Applications of Artificial Intelligence in Process Systems Engineering

Applications of Artificial Intelligence in Process Systems Engineering

Edited by

Jingzheng Ren
Assistant Professor, Department of Industrial and Systems Engineering,
The Hong Kong Polytechnic University, Hong Kong SAR, China

Weifeng Shen
School of Chemistry and Chemical Engineering, Chongqing University,
Chongqing, China

Yi Man
State Key Laboratory of Pulp and Paper Engineering, South China
University of Technology, Guangzhou, China

Lichun Dong
School of Chemistry and Chemical Engineering, Chongqing University,
Chongqing, China

ELSEVIER

Elsevier
Radarweg 29, PO Box 211, 1000 AE Amsterdam, Netherlands
The Boulevard, Langford Lane, Kidlington, Oxford OX5 1GB, United Kingdom
50 Hampshire Street, 5th Floor, Cambridge, MA 02139, United States

Notices
Knowledge and best practice in this field are constantly changing. As new research and experience
broaden our understanding, changes in research methods, professional practices, or medical
treatment may become necessary.

Practitioners and researchers must always rely on their own experience and knowledge in evaluating
and using any information, methods, compounds, or experiments described herein. In using
such information or methods they should be mindful of their own safety and the safety of others,
including parties for whom they have a professional responsibility.

To the fullest extent of the law, neither the Publisher nor the authors, contributors, or editors,
assume any liability for any injury and/or damage to persons or property as a matter of products
liability, negligence or otherwise, or from any use or operation of any methods, products,
instructions, or ideas contained in the material herein.

Library of Congress Cataloging-in-Publication Data
A catalog record for this book is available from the Library of Congress

British Library Cataloguing-in-Publication Data
A catalogue record for this book is available from the British Library

ISBN: 978-0-12-821092-5

For information on all Elsevier publications
visit our website at https://www.elsevier.com/books-and-journals

Publisher: Susan Dennis
Acquisitions Editor: Anita A Koch
Editorial Project Manager: Emerald Li
Production Project Manager: Kiruthika Govindaraju
Cover Designer: Mark Rogers

Typeset by SPi Global, India

Working together
to grow libraries in
developing countries

www.elsevier.com • www.bookaid.org

Contents

Contributors

Numbers in paraentheses indicate the pages on which the authors' contributions begin.

Mouloud Amazouz (207), CanmetENERGY-Natural Resources Canada (NRCan), Varennes, Canada

Masoud Babaei (255), Department of Control Engineering, Tarbiat Modares University, Tehran, Iran

Selim Ceylan (165), Ondokuz Mayıs University, Faculty of Engineering, Chemical Engineering Department, Samsun, Turkey

Zeynep Ceylan (165), Samsun University, Faculty of Engineering, Industrial Engineering Department, Samsun, Turkey

Lichun Dong (1), School of Chemistry and Chemical Engineering, Chongqing University, Chongqing, People's Republic of China

Jian Du (325), Institute of Chemical Process System Engineering, School of Chemical Engineering, Dalian University of Technology, Dalian, China

Mohamed El Koujok (207), CanmetENERGY-Natural Resources Canada (NRCan), Varennes, Canada

Xianlong Ge (497), School of Economics and Management; Key Laboratory of Intelligent Logistics Network, Chongqing Jiaotong University, Chongqing, China

Hakim Ghezzaz (207), CanmetENERGY-Natural Resources Canada (NRCan), Varennes, Canada

Mohsen Hadian (255), Department of Mechanical Engineering, University of Saskatchewan, Saskatoon, Canada

Rania M. Hathout (361), Department of Pharmaceutics and Industrial Pharmacy, Faculty of Pharmacy, Ain Shams University, Cairo, Egypt

Chang He (417), School of Materials Science and Engineering, Guangdong Engineering Centre for Petrochemical Energy Conservation, Sun Yat-sen University, Guangzhou, China

Mengna Hong (93, 119, 447), State Key Laboratory of Pulp and Paper Engineering, School of Light Industry and Engineering, South China University of Technology, Guangzhou, People's Republic of China

Yusha Hu (119), State Key Laboratory of Pulp and Paper Engineering, School of Light Industry and Engineering, South China University of Technology, Guangzhou, People's Republic of China

Priyanka Jha (187), Amity Institute of Biotechnology, Amity University, Kolkata, West Bengal, India

Yuanzhi Jin (1, 497), Department of Industrial and Systems Engineering, The Hong Kong Polytechnic University, Hong Kong SAR; School of Economics and Management, Chongqing Jiaotong University, Chongqing; Department of Computer Technology and Information Engineering, Sanmenxia Polytechnic, Sanmenxia, People's Republic of China

Vijay Kumar (187), Plant Biotechnology Lab, Division of Research and Development; Department of Biotechnology, Lovely Faculty of Technology and Sciences, Lovely Professional University, Phagwara, Punjab, India

Fuqiang Li (381), School of Management Science and Engineering, Dongbei University of Finance and Economics, Dalian, PR China

Jigeng Li (93, 119, 447), State Key Laboratory of Pulp and Paper Engineering, School of Light Industry and Engineering, South China University of Technology, Guangzhou, People's Republic of China

Feng Liu (381), School of Management Science and Engineering, Dongbei University of Finance and Economics, Dalian, PR China

Linlin Liu (325), Institute of Chemical Process System Engineering, School of Chemical Engineering, Dalian University of Technology, Dalian, China

Qilei Liu (325), Institute of Chemical Process System Engineering, School of Chemical Engineering, Dalian University of Technology, Dalian, China

Yi Man (1, 93, 119, 143, 447, 473), Department of Industrial and Systems Engineering, The Hong Kong Polytechnic University, Hong Kong SAR; State Key Laboratory of Pulp and Paper Engineering, School of Light Industry and Engineering, South China University of Technology, Guangzhou, People's Republic of China

Haitao Mao (325), Institute of Chemical Process System Engineering, School of Chemical Engineering, Dalian University of Technology, Dalian, China

Ardashir Mohammadzadeh (255), Department of Electrical Engineering, University of Bonab, Bonab, Iran

Sushreeta Paul (187), Amity Institute of Biotechnology, Amity University, Kolkata, West Bengal, India

Ahmed Ragab (207), CanmetENERGY-Natural Resources Canada (NRCan), Varennes, Canada

Jingzheng Ren (1), Department of Industrial and Systems Engineering, The Hong Kong Polytechnic University, Hong Kong SAR, People's Republic of China

Seyed Mohammad Ebrahimi Saryazdi (255), Department of Energy Engineering, Sharif University of Technology, Tehran, Iran

Weifeng Shen (1, 11, 39, 67), School of Chemistry and Chemical Engineering, Chongqing University, Chongqing, People's Republic of China

Tao Shi (1), Department of Industrial and Systems Engineering, The Hong Kong Polytechnic University, Hong Kong SAR, People's Republic of China

Yang Su (11), School of Chemistry and Chemical Engineering, Chongqing University, Chongqing, People's Republic of China

Zifei Wang (143, 473), State Key Laboratory of Pulp and Paper Engineering, School of Light Industry and Engineering, South China University of Technology, Guangzhou, People's Republic of China

Zihao Wang (39, 67), School of Chemistry and Chemical Engineering, Chongqing University, Chongqing, People's Republic of China

Ao Yang (1), Department of Industrial and Systems Engineering, The Hong Kong Polytechnic University, Hong Kong SAR, People's Republic of China

Zhaolong Yang (381), Hospital of Clinical Medicine Engineering Department, Deyang People's Hospital, Deyang, PR China

Huanhuan Zhang (447), State Key Laboratory of Pulp and Paper Engineering, School of Light Industry and Engineering, South China University of Technology, Guangzhou, People's Republic of China

Lei Zhang (325), Institute of Chemical Process System Engineering, School of Chemical Engineering, Dalian University of Technology, Dalian, China

Yang Zhang (93), State Key Laboratory of Pulp and Paper Engineering, School of Light Industry and Engineering, South China University of Technology, Guangzhou, People's Republic of China

Qiping Zhu (417), School of Materials Science and Engineering, Guangdong Engineering Centre for Petrochemical Energy Conservation, Sun Yat-sen University, Guangzhou; College of Chemistry and Bioengineering, Guilin University of Technology, Guilin, China

Chapter 1

Artificial intelligence in process systems engineering

Tao Shi[a], Ao Yang[a], Yuanzhi Jin[a], Jingzheng Ren[a], Weifeng Shen[b], Lichun Dong[b], and Yi Man[a]

[a]*Department of Industrial and Systems Engineering, The Hong Kong Polytechnic University, Hong Kong SAR, People's Republic of China,* [b]*School of Chemistry and Chemical Engineering, Chongqing University, Chongqing, People's Republic of China*

1 What is process system engineering

1.1 Background

In 1961, process system engineering (PSE) was first proposed in a special volume of AIChE Symposium Series. The term PSE was not widely accepted, however, until 1982 when the first international symposium on this topic took place in Kyoto, Japan. The first journal devoted to the PSE-related research was "Computers and Chemical Engineering" born in 1997. PSE is considered under the banner of chemical engineering process, which was developed from the concept of unit operation, mathematical models, and computer visualization. Of note is that chemical engineering differs so much from chemistry research since engineers have to take the responsibility for transferring the natural theory or microscale matters into the macroscale products. Therefore, in some sense, PSE is a multiscale systematical process beneficial for us humanity and a whole multiscale concept is shown in Fig. 1. The widely accepted definition of PSE is that an interdisciplinary research filed concerned with the development of the systematic procedures with the aid of mathematics and computer for the smart design, effective control, schedule distribution of product manufacture system, which is characterized by multiple scale [2].

Up to now, the PSE research deeply influenced by the engineering achievement and scientific theory is closely tied to mathematics, computer science, and software engineering. Much bigger data and high-performance computer greatly boost the investigation of PSE meanwhile connect the commercial enterprise and research institutions more closely. With the aid of real industrial data from the enterprise, the design of process model and simulation in advanced computer can be further improved by researchers. The improved

Applications of Artificial Intelligence in Process Systems Engineering
https://doi.org/10.1016/B978-0-12-821092-5.00010-3

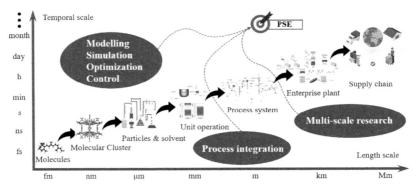

FIG. 1 Multiscale research flowsheet of PSE [1].

design could be continuously applied into the industrial adjustment and upgrading leading to an interesting cycle and feedback. As shown in Fig. 1, the major concerns in the PSE research are divided into the following sections like the modeling, simulation, optimization, and control, which are in multiscale process.

1.2 Challenges

Among these concerns there are many accomplishments in the past and a full review can be obtained by Klatt et al. [3]. It can be summarized that PSE strives to create representations and models to generate reasonable alternatives ranging from molecules to process superstructures and select the optimal solution from these candidates. Developing computationally efficient methods, accurate models, and simulation tools are all attributions in PSE area. PSE is relatively a young area from the viewpoint of discipline development, in which tough challenges stand there under the pressure of global competition and environmental protection. Overcoming these challenges eventually is rooted by innovative technology which is so closely to the improvement of manufacture efficiency, product yield, payback period, energy utilization, and so on [4].

First, the product design on molecular scale has received much more attention than ever before [5] in addition to the process design aiming to a more energy-efficient structure. During the product design, it is needed to make property predictions of pure components or mixtures involving thermodynamics, environment, health, safety, and so on. For example, evaluating the health-related and environmental property of new chemicals is so important in pharmaceutical engineering that researchers make great efforts to find out a green synthetic route with the aid of huge database and efficient computer [6]. Moreover, people pay more attention to the sustainability and safety during the product and process design.

How to deal with the coupling problem in the control structure of a process system in which multiinput and multioutput exist is also a big challenge. Traditionally, the proportion integral derivative (PID) control theory was the most classical and has been widely applied in most of the chemical and petrochemical industry embedded in the distributed control system. PID controllers take actions based on the linear combination of current and past deviations of the dynamic variables. As the process becomes more complex characterized by nonlinearity, multiple variables, and strong coupled function, classic control theory usually tends to a poor control quality [7]. Advanced control method is called for a solution in achieving the efficient control of such chemical processes.

Third, a big challenge lies in the process optimization which is of necessity not only to evaluate the impact of different factors but also manage the optimal design when considering the benefits of stakeholders. The high-level nonlinearity in chemical processes greatly increases the optimization solver difficulty as one added variable often represents more than 100,000 algebraic equations have to be calculated [3]. Multivariable problems combined with single-objective or multiobjective (e.g., economic, environmental, and safety indicators) optimization must be solved within a process design. Fortunately, the huge improvement in numerical modeling and process simulation has presented an effective method to solve such problem partially. Challenging requirements on efficient optimization algorithm to deal with nonlinearity, integer discrete variables, and constraints displayed by the model still exist [8, 9].

When the research area of PSE is expanded into the macroscale supply chain as shown in the Fig. 1, managers and investors have to consider the downstream logistics and product distribution activities meanwhile saving transportation cost and making more profit, which can be easily understood under the popularity of electrical commercial trade today. Each transaction online represents the logistics transportations and environmental influences offline. Therefore, it is necessary to introduce the life-cycle concept and carry out the supply chain optimization and inventory control in the integration of research, development, control, and operation at the business level [10, 11]. The life-cycle assessment integrated with the supply chain is an interesting topic in the PSE field. In summary, there is some challenges of multicriterion decision making in the PSE for enterprise and managers including the production schedule or plan if the industry itself wants to sell the products with little waste.

2 What is artificial intelligence?

2.1 Background

In 1956, the first artificial intelligence meeting in history, named Dartmouth Summer Research Project on Artificial Intelligence (AI), was initiated by John McCarthy and other experts in this area [12]. The meeting lasted for a month having a big wish for intelligent machines that could work perfectly like human beings. It is still a long way to go to achieve the big wish. However, some AI technologies specialized at one special skill have improved a lot since 2015 when parallel computing became much faster and more efficient with the aid of high-performance CPU and GPU. Those special skills in the AI machines like image identification, classification, logical calculation is obtained by learning and is more professional than human beings. One of the most important and popular ways to achieve the professional skills for AI is machine learning (ML). Basically, once the relative data is collected, next ML algorithms are applied to make feature extraction, model training, and evaluation. After the above learning procedure, effective models can be produced to make predictions and instruct the decision-making in the real world. It is summarized that the biggest advantage by ML is that trained AI models actively has some special skills and replaces the millions of lines of code lines to do us a favor in real world [13].

There are lots of different methods targeting on different problem solutions in the content of ML, which are generally divided into supervised learning, unsupervised learning, and reinforcement learning (RL) according to the label information in the training dataset and the dynamic decision-making process [14]. Of note is that the RL, a typical learning method with no prior label dataset, actively choose the next step (e.g., decision-making) to maximize long-term benefits according to the "good" or "bad" feedback in a dynamic environment with uncertainty. During the sequential decision-making process, reward and penalty are crucial elements to instruct the optimization [15]. Besides, ML methods branching of support vector machine based on the statistical learning theory and artificial neural network (ANN) based on the biological neural cells' connection have gain intensive attention in handling different problems, respectively. ANN architecture which looks like a thinking process of the human brain. The ANN model is composed of a large number of neural cells. Each node connects with each other and represents a specific output function which is called an activation function, and each connection between two nodes represents a weighted value for signal passing or calculations [16].

2.2 Deep learning

The first-generation ANN can only make some linear classification while cannot effectively finish the logical function classification [17]. The second-generation ANN was developed by Geoffrey Hinton who used multiple hidden layers to replace the feature layer in the ANN model and calculate the intermediate parameters by back-propagation algorithm [18]. As the number of hidden layers increases, the optimization function is more prone to local optimal solutions and the training of deep network are difficult to achieve. To overcome such problem, the method of hierarchical feature extraction and unsupervised pretraining are proposed by Hinton et al. [19]. Then, the third-generation ANN technology represented by deep learning was developed and recognized as a breakthrough with regards to handling the training of multilayer networks where "deep" referred to the number of layers in the network. Different architectures like deep neural network (DNN), recurrent neural network, and convolutional neural network are used in the deep learning study targeting on different applications in life [20]. The development of hardware and the explosion of datasets, the demand of image and text handling, greatly provide additional boosts to the effectiveness of training of DNN. Deep learning provides an efficient approach to obtain the intelligent model after training from the big data, which can be integrated into PSE to solve complex problems [15].

3 The AI-based application in PSE

As well-known, the application of AI could be extended in the PSE to achieve more intelligent computer-aided tools for chemical engineering. The investigation of AI in PSE is remarkable by some researchers in the late 1960s and early 1970s and then AI methods for chemical engineering problems have vigorously developed in the early 1980s [21]. The Adaptive Initial Design Synthesizer system for the chemical process synthesis is developed [22], which was arguably the first time to systematically describe the application of AI methods in chemical engineering including means-and-ends analysis, symbolic manipulation, and linked data structures. Afterwards, the AI technology has been widely extended in physics properties prediction, process modeling and simulation, dynamic control, process optimization, and fault detection and diagnosis, etc.

3.1 Physical properties prediction and product design

Physical properties (e.g., critical properties) and prediction models play an important role in chemical process and product design [23]. To improve the accuracy of prediction, an intelligent and automated quantitative structure property relationships (QSPR) model based on deep learning models is developed [24]. The DNN by using the combination of tree-long short-term memory network and back-propagation neural network (BPNN) can be used to model the proposed QSPR, and the results demonstrated that the proposed approach could well predict the critical temperature, pressure, and volume [24]. Furthermore, the proposed model could be extended to predict environmental properties such as octanol-water partition coefficients [25] and Henry's law constant [26]. Similarly, the hazardous property such as toxicity and flashpoint could be forecasted via ensemble neural network models. The investigation of physical properties prediction can realize high-throughput screening and further promote the development of product design based on the AI technique [27]. At the early stage, the BatchDesign-Kit system based on the state-transition network is used to find a new batch reaction process with lower health, safety, environmental impact and cost, which is explored by Linninger et al. [28]. Subsequently, Zhang et al. [13] develop a machine learning and computer-aided molecular design

(ML-CAMD) to establish and screen suitable fragrance molecules, which are widely used in modern industries. The results illustrated that the obtained molecule $C_9H_{18}O_2$ has the higher odor pleasantness compared with the existing product. Chai [29] extended the ML-CAMD to design crystallization solvent for an industrial crystallization process. A hybrid model is established for the determination of new food products [30]. In summary, the expanding of AI in physics properties prediction could effectively promote the development of product design and process synthesis.

3.2   Process modeling

Actually, the bottleneck of the chemical process requires a lot of time to perform the necessary detailed analyses (e.g., feed and product) in high-frequency optimization and process control. Such issues could be resolved via ANN models. Plehiers et al. [31] developed an ANN to model and simulate a steam-cracking process with a naphtha feedstock and a detailed composition of the steam cracker effluent and the computational illustrated that the presented method is applicable to any type of reactor without loss of performance. According to the study of George-Ufot [32], the forecasting of electric load could improve the energy efficiency and reduce the production cost. A hybrid framework including genetic algorithm (GA), particle swarm optimization, and BPNN is established, which displays good reliability and high accuracy in terms of electric load forecasting [33]. In addition, the prediction model based on the various neural network techniques could be used to forecast the energy consumption [34], chemical oxygen demand content [35], and throughput [36], respectively. The black-box model based on neural network can be employed to process optimization and control achieving, energy saving, emission reduction, and cleaner production.

3.3   Fault detection and diagnosis

Fault detection and diagnosis are very important in process control and monitoring because it could ensure the process safety and reliability (i.e., product quality) in chemical engineering processes [37, 38]. A back-propagation neural network-based methodology is proposed by Venkatasubramanian and Chan [39] to handle the fault diagnosis of the fluidized catalytic cracking process in oil refinery. Similarly, the proposed model could be used to fault diagnosis and process control [40]. The results illustrated that the established approach could diagnose novel fault combinations (not explicitly trained upon) and handle incomplete and uncertain data. Then, the neural network method is used to fault detection for reactor [41] and a fatty acid fractionation precut column pressure/temperature [42] sensors fault. An extended Deep Belief Networks is developed, which could avoid the valuable information in the raw data lost in the layer-wise feature compression in traditional deep networks [43].

3.4 Process optimization and scheduling

AI method for process optimization and scheduling has been attracted because it could increase the speed, reduce the time consumption of optimization, and achieve optimal matching. For instance, Qiu et al. [44] developed a data-driven model based on the radial basis function neural network combined with GA to solve the mixed integer nonlinear programing problem of distillation process. The separation of propylene/propane by using the externally heat-integrated distillation columns is employed to illustrate the proposed approach and the optimal solution could be quickly found in a wide space. A hybrid model via the combination of artificial neural network and GA is investigated by Khezri et al. [45], and the capability of the hybrid model is illustrated by means of a gas to liquids process. The hybrid model shows significant advantage because the computational time for optimization is greatly reduced from multiple days to just a few seconds and the relative error is in an accepted range. In addition, the flow shop scheduling [46, 47] optimization of cooling water system [48], dynamic vehicle routing scheduling [49] could be effectively solved via neural network models.

4 Summary

The application of AI in the above fields of PSE may be just one tip of the iceberg. With the continuous development and penetration of AI technology, we believe that AI could be applied more extensively in the future. In short term, maintaining AI's influence on society is conducive to the study of its deep applications in many fields such as logical reasoning and proof, natural language processing, intelligent information retrieval technology and expert systems, etc. In the long run, achieving the general AI machine better than human at most cognitive tasks is the final goal of scientists and engineers. Novel materials and high-performance computer have crucial influences on the process.

Acknowledgment

The authors would like to express their sincere thanks to the Research Committee of The Hong Kong Polytechnic University for the financial support of the project through a PhD studentship (project account code: RK3P).

References

[1] W. Marquardt, L. von Wedel, B. Bayer, Perspectives on lifecycle process modeling, in: AIChE Symposium Series, American Institute of Chemical Engineers, New York, 2000. 1998.
[2] I.E. Grossmann, A.W. Westerberg, Research challenges in process systems engineering, AICHE J. 46 (9) (2000) 1700–1703.
[3] K.-U. Klatt, W. Marquardt, Perspectives for process systems engineering—personal views from academia and industry, Comput. Chem. Eng. 33 (3) (2009) 536–550.

[4] I.E. Grossmann, I. Harjunkoski, Process systems engineering: academic and industrial perspectives, Comput. Chem. Eng. 126 (2019) 474–484.

[5] L.Y. Ng, F.K. Chong, N.G. Chemmangattuvalappil, Challenges and opportunities in computer-aided molecular design, Comput. Chem. Eng. 81 (2015) 115–129.

[6] H. Chen, et al., The rise of deep learning in drug discovery, Drug Discov. Today 23 (6) (2018) 1241–1250.

[7] X. Qian, et al., MPC-PI cascade control for the Kaibel dividing wall column integrated with data-driven soft sensor model, Chem. Eng. Sci. (2020) 116240.

[8] L.T. Biegler, I.E. Grossmann, Retrospective on optimization, Comput. Chem. Eng. 28 (8) (2004) 1169–1192.

[9] I.E. Grossmann, L.T. Biegler, Part II. Future perspective on optimization, Comput. Chem. Eng. 28 (8) (2004) 1193–1218.

[10] X. Tian, et al., Sustainable design of geothermal energy systems for electric power generation using life cycle optimization, AICHE J. 66 (4) (2020) e16898.

[11] N. Zhao, J. Lehmann, F. You, Poultry waste valorization via pyrolysis technologies: economic and environmental life cycle optimization for sustainable bioenergy systems, ACS Sustain. Chem. Eng. 8 (11) (2020) 4633–4646.

[12] J. McCarthy, et al., A proposal for the Dartmouth summer research project on artificial intelligence, August 31, 1955, AI Mag. 27 (4) (2006) 12.

[13] L. Zhang, et al., A machine learning based computer-aided molecular design/screening methodology for fragrance molecules, Comput. Chem. Eng. 115 (2018) 295–308.

[14] E. Alpaydin, Introduction to Machine Learning, MIT Press, 2020.

[15] J.H. Lee, J. Shin, M.J. Realff, Machine learning: overview of the recent progresses and implications for the process systems engineering field, Comput. Chem. Eng. 114 (2018) 111–121.

[16] T.G. Dietterich, Machine-learning research, AI Mag. 18 (4) (1997) 97.

[17] F. Rosenblatt, The perceptron: a probabilistic model for information storage and organization in the brain, Psychol. Rev. 65 (6) (1958) 386.

[18] D.E. Rumelhart, G.E. Hinton, R.J. Williams, Learning Internal Representations by Error Propagation, California Univ San Diego La Jolla Inst for Cognitive Science, 1985.

[19] G.E. Hinton, S. Osindero, Y.-W. Teh, A fast learning algorithm for deep belief nets, Neural Comput. 18 (7) (2006) 1527–1554.

[20] Y. LeCun, Y. Bengio, G. Hinton, Deep learning, Nature 521 (7553) (2015) 436–444.

[21] D.F. Rudd, G.J. Powers, J.J. Siirola, Process Synthesis, Prentice-Hall, 1973.

[22] J.J. Siirola, D.F. Rudd, Computer-aided synthesis of chemical process designs. From reaction path data to the process task network, Ind. Eng. Chem. Fundam. 10 (3) (1971) 353–362.

[23] V. Venkatasubramanian, The promise of artificial intelligence in chemical engineering: is it here, finally? AICHE J. 65 (2) (2019) 466–478.

[24] Y. Su, et al., An architecture of deep learning in QSPR modeling for the prediction of critical properties using molecular signatures, AICHE J. 65 (9) (2019) e16678.

[25] Z. Wang, et al., Predictive deep learning models for environmental properties: the direct calculation of octanol-water partition coefficients from molecular graphs, Green Chem. 21 (16) (2019) 4555–4565.

[26] Z. Wang, et al., A novel unambiguous strategy of molecular feature extraction in machine learning assisted predictive models for environmental properties, Green Chem. 22 (12) (2020) 3867–3876.

[27] L. Zhang, et al., Chemical product design—recent advances and perspectives, Curr. Opin. Chem. Eng. 27 (2020) 22–34.

[28] A.A. Linninger, et al., Generation and assessment of batch processes with ecological considerations, Comput. Chem. Eng. 19 (1995) 7–13.

[29] S. Chai, et al., A grand product design model for crystallization solvent design, Comput. Chem. Eng. 135 (2020) 106764.

[30] X. Zhang, et al., Food product design: a hybrid machine learning and mechanistic modeling approach, Ind. Eng. Chem. Res. 58 (36) (2019) 16743–16752.

[31] P.P. Plehiers, et al., Artificial intelligence in steam cracking modeling: a deep learning algorithm for detailed effluent prediction, Engineering 5 (6) (2019) 1027–1040.

[32] G. George-Ufot, Y. Qu, I.J. Orji, Sustainable lifestyle factors influencing industries' electric consumption patterns using fuzzy logic and DEMATEL: the Nigerian perspective, J. Clean. Prod. 162 (2017) 624–634.

[33] Y. Hu, et al., Short term electric load forecasting model and its verification for process industrial enterprises based on hybrid GA-PSO-BPNN algorithm—a case study of papermaking process, Energy 170 (2019) 1215–1227.

[34] C. Chen, et al., Energy consumption modelling using deep learning embedded semi-supervised learning, Comput. Ind. Eng. 135 (2019) 757–765.

[35] A.R. Picos-Benítez, et al., The use of artificial intelligence models in the prediction of optimum operational conditions for the treatment of dye wastewaters with similar structural characteristics, Process Saf. Environ. Prot. 143 (2020) 36–44.

[36] A. Sagheer, M. Kotb, Time series forecasting of petroleum production using deep LSTM recurrent networks, Neurocomputing 323 (2019) 203–213.

[37] L. Ming, J. Zhao, Review on chemical process fault detection and diagnosis, in: 2017 6th International Symposium on Advanced Control of Industrial Processes (AdCONIP), IEEE, 2017.

[38] N.M. Nor, C.R.C. Hassan, M.A. Hussain, A review of data-driven fault detection and diagnosis methods: applications in chemical process systems, Rev. Chem. Eng. 1 (2019) (ahead-of-print).

[39] V. Venkatasubramanian, K. Chan, A neural network methodology for process fault diagnosis, AICHE J. 35 (12) (1989) 1993–2002.

[40] L. Ungar, B. Powell, S. Kamens, Adaptive networks for fault diagnosis and process control, Comput. Chem. Eng. 14 (4–5) (1990) 561–572.

[41] A. Ahmad, M. Hamid, K. Mohammed, Neural networks for process monitoring, control and fault detection: application to Tennessee Eastman plant, in: Proceedings of the Malaysian Science and Technology Congress, Melaka, Malaysia, Universiti Teknologi Malaysia, Johor Bahru, Malaysia, 2001.

[42] M.R. Othman, M.W. Ali, M.Z. Kamsah, Process fault detection using hierarchical artificial neural network diagnostic strategy, J. Teknol. 46 (1) (2007) 11–26.

[43] Y. Wang, et al., A novel deep learning based fault diagnosis approach for chemical process with extended deep belief network, ISA Trans. 96 (2020) 457–467.

[44] P. Qiu, et al., Data-driven analysis and optimization of externally heat-integrated distillation columns (EHIDiC), Energy 189 (2019) 116177.

[45] V. Khezri, et al., Hybrid artificial neural network–genetic algorithm-based technique to optimize a steady-state gas-to-liquids plant, Ind. Eng. Chem. Res. 59 (18) (2020) 8674–8687.

[46] F. Liu, et al., On the robust and stable flowshop scheduling under stochastic and dynamic disruptions, IEEE Trans. Eng. Manage. 64 (4) (2017) 539–553.

[47] Z. Zeng, et al., Multi-object optimization of flexible flow shop scheduling with batch process—consideration total electricity consumption and material wastage, J. Clean. Prod. 183 (2018) 925–939.

[48] Q. Zhu, et al., Model reductions for multiscale stochastic optimization of cooling water system equipped with closed wet cooling towers, Chem. Eng. Sci. 224 (2020) 115773.

[49] W. Joe, H.C. Lau, Deep reinforcement learning approach to solve dynamic vehicle routing problem with stochastic customers, in: Proceedings of the Thirtieth International Conference on Automated Planning and Scheduling, 2020, pp. 394–402.

Chapter 2

Deep learning in QSPR modeling for the prediction of critical properties

Yang Su and Weifeng Shen

School of Chemistry and Chemical Engineering, Chongqing University, Chongqing, People's Republic of China

1   Introduction

The chemical process and product design rely heavily on physical properties (e.g., critical properties) and prediction models [1, 2]. To investigate relationships between molecular structures and properties, plenty of mathematical models have been developed [3]. Most prediction models are based on semi-empirical quantitative structure property relationships (QSPRs) including group contribution (GC) methods and topological indices (TIs).

In GC methods, any compound can be divided into fragments (e.g., atoms, bonds, groups of both atoms, and bonds). Each fragment has a partial value called a contribution, and the final property value is given by summing the fragmental contributions. A large variety of these models has been designed differing in the field of their applicability and in the set of experimental data. For example, GC methods reported by Lydersen [4], Klincewicz and Reid [5], Joback and Reid [6], Constantinou and Gani [7], and Marrero and Gani [8] are generally suitable to obtain values of physical properties, because these methods provide the advantage of quick estimation without substantial computational work. As alternative approaches, topological indices (TIs) are used to estimate properties similar to the way of GC methods. In TI methods, molecular topology is characterized depending on standard molecular graph properties such as vertex degrees, connectivity, atomic types, etc. Additionally, one of the main advantages is that TI methods can make a distinction between two similar structures from a more holistic perspective than GC methods [9].

Another method named 'signature molecular descriptor' combining the advantages of GC and TI methods is developed by Faulon et al. [10, 11] Similar to TI methods, chemical structures is conceived of as chemical graphs. The

Applications of Artificial Intelligence in Process Systems Engineering
https://doi.org/10.1016/B978-0-12-821092-5.00012-7

11

signature descriptor retains all the structural and connectivity information of every atom in a molecule, rather than ascribe various numerical values to a complete molecular graph [9]. Meanwhile, the signature descriptor has the ability to represent molecular substructures similar to GC methods. Faulon et al. [12] also introduced a canonical form of molecular signatures to solve molecular graph isomorphism which provides a holistic picture depicting molecular graphs and also holds the sub-structural information of a molecule. Nevertheless, we found that the previous researches have few attempts to use the canonical molecular signature for QSPR modeling. To the best of our knowledge, the main reason is that the canonical molecular signature is not represented in a numeric form, and it cannot be employed within the common-used mathematical models for QSPRs.

For the property estimation, most above-mentioned QSPR models, based on the specific rules such as a certain set of molecular substructures or an array of molecular graph-theoretic properties, are often formulated by multiple linear regressions (MLRs). Facts proved the MLR techniques have strong ability to correlate QSPRs; however, their encoding rules and mapping functions are defined a priori (i.e., mathematical formulations are not adaptive to the different regression tasks). Moreover, the MLRs cannot be applied with the canonical molecular signatures for QSPR modeling. On the other hand, an alternative technique, the neural network, has been used to learn molecular structures and correlate physical properties or activities [13]. A variety of molecular descriptors (e.g., topological characteristics, frequency of molecule substructures, and microscopic data of molecules) are fed to artificial neural networks (ANNs). With the limitation of the computing capability and development platform at that period, most researchers adopted feedforward neural networks (FNNs) with static computing graphs in their studies [14–32].

Although these methods are well-used or precise in properties prediction, the molecular features are chosen manually as the input for above-mentioned models. For example, the splitting rules of molecular groups are pre-determined manually in the GC methods, or the well-chosen descriptors are input the ANNs. With the number of various properties and product designs has been increasing, some properties/activities may need to be correlated with more molecular features or calculated by more complex mathematical models. It is therefore a challenge to pick out relevant features of molecules from massive data in the classical QSPR modeling.

Recently, many researchers were encouraged to study deep learning in artificial intelligence with improvements of computing performance. The deep learning is a much more intelligent technique that can capture the valuable features automatically. This advantage enables deep neural networks (DNNs) to formulate models from a great variety of Big Data. As such, some new information carriers (e.g., graphs, images, text, and 3D models) could be used to represent molecular structures in the QSPR modeling with DNNs. Lusci et al. [33] utilized the recurrent neural networks (RNNs) to present a molecular graph by considering molecules as undirected graphs (UGs) and proposed an approach for mapping aqueous solubility to molecular structures. Goh et al. [34] developed a deep RNN "SMILES2vec" that automatically learns features from simplified molecular-input line-entry system [35] (SMILES) to correlate properties without the aid of additional explicit feature engineering. Goh et al. [36] also developed a deep CNN for the prediction of chemical properties, using just the images of 2D drawings of molecules without providing any additional explicit chemistry knowledge such as periodicity, molecular descriptors, and fingerprints. These creative works [34, 36] demonstrate the plausibility of using DNNs to assist in computational chemistry researches. The neural networks based on the long short-term memory (LSTM) units suggested by Hochreiter et al. [37] also have been adopted in the quantitative structure–activity relationship (QSAR) researches. Altae-Tran et al. [38] proposed a new deep learning architecture based on the iterative refinement LSTM to improve learning of meaningful distance metrics over small-molecules. The Tree-LSTM introduced by Tai et al. [39] is able to capture the syntactic properties of natural languages and two natural extensions were proposed depending on the basic LSTM architecture, which outperform other RNNs in their experiments. We noticed that the new neural network Tree-LSTM might be possible to depict the canonical molecular signature.

Motivated by the preceding researches, in this contribution, we focus on developing a deep learning approach that can learn QSPRs automatically and cover a wider range of substances for better predictive capabilities. A Python-based implementation with Faulon's algorithm [12] is achieved to convert molecules into canonical signatures for depicting molecular graphs and an in-house encoding approach is developed to parse the signatures into tree data-structures conveniently. The Tree-LSTM network and BPNN are incorporated into the DNN for modeling QSPR, among which, the Tree-LSTM mimics the tree structures of canonical signatures and outputs a feature vector that is used to correlate properties within a BPNN. As such, there is no need to convert molecules to bitmap images for training CNNs and to treat molecules as linear languages for training RNNs. Then, the novelty of the proposed approach is that the canonical molecular signatures are used as templates to generate the topological structures of Tree-LSTM networks. In this sense, the contribution of this work is to propose an intelligent strategy of QSPR modeling based on deep learning that can extract the valuable features from molecular structures automatically. An important type of properties in process and product designs, critical properties, is used as a case study to clarify the main details of the deep learning architecture, which highlights the outperformance of the implemented QSPR modeling strategies within the proposed DNN.

2 Methodology

In this section, the technical details respected to the deep learning architecture for modeling QSPR will be introduced. The proposed deep learning architecture incorporates multiple techniques that including canonical molecular signatures, word embedding, Tree-LSTM network, BPNN, etc. The proposed architecture consisting of eight steps is illustrated in Fig. 1. Step 1 mainly involves the data acquisition of molecular structures, where the SMILES expressions are captured from open access databases. The second step is the embedding stage, where the vectors representing the substrings of chemical bonds are generated and collected into a dictionary with a widely used word-embedding algorithm. The third step is focused on the canonization of a molecule, where the molecular structures are transformed into the canonical molecular signatures as the templates for formulating the Tree-LSTM network. Step 4 refers to the mapping stage, where the adaptive structure of the Tree-LSTM network is obtained by the recursive algorithm from the canonical signature. In other words, the Tree-LSTM network is self-adaptive to a molecule. Step 5 involves the inputting vectors of each substrings corresponding to each node. The Tree-LSTM

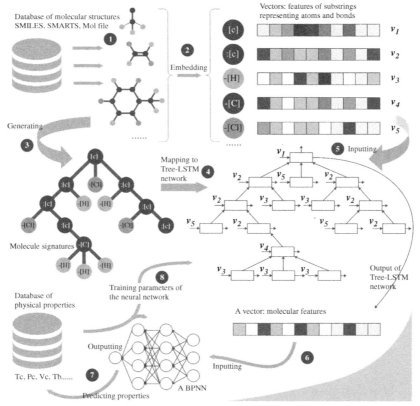

FIG. 1   Schematic diagram of technical architecture for deep learning in the prediction of physical properties.

network will be calculated from the lowest leaf node to the root node in this step. Finally, a vector representing a molecule is given from the root node. Step 6 is focused on the correlation stage of a property, where the vector representing a molecule is input into a BPNN to compute a scalar output for the property prediction. Step 7 is the comparison stage, where the tolerance between the predicted value and the experimental value is calculated. Step 8 is the feedback stage, where the adjustable parameters in the Tree-LSTM network and the BPNN are corrected for reducing the tolerance in step 7. The training process of the proposed DNN is the iterative loop within steps 5, 6, 7, and 8.

2.1 The signature molecular descriptor

The canonical molecular signature is employed to depict molecules in this work. One reason is that a computer program can generate signatures automatically. Another important reason is that the canonical molecular signature provides a method to distinguish molecular structures for isomorphism. This also transforms the molecules with a uniform form for mapping to the neural network model.

To introduce canonical molecular signatures, atomic and molecular signatures have to be defined. An atomic signature is a subgraph originated at a specific root atom and includes all atoms/bonds extending out to the predefined distance, without backtracking. The predefined distance is a user-specified parameter called the signature height h, and it determines the size of the local neighborhood of atoms in a molecule. It means that specified a certain root atom in a chemical graph, its atomic signature represents all of the atoms that are within a certain distance h, from the root. The atomic signature of atom x in height h given as $^h\sigma_{G(x)}$, is a representation of the subgraph of the 2D graph $G = (V, E)$ containing all atoms that are at distance h from x. It is noted that V, E corresponds to the vertex (atom) set and edge (bond) set, respectively. Acetaldoxime (CAS No. 107-29-9) is taken as an example to provide atomic signatures shown in Fig. 2. The carbon atom numbered by 0 (C0) is given as the root atom, and it is single-bonded to three hydrogen atoms and another carbon atom numbered by 1 (C1). Thus, the atomic signature for this root atom at height 1 is [C]([C][H][H][H]), the other atomic signatures is shown in Fig. 2B.

In Faulon's theory [8], the molecular signature shown in Fig. 2C is a linear combination of all the atomic signatures and is defined as Eq. (1).

$$^h\sigma(G) = \sum_{x \in V} {}^h\sigma_{G(X)} \tag{1}$$

In a given compound, any atomic signature can appear more than once. For example, the atomic signature [H]([C]) occurs four times in acetaldoxime. When the height of atomic signatures reaches the maximum value, the molecular graph can be reconstructed from any of the atomic signatures. Consequently, as long as graph canonization is concerned, there is no need to record

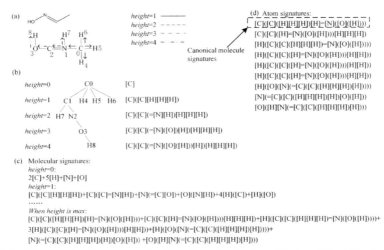

FIG. 2 Signature descriptors generating from the molecule of acetaldoxime. (Note: C0 is the root. An atomic signature begins from this root atom numbered by 0, steps forward a predetermined height, and records all the atoms encountered on the path are connected to the root atom. The process is repeated for all atoms in the molecule at a certain height, and then the molecular signature will be achieved by a linear combination of all the atomic signatures. Acetaldoxime is used as an example, we present the heights from the root atom in: (A) the molecular structure; (B) tree form and atomic signatures of different heights from a certain atom; (C) the molecular signatures from height $=0$ to height $=1$; (D) the canonical molecular signature in the *red rectangle* is the lexicographically largest atomic signature. The atoms found further down the tree of a branch point atom are marked by nested parenthesis. Single bonds between atoms are omitted in atomic signatures. In addition, other bond-types are presented as follows ("$=$"is double bond; "#" is a triple bond; ":" is an aromatic bond).)

all atomic signatures. The lexicographically largest atomic signature suffices to represent the graph in a unique manner [10]. For example, acetaldoxime has nine atomic signatures at the maximum height as shown in Fig. 2D, and each of them is able to describe the complete molecular structure. If these nine signatures are sorted in decreasing lexicographic order (a canonical order), the lexicographically largest one can be defined as the canonical molecular signature that could be encoded and then mapped to the Tree-LSTM network.

2.2 Data preparation: Molecules encoding and canonizing

In this work, SMILES expressions that used for depicting molecular structures are gathered from PubChem database [40]. We developed a program based on RDKit [41] for parsing and preserving the canonical molecular signature. The program implements Faulon's algorithm to generate and canonize atomic signatures, which can translate SMILES expressions to molecular graphs before canonizing molecular structures. There exist two rules for coding molecular structures in this program, one is the canonical string encoding a canonical molecular signature, and the other is the developed in-house coding method. The canonical molecular signature is used to determine the root atom in different molecules. However, it is difficult to reproduce the molecular structures and feed into the neural network from a molecular signature represented by a canonical string. When training the neural networks, one needs a more straightforward and simpler expression for parsing a molecule as a tree data-structure. As such, a specified in-house coding method is proposed. Again, taking acetaldoxime as an example, the codes of atomic signatures from height 0 to 4 are demonstrated in Table 1, which is started from the C0 in Fig. 2B.

The substring shown in Fig. 3 represents the atom carbon in the atomic signature of height 1. The first character in the code involves the layer position (height) of this atom in the tree form of a molecule, and the character "S" refers to the layer position equals 0. The second character "0" is the initial position index of the atom in a molecule. The third and fourth characters represent the number of neighbor atoms and children atoms, respectively. The next character "C" is the SMILES expression of an atom. The sixth character is the initial position index of parent atom. When an atom is a root atom of a signature structure, the sixth character "S" represents the atom has no superior atom. The seventh character is the valence of this atom. The eighth character represents

TABLE 1 Atomic signatures coding started from C0 in Fig. 2B.

Height	Start index	Code
0	0	S,0,4,4,C,S,4,S,N|
1	0	S,0,4,4,C,S,4,S,N|1,1,3,2,C,0,4,1,N|1,4,1,0,[H],0,1,1, N|1,5,1,0,[H],0,1,1,N|1,6,1,0,[H],0,1,1,N|
2	0	S,0,4,4,C,S,4,S,N|1,1,3,2,C,0,4,1,N|1,4,1,0,[H],0,1,1, N|1,5,1,0,[H],0,1,1,N|1,6,1,0,[H],0,1,1,N|2,2,2,1,N,1,3,2, E|2,7,1,0,[H],1,1,1,N|
3	0	S,0,4,4,C,S,4,S,N|1,1,3,2,C,0,4,1,N|1,4,1,0,[H],0,1,1, N|1,5,1,0,[H],0,1,1,N|1,6,1,0,[H],0,1,1,N|2,2,2,1,N,1,3,2, E|2,7,1,0,[H],1,1,1,N|3,3,2,1,O,2,2,1,N|
4	0	S,0,4,4,C,S,4,S,N|1,1,3,2,C,0,4,1,N|1,4,1,0,[H],0,1,1, N|1,5,1,0,[H],0,1,1,N|1,6,1,0,[H],0,1,1,N|2,2,2,1,N,1,3,2, E|2,7,1,0,[H],1,1,1,N|3,3,2,1,O,2,2,1,N|4,8,1,0,[H],3,1,1, N|

Note: The atomic signature only contains one atom that expressed by a short string when its height is zero. If the height of an atomic signature is greater than zero, the atomic signature consists of more than one atom. Every atom and the information are represented by a substring between two vertical bars in a line of code.

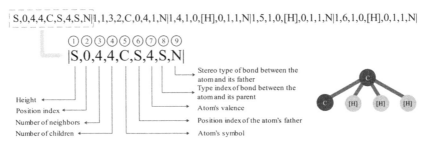

FIG. 3 Signature coding for the molecule of acetaldoxime when height is one.

the type of bond between this atom and its parent atom. The last character indicates isomeric type of the bond between this atom and its parent atom.

2.3 Data preparation: Atom embedding from chemical bonds

As the input of Tree-LSTM networks, atoms and bonds need to be translated and represented in form of vectors. Word embedding has been widely applied in natural language processing, several known program in the field has been developed, such as "Word2vec" [40]. Inspired by this method, we proposed a simple approach to generate vector representations of atoms (see Fig. 4) by breaking a chemical bond string into two smaller particles.

As we all know, chemical bonds are frequently represented in form of "A-B", "A" and "B" represent atoms, and "-" represents chemical bond types between two atoms. The strings as "A-B" are extracted from a data set of molecular structures, and then it is split into two part, "A" and "-B", as the samples to train the embedding neural network. For this application, the skip-gram algorithm [42] is employed. As such, the substrings "A" and "-B" can be mapped into vectors for expressing each node in the Tree-LSTM network. In other

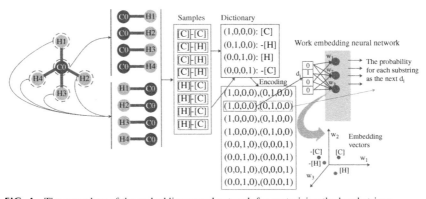

FIG. 4 The procedure of the embedding neural network for vectorizing the bond-strings.

words, a molecule is considered as a sentence in the embedding algorithm, and "A" or "-B" is equivalent to a word.

Here, the methane molecule including five atoms is taken as an example shown in Fig. 4. Every atom is considered as the starting point to record its connected bonds and atoms. A dictionary is extracted from the samples of chemical bonds. The substrings "A" and "-B" are represented by some initial vectors, for example, one-hot codes. Each initial vector is employed to train the embedding neural network. Based on these training samples, the neural network will output probabilities representing that each substring of the dictionary is the next substring. After training completed, the weights of neurons in the embedding network are formed into target vectors.

2.4 Deep neural network

A DNN combining Tree-LSTM and BPNN is developed in this work. The Tree-LSTM neural network is employed for depicting molecular tree data-structures with the canonical molecular signatures while the BPNN is used to correlate properties.

The Child-sum Tree-LSTM can be used to the dependency tree while the N-ary Tree-LSTM is applied to the constituency tree [39], and the mathematical models of these two Tree-LSTM models are listed in Table 2. The gating vectors and memory cell updates of the Tree-LSTM are dependent on the states of child units, which is different from the standard LSTM. Additionally, instead of a single forget gate, the Tree-LSTM unit contains one forget gate f_{jk} for each child k. This allows the Tree-LSTM to incorporate information selectively from each child. Since the components of the Child-Sum Tree-LSTM unit are calculated from the sum of child hidden states h_k, the Child-Sum Tree-LSTM is well suited for trees with high branching factor or whose children are unordered. The vector $\tilde{h_j}$ is the sum of the hidden states of all sub nodes under the current node j in the Child-sum Tree-LSTM model. The N-ary Tree-LSTM model can be utilized in the tree structure where the branching factor is at most N and where children are ordered from 1 to N. For any node j, the hidden state and memory cell of its kth child are written as h_{jk} and c_{jk}, respectively. The introduction of separate parameter matrices for each child k allows the N-ary Tree-LSTM model to learn more fine-grained conditioning on the states of a unit's children than those of Child-Sum Tree-LSTM.

The performance evaluation of two Tree-LSTM models on semantic classification indicated that both Tree-LSTM models are superior to the sequential LSTM model and is able to provide better classification capability [36]. Therefore, the N-ary Tree-LSTM network is employed in this work to depict molecules, and the input variables are vectors converted by the embedding algorithm. In the QSPR model, the variable x_j is the input vector representing a substring of a bond ("A" or "-B"), and the vector h_j is the output vector representing a molecular structure. The vector h_j is finally associated with the

TABLE 2 The transition equations of Child-sum Tree-LSTM and N-ary Tree-LSTM [36].

Child-sum Tree-LSTM		N-ary Tree-LSTM	
$\tilde{h}_j = \sum_{k \in C(j)} h_k$	(2)	$-^a$	
$i_j = \sigma\left(W^{(i)} x_j + U^{(i)} \tilde{h}_j + b^{(i)}\right)$	(3)	$i_j = \sigma\left(W^{(i)} x_j + \sum_{l=1}^{N} U_l^{(i)} h_{jl} + b^{(i)}\right)$	(9)
$f_{jk} = \sigma(W^{(f)} x_j + U^{(f)} h_k + b^{(f)})$	(4)	$f_{jk} = \sigma\left(W^{(f)} x_j + \sum_{l=1}^{N} U_{kl}^{(f)} h_{jl} + b^{(f)}\right)$	(10)
$o_j = \sigma\left(W^{(o)} x_j + U^{(o)} \tilde{h}_j + b^{(o)}\right)$	(5)	$o_j = \sigma\left(W^{(o)} x_j + \sum_{l=1}^{N} U_l^{(o)} h_{jl} + b^{(o)}\right)$	(11)
$u_j = \tanh\left(W^{(u)} x_j + U^{(u)} \tilde{h}_j + b^{(u)}\right)$	(6)	$u_j = \tanh\left(W^{(u)} x_j + \sum_{l=1}^{N} U_l^{(u)} h_{jl} + b^{(u)}\right)$	(12)
$c_j = i_j \cdot u_j + \sum_{k \in C(j)} f_{jk} \cdot c_k$	(7)	$c_j = i_j \cdot u_j + \sum_{l=1}^{N} f_{jl} \cdot c_{jl}$	(13)
$h_j = o_j \cdot \tanh(c_j)$	(8)	$h_j = o_j \cdot \tanh(c_j)$	(14)

^{a}Note: "–" represents null since $\tilde{h}_j$ is not involved in the N-ary Tree-LSTM unit.

properties by the BPNN. The BPNN involves an input layer, a hidden layer and an output layer. For other variables and functions in Table 2, $W^{(i,o,u,f)}$, $U^{(i,o,u,f)}$, $b^{(i,o,u,f)}$ are parameters that need to be learned, and σ represents the activation function sigmoid. For example, the model can learn parameters $W^{(i)}$ such that the components of the input gate i_j have values close to 1 (i.e., "open") when an important atom is given as input, and values close to 0 (i.e., "closed") when the input is a less important atom. Taking acetaldoxime as an example again, the computing graph of the neural network is presented in Fig. 5. It can be observed that the Tree-LSTM network mimics the topological structure of the acetaldoxime molecule. That is, if other molecular structures are learned, the Tree-LSTM network can vary the computing graph automatically. The BPNN accepts the output vectors from the Tree-LSTM network and correlates them with the property values. In this way, a DNN is built based on the Tree-LSTM network and BPNN.

Moreover, in this work, the aim of the DNN is to predict a numeric value instead of classification. Hereby, there is no need to employ the activation function "softmax" [43]. The regularization technique "dropout" [44] is introduced to the BPNN for reducing overfitting. Huber loss [45] is adopted as the loss function in the training process, which is different from the frequently used classification scheme of Tree-LSTM network. The information about the DNN is provided in Tables 3 and 4.

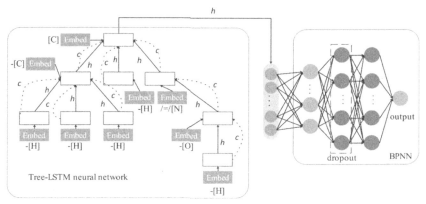

FIG. 5   The computational graph of the neural network describing the molecule acetaldoxime and predicting properties.

The regularization technique "dropout" is used to reduce overfitting in the proposed DNN. The "dropout" is easily implemented by randomly selecting nodes of a neural network to be dropped-out with a given probability (e.g., 20%) in each weight update cycle. With the cross validation, the expected probability is located between 5% and 25%.

2.5   Model training and evaluation

The Tree-LSTM network has a dynamic computational graph that is a mutable directed graph with operations as vertices and data as edges. Hence, this neural network is implemented and trained in the deep learning framework PyTorch [46]. The Adam algorithm [47] is employed to train the DNN with a learning rate of 0.02 for the first 200 epochs, and subsequent epochs with 0.0001 in learning rate. Early stopping and batch normalization are utilized to decrease overfitting. The training process proceeded by monitoring the loss of test set and it will not finish until there is no improvement in the testing loss within continuous

TABLE 3 The structural parameters of the DNN.

Names of the DNN structural parameters	Values
Shape of embedding vectors	(50,1)
Shape of parameters of Tree-LSTM	(128,128)
Shape of output vectors of Tree-LSTM	(128,1)
Layer number of the BPNN	3

TABLE 4 The hyper parameters of training the DNN.

Names of the hyper parameters	Values
Learning rate	0.02 (the first 200 epochs); 0.0001 (others)
L2 weight decay	0.00001
Batch size of training set	200
Batch size of testing set	200

50 epochs. Finally, the model with the lowest testing loss will be saved as the final model. To evaluate the correlative and predictive capacities of proposed deep learning architecture, the critical properties of pure compounds are adopted as case studies. It is acceptable that critical properties play vital roles in predicting phase behavior; however, the experimental measurements of critical properties are time-consuming, costly, and tough especially for large molecules that are easily decomposed. Moreover, several frequently used methods for the estimation of critical properties can be employed to compare with the learned DNN model. The values of critical properties are sourced from the Yaws' handbook [48] and the molecular structures of the relevant substances are gathered from PubChem database [40].

The following statistical metrics will be employed to evaluate the performance of the learned DNN models. Standard deviation (s), used to measure the amount of variation or dispersion of a data set, is given by Eq. (15).

$$s = \sqrt{\frac{\sum_{i=1}^{N} (x_i - \bar{x})^2}{N}} \tag{15}$$

Average absolute error (AAE) is the measure of deviation of predicted property values from the experimentally measured property values, and it is obtained via Eq. (16).

$$AAE = \frac{\sum_{i=1}^{N} |x_i^{exp} - x_i^{prep}|}{N} \tag{16}$$

Average relative error (ARE) provides an average of relative error calculated with respect to the experimentally measured property values, and it is expressed as Eq. (17).

$$ARE = \frac{1}{N} \sum_{i=1}^{N} \frac{|x_i^{exp} - x_i^{prep}|}{x_i^{exp}} \times 100 \tag{17}$$

3 Results and discussion

The embedding vectors representing the bond-substrings are presented at first. The DNN's capability of correlation and prediction on the data set of critical properties is evaluated in the section, and it is compared with two classical GC methods involving Joback and Reid (JR) method, Constantinou and Gani (CG) method.

3.1 Embedding of bond substrings

The input vectors of the Tree-LSTM network are translated from the substrings of the chemical bonds by the embedding neural network. After training, 106 substrings are extracted from the chemical bonds of 11,052 molecules, and then they are converted to 50-dimensional real-valued vectors as the input data representing substring of every node in the Tree-LSTM network. This is contrasted to the more dimensions required for sparse word representations, such as a one-hot encoding. These 50-dimensional vectors have been reduced to two dimensions by t-SNE algorithm [49] (see Fig. 6) for understanding easily.

The approach generates vector representing bond-substrings by the embedding neural network [50, 51] as shown in Fig. 7. The input vectors of the Tree-LSTM network are translated from the substrings of the chemical bonds by the embedding algorithm. In the training process, 106 substrings shown in Table 5 are extracted from the chemical bonds of 11,052 molecules, and then they are converted to 106 vectors by the embedding algorithm. These 50-dimensional vectors will be used as input data for every node of the Tree-LSTM network.

3.2 The DNN performance

The key idea behind the new deep learning architecture is to distinguish molecular structures by signature descriptors and to simulate molecular structures by a Tree-LSTM network. The QSPR models are obtained by training the DNN. The substances in the training and test sets are not screened carefully, which contains several small molecules and inorganic acids. Actually, these substances should be excluded from the modeling of the group contribution method, because they may cause deviation in the prediction. Finally, they are kept as some noise to the DNN. The predicting capabilities of the learned models are validated by a test set including independent compounds never used in training. The results of training and testing demonstrate that the Tree-LSTM network is capable of correlating physical properties and molecular structures (see Table 6).

The distributions of the standard deviation, average absolute error and average relative error are presented in Table 6 for three critical properties of training and test sets, respectively. The number of data points for the average relative error that is $<5\%$ and $>10\%$ are also presented. The residuals ($x^{exp}\text{-}x^{pred}$) of data points are plotted in the form of residual distribution plots in Fig. 8. Also, the

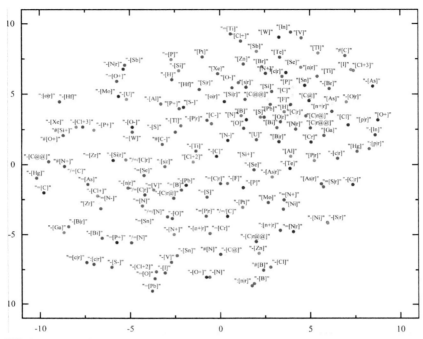

FIG. 6 The two-dimensional vectors resulting from the dimension reduction of 50-dimensional embedding vectors.

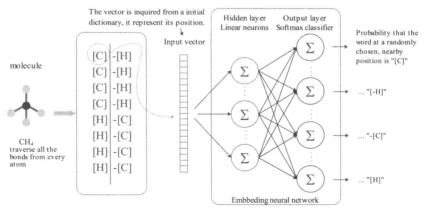

FIG. 7 The embedding neural network for vectorizing the bond-strings.

TABLE 5 The list of substrings captured from chemical bonds.

[c|r]	-[Cl]	[Si]	-[B]	-[Al]	[Tl]
-[H]	[Br]	[As]	[N+]	[Al]	-[P]
:[c|r]	-[Br]	-[As]	=[N+]	:[n|r]	[Cl]
[H]	=[C]	-[Hg]	-[N+]	[n|r]	#[Si+]
-[c|r]	#[C]	[Hg]	=[Se]	-[n|r]	[S|r]
[C]	#[N]	-[Ni]	=[Sn]	-[In]	-[Si]
-[C]	=[O]	[Ni]	[Sn]	[In]	[Si+]
[F]	[B]	-[Pb]	=[Ti]	-[Ga]	[Cl+]
-[F]	#[B]	[Pb]	[Ti]	[Ga]	-[O-]
[I]	[C|r]	=[Pb]	-[Zn]	:[o|r]	-[O+]
-[I]	-[C|r]	=[C|r]	[Zn]	[o|r]	[O+]
[N]	=[S]	[Se]	=[Zr]	=[B]	#[C-]
-[N]	=[N]	-[Se]	[Zr]	-[Cl+2]	=[O+]
[O]	-[O|r]	/=\[C]	#[N+]	[Cl+2]	/=\[N]
-[O]	[O|r]	/=/[C]	[Xe]	-[S-]	=[P]
[S]	-[N|r]	/=/[N]	-[Xe]	[S-]	-[Tl]
-[S]	[N|r]	[O-]	[C-]	:[s|r]	[s|r]
[P]	-[S|r]	-[Cl+]	#[O+]		

predicted values of these compounds by the proposed DNN in comparison with the experimental data are shown in Fig. 9.

3.3   Comparisons with two existing methods

Taken as examples, two existing GC methods for the estimation of critical properties are compared with the proposed DNN method, which involves JR method and CG method. The available performance data are provided by Poling et al. [52] We have to admit that the completely equitable comparison with other existing methods of property predictions is impossible since every method might be regressed from different data sources.

For the critical temperature (see Table 7), the JR method based on the experimental boiling points exhibits more accuracy than other GC methods; however,

TABLE 6 Global comparison of critical properties between training and test sets.

Properties	Data points		s		AAE		ARE (%)		# Err <5%		# Err > 10%	
	Train	Test	Train	Test	Train	Test	Train	Test	Train	Test	Train	Test
T_c (K)	1432	360	145.89	166.09	22.48	23.77	4.23	5.29	1104	266	109	36
P_c ($\times 10^5$ Pa)	1380	346	161.80	139.55	1.34	3.18	3.81	8.29	1104	177	98	89
V_c ($\times 10^{-6}$ m³/mol)	1440	361	199.18	169.73	7.10	19.92	1.97	6.15	1361	245	19	59

Note: s, standard deviation; AAE, average absolute error; ARE, average relative error.

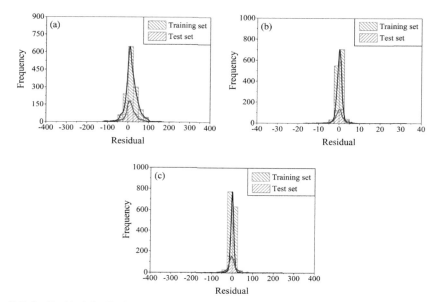

FIG. 8 Residual distribution plots for the training and test sets of: (A) T_c; (B) P_c; (C) V_c.

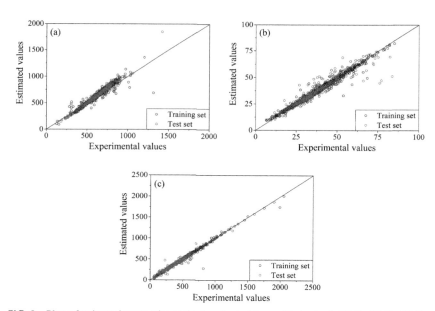

FIG. 9 Plots of estimated vs experimental values for training and test sets of: (A) T_c; (B) P_c; (C) V_c.

TABLE 7 The comparisons among DNN and GC methods in predicting critical temperature.

Methods	Substances	AAE^a	ARE^a	# Err < 5%[b]	# Err > 10%[c]
JR [48] (Exp. $T_b)^d$	352[e]	6.65	1.15	345	0
	290[f]	6.68	1.10	286	0
JR [48] (Est. $T_b)^g$	352[e]	25.01	4.97	248	46
	290[f]	20.19	3.49	229	18
DNN[h]	352[e]	15.39	2.92	299	15
	290[f]	13.92	2.31	265	7
CG (1st) [48]	335[e]	18.48	3.74	273	28
	286[f]	13.34	2.25	254	4
CG (2nd) [48]	108[e]	17.69	13.61	274	29
	104[f]	12.49	2.12	254	6
DNN[i]	452[e]	26.59	5.87	343	51
	335[f]	15.98	2.62	294	11

Note:
[a]AAE is average absolute error; ARE is average relative error.
[b]The number of substances for which the ARE was <5% (# Err< 5%).
[c]The number of substances for which the ARE was >10% (# Err> 10%).
[d]The values of estimation is based on the experimental values of normal boiling point.
[e]The number of substances in the list provided by Poling et al. [50] with data that could be tested with the method in the current line.
[f]The number of substances in the list provided by Poling et al. [50] having three or more carbon atoms with data that could be tested with the method in the current line.
[g]The values of estimation is based on the estimation values of normal boiling point.
[h]The number of substances is kept consistent with the JR method.
[i]The number of all the substances that could be predicted by DNN.

the accuracy of the JR method based on the estimated boiling points shows a marked decline [52]. To make the comparison as fair as possible, the substances from the same list provided by Poling et al. [52] are chosen to predict the critical temperature (T_c) using the proposed DNN. It can be seen from Table 7 that the DNN shows higher performance than JR method (Est. Tb). It is noticed that the CG method involves groups in two orders, and the second order partially overcomes the limitation of the single order that cannot distinguish special molecular structures. Hereby, the number of substances estimated by the CG (2nd) method shown in Table 7 is actually apart from the substances estimated by the CG (1st) method. Although the list of 335 compounds within the CG method is not ascertained, it can be concluded that the accuracy of the learned DNN model is close to the CG method. When the learned DNN model is evaluated with all substances in the list provided by Poling et al. [52], a decline in precision can be observed but the resulting ARE is still close to the others. Hereby, the DNN method can predict some substances that these GC methods cannot estimate, and the accuracy is close to the CG method for the critical temperature when the amount of substances engaged in the comparison is approximate. Moreover, the DNN method also provides better precision when only predicting molecules with more than three carbon atoms.

For critical pressure (see Table 8), the estimations with the learned DNN model are more accurate than all other methods. It also proves that the method can correlate properties with more substances and has better accuracy for predicting the critical pressure (P_c). Furthermore, for the estimation of the critical volume (V_c), as indicated in Table 9, the estimation of the critical volume with the DNN method reaches precision close to other methods.

Actually, in total 468 substances are provided by Poling et al. [52] It can be observed that the number of substances estimated by the CG method and JR

TABLE 8 The comparisons among DNN and GC methods in predicting critical pressure.

Methods	Substances	AAE^a	ARE^a	# Err $<5\%^b$	# Err $>10\%^c$
JR [48]	328[d]	2.19	5.94	196	59
	266[e]	1.39	4.59	180	30
DNN[f]	328[d]	1.46	4.03	248	23
	266[e]	1.21	3.94	206	19
CG (1st) [48]	316[d]	2.88	7.37	182	52
	263[e]	1.80	5.50	156	32
CG (2nd) [48]	99[d]	2.88	7.37	187	56
	96[e]	1.80	5.50	160	36
DNN[g]	450[d]	2.66	5.43	314	58
	335[e]	1.33	4.40	241	26

Note:
[a]AAE is average absolute error; ARE is average relative error.
[b]The number of substances for which the ARE was less than 5% (# Err< 5%).
[c]The number of substances for which the ARE was greater than 10% (# Err> 10%).
[d]The number of substances in the list provided by Poling et al. [50] with data that could be tested with the method in the current line.
[e]The number of substances in the list provided by Poling et al. [50] having 3 or more carbon atoms with data that could be tested with the method in the current line.
[f]The number of substances is kept consistent with the JR method.
[g]The number of all the substances could be predicted by DNN.

TABLE 9 The comparisons among DNN and GC methods in predicting critical volume.

Methods	Substances	AAE^a	ARE^a	# Err < 5%^{b}	# Err > 10%^{c}
JR [48]	236^d	12.53	3.37	189	13
	185^e	13.98	3.11	148	9
DNN^f	236^d	10.07	2.99	197	13
	185^e	11.20	2.69	157	10
CG (1st) [48]	220^d	15.99	4.38	160	18
	180^e	16.68	4.57	159	22
CG (2nd) [48]	76^d	16.5	3.49	136	10
	72^e	17.4	3.70	134	15
DNN^g	402^d	15.05	4.84	301	56
	230^e	17.38	4.20	236	31

Note:
^{a}AAE is average absolute error; ARE is average relative error.
^{b}The number of substances for which the ARE was <5% (# Err< 5%).
^{c}The number of substances for which the ARE was >10% (# Err> 10%).
^{d}The number of substances in the list provided by Poling et al. [50] with data that could be tested with the method in the current line.
^{e}The number of substances in the list provided by Poling et al. [50] having three or more carbon atoms with data that could be tested with the method in the current line.
^{f}The number of substances is kept consistent with the JR method.
^{g}The number of all the substances could be predicted by DNN.

method is less than the learned DNN model. In other words, the critical properties of some substances cannot be predicted using these two existing GC methods. The reason is that the GC methods are limited by the types and segmentation rules of groups while the DNN method is not subject to them. Hereby, the DNN can predict more compounds and achieve a decent precision while it performs the acceptable precision on the substances provided by Poling et al. [52]

Another important fact has also to be considered is the compounds exemplified in Tables 7–9 has been involved more or less in the regression samples of the JR method, the CG method, and the DNN. Although the above-mentioned comparison can evaluate the predictive capability of the DNN differed from those two existing GC methods, the extrapolation ability is also necessary to be evaluated. There is no program available to us to estimate properties by the CG method, Fig. 10 only shows the comparison of extrapolation abilities

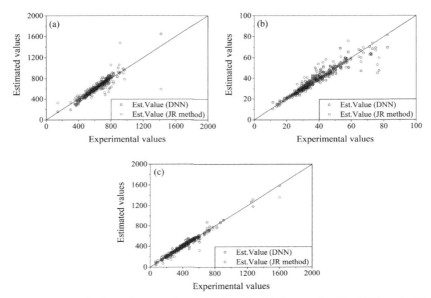

FIG. 10 Plots of estimated vs experimental values of critical properties (the DNN vs the JR method) of: (A) T_c; (B) P_c; (C) V_c.

between the JR method and the DNN according to the substances of test sets shown in Table 7.

Some works based on other artificial intelligence methods for optimization still used the manually selected molecular features. The completely fair comparisons are not practical among different approaches of QSPR modeling, because of the different data, properties, and training theories. The statistical evaluations between DNN and other methods are provided in Table 10. Joback and Reid [6] developed a useful group-contribution method to estimate critical properties. Lee and Chen [53] used two ANNs and the Joback model to correlate critical properties.

The number of substances employed in this work (i.e., 1767–1784 compounds sourced from DIPPR database [54]) is larger than those of the other two studies (i.e., 321–421 compounds), thus it is more difficult for DNN to provide better training results than other works may attribute to the data set including more inaccurate experimental and predictive values than the other two methods. Another reason is that the anti-overfitting techniques involving batch normalization and early stopping are employed in DNN. However, of note is that the DNN models can provide the better test results. On the other hand, some substances could not be estimated in group contributions when molecular groups are not encountered.

TABLE 10 The comparison among this work and the other two methods in predicting critical properties.

Methods	Properties	Substances		Correlation coefficient		Average relative error (%)	
		Train	Test	Train	Test	Train	Test
This work (DNN)	Critical temperature	1427	357	0.957	0.967	3.85	3.31
	Critical pressure	1424	357	0.969	0.977	6.26	5.43
	Critical volume	1413	354	0.993	0.991	5.41	4.01
Joback [4, 5] (based on Est. Tb)	Critical temperature[a]	409	352	–	–	0.8	4.97
	Critical pressure[a]	392	328	–	–	5.2	5.94
	Critical volume[a]	310	236	–	–	2.3	3.37
Lee and Chen [6]	Critical temperature	292	29	–	–	3.1	7.7
	Critical pressure	292	29	–	–	3.4	7.1
	Critical volume	292	29	–	–	6.4	14.1

[a]The train set cited from the original literature of Joback while the test set is cited from Poling et al.

3.4 Distinction of isomers

Signature descriptors have the ability to distinguish isomers. Table 11 exhibits the estimations through the DNN method as opposed to experimental values and other GC methods. Apparently, the JR method cannot recognize isomers, although it is able to predict more accurate according to the experimental boiling point. The CG method with the second order of groups can obtain decent prediction for isomers, and the DNN method can achieve similar results.

4 Conclusions

In this work, a deep learning architecture is developed and the prediction of physical properties from the holistic molecular structure is achieved in

TABLE 11 Experimental and estimated critical temperature values of isomeric trimethylpentane and methyl propanol.

Compounds	CAS no.	Exp. value (K)	JR method (Est. T_b) Est. value (K)	JR method (Exp. T_b) Est. value (K)	CG method Est. value (K)	DNN Est. value (K)
2,2,3-Trimethylpentane	564-02-3	563.40	557.09	563.31	562.10	563.42
2,2,4-Trimethylpentane	540-84-1	543.90	557.09	547.71	540.33	544.98
2,3,3-Trimethylpentane	560-21-4	573.50	557.09	570.55	577.45	576.88
2,3,4-Trimethylpentane	560-21-4	566.30	556.23	564.26	581.37	565.84
2-Methyl-1-propanol	78-83-1	547.78	548.34	546.11	543.32	552.96
2-Methyl-2-propanol	75-65-0	506.20	548.34	509.38	497.46	500.57

following four steps. First, an embedding neural network is used to generate the vector representations of bond-substrings. Then, a canonization algorithm is employed to convert the molecules to uniform data-structures for providing templates to the Tree-LSTM neural network. Next, the computational graph of the Tree-LSTM network accepts a vector of bond-string on its each node, which is self-adaptive to various molecular structures. Finally, a vector outputting from the root node of the Tree-LSTM network is introduced to a BPNN to generate predictive property values. The proposed DNN does not rely on the well-chosen descriptors to correlate properties, it could learn some valuable features of molecule and achieve an acceptable precision of a specific property for more substances with less human effort.

The proposed approach neither counts the frequencies of molecular substructures nor calculates any TIs, instead, provides a way to build the QSPR models from the text-type descriptor, the canonical signature representing molecular graphs. Hence, the strategy has a capability to capture the relevant molecular features for QSPR modeling automatically. Furthermore, those parameters involved in the learned DNN model are not the contribution value of each group in GC methods but tensors containing potential information.

For validating the effectiveness of the proposed deep learning architecture, critical properties are taken as case studies to train and test the QSPR models built from the proposed DNN combining Tree-LSTM and back-propagation neural network (BPNN). It has been proven that these QSPR models provide more accurate prediction and cover more diverse molecular structures. Moreover, the DNN behaves a better ability in distinguishing isomers. We admit that the data used to train the model is still far from enough. This signifies that there needs to be more data to capture the delicate relationships that may exist between molecular structures and physical properties.

In a word, the wide applicability of the proposed architecture highlights the significance of deep learning providing an intelligent tool to predict properties in the design or synthesis of chemical, pharmaceutical, bio-chemical products, and processes. It is worth mentioning that the proposed strategy could be widely applied for the estimation of other properties of pure compounds, such as environment-related properties and safety-related properties.

Acknowledgments

The chapter is reprinted from AIChE Journal, 2019, 65 (9), e16678, Yang Su, Zihao Wang, Saimeng Jin, Weifeng Shen, Jingzheng Ren, Mario R. Eden, An architecture of deep learning in QSPR modeling for the prediction of critical properties using molecular signatures, by permission of John Wiley & Sons.

References

[1] W. Shen, L. Dong, S. Wei, J. Li, H. Benyounes, X. You, V. Gerbaud, Systematic design of an extractive distillation for maximum-boiling azeotropes with heavy entrainers, AICHE J. 61 (11) (2015) 3898–3910.

[2] A. Yang, W. Shen, S. Wei, L. Dong, J. Li, V. Gerbaud, Design and control of pressure-swing distillation for separating ternary systems with three binary minimum azeotropes, AICHE J. 65 (4) (2019) 1281–1293.

[3] G.M. Kontogeorgis, R. Gani, Introduction to computer aided property estimation, in: G.M. Kontogeorgis, R. Gani (Eds.), Computer Aided Property Estimation for Process and Product Design, Elsevier, Amsterdam, the Netherlands, 2004, pp. 3–26.

[4] A.L. Lydersen, Estimation of Critical Properties of Organic Compounds, University of Wisconsin College of Engineering, Madison, WI, 1995 (Eng. Exp. Stn. Rep).

[5] K.M. Klincewicz, R.C. Reid, Estimation of critical properties with group contribution methods, AICHE J. 30 (1) (1984) 137–142.

[6] K.G. Joback, R.C. Reid, Estimation of pure-component properties from group-contributions, Chem. Eng. Commun. 57 (1–6) (1987) 233–243.

[7] L. Constantinou, R. Gani, New group contribution method for estimating properties of pure compounds, AICHE J. 40 (10) (1994) 1697–1710.

[8] J. Marrero, R. Gani, Group-contribution based estimation of pure component properties, Fluid Phase Equilib. 183-184 (2001) 183–208.

[9] N.D. Austin, N.V. Sahinidis, D.W. Trahan, Computer-aided molecular design: an introduction and review of tools, applications, and solution techniques, Chem. Eng. Res. Des. 116 (2016) 2–26.

[10] J.L. Faulon, D.P. Visco, R.S. Pophale, The signature molecular descriptor. 1. Using extended valence sequences in QSAR and QSPR studies, J. Chem. Inf. Comput. Sci. 43 (3) (2003) 707–720.

[11] J.L. Faulon, C.J. Churchwell, D.P. Visco, The signature molecular descriptor. 2. Enumerating molecules from their extended valence sequences, J. Chem. Inf. Comput. Sci. 34 (3) (2003) 721–734.

[12] J.L. Faulon, M.J. Collins, R.D. Carr, The signature molecular descriptor. 4. Canonizing molecules using extended valence sequences, J. Chem. Inf. Comput. Sci. 44 (2) (2004) 427–436.

[13] S. Borman, Neural network applications in chemistry begin to appear, Chem. Eng. News 67 (17) (1989) 24–28.

[14] N. Bodor, A. Harget, M.J. Huang, Neural network studies. 1. Estimation of the aqueous solubility of organic compounds, J. Am. Chem. Soc. 113 (25) (1991) 9480–9483.

[15] T. Aoyama, Y. Suzuki, H. Ichikawa, Neural networks applied to structure-activity relationships, J. Med. Chem. 33 (3) (1990) 905–908.

[16] L.M. Egolf, P.C. Jurs, Prediction of boiling points of organic heterocyclic compounds using regression and neural network techniques, J. Chem. Inf. Comput. Sci. 33 (4) (1993) 616–625.

[17] D.B. Kireev, ChemNet: a novel neural network based method for graph/property mapping, J. Chem. Inf. Comput. Sci. 35 (2) (1995) 175–180.

[18] J. Devillers, Neural Networks in QSAR and Drug Design, Principles of QSAR and Drug Design, second ed., Academic Press, San Diego, 1996.

[19] A.P. Bünz, B. Braun, R. Janowsky, Application of quantitative structure-performance relationship and neural network models for the prediction of physical properties from molecular structure, Ind. Eng. Chem. Res. 37 (8) (1998) 3043–3051.

[20] B. Beck, A. Breindl, T. Clark, QM/NN QSPR models with error estimation: vapor pressure and logP, J. Chem. Inf. Comput. Sci. 40 (4) (2000) 1046–1051.

[21] G. Espinosa, D. Yaffe, Y. Cohen, A. Arenas, F. Giralt, Neural network based quantitative structural property relations (QSPRs) for predicting boiling points of aliphatic hydrocarbons, J. Chem. Inf. Comput. Sci. 40 (3) (2000) 859–879.

[22] X. Yao, X. Zhang, R. Zhang, M. Liu, Z. Hu, B. Fan, Prediction of enthalpy of alkanes by the use of radial basis function neural networks, Comput. Chem. 25 (5) (2001) 475–482.

[23] D. Yaffe, Y. Cohen, Neural network based temperature-dependent quantitative structure property relations (QSPRs) for predicting vapor pressure of hydrocarbons, J. Chem. Inf. Comput. Sci. 41 (2) (2001) 463–477.

[24] D. Yaffe, Y. Cohen, G. Espinosa, A. Arenas, F. Giralt, Fuzzy ARTMAP and back-propagation neural networks based quantitative structure-property relationships (QSPRs) for octanol-water partition coefficient of organic compounds, J. Chem. Inf. Comput. Sci. 42 (2) (2002) 162–183.

[25] T.A.A. And, R.S. George, Artificial neural network investigation of the structural group contribution method for predicting pure components auto ignition temperature, Ind. Eng. Chem. Res. 42 (22) (2003) 5708–5714.

[26] T.L. Chiu, S.S. So, Development of neural network QSPR models for Hansch substituent constants. 1. Method and validations, J. Chem. Inf. Comput. Sci. 44 (1) (2004) 147–153.

[27] J.S. Torrecilla, F. Rodríguez, J.L. Bravo, G. Rothenberg, K.R. Seddon, I. Lopez-Martin, Optimising an artificial neural network for predicting the melting point of ionic liquids, Phys. Chem. Chem. Phys. 10 (38) (2008) 5826–5831.

[28] F. Gharagheizi, R.F. Alamdari, M.T. Angaji, A new neural network-group contribution method for estimation of flash point temperature of pure components, Energy Fuel 22 (3) (2008) 1628–1635.

[29] F. Gharagheizi, New neural network group contribution model for estimation of lower flammability limit temperature of pure compounds, Ind. Eng. Chem. Res. 48 (15) (2009) 7406–7416.

[30] R. Wang, J. Jiang, Y. Pan, H. Cao, Y. Cui, Prediction of impact sensitivity of nitro energetic compounds by neural network based on electrotopological-state indices, J. Hazard. Mater. 166 (2009) 155–186.

[31] A. Guerra, N.E. Campillo, J.A. Páez, Neural computational prediction of oral drug absorption based on CODES 2D descriptors, Eur. J. Med. Chem. 45 (2010) 930–940.

[32] M. Bagheri, T.N.G. Borhani, G. Zahedi, Estimation of flash point and autoignition temperature of organic sulfur chemicals, Energy Convers. Manag. 58 (2012) 185–196.

[33] A. Lusci, G. Pollastri, P. Baldi, Deep architectures and deep learning in chemoinformatics: the prediction of aqueous solubility for drug-like molecules, J. Chem. Inf. Comput. Sci. 53 (7) (2013) 1563–1575.

[34] G.B. Goh, N.O. Hodas, C. Siegel, A. Vishnu, Smiles2vec: an interpretable general-purpose deep neural network for predicting chemical properties, arXiv preprint (2017) (1712.02034).

[35] D. Weininger, SMILES, a chemical language and information system. 1. Introduction to methodology and encoding rules, J. Chem. Inf. Comput. Sci. 28 (1) (1988) 31–36.

[36] G.B. Goh, C. Siegel, A. Vishnu, N.O. Hodas, N. Baker, Chemception: a deep neural network with minimal chemistry knowledge matches the performance of expert-developed QSAR/QSPR models, arXiv preprint (2017) (1706.06689).

[37] S. Hochreiter, J. Schmidhuber, Long short-term memory, Neural Comput. 9 (8) (1997) 1735–1780.

[38] H. Altae-Tran, B. Ramsundar, A.S. Pappu, V. Pande, Low data drug discovery with one-shot learning, ACS Cent. Sci. 3 (4) (2017) 283–293.

[39] K.S. Tai, R. Socher, C.D. Manning, Improved semantic representations from tree-structured long short-term memory networks, Comput. Sci. 5 (1) (2015) 36.

[40] S. Kim, P.A. Thiessen, E.E. Bolton, J. Chen, G. Fu, A. Gindulyte, L. Han, J. He, S. He, B.A. Shoemaker, J. Wang, B. Yu, J. Zhang, S.H. Bryant, PubChem substance and compound databases, Nucleic Acids Res. 44 (D1) (2015) D1202–D1213.

[41] G. Landrum, Rdkit: Open-Source Cheminformatics Software, 2017, Available at: http://www.rdkit.org/. https://github.com/rdkit/rdkit.

[42] C. McCormick, Word2Vec Tutorial-the Skip-Gram Model, 2016, Available at: http://www.mccormickml.com.

[43] G. Klambauer, T. Unterthiner, A. Mayr, S. Hochreiter, Self-normalizing neural networks, in: I. Guyon, U.V. Luxburg, S. Bengio, H. Wallach, R. Fergus, S. Vishwanathan, R. Garnett (Eds.), Advances in Neural Information Processng Systems 30, Curran Associates, Inc, New York, 2017, pp. 971–980.

[44] G.E. Hinton, N. Srivastava, A. Krizhevsky, I. Sutskever, R.R. Salakhutdinov, Improving neural networks by preventing co-adaptation of feature detectors, Comput. Sci. 3 (4) (2012) 212–223.

[45] P.J. Huber, Robust estimation of a location parameter, Ann. Math. Stat. 35 (1) (1964) 73–101.

[46] A. Paszke, S. Gross, S. Chintala, G. Chanan, Pytorch: Tensors and Dynamic Neural Networks in Python With Strong GPU Acceleration, 2019, Available at: https://pytorch.org.

[47] D.P. Kingma, J. Ba, Adam: a method for stochastic optimization, arXiv Preprint (2017), 1412.6980v5.

[48] C.L. Yaws, C. Gabbula, Yaws' Handbook of Thermodynamic and Physical Properties of Chemical Compounds, Knovel, New York, 2003.

[49] L.V.D. Maaten, G. Hinton, Visualizing data using t-sne, J. Mach. Learn. Res. 9 (2605) (2008) 2579–2605.

[50] T. Mikolov, I. Sutskever, K. Chen, G. Corrado, J. Dean, Distributed representations of words and phrases and their compositionality, arXiv eprint (2013). 1310.4546.

[51] R. Lebret, R. Collobert, Word emdeddings through Hellinger PCA, arXiv eprint (2013). 1312.5542.

[52] B.E. Poling, J.M. Prausnitz, J.P. O'connell, The Properties of Gases and Liquids, fifth ed., McGraw-Hill, New York, 2001.

[53] M.J. Lee, J.T. Chen, Fluid property predictions with the aid of neural networks, Ind. Eng. Chem. Res. 32 (5) (1993) 995–997.

[54] Design Institute for Physical Properties, Sponsored by AIChE, DIPPR Project 801 - Full Version. Design Institute for Physical Property Research/AIChE, 2017, Retrieved from https://app.knovel.com/.

Chapter 3

Predictive deep learning models for environmental properties

Zihao Wang and Weifeng Shen
School of Chemistry and Chemical Engineering, Chongqing University, Chongqing, People's Republic of China

1 Introduction

As one of the cornerstones for the sustainable development [1], environmental benefit drives chemical process technology and environmental science toward environmentally friendly technology [2–7]. The environmental impact is an indispensable factor that should be considered in the molecular design, chemical synthesis, and solvent selection [8–12]. As an essential environmental property, the lipophilicity refers to the affinity of a compound for lipids and provides valuable information about the absorption, distribution, and metabolism of compounds [13–15].

Usually, the lipophilicity of a compound is measured as its partition coefficient between lipid and aqueous phases. Octanol has been widely accepted as a token representing cell membranes, tissue, and lipids [13] and the partition coefficient between octanol and aqueous phases is frequently adopted as a measure for the lipophilicity of organic chemicals [15–17] as well as a physicochemical criterion for solvent selection [18–20]. This partition coefficient could be further employed to predict various indicators of toxicity [e.g., 50% effective concentration (EC50) and 50% lethal concentration (LC50)] [21,22].

The octanol-water partition coefficient (K_{OW}) describes the distribution of a substance between the octanol and aqueous phases in a two-phase octanol-water system at equilibrium [23]. It can be ideally measured by experiments, but the existing database is not enough for many compounds of interest. Additionally, the experimental determinations are not always feasible for those compounds with low water solubility. In this context, various methods that rely on property estimation were developed and continue to be proposed to solve the problems of generality and accuracy.

Many researchers reviewed the existing models on predicting log K_{OW} and they highlighted the frontiers and prospects of developing prediction

Applications of Artificial Intelligence in Process Systems Engineering
https://doi.org/10.1016/B978-0-12-821092-5.00015-2

39

approaches in this respect [24–26]. Additionally, some investigators elaborated the state-of-the-art and assessed the performances of the representative log K_{OW} predictive models [27,28]. A large number of studies have focused on developing empirical relationship models in which K_{OW} is described as a function of molecular physicochemical properties [14,29–33]. These empirical methods can be efficient in the computation but heavily rely on correlated properties which are not always available. In contrast, the quantity structure-property relationship (QSPR) methods, such as group contribution (GC) methods and topological methods, are more readily implemented since only structural information needs to be provided [34].

Prior to the GC methods, atom and fragment contribution methods laid a solid foundation for the prediction of properties [16,35,36]. A molecule can be divided into atoms and fragments without any ambiguity and as such these methods achieved success in property estimation. However, molecules are much more than a collection of atoms [28]. In this regard, as extensions of atom and fragment contribution methods, modified GC methods have been put forward with the purpose of predicting properties for organic chemicals [37–40]. In the GC methods, various groups (e.g., substructures containing atoms and bonds) can be defined, and the target property value of a compound is given by summarizing the contributions of groups. A typical example is the three-level GC method proposed by Marrero and Gani [15,39] and it was applied to estimate K_{OW}. This method showed a better predictive accuracy regarding a large quantity of organic compounds. On the other hand, topology is an unambiguous feature and topological properties can be directly derived from molecular structures [28]. Therefore, different topological characteristics have been extensively adopted as descriptors to develop QSPR models and correlate properties [41–44].

Although the GC and topological methods have revealed satisfactory performance in property estimation, a few shortcomings have limited their extensive applications, such as the limited discriminative power in isomers and the inadequate consideration in the holistic molecular structures [45]. To overcome these shortcomings, signature molecular descriptors were introduced which could capture a whole picture with connectivity information of each atom for a molecule [46,47]. This signature descriptor was developed specifically for molecular structures and it could be a potential tool for QSPR modeling without the need for calculation and selection of other numerical descriptors [45]. Moreover, it was further detailed and applied in QSPR researches [48,49].

Meanwhile, a major expansion has appeared in the field of QSPR due to the advent of artificial intelligence (AI). Artificial neural networks have been extensively employed to determine the correlations between molecular structures and properties [41,50–54]. In this respect, based on the long short-term memory (LSTM) [55], the traditional LSTM network is usually structured as a linear chain and this exhibited the superior representational power and effectiveness. However, some types of data (such as text) are better represented as the tree structures. In this context, an advanced Tree-structured LSTM (Tree-LSTM) network was put forward as a variant of the LSTM network to capture the syntactic properties of natural languages [56]. With regard to the complex and various molecular structures, the Tree-LSTM network is supposed as an attractive option in representing the relationships of the atoms or groups.

Recently, a deep learning approach for predicting the properties of chemical compounds was proposed, in which the Tree-LSTM network was successfully implemented with the purpose of expressing and processing chemical structures [57]. Additionally, taking the critical properties as examples, it proved that the proposed deep learning approach is suitable for a more diverse range of molecular structures and enables users to achieve more accurate predictions.

Based on the state-of-the-art, there are still three issues to be solved in K_{OW} estimation for organic compounds, and they are:

(i) Human intervention was involved in the feature selection of the molecular structures during the model development, which caused the omission of the important molecular information.
(ii) Too many topological features or physicochemical descriptors have been adopted, increasing the complexity of models and decreasing the computational efficiency.
(iii) The ability of differentiating structural isomers and stereoisomers is limited in the reported QSPR models, which constrains the application scope of the predictive model.

In order to overcome the three challenges and motivated by the successful deep learning approach in property estimation, a QSPR model was developed in this research to accurately predict K_{OW} values for organic compounds and provide the valuable environmental information for guiding the selection and development of the important chemicals including green solvents. In this model, the automatic feature selection for molecules was achieved by coupling the canonical molecular signature and deep neural network (DNN).

2 Methodology

A DNN model, which couples the Tree-LSTM network and back-propagation neural network (BPNN), was developed in this study based on the deep learning approach. It was built to specialize in the determination of the correlation between molecular structures and log K_{OW} values of organic compounds. The process of developing a reliable QSPR model with the DNN model is comprised of the following five basic steps, as illustrated in the Fig. 1.

Step 1: Data collection. The experimentally measured log K_{OW} values and simplified molecular-input line-entry system (SMILES) strings of compounds were collected since they are necessary for developing a QSPR

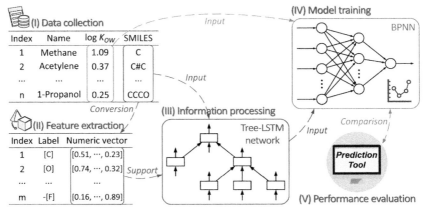

FIG. 1 The schematic diagram of the process for developing a QSPR model with the deep learning approach.

model. Herein, the SMILES strings sufficed for representing the basic molecular structural information.

Step 2: Feature extraction. The SMILES strings of compounds were utilized to generate a list of numeric vectors based on a proposed atom embedding algorithm which was implemented with the atomic signatures. The vectors are able to describe molecular structures and represent their features.

Step 3: Information processing. The SMILES strings were converted to canonical molecular signatures with the theory of the canonizing molecular graph [58]. On this basis, these signatures were mapped on the Tree-LSTM networks with the aim of creating vectors as inputs for the BPNN.

Step 4: Model training. After receiving the inputs from the Tree-LSTM networks, the BPNN supported the correlation process, and it was repeatedly run to learn a satisfactory QSPR model. In the training process, parameters were updated to optimize the parameters of the DNN model and finally the QSPR model with better performance was preserved for log K_{OW} estimation.

Step 5: Performance evaluation. Based on the developed QSPR model, the generalization ability was assessed by the predictive performance of an external dataset. And the external competitiveness of the QSPR model was evaluated by comparing to an authoritative predictive model.

All the above steps for obtaining the QSPR model to predict the K_{OW} were achieved with a series of programs which were written in the Python language and successfully tested on Windows platforms.

2.1 Data acquisition and processing

The dimensionless K_{OW} values span over 10 orders of magnitude and therefore the decimal logarithm of K_{OW} (log K_{OW}) was frequently adopted in property estimation. A large number of experimentally measured log K_{OW} values of chemical compounds were collected [59], and all the experimental values were originated from references to guarantee the reasonability of the predictive model. To investigate the QSPR model for organic compounds, a number of irrelevant compounds were eliminated. The excluded irrelevant compounds involve the inorganic compounds (e.g., carbon dioxide, sulfur hexafluoride, and hydrazine), metal-organic compounds (i.e., the organic compounds containing metal atoms such as sodium, chromium or/and stannum) and mixtures consisting two or more compounds. Hence, the remaining 10,754 pure organic compounds were assembled for the model development.

As a large dataset was collected, the data cleaning is essential to be carried out by detecting and removing outliers, which contain gross errors. Accordingly, the Pauta criterion [60], also referred to as the three-sigma rule, was applied for the cleaning process. It describes that 99.73% of all values of a normally distributed parameter fall within three times the standard deviation (σ) of the average (μ). Any error beyond this interval is not a random error but a gross error. Accordingly, data points which include gross error are regarded as outliers and should be excluded from the sample data. The data cleaning process with Pauta criterion is graphically illustrated in the Fig. 2.

As a result, 86 out of 10,754 organic compounds (about 0.8% of the dataset) were detected as outliers based on their experimental values and they were

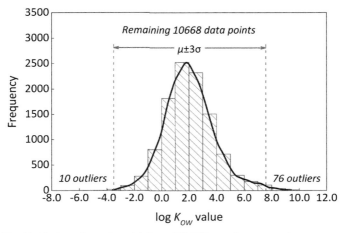

FIG. 2 The distribution of experimental data of 10,754 organic compounds.

removed from the dataset. The remaining 10,668 organic compounds were preserved as the final dataset for developing a QSPR model to predict log K_{OW}. The dataset of compounds spans a wide class of molecular structures including aliphatic and aromatic hydrocarbons, alcohols, and phenols, heterocyclic compounds, amines, acids, ketones, esters, aldehydes, ethers, and so on. In order to demonstrate the chemical diversity of the dataset, the corresponding counts of different types were detailed in Table 1, and their distributions in the training, test, and external sets were also provided. Since the subsets were divided with a random selection routine, proportions of different types of compounds in each subset approximate corresponding proportions for the compounds of subset in the entire dataset.

In addition to the experimental values, the information of molecular structures is also indispensable in developing a QSPR model. SMILES [61] is a chemical language representing structural information in the text form and it is widely applied in the chemo-informatics software because it can be employed to build molecular two-dimensional or three-dimensional structures. Moreover, one can manually provide the SMILES string of any compound after simply learned the encoding rules.

The open chemistry database, PubChem [62], contains the largest collection of publicly available chemical information and provides two types of SMILES

TABLE 1 The detailed analysis for the types of compounds in the entire dataset and three disjoint subsets.

	Training set	Test set	External set	Entire dataset
Aliphatic and aromatic hydrocarbons	731	80	114	925
Alcohols and phenols	523	67	67	657
Heterocyclic compounds	2074	275	243	2592
Amines	1596	188	192	1976
Acids	988	104	110	1202
Ketones	782	100	116	998
Esters	521	64	61	646
Aldehydes	49	7	11	67
Ethers	134	21	13	168
Others	1136	161	140	1437
Total	*8534*	*1067*	*1067*	*10,668*

strings (i.e., canonical SMILES and isomeric SMILES) for tens of millions of compounds. The canonical SMILES strings are available for all the compounds, whereas the isomeric SMILES strings which contain isomeric information are only provided for isomers. With respect to the dataset applied in this study, the SMILES string of each compound was derived from the PubChem according to its chemical abstracts service registry number (CASRN). During the SMILES acquisition, the isomeric SMILES string was collected if available. Otherwise, the canonical SMILES string was adopted. Eventually, the experimental values and SMILES strings of 10,668 organic compounds were adopted as the inputs for developing the QSPR model.

2.2   Tree structures in information processing

The signature molecular descriptor was introduced specifically for describing molecular structures, and all the connectivity information for every atom in a molecule was retained. Additionally, it can be theoretically applied to represent any organic compound which means that it is able to cover various molecular structures without limitation.

Herein, taking 1-propanol (CASRN: 71-23-8; SMILES: CCCO) as an example. When a root atom was specified in the molecule, a tree spanning all atoms and bonds of the molecule was constructed (refer to Fig. 3A), and the signatures were generated relying on the theory of canonizing molecular graph [58].

Up to a point, the syntactic property of natural languages is analogous to the connectivity information for atoms in a molecule. The former one is able to be captured by the Tree-LSTM network while the later one can be expressed with a signature. In addition, the tree structure of Tree-LSTM network (refer to Fig. 3B) is similar to the signature tree displayed in the Fig. 3A. Therefore, it was assumed that the molecular structural information can be processed and transmitted by coupling the signatures and Tree-LSTM network, and this was proven to be practical [47].

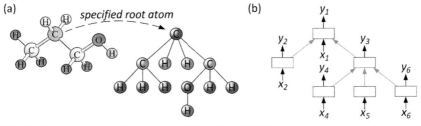

FIG. 3   The tree structures in expressing information of (A) the signature tree for the 1-propanol molecule and (B) the Tree-LSTM network.

2.3 Signature molecular descriptor and encoding rules

The structural information of molecules was extracted from the SMILES strings and expressed by atomic and molecular signatures with text form in this study. The atomic signatures can represent the substructure of a molecule while the molecular signatures describe the whole molecular structure. To specify atomic features, atoms were converted to strings relying on encoding rules which refer the regulations defined in SMARTS [63] (a straightforward extension of SMILES for describing molecular substructures). In order to be well applied in this task, some new definitions were made as a complement of the encoding rules. Herein, RDKit [64] was adopted as an auxiliary tool for implementing the encoding rules by identifying the element symbols of atoms, the types of chemical bonds, the types of chirality centers and so forth.

Atomic signature of height 1, also called 1-signature, contains only the root atom and its chemical bonds along with connected atoms (refer to Fig. 4) [47]. The 1-signature of each atom in molecules were generated with encoding rules, and subsequently a series of substrings representing molecular features were extracted with adopting the atom embedding program [57]. During the embedding process, each substring was assigned a numeric vector for distinction and adopted as the label for this vector. In spite of that these vectors were only used to represent molecular features. The structural information of molecules and atom connectivity will be totally preserved with the aid of the combination of signatures and the Tree-LSTM networks. For illustrative purpose, all the symbols involving in the labels of molecular features are listed and explained in Table 2.

The molecular signature was defined as the linear combination of atomic signatures covering all the atoms and bonds [47]. However, the molecular signatures involve redundant and duplicated information. Accordingly, canonical molecular signature, the lexicographically largest atomic signature, which suffices to represent the molecular graph, was introduced to simplify the molecular signature [58]. Herein, to be used in conjunction with the Tree-LSTM network, the canonical molecular signature of each compound was generated in a unique manner for describing the molecular structure. For instance, the canonical molecular signature for 1-propanol (CASRN: 71-23-8; SMILES: CCCO) is

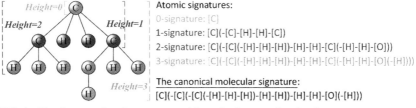

Atomic signatures:
0-signature: [C]
1-signature: [C](-[C]-[H]-[H]-[C])
2-signature: [C](-[C](-[H]-[H]-[H])-[H]-[H]-[C](-[H]-[H]-[O]))
3-signature: [C](-[C](-[H]-[H]-[H])-[H]-[H]-[C](-[H]-[H]-[O](-[H])))

The canonical molecular signature:
[C](-[C](-[C](-[H]-[H]-[H])-[H]-[H])-[H]-[H]-[O]-[H]))

FIG. 4 The signature descriptors generated from the 1-propanol molecule.

TABLE 2 The explanations and examples for symbols involved in the labels of molecular features.

Symbol	Explanation	Example
[A]	Atom in aliphatic compound	[C]—carbon atom in an aliphatic compound
[a]	Atom in aromatic compound	[c]—carbon atom in an aromatic compound
|r	Atom in a ring	[C|r]—carbon atom in a ring
+ (inside [])	Atom with a positive charge	[N+]—nitrogen atom with a positive charge
− (inside [])	Atom with a negative charge	[N-]—nitrogen atom with a negative charge
− (outside [])	Single bond	-[C]—carbon atom with a single bond
=	Double bond	=[C]—carbon atom with a double bond
#	Triple bond	#[C]—carbon atom with a triple bond
:	Aromatic bond	:[c]—carbon atom with an aromatic bond
/=\	Atoms in same side	/=\[C]—carbon atom in same side of connected atom
/=/	Atoms in opposite side	/=/[C]—carbon atom in opposite side of connected atom
*	Atom is a r-chirality center	[C*]—carbon atom is a r-chirality center
**	Atom is a s-chirality center	[C**]—carbon atom is a s-chirality center

represented as [C](-[C](-[C](-[H]-[H]-[H])-[H]-[H])-[H]-[H]-[O](-[H])) relying on the canonizing algorithm [47] and proposed encoding rules.

The molecular features chosen in the QSPR model rely on the molecular structure of the compound (refer to Fig. 5). First, the canonical molecular signature was generated for a compound, and then the Tree-LSTM network for this compound was built according to the signature tree for mapping the molecular structure. Afterwards, the numeric vectors representing molecular features were fed to the nodes of the Tree-LSTM network. Finally, a vector generated in the Tree-LSTM network was introduced to the BPNN for training the predictive model.

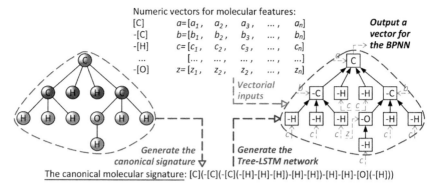

FIG. 5 The way of selecting molecular features for presenting the compound during predictions.

2.4 Structural features and parameters of DNN

In the DNN model, the Tree-LSTM network was utilized in conjunction with the BPNN to develop a QSPR model for predicting log K_{OW}. The Tree-LSTM network was employed to describe molecular tree structures with canonical molecular signatures while the BPNN was used to correlate structures and properties. Back-propagation (BP) algorithm is a supervised learning procedure in the machine learning process, and it was commonly used to train the DNN [65–67]. In this study, the BPNN was built with three layers including one input layer, one hidden layer, and one output layer. The topological structure of the fully connected three-layer neural network is graphically presented in Fig. 6. The input layer receives the vectors produced by Tree-LSTM network and the output layer gives the predicted log K_{OW} values. As single layers of linear neurons, the hidden layer take in a set of weighted inputs from the input layer and produce an output for the output layer.

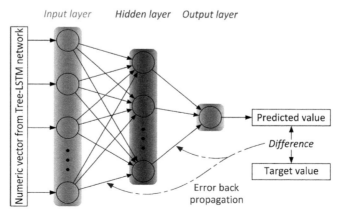

FIG. 6 The structure of the fully connected BPNN model for log K_{OW} prediction.

As an open-source deep learning library for Python, PyTorch [68] mainly supported the development of the DNN model in this research. Huber loss is a common loss function which is characterized by rapid convergence and inclusiveness to outliers because it combines the advantages of two basic loss functions, that is, the mean square error and the mean absolute error. Therefore, the Huber loss [69] was adopted as the loss function in this research to evaluate the model performance during the training process. Additionally, Adam algorithm [70] was employed for optimizing the predictive model by minimizing the loss function due to the attractive benefits that it is computationally efficient and suitable for tasks with a large dataset.

A machine learning model is parameterized, and it refers to numerous variables that can be classified into two types: model parameters and model hyperparameters. The model parameters, such as weights and biases, are learned from the given dataset and updated with the BP algorithm by calculating the gradient of the loss function during the model training. In contrast, with the purpose of controlling the learning process efficiently, model hyper-parameters were specified before the training activates.

In order to achieve the better performance as well as make the DNN model specialize in the prediction of log K_{OW}, hyper-parameters were specified and detailed as follows:

  (i)   The hidden layer of the BPNN has 32 neurons.
 (ii)   The batch size of the set for training, the number of training examples utilized in one iteration, is set as 250.
(iii)   The Learning rate is set as 1.00E-03 to control the rate of convergence.
(iv)   The weight decay rate is set as 1.00E-06 to alleviate the problem of the over-fitting.

The algorithm of model development with the Tree-LSTM network and BPNN is illustrated in Fig. 7. For supporting the development of the predictive model, molecular features were firstly extracted from the molecules of the collected dataset. Afterwards, the signature trees of compounds were generated for further mapping to the Tree-LSTM networks. Therefore, the vectors of molecular features can be inputted into the Tree-LSTM networks, and a vector was generated as an input for the BPNN. Within the BPNN, the properties were correlated to the molecular structures, and the QSPR model was obtained after massive training and testing. Afterwards, the QSPR model was evaluated with an external set, discussed on its Applicability Domain and compared with the reported model to investigate its performance. As such, an accurate and reliable QSPR model was generated for predicting the log K_{OW} of organic compounds.

The algorithm of the proposed model for predicting with the Tree-LSTM network and BPNN is illustrated in Fig. 8. During predicting with the developed QSPR model, the molecular structure of a new compound is used to generate the signature tree which can be mapped to the Tree-LSTM network. Afterwards, using the vectors which were generated during the model development, the

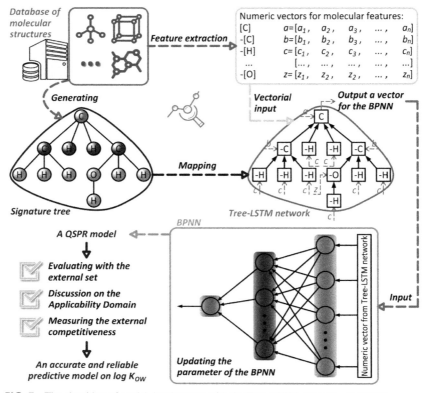

FIG. 7 The algorithm of model development with the Tree-LSTM network and BPNN.

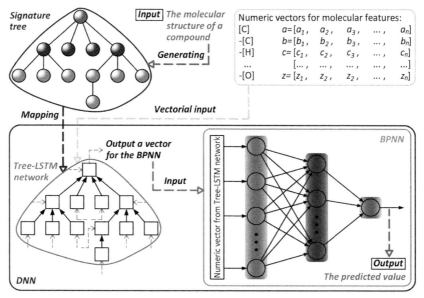

FIG. 8 The algorithm of the model for predicting with the Tree-LSTM network and BPNN.

Tree-LSTM network outputs a vector integrated the features of the molecular structure for the BPNN. Relying on the parameters and hyper-parameters of the BPNN determined during model development, the BPNN makes a numeric prediction and outputs a predicted value for the log K_{OW} of the compound.

3   Results and discussion

3.1   List of molecular features

To ensure that molecular structures can be introduced to the Tree-LSTM network, all the atoms and bonds of a molecule need to be expressed in the form of numerical vectors. Initially, the 1-signatures of every atom in the 10,668 chemical compounds were obtained relying on the aforementioned encoding rules and signature descriptors. Subsequently, 1-signatures were taken as samples to produce substring vectors with the atom embedding program. Finally, as exhibited in Table 3, 87 types of molecular features were extracted and assembled as a list for representing molecular structures in the Tree-LSTM networks.

The detailed count for the number of compounds in the training set presenting each molecular feature is shown in Table 4. Such information is important for the future use of the predictive model to estimate the level of confidence on individual predictions considering the number of training compounds associated to each molecular feature.

3.2   Training process

A network is said to generalize when it appropriately predicts the properties of compounds which are not in the set for training [71]. To measure the generalization ability of a QSPR model, an external dataset which is not used for training the QSPR model should be employed to evaluate the model performance. Consequently, the entire dataset containing 10,668 organic compounds was divided into three disjoint subsets (i.e., a training set, a test set, and an external set) by a random selection routine according to their corresponding proportions (80%, 10%, and 10%). The training set (8534 compounds), test set (1067 compounds), and external set (1067 compounds) were used to build and optimize the QSPR model, determine the timing for stopping training, and measure the external predictive performance of the final model, respectively.

In the DNN model, the canonical molecular signatures were fed into the Tree-LSTM networks to describe molecular tree structures, and subsequently the BPNN was used to correlate molecular structures and log K_{OW} values. In the training process, the Huber loss function was minimized with the Adam algorithm to receive a better QSPR model which can provide more accurate predictions. The learning process proceeded epoch by epoch, and the losses on training set and test set were calculated at the end of each epoch to evaluate the model performance and determine the timing for stopping training. Once

TABLE 3 The labels of molecular features classified by the chemical elements for representing molecular structures.

Chemical elements	Labels of molecular features
Carbon (C)	[C]; -[C]; =[C]; #[C]; [C|r]; -[C|r]; =[C|r]; [C*]; -[C*]; [C|r*]; -[C|r*]; [C**]; -[C**]; [C|r**]; -[C|r**]; [c|r]; -[c|r]; =[c|r]; :[c|r]; / =\[C]; /=/[C]; /=\[C|r]; /=/[C|r]; /=/[c|r]
Oxygen (O)	[O]; -[O]; =[O]; [O|r]; -[O|r]; [o|r];: [o|r]; [O-]; -[O-]
Nitrogen (N)	[N]; -[N]; =[N]; #[N]; [N|r]; -[N|r]; =[N|r]; [n|r]; -[n|r];: [n|r]; [N +]; -[N+]; =[N+]; #[N+]; [N-]; =[N-]; [N+|r]; -[N+|r]; =[N+|r]; [n +|r]; -[n+|r]; =[n+|r];: [n+|r]; /=\[N]; /=/[N]; /=\[N+]
Phosphorus (P)	[P]; -[P]; =[P]; [P|r]; -[P|r]; =[P|r]; [P+]; -[P+]; [P+|r]; -[P+|r]
Sulfur (S)	[S]; -[S]; =[S]; [S|r]; -[S|r]; =[S|r]; [s|r];: [s|r]
Others	[H]; -[H]; [F]; -[F]; [Cl]; -[Cl]; [Br]; -[Br]; [I]; -[I]

TABLE 4 The frequencies of compounds presenting each molecular feature in model training.

Molecular feature	Frequency	Molecular feature	Frequency	Molecular feature	Frequency
[C]	5628	-[C]	6303	=[C]	381
#[C]	46	[C|r]	190	-[C|r]	2211
=[C|r]	526	[C*]	75	-[C*]	93
[C|r*]	5	-[C|r*]	638	[C**]	279
-[C**]	130	[C|r**]	22	-[C|r**]	618
[c|r]	541	-[c|r]	6201	=[c|r]	9
:[c|r]	6950	/=\[C]	102	/=/[C]	176
/=\[C|r]	15	/=/[C|r]	20	[O]	5
-[O]	4466	=[O]	5753	-[O|r]	677
[o|r]	3	:[o|r]	300	-[O-]	819
-[N]	3920	=[N]	153	#[N]	306
-[N|r]	1232	=[N|r]	217	-[n|r]	597
:[n|r]	2238	-[N+]	753	=[N+]	17
#[N+]	3	=[N-]	8	=[N+|r]	1

TABLE 4 The frequencies of compounds presenting each molecular feature in model training—cont'd

Molecular feature	Frequency	Molecular feature	Frequency	Molecular feature	Frequency
-[n+|r]	2	:[n+|r]	60	/=\[N]	42
/=/[N]	50	/=\[N+]	2	-[P]	186
-[P|r]	13	-[P+]	12	-[P+|r]	1
-[S]	866	=[S]	221	-[S|r]	291
[s|r]	1	:[s|r]	265	[H]	23
-[H]	8510	-[F]	662	[Cl]	1417
-[Cl]	705	[Br]	345	-[Br]	64
-[I]	112				

the training process is activated, it does not terminate until there is no decrease in the loss on the test set within 20 consecutive epochs. Furthermore, early stopping was used to prevent the problem of the over-fitting and improve the performance of model on data outside of the training set. The QSPR model was preserved by storing the topological structure of the DNN on each epoch during the training process, and the final QSPR model can be rebuilt with its corresponding structural parameters of the DNN.

The training was terminated at the 130th epoch for the model developed based on isomeric SMILES (represented as ISO-DNN model), and the tendency in losses on training and test sets is displayed in the Fig. 9. Since the 110th epoch, the loss on the training set significantly decreased while the loss on the test set kept at the same level in 20 consecutive epochs which means that the model was over-fitting during these epochs. Accordingly, the QSPR model obtained in the 110th epoch was considered to be the global optimum model and it was saved as the final model for log K_{OW} prediction to prevent an over-fitting model.

A comprehensive investigation found that 1663 out of the 10,668 compounds contain isomeric features in developing the QSPR model and they were described with the isomeric features, while the remaining 9005 compounds described with the canonical SMILES strings contain no isomeric features.

In order to evaluate the impact of isomeric features on the predictive accuracy of the model, we replace all the isomeric SMILES strings with the canonical SMILES strings, and on this basis a new QSPR model was developed with

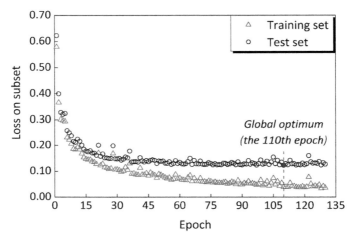

FIG. 9 The tendency in losses on training and test sets during training for the ISO-DNN model.

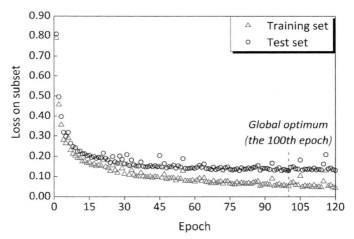

FIG. 10 The tendency in losses on the training and test sets during training for the CAN-DNN model.

the same training, test and external sets used for developing the ISO-DNN model. The QSPR model obtained in the 100th epoch was considered to be the global optimum model (refer to Fig. 10) and it was saved as the final model (represented as CAN-DNN model) for log K_{OW} prediction.

Furthermore, as summarized in Table 5, the predictive accuracy of the two model was quantified using three evaluation indexes, the root mean squared error $(RMSE)$, average absolute error (AAE), and determination coefficient (R^2), according to the research of Chirico et al. [72] These evaluation indexes

TABLE 5 The statistics results for the ISO-DNN and CAN-DNN models in log K_{OW} prediction.

	N^a	$RMSE^b$	MAE^c	R^{2d}
ISO-DNN	10,668	0.3386	0.2376	0.9606
CAN-DNN	10,668	0.3707	0.2634	0.9552

^{a}The number of data points.

$^b RMSD = \sqrt{\sum_{n=1}^{N}\left(x_n^{exp} - x_n^{pre}\right)^2 / N}.$

$^c MAE = \frac{1}{N}\sum_{n=1}^{N}\left|x_n^{exp} - x_n^{pre}\right|.$

$^d R^2 = 1 - \left[\sum_{n=1}^{N}\left(x_n^{exp} - x_n^{pre}\right)^2 / \sum_{n=1}^{N}\left(x_n^{exp} - \mu\right)^2\right]$ (where $\mu = \frac{1}{N}\sum_{n=1}^{N}x_n^{exp}$).

are related to the experimental values (x^{exp}), predicted values (x^{pre}) and the number of data points (N). The results show that the predictive accuracy of the CAN-DNN model is lower than that of the ISO-DNN model. It proves that the isomeric features involving in model development are conducive to improve the accuracy of the predictive model.

Additionally, the impact of isomeric features on the predictive accuracy of the model was investigated by comparing the predictive accuracy of the KOW-WIN model [59] and the developed ISO-DNN model performing on the 1663 isomeric compounds and the remaining 9005 compounds. As displayed in Table 6, the predictive accuracy on the 1663 isomeric compounds is much lower than that on the other 9005 compounds with the KOWWIN model. However, using the ISO-DNN model, the predictive accuracy on the 1663 isomeric compounds is close to that on the other 9005 compounds. Meanwhile, comparing with the KOWWIN model, the improvement in the predictive accuracy for

TABLE 6 The statistics results for the KOWWIN and ISO-DNN models in log K_{OW} prediction for isomeric compounds and other compounds.

	Isomeric compounds		Other compounds	
	KOWWIN	ISO-DNN	KOWWIN	ISO-DNN
N	1663	1663	9005	9005
RMSE	0.5563	0.3957	0.3928	0.3269
MAE	0.4060	0.2546	0.2857	0.2345
R^2	0.9352	0.9619	0.9480	0.9603

the 1663 isomer compounds is markedly higher than the other 9005 compounds using the ISO-DNN model. It also demonstrated that the predictive accuracy on log K_{OW} is significantly improved with the participation of the isomeric features. Therefore, the isomeric features are considered beneficial to improve the predictive accuracy of the model.

3.3 Generalization ability

The generalization ability acts as an important indicator to evaluate the predictive performance of a model in machine learning. The problem of over-fitting in the DNN model is bound to cause the loss of generalization ability and low external prediction accuracy [73]. The traditional models for property prediction were developed with the entire dataset for training, and therefore the generalization ability was unable to be guaranteed around two decades ago. However, efforts have been made to avoid the publication of models without external validations and the generalization ability of predictive models has been highly improved in the last decade [74,75].

As stated, early stopping was used in this study to avoid over-fitting and ensure that the final QSPR model has satisfactory generalization ability. In addition, the external set was employed to measure the generalization ability of the final QSPR model. The training and external sets were applied in the final QSPR model, and the predicted values were compared against the original experimental values. The predictive performance is visualized in the Fig. 11 with the scatter graphs of predicted values versus experimental values. It is observed that the distribution and predictive accuracy of the external set are similar to those of the training set. It demonstrated that the developed QSPR model has the satisfactory generalization ability in predicting log K_{OW} values for organic compounds.

Apart from the training and test sets, an external set was adopted as the additional test set to evaluate the predictivity of the final model in this research. The external validation indices (i.e., $RMSE$, MAE, and R^2) have been calculated to measure the performance of the predictive model as summarized in Table 7.

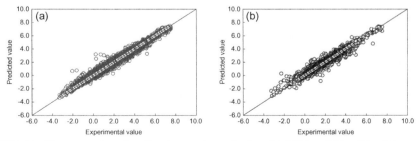

FIG. 11 The scatter plots of predicted—experimental value with DNN model for (A) training set and (B) external set.

TABLE 7 The statistics results for the ISO-DNN model on the training, test, and external sets.

	N	RMSE	MAE	R^2
Training set	8534	0.2836	0.2101	0.9741
Test set	1067	0.5349	0.3596	0.9091
External set	1067	0.4656	0.3355	0.9285

As it turns out, the *RMSE* and *MAE* of the training set is lower than those of the external set, and the R^2 of the training set is closer to 1.0000. It indicated that the developed QSPR model can make more accurate predictions for the training set. In view of that the similar results of these external validation indices between the training and external sets, the developed predictive model is acceptable, and it has satisfactory applicability.

3.4 Applicability domain

The predictions can be considered reliable for the compounds which fall in the applicability domain (AD) of the predictive model. The Williams plot is a recommended leverage approach for AD investigation which provides a graphical detection of both the response outliers and the structurally influential outliers in a predictive model. In this research, the AD of the developed predictive model is visualized with the Williams plot which is displayed with the plot of standardized residuals versus hat values as exhibited in Fig. 12. Moreover, the

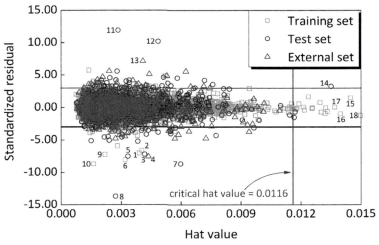

FIG. 12 The Applicability Domain of the developed QSPR model.

way for calculating the standardized residual, hat value, and critical hat value can be found in the published works [14, 76].

In the Williams plot, the compounds with standardized residuals greater than 3 standard deviation units are identified as response outliers. Part of the response outliers of the developed predictive model were marked with numbers and detailed as follows: cephaloridine (1), methyl 3-[5-acetyl-2-[2-[[3-ethyl-5-[(3-ethyl-4-methyl-5-oxopyrrol-2-yl) methylidene]-4-methylpyrrol-2-ylidene] methyl]-3-methyl-4-oxo-1H-cyclopenta[*b*]pyrrol-6-ylidene]-4-methyl-3,4-dihydropyrrol-3-yl]propanoate (2), sarmoxicillin (3), (4aR,6R,7R,7aS)-6-[3,2-d][1,3,2] dioxaphosphinin-7-ol (4), prolylphenylalanine (5), thienylglycine (6), 8-thiomethyl cyclic AMP (7), 1-bromo-4-[2-[2-(4-methoxyphenoxy)ethoxy]ethoxy]-2,5-dimethylbenzene (8), N,N-diethyl-3-methoxy-4-(2-hydroxy-5-s-butylphenylazo)benzenesulfonamide (9), 1-butyl-5-[[3-[(2-chlorophenyl) methyl]-4-oxo-2-sulfanylidene-1,3-thiazolidin-5-ylidene]methyl]-4-methyl-2-oxo-6-(4-phenylpiperazin-1-yl)pyridine-3-carbonitrile (10), (E)-4-(dimethylamino)- 4-oxobut-2-enoic acid (11), (E)-2-cyano-3-amino-3-(isopropylamino) propenoic acid methyl ester (12) and [4-[2-(diaminomethylidene)hydrazinyl] phenyl]iminourea (13). Moreover, the hat value of a compound greater than the critical hat value indicates the compound is outside of the model's structural AD and it could lead to unreliable predictions. As it turns out, 22 compounds were detected as structurally influential outliers and part of them were marked with numbers and exhibited as follows: mellitic acid (14), perfluoromethylcyclohexylpiperidine (15), 2-nitrostrychnidin-10-one (16), strychnine (17), and perfluorocyclohexane (18). The way for calculating the standardized residual, hat value, and critical hat value can be found in the published works [14,76].

3.5 External competitiveness

The satisfactory predictive capability of a new QSPR model needs to be proven by its external competitiveness. As an authoritative log K_{OW} prediction tool relying on an atom and fragment contribution method, the KOWWIN program was developed and maintained by the United State Environmental Protection Agency and the Syracuse Research Corporation [77]. In this research, a comparison between the KOWWIN and developed QSPR model (represented as ISO-DNN model) was conducted to measure their predictive capabilities. It should be noted that it is not always possible to compare the performance of different models unless they are evaluated using the same dataset. Therefore, a dataset of KOWWIN predicted values was collected [59]. For a fair comparison, the samples in this dataset are consistent with the whole dataset adopted in this study. The overall predictive capabilities of the ISO-DNN and KOWWIN models were further assessed and compared in the form of scatter graphs (refer to Fig. 13). Overall, the data points in the Fig. 13B were closer to the diagonal line (predicted value equals experimental value) when compared to the data

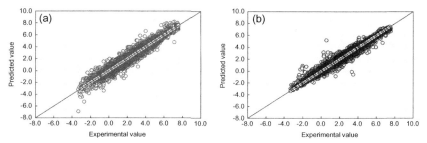

FIG. 13 The scatter plots of predicted—experimental value with (A) KOWWIN model and (B) ISO-DNN model.

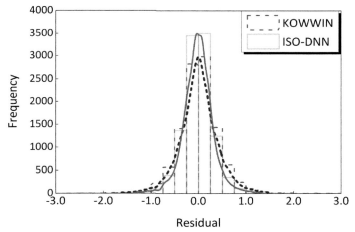

FIG. 14 The residual distributions of log K_{OW} estimation with KOWWIN and ISO-DNN models.

points in Fig. 13A. This suggests that the ISO-DNN model has better predictive accuracy in log K_{OW} estimation, although some data points still exhibited relatively large deviations.

The scatter plots can only appear to indicate a better predictive accuracy of the ISO-DNN model. It is more persuasive if the external competitiveness of the developed QSPR model can be demonstrated from perspective of statistics. The residual (experimental value minus predicted value) of each compound in the dataset was calculated using the KOWWIN and ISO-DNN models. The residual distributions in Fig. 14 show that the residuals produced by the ISO-DNN model are more densely gathered around the zero value in contrast with those obtained using the KOWWIN model. This indicates that the residual distribution of the ISO-DNN model has a lower standard deviation. According to the analysis, it was further proven that the develop QSPR model enables a more accurate prediction for log K_{OW} from a statistical point of view.

TABLE 8 The statistical results of the KOWWIN and ISO-DNN models in log K_{OW} prediction.

Predictive model	N	RMSE	MAE	R^2
KOWWIN	10,668	0.4224	0.3045	0.9451
ISO-DNN	10,668	0.3386	0.2376	0.9606

Furthermore, the model performance was also quantified using the evaluation indexes. From Table 8, it can be observed that both RMSD and AAE present lower values in the ISO-DNN model for the investigated compounds, and the R^2 for the ISO-DNN model is closer to 1.0000. The results demonstrate that the developed QSPR model has a better agreement between the experimental and predicted log K_{OW} values for organic compounds.

3.6 Discriminative power in isomers

In chemistry, isomers are these molecules with identical formulae, but distinct structures and they do not necessarily have similar properties. The isomers include structural isomers whose atoms and functional groups are joined together in different ways and stereoisomers that differ in three-dimensional orientations of their atoms and functional groups in space.

Although the KOWWIN program exhibits the strong ability in the log K_{OW} prediction, it is found that, under most circumstances, structural isomers can be differentiated but stereoisomers cannot. Because the KOWWIN program was developed relying on an atom and fragment contribution method, its discriminative power was limited in isomers. In contrast, the developed DNN-based QSPR model is able to differentiate the structural isomers and stereoisomers and account for stereochemistry due to the interaction between the canonical molecular signatures and Tree-LSTM networks in the processing and transmitting structural information for molecules.

Two pairs of structural isomers and two pairs of stereoisomers were extracted from the investigated dataset and they were taken as examples for illustrating the capabilities of two models in discriminating isomers. The experimental values, KOWWIN predicted values, and ISO-DNN predicted values of these structural isomers and stereoisomers are summarized in Tables 9 and 10, respectively.

Regarding the isomers as shown in Tables 9 and 10, both compounds in each pair of isomers have different experimental values, while they were given the same KOWWIN predicted value since the predictive model does not have obvious advantages over differentiating isomers. In contrast, the ISO-DNN model has greater discriminative power, and the predicted values were assigned

TABLE 9 The capabilities of the KOWWIN and ISO-DNN models in distinguishing the structural isomers.

Chemical name	CASRN	$x^{exp a}$	$x^{pre-K b}$	$x^{pre-D c}$	$\Delta x_K{}^d$	$\Delta x_D{}^e$
Pyridazine-4-carboxamide	88511-47-1	−0.96	−1.31	−0.75	0.35	0.21
Pyrimidine-5-carboxamide	40929-49-5	−0.92	−1.31	−0.63	0.39	0.29
Thiophene-2-carboxylic acid	527-72-0	1.57	1.69	1.61	0.12	0.04
Thiophene-3-carboxylic acid	88-13-1	1.50	1.69	1.49	0.19	0.01

[a]Experimental value.
[b]KOWWIN predicted value.
[c]ISO-DNN predicted value.
[d]$\Delta x_K = |x^{exp} - x^{pre-K}|$.
[e]$\Delta x_D = |x^{exp} - x^{pre-D}|$.

TABLE 10 The capabilities of the KOWWIN and ISO-DNN models in distinguishing the stereoisomers.

Chemical name	CASRN	x^{exp}	x^{pre-K}	x^{pre-D}	Δx_K	Δx_D
(2R,6R)-2,6-dimethylcyclohexan-1-ol	39170-83-7	2.10	2.47	2.55	0.37	0.45
(2R,6S)-2,6-dimethylcyclohexan-1-ol	39170-84-8	2.37	2.47	2.38	0.10	0.01
(1R)-cis-(alphaS)-cypermethrin	65731-84-2	6.05	6.38	5.78	0.33	0.27
(1R)-trans-(alphaS)-cypermethrin	65732-07-2	6.06	6.38	6.05	0.32	0.01

depending on the distinct structures of structural isomers and stereoisomers. Meanwhile, the predictive accuracy was guaranteed. As it turns out, the strong discriminative power of the developed ISO-DNN model can be attributed to the interaction between the canonical molecular signatures and Tree-LSTM networks.

4 Conclusions

In this study, a QSPR model under the deep learning approach was developed for accurately and reliably predicting the octanol-water partition coefficients for organic compounds and providing valuable environmental information for guiding selection and development of important chemicals including green solvents. Prior to the learning process, molecular features were extracted to describe structural information and connectivity. The canonical molecular signatures were produced relying on the theory of canonizing molecular graph and mapped on the Tree-LSTM networks to generate input parameters in preparation for obtaining a QSPR model. Afterwards, the learning process was performed by the built DNN model combining the Tree-LSTM network and the BPNN, and the final QSPR model for the log K_{OW} estimation was determined after the massive training and test. The evaluations were finally carried out for exhibiting better predictive accuracy and external competitiveness of the DNN-based QSPR model in contrast with an authoritative log K_{OW} prediction tool. Moreover, the developed QSPR model revealed greater discriminative power in the structural isomers and stereoisomers.

Differing from the traditional property prediction models, the developed deep-learning-assisted model avoids the human intervention in the feature selection of molecular structures. Meanwhile, by coupling the canonical molecular signatures and Tree-LSTM networks, the molecular features were automatically extracted from chemical structures which circumvents numerous topological features and physicochemical descriptors. Therefore, the deep learning approach was successfully implemented to develop a QSPR model for predicting the log K_{OW} for the organic compounds which provide valuable information in the absorption, distribution, and metabolism of chemicals and could be further employed to predict various indicators of toxicity. It proved that the deep learning approach can serve as a promising and intelligent approach to develop property prediction models with high predictive accuracy and a wide application scope. The established deep learning-based models can be further used in the computer-aided molecular design framework. Although this research focused on predicting the log K_{OW} for measuring the lipophilicity of organic chemicals, the proposed approach can be further popularized to some other environmentally important properties such as water solubility and bioconcentration factor, which exhibits its vital potentials in the development of green chemistry and engineering.

Acknowledgments

The chapter is reproduced from Green Chemistry, 2019, 21, 4555-4565, Zihao Wang, Yang Su, Weifeng Shen, Saimeng Jin, James H. Clark, Jingzheng Ren, Xiangping Zhang, Predictive deep learning models for environmental properties: the direct calculation of octanol-water partition coefficients from molecular graphs, by permission of The Royal Society of Chemistry.

References

[1] J.H. Clark, Green chemistry: today (and tomorrow), Green Chem. 8 (2006) 17–21.

[2] D. Prat, A. Wells, J. Hayler, H. Sneddon, C.R. McElroy, S. Abou-Shehada, P.J. Dunne, CHEM21 selection guide of classical- and less classical-solvents, Green Chem. 18 (2016) 288–296.

[3] S. Jin, A.J. Hunt, J.H. Clark, C.R. McElroy, Acid-catalysed carboxymethylation, methylation and dehydration of alcohols and phenols with dimethyl carbonate under mild conditions, Green Chem. 18 (2016) 5839–5844.

[4] S. Jin, Y. Tian, C.R. McElroy, D. Wang, J.H. Clark, A.J. Hunt, DFT and experimental analysis of aluminium chloride as a Lewis acid proton carrier catalyst for dimethyl carbonate carboxymethylation of alcohols, Catal. Sci. Technol. 7 (2017) 4859–4865.

[5] S. Jin, F. Byrne, C.R. McElroy, J. Sherwood, J.H. Clark, A.J. Hunt, Challenges in the development of bio-based solvents: a case study on methyl (2,2-dimethyl-1,3-dioxolan-4-yl) methyl carbonate as an alternative aprotic solvent, Faraday Discuss. 202 (2017) 157–173.

[6] R. Luque, J.A. Menendez, A. Arenillas, J. Cot, Microwave-assisted pyrolysis of biomass feedstocks: the way forward? Energy Environ. Sci. 5 (2012) 5481–5488.

[7] C.S.K. Lin, L.A. Pfaltzgraff, L. Herrero-Davila, E.B. Mubofu, S. Abderrahim, J.H. Clark, A.K. Apostolis, K. Nikolaos, S. Katerina, D. Fiona, T. Samarthia, M. Zahouily, B. Robert, L. Rafael, Food waste as a valuable resource for the production of chemicals, materials and fuels. Current situation and global perspective, Energy Environ. Sci. 6 (2013) 426–464.

[8] J.H. Clark, Green chemistry: challenges and opportunities, Green Chem. 1 (1999) 1–8.

[9] W. Shen, L. Dong, S. Wei, J. Li, H. Benyounes, X. You, V. Gerbaud, Systematic design of an extractive distillation for maximum-boiling azeotropes with heavy entrainers, AICHE J. 61 (2015) 3898–3910.

[10] A. Jayswal, X. Li, A. Zanwar, H.H. Lou, Y. Huang, A sustainability root cause analysis methodology and its application, Comput. Chem. Eng. 35 (2011) 2786–2798.

[11] Y. Hu, Y. Su, S. Jin, I.L. Chien, W. Shen, Systematic approach for screening organic and ionic liquid solvents in homogeneous extractive distillation exemplified by the tert-butanol dehydration, Sep. Purif. Technol. 211 (2019) 723–737.

[12] A. Yang, H. Zou, I. Chien, D. Wang, S. Wei, J. Ren, W. Shen, Optimal design and effective control of triple-column extractive distillation for separating ethyl acetate/ethanol/water with multi-azeotrope, Ind. Eng. Chem. Res. 58 (2019) 7265–7283.

[13] S. Neidle, Cancer Drug Design and Discovery, Elsevier, New York, 2011, pp. 131–154.

[14] A. Rybinska, A. Sosnowska, M. Grzonkowska, M. Barycki, T. Puzyn, Filling environmental data gaps with QSPR for ionic liquids: modeling n-octanol/water coefficient, J. Hazard. Mater. 303 (2016) 137–144.

[15] J. Marrero, R. Gani, Group-contribution-based estimation of octanol/water partition coefficient and aqueous solubility, Ind. Eng. Chem. Res. 41 (2002) 6623–6633.

[16] T. Cheng, Y. Zhao, X. Li, F. Lin, Y. Xu, X. Zhang, Y. Li, R. Wang, Computation of octanol-water partition coefficients by guiding an additive model with knowledge, J. Chem. Inf. Model. 47 (2007) 2140–2148.

[17] M. Turchi, Q. Cai, G. Lian, An evaluation of in-silico methods for predicting solute partition in multiphase complex fluids—A case study of octanol/water partition coefficient, Chem. Eng. Sci. 197 (2019) 150–158.

[18] F.P. Byrne, S. Jin, G. Paggiola, T.H.M. Petchey, J.H. Clark, T.J. Farmer, A.J. Hunt, C.R. McElroy, J. Sherwood, Tools and techniques for solvent selection: green solvent selection guides, Sustain. Chem. Process. 4 (2016) 7.

[19] F.P. Byrne, B. Forier, G. Bossaert, C. Hoebers, T.J. Farmer, A.J. Hunt, A methodical selection process for the development of ketones and esters as bio-based replacements for traditional hydrocarbon solvents, Green Chem. 20 (2018) 4003–4011.

[20] M. Tobiszewski, S. Tsakovski, V. Simeonov, J. Namieśnik, F. Pena-Pereira, A solvent selection guide based on chemometrics and multicriteria decision analysis, Green Chem. 17 (2015) 4773–4785.

[21] M.D. Ertürk, M.T. Saçan, Assessment and modeling of the novel toxicity data set of phenols to *Chlorella vulgaris*, Ecotoxicol. Environ. Saf. 90 (2013) 61–68.

[22] S. Bakire, X. Yang, G. Ma, X. Wei, H. Yu, J. Chen, H. Lin, Developing predictive models for toxicity of organic chemicals to green algae based on mode of action, Chemosphere 190 (2018) 463–470.

[23] S.G. Machatha, S.H. Yalkowsky, Comparison of the octanol/water partition coefficients calculated by ClogP, ACDlogP and KowWin to experimentally determined values, Int. J. Pharm. 294 (2005) 185–192.

[24] M. Reinhard, A. Drefahl, Handbook for Estimating Physicochemical Properties of Organic Compounds, Wiley, New York, 1999.

[25] C. Nieto-Draghi, G. Fayet, B. Creton, X. Rozanska, P. Rotureau, J.C. Hemptinne, P. Ungerer, B. Rousseau, C. Adamo, A general guidebook for the theoretical prediction of physicochemical properties of chemicals for regulatory purposes, Chem. Rev. 115 (2015) 13093–13164.

[26] J.C. Dearden, P. Rotureau, G. Fayet, QSPR prediction of physico-chemical properties for REACH, SAR QSAR Environ. Res. 24 (2013) 279–318.

[27] R. Mannhold, G.I. Poda, C. Ostermann, I.V. Tetko, Calculation of molecular lipophilicity: state-of-the-art and comparison of log P methods on more than 96,000 compounds, J. Pharm. Sci. 98 (2009) 861–893.

[28] R. Mannhold, H. van de Waterbeemd, Substructure and whole molecule approaches for calculating log P, J. Comput. Aided Mol. Des. 15 (2001) 337–354.

[29] C.W. Cho, S. Stolte, Y.S. Yun, Validation and updating of QSAR models for partitioning coefficients of ionic liquids in octanol-water and development of a new LFER model, Sci. Total Environ. 633 (2018) 920–928.

[30] B. Admire, B. Lian, S.H. Yalkowsky, Estimating the physicochemical properties of polyhalogenated aromatic and aliphatic compounds using UPPER. Part 2. Aqueous solubility, octanol solubility and octanol-water partition coefficient, Chemosphere 119 (2015) 1441–1446.

[31] C.C. Bannan, G. Calabró, D.Y. Kyu, D.L. Mobley, Calculating partition coefficients of small molecules in octanol/water and cyclohexane/water, J. Chem. Theory Comput. 12 (2016) 4015–4024.

[32] K.B. Hanson, D.J. Hoff, T.J. Lahren, D.R. Mount, A.J. Squillace, L.P. Burkhardb, Estimating *n*-octanol-water partition coefficients for neutral highly hydrophobic chemicals using measured n-butanol-water partition coefficients, Chemosphere 218 (2019) 616–623.

[33] E. Wyrzykowska, A. Rybińska-Fryca, A. Sosnowska, T. Puzyn, Virtual screening in the design of ionic liquids as environmentally safe bactericides, Green Chem. 21 (2019) 1965–1973.

[34] G.M. Kontogeorgis, R. Gani, Computer Aided Property Estimation for Process and Product Design, Elsevier, Amsterdam, 2004, pp. 3–26.

[35] W.M. Meylan, P.H. Howard, Atom/fragment contribution method for estimating octanol-water partition coefficients, J. Pharm. Sci. 84 (1995) 83–92.

[36] A.J. Leo, Calculating log P_{oct} from structures, Chem. Rev. 93 (1993) 1281–1306.

[37] K.G. Joback, R.C. Reid, Estimation of pure-component properties from group-contributions, Chem. Eng. Commun. 57 (1987) 233–243.

[38] L. Constantinou, R. Gani, New group contribution method for estimating properties of pure compounds, AICHE J. 40 (1994) 1697–1710.

[39] J. Marrero, R. Gani, Group-contribution based estimation of pure component properties, Fluid Phase Equilib. 183 (2001) 183–208.

[40] S. Jhamb, X. Liang, R. Gani, A.S. Hukkerikar, Estimation of physical properties of amino acids by group-contribution method, Chem. Eng. Sci. 175 (2018) 148–161.

[41] I.V. Tetko, V.Y. Tanchuk, A.E.P. Villa, Prediction of n-octanol/water partition coefficients from PHYSPROP database using artificial neural networks and E-state indices, J. Chem. Inf. Model. 41 (2001) 1407–1421.

[42] V.K. Gombar, K. Enslein, Assessment of *n*-octanol/water partition coefficient: when is the assessment reliable? J. Chem. Inf. Model. 36 (1996) 1127–1134.

[43] J.J. Huuskonen, D.J. Livingstone, I.V. Tetko, Neural network modeling for estimation of partition coefficient based on atom-type electrotopological state indices, J. Chem. Inf. Model. 40 (2000) 947–955.

[44] J.I. García, H. García-Marín, J.A. Mayoral, P. Pérez, Quantitative structure-property relationships prediction of some physico-chemical properties of glycerol based solvents, Green Chem. 15 (2013) 2283–2293.

[45] N.D. Austin, N.V. Sahinidis, D.W. Trahan, Computer-aided molecular design: an introduction and review of tools, applications, and solution techniques, Chem. Eng. Res. Des. 116 (2016) 2–26.

[46] D.P. Visco, R.S. Pophale, M.D. Rintoul, J.L. Faulon, Developing a methodology for an inverse quantitative structure-activity relationship using the signature molecular descriptor, J. Mol. Graph Model. 20 (2002) 429–438.

[47] J.L. Faulon, D.P. Visco, R.S. Pophale, The signature molecular descriptor. 1. Using extended valence sequences in QSAR and QSPR studies, J. Chem. Inf. Model. 43 (2003) 707–720.

[48] N.G. Chemmangattuvalappil, M.R. Eden, A novel methodology for property-based molecular design using multiple topological indices, Ind. Eng. Chem. Res. 52 (2013) 7090–7103.

[49] N.G. Chemmangattuvalappil, C.C. Solvason, S. Bommareddy, M.R. Eden, Reverse problem formulation approach to molecular design using property operators based on signature descriptors, Comput. Chem. Eng. 34 (2010) 2062–2071.

[50] U. Safder, K.J. Nam, D. Kim, M. Shahlaei, C.K. Yoo, Quantitative structure-property relationship (QSPR) models for predicting the physicochemical properties of polychlorinated biphenyls (PCBs) using deep belief network, Ecotoxicol. Environ. Saf. 162 (2018) 17–28.

[51] A. Eslamimanesh, F. Gharagheizi, A.H. Mohammadi, D. Richona, Artificial neural network modeling of solubility of supercritical carbon dioxide in 24 commonly used ionic liquids, Chem. Eng. Sci. 66 (2011) 3039–3044.

[52] A. Lusci, G. Pollastri, P. Baldi, Deep architectures and deep learning in chemoinformatics: the prediction of aqueous solubility for drug-like molecules, J. Chem. Inf. Model. 53 (2013) 1563–1575.

[53] F. Gharagheizi, A. Eslamimanesh, F. Farjood, A.H. Mohammadi, D. Richon, Solubility parameters of nonelectrolyte organic compounds: determination using quantitative structure-property relationship strategy, Ind. Eng. Chem. Res. 50 (2011) 11382–11395.

[54] J. Zheng, Y. Zhu, M. Zhu, G. Sun, R. Sun, Life-cycle assessment and techno-economic analysis of the utilization of bio-oil components for the production of three chemicals, Green Chem. 20 (2018) 3287–3301.

[55] S. Hochreiter, J. Schmidhuber, Long short-term memory, Neural Comput. 9 (1997) 1735–1780.

[56] K.S. Tai, R. Socher, C.D. Manning, Improved Semantic Representations From Tree-Structured Long Short-Term Memory Networks, 2015. arXiv: 1503.00075.

[57] Y. Su, Z. Wang, S. Jin, W. Shen, J. Ren, M.R. Eden, An architecture of deep learning in QSPR modeling for the prediction of critical properties using molecular signatures, AICHE J. 65 (9) (2019) e16678.

[58] J.L. Faulon, M.J. Collins, R.D. Carr, The signature molecular descriptor. 4. Canonizing molecules using extended valence sequences, J. Chem. Inf. Model. 44 (2004) 427–436.

[59] KOWWIN Data. http://esc.syrres.com/interkow/KowwinData.htm. (Accessed 28 November 2018).

[60] L. Li, Z. Wen, Z. Wang, Outlier detection and correction during the process of groundwater lever monitoring base on Pauta criterion with self-learning and smooth processing, in: Theory, Methodology, Tools and Applications for Modeling and Simulation of Complex Systems, Springer, Singapore, 2016, pp. 497–503.

[61] D. Weininger, SMILES, a chemical language and information system. 1. Introduction to methodology and encoding rules, J. Chem. Inf. Model. 28 (1988) 31–36.

[62] S. Kim, J. Chen, T. Cheng, A. Gindulyte, J. He, S. He, Q. Li, B.A. Shoemaker, P.A. Thiessen, B. Yu, L. Zaslavsky, J. Zhang, E.E. Bolton, PubChem 2019 update: improved access to chemical data, Nucleic Acids Res. 47 (2019) D1102–D1109.

[63] Daylight Chemical Information Systems, Inc. http://www.daylight.com/. (Accessed 5 December 2018).

[64] RDKit, Open-Source Cheminformatics Software. http://www.rdkit.org/. (Accessed 5 December 2018).

[65] D.E. Rumelhart, G.E. Hinton, R.J. Williams, Learning representations by back-propagating errors, Cognitive Model. 5 (1988) 1.

[66] M.A. Nielsen, Neural Networks and Deep Learning, Determination Press, California, 2015.

[67] Y. LeCun, Y. Bengio, G. Hinton, Deep learning, Nature 521 (2015) 436–444.

[68] PyTorch. https://pytorch.org/. (Accessed 5 December 2018).

[69] P.J. Huber, Robust estimation of a location parameter, Ann. Math. Stat. 35 (1964) 73–101.

[70] D.P. Kingma, J. Ba, Adam: A Method for Stochastic Optimization, 2014. arXiv:1412.6980.

[71] N.J. Nilsson, Artificial Intelligence: A New Synthesis, Morgan Kaufmann, California, 1998.

[72] P. Gramatica, External evaluation of QSAR models, in addition to cross-validation: verification of predictive capability on totally new chemicals, Mol. Inform. 33 (2014) 311–314.

[73] F. Melnikov, J. Kostal, A. Voutchkova-Kostal, J.B. Zimmerman, P.T. Anastas, Assessment of predictive models for estimating the acute aquatic toxicity of organic chemicals, Green Chem. 18 (2016) 4432–4445.

[74] N. Chirico, P. Gramatica, Real external predictivity of QSAR models: how to evaluate it? Comparison of different validation criteria and proposal of using the concordance correlation coefficient, J. Chem. Inf. Model. 52 (2012) 2044–2058.

[75] A. Tropsha, Best practices for QSAR model development, validation, and exploitation, Mol. Inform. 29 (2010) 476–488.

[76] P.P. Roy, S. Kovarich, P. Gramatica, QSAR model reproducibility and applicability: a case study of rate constants of hydroxyl radical reaction models applied to polybrominated diphenyl ethers and (benzo-)triazoles, J. Comput. Chem. 32 (2011) 2386–2396.

[77] EPI SuiteTM-Estimation Program Interface. https://www.epa.gov/tsca-screening-tools/epi-suitetm-estimation-program-interface. (Accessed 3 February 2019).

Chapter 4

Automated extraction of molecular features in machine learning-based environmental property prediction

Zihao Wang and Weifeng Shen
School of Chemistry and Chemical Engineering, Chongqing University, Chongqing, People's Republic of China

1 Introduction

Environmental properties of compounds play a crucial role in many fields such as sustainable chemistry [1–3], process design [4, 5], environmental remediation, and evaluation of chemicals' environmental behaviors [6–8]. Environmental benefits drive the development of green solvents, chemical synthesis, and molecular design toward eco-friendly technology [9–11], because environmental properties provide valuable information on the absorption, distribution, and metabolism of compounds and direct the treatment of organic pollutants which may pose serious threats to humans and wildlife [12]. However, reliably measuring the environmental properties for compounds is a costly task and sometimes tedious, especially for those compounds with very low vapor pressure, low aqueous solubility, or high risk. Therefore, different approaches have been proposed in the open literature to predict properties for various types of chemical compounds.

Empirical relationship method is one of the popular approaches for property estimation, in which different physicochemical properties (e.g., critical temperature, vapor pressure, and aqueous solubility) serve as input parameters to calculate target properties of compounds [7, 13, 14]. For instance, Gharagheizi et al. [14] developed a fairly accurate empirical model to predict Henry's law constant values of organic compounds relying on several basic properties (e.g., normal boiling point temperature and critical pressure). This model can be easily applied for rapid estimation and it exhibits an absolute average deviation of about 10% with respect to 1816 organic compounds. However, empirical

Applications of Artificial Intelligence in Process Systems Engineering
https://doi.org/10.1016/B978-0-12-821092-5.00005-X

67

relationship approaches heavily depend on the availability and accuracy of the required input properties. Thus, it is not practical to use if one of the inputs is unavailable (or cannot be estimated).

Another popular type of the predictive tools has focused on the application of quantitative structure-property relationship (QSPR) models, in which the physicochemical properties are supposed to be related to molecular structures. A number of studies have made great contributions in this regard [15–21]. In addition, several QSPR models were put forward based on group contribution (GC) methods [6, 12, 22, 23]. In such models, molecules of interest are divided into various groups (e.g., atoms and substructures containing atoms and chemical bonds), and each group is assigned a specific contribution value. Afterwards, the target property of a compound can be given by summarizing the contributions of groups. The GC methods therefore are regarded as multiple linear mathematical models. Whereas the same groups in different GC methods have distinct contribution values and the definitions of groups are not entirely the same. Thus, different GC methods work in a similar way though exhibit different results. A classic GC method is the three-level GC estimation approach proposed by Marrero and Gani [24], in which a total of 370 kinds of groups were defined for recognizing molecular structures. Attributed to its superior performance, the three-level GC method has been extensively applied for estimating various physicochemical properties such as critical properties, standard enthalpy of vaporization, and the octanol-water partition coefficient [25–27].

GC methods are characterized by simple models, quick and fairly reliable estimations. Based on a comprehensive literature review, three typical shortcomings of the traditional GC methods have often emerged in applications:

(a) Difficulties in understanding the definitions and structures of complex groups;
(b) Computational error and time consumption due to complexities in recognizing groups and calculating the property;
(c) Multiple predicted values for compounds resulting from different feasible strategies in the structure recognition.

With the rapid development of artificial intelligence and computational power, many QSPR models have been investigated with the aid of artificial neural networks (ANNs) and have gained popularity for estimating physicochemical properties [28–35]. Because of their ability to model and reproduce nonlinear processes, ANN-GC hybrid models have presented accurate predictive tools mainly with the aim of alleviation of the problems in the traditional GC methods [36–38]. As such, the assistance of computer-aided technologies enables the ANN-GC models to readily correct these shortcomings (a) and (b). Meanwhile, the shortcoming (c) can also be removed by predefining the priority rules of available strategies or developing new GC methods. However, the priority rules or the defined groups need to be updated when new chemicals and chemical structures are introduced.

In this research, a novel strategy is proposed to rapidly recognize molecular structures, extract molecular features, and transfer features into identifiers according to encoding rules. The feature extraction algorithm for accomplishing these works is developed and is introduced in detail. Moreover, using the proposed strategy, a QSPR model is developed to predict property values for organic compounds in water, based on their experimental data, and molecular structures. For this, adopting machine learning algorithms, a simple four-layer ANN is constructed to generate a predictive model which is expected to exhibit the following features:

(a) Using fewer molecular descriptors to achieve more accurate predictions compared to the available models in the literature;
(b) Avoiding various feasible strategies in structure recognition to prevent the multiple predicted values for compounds;
(c) Enhancing the generality of the model with respect to the types of the organic compounds.

2 Methodology

Herein, a strategy is proposed to rapidly recognize molecular structures and to extract molecular features without ambiguity followed by a neural network specially built for producing a predictive model to estimate physicochemical properties of the compounds of interest. Henry's law constant (HLC) for compounds in water is employed as a case study in this research. It is the air-water partition coefficient which describes the equilibrium distribution of a chemical between air and water, and it can be expressed as the ratio of partial pressure above water to the amount of dissolved gas in water [39, 40]. The HLC is an indicator of the chemical's volatility. It is important in describing the distribution and transport of chemicals between aquatic ecosystems and the atmosphere, which determines the fate of chemicals in environment. Compounds displaying higher HLC values, especially the lower molecular weight compounds, are more likely to volatilize from aqueous solutions, they must be handled carefully to improve air quality and avoid short-term and long-term adverse health effects. Fig. 1 illustrates the procedure of model development as follows:

> **Stage 1: Data collection.** The experimental data [41] of organic compounds is essential for the development of a QSPR model. In addition, simplified molecular-input line-entry system (SMILES) string is also treated as a key parameter to the presented model, which expresses fundamental information of molecular structures.
> **Stage 2: Feature extraction.** To ensure that the information of molecular structures can be processed by the neural network, molecular features are extracted with the proposed strategy and later converted to numeric vectors which are generated in a unique manner relying on built-in encoding rules.

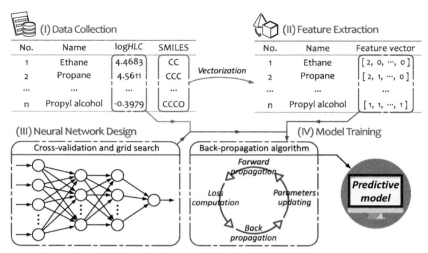

FIG. 1 The procedure for developing a predictive model to predict the logHLC values of organic compounds.

In this way, the molecular information can be introduced to the neural network and be correlated to the value of the target property.

Stage 3: Neural network design. On the basis of the experimental data and molecular feature vectors, a fully connected neural network is constructed to develop the predictive model. The structural parameters required in the design of neural network are optimized using cross-validation and grid search in order to provide stability and reliability in model training.

Stage 4: Model training. Having received the feature vectors describing molecular structures, the neural network establishes a complex mathematical model and then produces the estimated property values. The training process runs repeatedly aiming to obtain a better predictive model, which could provide more accurate predictions for HLC values of organic compounds in water.

All the above steps are achieved with a series of programs written in Python. The program has been run successfully on a desktop computer with Intel Core i3-8100 processor under Windows 10 operating system.

2.1 Data collection

To ensure the reliability of the predictive model, the experimental HLC values at 298.15 K are gathered from one of the most reliable and comprehensive databases [41]. The HLC is commonly reported in units of atm·m^3/mol (mole fraction basis) but here it is represented as its decimal logarithmic form (logHLC) because it spans over many orders of magnitude with regard to the collected massive samples. In this research, a number of irrelevant compounds (e.g., inorganic compounds and ionic compounds) and the compounds provided with estimated HLC values have been discarded. Therefore, the model is applicable only to organic compounds and its reliability is significantly improved. As a consequence, the HLC values of 2566 diverse organic compounds in water are kept and assembled as the dataset for developing the predictive model. The compounds span a wide class of molecular structures including aliphatic and aromatic hydrocarbons, alcohols and phenols, heterocyclic compounds, amines, acids, ketones, esters, aldehydes, ethers, and so on. The distribution of the treated logHLC values is displayed in Fig. 2.

The other input for the development of the QSPR model is the information of molecular structures. The SMILES is a specification in the form of a line notation for describing the structure of chemical species, and it can be used to build two- or three-dimensional structure of a molecule [42, 43]. As a chemical language, the SMILES string is sufficient to provide structural information for molecules required in model development. Thus, SMILES strings have been widely employed in the literature for developing QSPR-based models and cheminformatics software. Additionally, having learned the simple encoding rules of the SMILES strings, one can readily and correctly give the SMILES string of a compound from its molecular structure.

PubChem [44] is a massive open repository which provides over 200 million kinds of compounds with chemical information such as molecular formula, SMILES string, and so forth. It should be noted that there are two types of SMILES strings, canonical SMILES, and isomeric SMILES. The former one

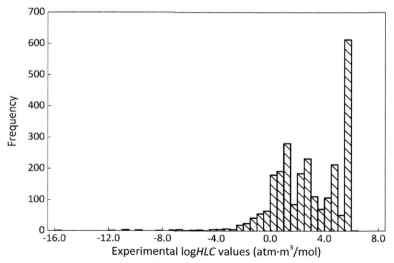

FIG. 2 The distribution of the collected experimental logHLC values for 2566 organic compounds.

is available for all the existing compounds, whereas the latter one is only provided for isomers since the isomeric SMILES strings contain isomeric information of molecules.

The SMILES strings for these investigated compounds have been collected from the PubChem database. In order to preserve the isomeric information, the isomeric SMILES string has been adopted if it is available for a given compound; otherwise, the canonical SMILES string is employed. Therefore, the experimental data and SMILES strings of the investigated 2566 organic compounds have been prepared for the correlation of molecular structures and properties.

2.2 Feature extraction

To be provided to the neural network, all types of data need to be translated into the numeric form contained in vectors. Accordingly, the molecular information of each compound needs to be converted and included in a numeric vector. For this purpose, an unambiguous strategy is proposed and programed to rapidly recognize molecular structures and extract molecular features. In the proposed strategy, each molecular feature represents a molecular substructure that only contains single nonhydrogen atom accompanied with its connected hydrogen atoms and chemical bonds. Therefore, only one strategy is feasible in subdividing a molecule into several substructures, and it avoids scattered predicted values. These features are created with built-in encoding rules in which various traditional chemical information (such as type of the non-hydrogen atom, number of hydrogen atoms, and formal charge [45]) of substructures is taken into consideration. In addition, the types of chemical bonds between the substructure and its connected substructures in the molecule are considered in the encoding rules for creating molecular features, and meanwhile, stereoisomers are also identified by the encoding rules and the stereo-centers are recorded in the molecular features. In this way, molecular features are extracted with the encoding rules and similar substructures can be distinguished to the greatest extent. On this basis, molecular structures have been converted to numeric vectors according to the frequency of each feature. Therefore, similar to GC-based methods, the proposed strategy is characterized by high interpretability as molecule are the combination of fragments.

In order to preserve molecular information and specify distinct molecular features, the RDKit cheminformatics tool has been adopted for implementing the encoding rules to present the features with identifiers. The extraction algorithm of molecular features is developed to automatically encode features using character strings and generate feature vectors based on SMILES strings. The definitions of the characters incorporated in identifiers are provided in Table 1.

The procedure for the feature extraction and vectorization of molecular structures is comprised of the following three steps as depicted in Fig. 3.

TABLE 1 The explanations and examples for the characters involved in identifiers.

Symbol	Explanation	Example
A	Atom in aliphatic compound	"[C]" Carbon atom in aliphatic compound
a	Atom in aromatic compound	"[c]" Carbon atom in aromatic compound
Hn	Atom with n hydrogen atoms	"[CH₃]" Carbon atom with three hydrogen atoms
+ (inside [ ])	Atom with a positive charge	"[N+]" Nitrogen atom with a positive charge
– (inside [ ])	Atom with a negative charge	"[N-]" Nitrogen atom with a negative charge
– (outside [ ])	Single bond not in rings	"[C]-" Carbon atom with a single bond not in rings
.	Single bond within rings	"[C]." Carbon atom with a single bond within rings
=	Double bond not in rings	"[C]=" Carbon atom with a double bond not in rings
:	Double bond within rings	"[C]:" Carbon atom with a double bond within rings
#	Triple bond not in rings	"[C]#" Carbon atom with a triple bond not in rings
*	Aromatic bond	"[c]*" Carbon atom with an aromatic bond
@	Atom is a R-chirality center	"[C@]" Carbon atom is a R-chirality center
@@	Atom is a S-chirality center	"[C@@]" Carbon atom is a S-chirality center
|	Separator	–

Step 1: The molecular features are extracted from the molecular structures of organic compounds of interest which have been already expressed with identifiers using the predefined encoding rules. The process covers all the atoms and chemical bonds in a molecule to acquire the information of molecular structures without omissions.

Step 2: The molecular features represented with identifiers are assembled into a list and the duplicates are removed to ensure that each feature only

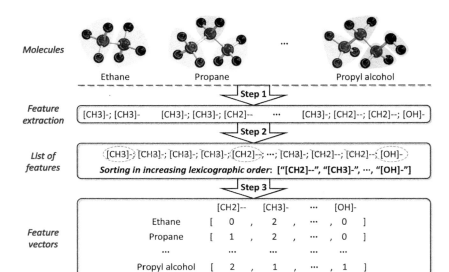

FIG. 3 The procedure of the feature extraction and vectorization for molecular structures.

appears once in the list. Then, all the remaining molecular features in the list are sorted in increasing lexicographic order (according to the Python function of "sorted") to fix the location of every feature in the list.

Step 3: For any individual compound, the feature extraction is performed again following step 1. Afterwards, the frequency of each feature in the molecule is assigned to the numeric vector according to its corresponding location in the feature list. Therefore, the final vectors include the required molecular information for all of the compounds presented in the database.

In this way, molecular features are extracted and molecular vectors are generated. Attributed to the chemical information incorporated in molecular features, the proposed strategy is able to differentiate isomers, and whereas, part of structural isomers cannot be distinguished. Therefore, plane of best fit (PBF) [46], a rapid and amenable method for describing the 3D character of molecules is employed to retain the molecular information omitted in the proposed feature extraction strategy. The proposed strategy is improved with the introduction of PBF, and both structural and geometric isomers are well identified.

2.3 Neural network design

The input parameters to the developed predictive model are transferred from the first layer of neural network (the input layer) to its last layer (the output layer) through specific mathematical relations (neurons) and, accordingly, results in the predicted HLC values.

Layer is the basis to determine the architecture of a neural network. In this research, the neural network has been built with four layers including one input layer, two hidden layers and one output layer. The number of neurons in the input layer matches the number of numeric values in the input vector so that all extracted molecular features are completely loaded. In addition, the output layer only contains one neuron for producing predicted values for the target property. The network is fully connected which means that each neuron in a layer is connected to all neurons in the previous layer, as shown in Fig. 4.

The four-layer neural network has been developed using Python as follows:

(i) PyTorch is an open-source machine learning library for Python which is rising in popularity, and it is used to build different structures of the neural network in a flexible way [47];

(ii) Root mean square error measures the differences between predicted and experimental values, and it is adopted as the loss function to quantify the performance of the developed model;

(iii) Adam algorithm [48] is an optimization method to update the weights and biases of the neural network, and it is applied to optimize the predictive model because of its high computational efficiency;

(iv) Back-propagation algorithm, a supervised learning procedure commonly used to train neural networks, is employed to update the weights and biases of neurons by calculating the gradient of the loss function.

The parameters of the neural networks are generally divided into two categories: model parameters and hyperparameters. Model parameters (e.g., weights and biases) are automatically tuned or optimized by calculating the gradient of the loss function during training. On the other hand, model hyperparameters are commonly set by the operators in advance before the neural network is functional. With the aim of efficiently controlling the training process and

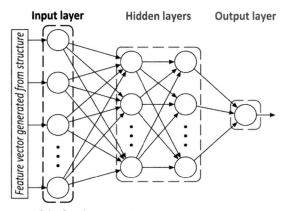

FIG. 4 The structure of the four-layer neural network.

generating a robust model, the hyperparameters are herein optimized using the approaches of cross-validation and grid search. Learning rate is set to 1.00E-03 to control the rate of convergence for the neural network. Activation functions map inputs to outputs and enhance the ability of neural networks in handling complex tasks. Two types of activation functions, "sigmoid" and "softplus" (corresponding equations are provided in Table 2), are introduced to hidden layers 1 and 2, respectively.

The application of the developed tool in making predictions for new compounds is illustrated in Fig. 5. The molecular structure (SMILES notation) of a new compound is used to generate the molecular feature vector. Relying on the parameters and hyperparameters of the ANN framework determined in the model development, the predictive model (predictive tool) makes a numeric prediction and outputs a predicted value for the logHLC of the compound.

2.4 Model training

During iterative training of the neural network, one epoch represents one forward pass (regression process from input layer to output layer) and one backward pass (back-propagation process from output layer to input layer) for all the data of a dataset. As per the batch size of training set, an epoch is divided into several iterations. The weights and biases of neurons are updated after every iteration completed so that the model can be optimized multiple times during one epoch.

The predictability of the neural network is generally verified with an external dataset which is not involved in the training of neural network. Herein, the collected dataset (including the HLC values measured in water for 2566 compounds) has been divided into two subsets: a modeling set and a test set, holding 80% and 20% of the whole dataset by using a random selection routine or k-means clustering method (i.e., random sampling and cluster sampling). Data points might be distributed very nonuniformly in the input space, and therefore, adopting the k-means clustering in the data partitioning would lead to better training, validation, and test sets than simply using randomization. The modeling and test sets are employed to, respectively, build the predictive model and evaluate the predictability of the developed model. The best set of

TABLE 2 The equations for activation functions in the ANN model.

Activation function	Equation
Sigmoid	$f(x) = \frac{1}{1+e^{-x}}$
Softplus	$f(x) = \ln(1 + e^{-x})$

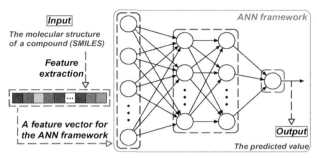

FIG. 5 The application of the developed tool in making predictions for new compounds.

hyperparameters are determined by the fivefold cross-validation. In the five-fold cross-validation, the dataset is equally partitioned into five subsets and the model training is carried out five times. During each training process, one of the five subsets is regarded as the validation set and the remaining four subsets are assigned to the training set. Therefore, each subset is used for training four times and for validation once. After training five times, the model performance is finally evaluated with the results from five independent validation sets.

During training the neural network, the error in the validation set is compared with that in the training set. Usually where both learning curves meet the tolerance is the point at which training should stop. The error is measured with the adopted loss function, and the tolerance is set to $1.00E-03$.

3 Results and discussion

3.1 Feature vector

58 Types of molecular features have been extracted from the molecular structures relying on the proposed strategy, and they are summarized in Table 3. These features are represented by identifiers involving various chemical information. For instance, the molecular feature "[CH0]-#" indicates an aliphatic carbon atom attached with zero hydrogen atoms, a single bond, and a triple bond.

Afterwards, the extracted molecular features are sorted in increasing lexicographic order as mentioned earlier. On this basis, the frequency of each feature appeared in a molecule is computed. Integers represented the frequencies of features are assigned in the corresponding locations of a numeric vector. In this way, the numeric vector containing 58 nonnegative integers is generated to describe the structural information of the molecule. Three small molecules (ethane, propane, and 1-propanol) are taken as examples to illustrate the production of vectors as shown in the Fig. 3. Once the numeric vectors have been prepared, they act as input parameters for the neural network to correlate the relationship between structures and properties.

TABLE 3 The molecular features represented by identifiers for describing molecular structures.

Chemical element	Molecular features expressed with identifiers
Carbon (C)	[CH0]-#; [CH0]----; [CH0]--..; [CH0]--=; [CH0]-...; [CH0]-.; [CH0]=..; [CH0]=..; [CH0]==; [CH0]@]--..; [CH0]@]-...; [CH1]#; [CH1]---; [CH1]-..; [CH1]-=; [CH1]...; [CH1].; [CH1]@@]---; [CH1]@@]-..; [CH1]@]---; [CH1]@]-..; [CH1]@]...; [CH2]--; [CH2]..; [CH2]]=; [CH3]-; [CH4]; [cH0]***; [cH0]**.; [cH0]-**; [cH0]=**; [cH1]**
Oxygen (O)	[OH0]-]-; [OH0]--; [OH0]..; [OH0]=; [OH1]-; [oH0]**
Nitrogen (N)	[NH0]+]--=; [NH0]#; [NH0]---; [NH0]-=; [NH1]--; [NH1]..; [NH2]-; [nH0]]**; [nH1]**
Phosphorus (P)	[PH0]---=
Sulfur (S)	[SH0]--; [SH0]--==; [SH0]..; [SH0]=; [SH1]-; [sH0]**
Halogen	[FH0]-; [ClH0]-; [BrH0]-; [IH0]-

3.2 Training process

The numeric vectors characterizing molecular information are introduced as input parameters to the neural network. The number of neurons in the input layer is equivalent to the number of numeric values in the feature vector, and thus, all the molecular information can be completely loaded to the neural network. During training of the model presented in this research, the loss function of training set has been minimized by the optimizer to search for a fairly accurate predictive model to describe the relationship between the molecular structures and the target property. Once a batch passes through the neural network, the molecular information traverses all the neurons from the first to last layer, and the neural network produces predicted values for this batch. Subsequently, the deviations between experimental and predicted values are calculated, and then the weights and biases of the neurons are updated from the output layer to the input layer with the back-propagation algorithm.

In order to improve the robustness of the neural network, the numbers of neurons in two hidden layers are optimized using the fivefold cross-validation and grid search method. Four models are investigated considering two different input vectors (i.e., feature vector and feature vector supplemented with PBF) and two different sampling methods (random sampling and cluster sampling). As highlighted in Fig. 6, the optimal set of structural parameters for each model

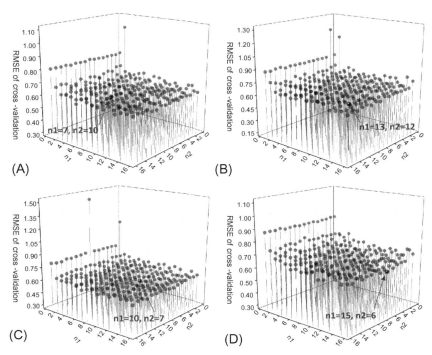

FIG. 6 The optimization and determination for the numbers of neurons in two hidden layers (n1 and n2) trained using (A) feature vector under random sampling (Scheme 1); (B) feature vector under cluster sampling (Scheme 2); (C) feature vector supplemented with PBF under random sampling (Scheme 3); (D) feature vector supplemented with PBF under cluster sampling (Scheme 4).

is determined by the lowest loss function value, and for four discussed schemes, the numbers of neurons in hidden layers are 7 and 10, 13 and 12, 10 and 7, and 15 and 6, respectively.

The learning curves of these models with optimal sets of structural parameters are provided in Fig. 7, which compare the errors in the training and validation sets for predictive models trained with different input vectors and dataset dividing methods.

It is worth mentioning that the number of cluster centers in cluster sampling is optimized by calculating Calinski-Harabasz (CH) index [49], and the results show that the clustering is better when four cluster centers are given, as shown in Fig. 8.

In order to directly compare model performance with the number of clusters, different numbers of cluster centers are adopted in the cluster sampling of model development based on Scheme 4. Analogously, the structures of these predictive models are optimized with the fivefold cross-validation as shown in Fig. 9.

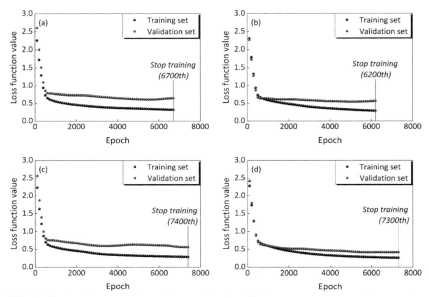

FIG. 7   The learning curves for training model using (A) feature vector under random sampling (Scheme 1); (B) feature vector under cluster sampling (Scheme 2); (C) feature vector supplemented with PBF under random sampling (Scheme 3); (D) feature vector supplemented with PBF under cluster sampling (Scheme 4).

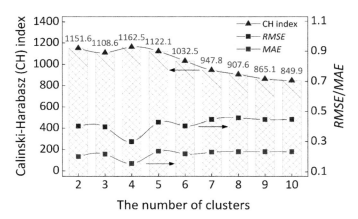

FIG. 8   The optimization for the number of cluster centers in cluster sampling and the effect of cluster centers on model performance.

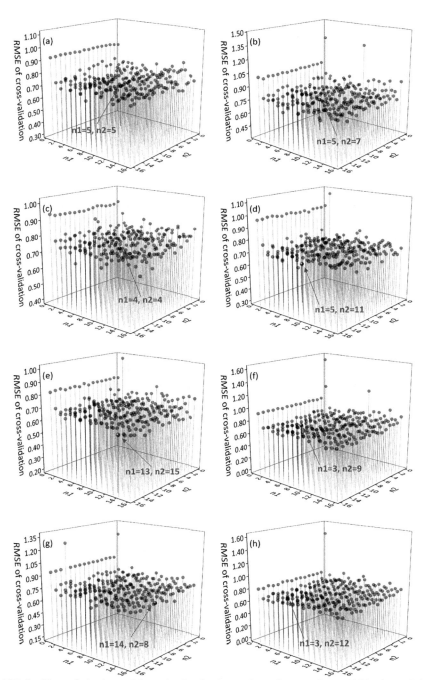

FIG. 9 The optimization and determination for the numbers of neurons in two hidden layers (n1 and n2) based on Scheme 4 trained with different numbers of cluster centers: (A) two; (B) three; (C) five; (D) six; (E) seven; (F) eight; (G) nine; (H) ten.

Relying on the optimized ANN structures, the predictive models are developed using different number of clusters in sampling for data partitioning, and the model performance is provided in Table 4. The statistical analysis is carried out with three indicators based on the experiment value (x^{exp}), predicted value (x^{pre}), and number of data points (N). The first is the root mean squared error ($RMSE$) which measures the standard deviation of differences between estimated and experimental values. The second is mean absolute error (MAE) which indicates the magnitude of differences between estimated and experimental values. Another is coefficient of determination (R^2) which provides information about the quality of the model fit.

Moreover, the predictive performance (indicated with $RMSE$ and MAE) is directly compared with the number of clusters as shown in Fig. 8, along with the CH index which is used to measure the performance of clustering. It is observed that the cluster performance is better (indicated with a greater CH index) when four cluster centers are assigned in sampling. Meanwhile, the model accuracy is also better (indicated with a smaller $RMSE$ or MAE) in the case of four cluster centers. As the number of cluster increases, the variation of CH index is basically contrary to that of $RMSE$ or MAE, demonstrating that the optimization criterion (CH index) used is also optimal for the model performance.

TABLE 4 The statistical analysis for the whole dataset in logHLC prediction adopting different numbers of clusters in Scheme 4.

Number of clusters	N^a	$RMSE^b$	MAE^c	R^{2d}
2	2566	0.4009	0.2000	0.9740
3	2566	0.3952	0.2167	0.9747
4	2566	0.2981	0.1544	0.9856
5	2566	0.4274	0.2360	0.9704
6	2566	0.4016	0.2182	0.9739
7	2566	0.4465	0.2297	0.9677
8	2566	0.4576	0.2331	0.9661
9	2566	0.4454	0.2334	0.9679
10	2566	0.4470	0.2329	0.9676

^{a}The number of data points.

$^b RMSD = \sqrt{\sum_{n=1}^{N} (x_n^{exp} - x_n^{pre})^2 / N}.$

$^c MAE = \frac{1}{N}\sum_{n=1}^{N} |x_n^{exp} - x_n^{pre}|.$

$^d R^2 = 1 - \left[\sum_{n=1}^{N} (x_n^{exp} - x_n^{pre})^2 / \sum_{n=1}^{N} (x_n^{exp} - \mu)^2\right]$ (where $\mu = \frac{1}{N}\sum_{n=1}^{N} x_n^{exp}$).

3.3   Model performance

The problem of overfitting in the neural network eventually leads to the loss of the model's predictability. In the traditional methods for property prediction, the whole dataset is employed to train and test the predictive models. Thus, these traditional prediction models may have weak predictability in predicting properties for new compounds of interest.

The four-layer fully connected neural network with given parameters is in fact a correlation between molecular structures and logHLC values. Based on the modeling and test sets, the predictability of the developed predictive model is measured by estimating logHLC values from the molecular structures. The numeric vectors of structure features of the independent compounds (not used in training and validation steps) are fed into the neural network and the estimated values are given based on the developed model.

The statistical parameters for four predictive models in Table 5 reveal that these models have satisfactory predictability and can make accurate prediction on the new data. In comparison, the model trained using feature vector supplemented with PBF under cluster sampling (i.e., Scheme 4) is significantly better than others, which indicates that introducing PBF descriptor as input and adopting k-means clustering in sampling lead to better predictive performance. The

TABLE 5 The statistical analysis for the subsets and whole dataset in logHLC prediction using different input vectors and sampling methods.

Predictive model	Dataset	N	RMSE	MAE	R^2
Scheme 1	Modeling set	2052	0.3197	0.1686	0.9824
	Test set	514	0.6469	0.2553	0.9453
	Whole dataset	2566	0.4069	0.1860	0.9732
Scheme 2	Modeling set	2052	0.2886	0.1579	0.9866
	Test set	514	0.5683	0.2558	0.9467
	Whole dataset	2566	0.3623	0.1775	0.9787
Scheme 3	Modeling set	2052	0.2875	0.1535	0.9858
	Test set	514	0.5619	0.2410	0.9587
	Whole dataset	2566	0.3596	0.1710	0.9791
Scheme 4	Modeling set	2052	0.2592	0.1399	0.9888
	Test set	514	0.4188	0.2121	0.9741
	Whole dataset	2566	0.2981	0.1544	0.9856

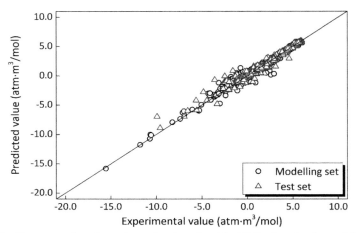

FIG. 10 The scatter plot of experimental and predicted logHLC values for the modeling and test sets.

logHLC values calculated with Scheme 4 versus the corresponding experimental data is illustrated in Fig. 10.

3.4 Comparison with reported models

The ultimate objective of developing the predictive model is to accurately estimate logHLC values in water for organic compounds. Although a number of predictive models have been reported in the literature for this purpose, different models need be compared using the same experimental dataset.

Herein, the performance of the developed neural-network-assisted predictive model based on feature vector supplemented with PBF under cluster sampling (i.e., Scheme 4) in this research (represented as NN model) is compared with a few of available models in the literature. An empirical relationship method [14] (represented as ER model) is picked in contrast with the NN model to measure the predictive power of the developed model. A comprehensive comparison shows that over 80% compounds (1475 out of 1816) used in the ER model are included in the development of the NN model, which proves that both models have employed similar datasets.

As far as the 1475 organic compounds are concerned, both models exhibit satisfactory predictive accuracy. As displayed in Fig. 11, it is clear that the NN model produced relatively small deviations. In other words, the NN model has a better agreement between the predicted and experimental values in terms of the overlapped 1475 organic compounds.

From the view of statistics, residual (experimental value minus estimated value) of each compound is calculated to compare the residual distribution plots of both ER and NN models, as illustrated in Fig. 12. With respect to the residual

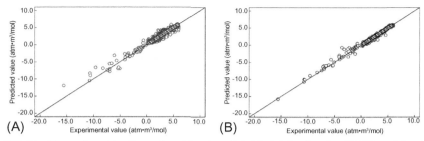

FIG. 11 The scatter plots of experimental and predicted logHLC values for (A) ER model and (B) NN model.

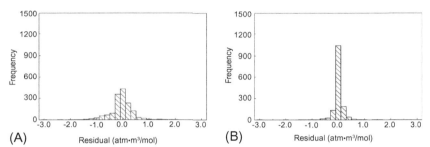

FIG. 12 The residual distributions of logHLC values estimated with (A) ER model and (B) NN model.

distribution, the residuals produced by the NN model are more densely gathered around the zero value which indicates that the proposed NN model perfectly estimated the logHLC values for more compounds than the conventional ER model.

On the other hand, several statistical indicators such as *RMSE*, *MAE*, and R^2 are analyzed based on the same data subset as shown in Table 6. For the overlapped 1475 organic compounds, the *RMSE* and *MAE* of the NN model are significantly lower than those of the ER model which means that the NN model generated smaller errors in predicting logHLC. Meanwhile, the R^2 of the NN model is closer to 1.0000 which donates that the predicted values given by the NN model are better fitted with the experimental values. All these statistical results further confirmed the conclusion drew with the residual distribution, and it demonstrated the stronger predictive capability of the NN model.

Except for the empirical relationship method, a hybrid method coupling the GC method and the neural network was proposed to develop a predictive model (represented as GN model) for estimating logHLC values of organic compounds [36]. In the GN model, 107 functional groups are extracted from 1940 compounds, and on this basis, a four-layer neural network was built to produce a

TABLE 6 The comparison for statistical results of the ER and NN models in logHLC prediction.

Predictive model	N	$RMSE$	MAE	R^2
ER model (Gharagheizi et al. [14])	1475	0.4400	0.2898	0.9660
NN model (this research)	1475	0.2124	0.1069	0.9921

nonlinear model for property estimation. In comparison with previous studies, it covered a lager dataset and showed a lower *RMSE* value.

Although different approaches for structure representation are used in the GN and NN models, the neural network is adopted as a tool to develop predictive models for estimating logHLC. Accordingly, it is necessary to evaluate the predictive performance of NN model in contrast with the GN model. A thorough comparison reveals that over 80% organic compounds (1567 out of 1940) employed in the GN model are used to develop the predictive model in this research. In terms of the overlapped 1567 compounds, the predictive capabilities of both GN and NN models are visualized in Fig. 13 with the scatter plots of estimated values versus experimental values.

From the scatter plots, it is observed that both models exhibit satisfactory predictive accuracy although some data points represent relatively large deviations. In this regard, it is hard to conclude that which model is better in estimating logHLC for organic compounds. Thus, analysis is carried out in the statistical perspective to evaluate the predictive performance of both models. From the residual distribution plots displayed in Fig. 14, almost all the residuals produced by GN and NN models are within ±0.5 log units from the zero value. Accordingly, with respect to the overlapped 1567 compounds, both models made accurate estimation for logHLC. However, there are no obvious differences between the distributions of the residuals produced by the GN and NN

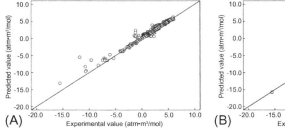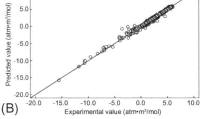

FIG. 13 The scatter plots of experimental and predicted logHLC values for (A) GN model and (B) NN model.

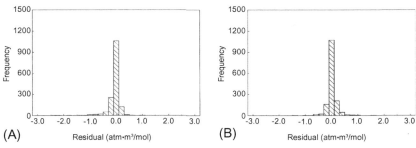

FIG. 14   The residual distributions of logHLC values estimated with (A) GN model and (B) NN model.

TABLE 7 The comparison for statistical results of the GN and NN models in logHLC prediction.

Predictive model	N	$RMSE$	MAE	R^2
GN model (Gharagheizi et al. [36])	1567	0.3283	0.1356	0.9822
NN model (this research)	1567	0.2187	0.1123	0.9921

models. Thus, their predictive performances are further quantified with several statistical indicators as shown in Table 7.

The consistency of these indicators for both models suggested that they have similar predictive performance in this task. Diving into this situation, NN model has a slightly lower $RMSE$ and MAE together with a bit higher R^2. These subtle differences in statistical results prove that the NN method is slightly better than GN method in predictive accuracy for the (overlapped 1567) organic compounds.

From the above, the developed neural-network-assisted predictive model based on feature vector supplemented with PBF under cluster sampling (i.e., the NN model) exhibits a distinct advantage over the empirical model (i.e., the ER model) in the predictive accuracy and application scope. On the other hand, the NN model is slightly better than the GC-based neural network model (i.e., the GN model) with the aid of the proposed feature extraction algorithm. Nevertheless, the GN model extracted 107 functional groups to develop the predictive model, whereas the NN model only adopted 58 molecular features. In other words, the NN model achieved a higher predictive accuracy with fewer molecular descriptors.

Moreover, the predictability of the developed model is further evaluated with compounds outside the adopted dataset. Totally, 373 compounds that are not included in the employed dataset are gathered from the literature [14] to validate the performance of the predictive model. Despite experimental HLC values and chemical names are provided, the necessary SMILES notations of compounds are unavailable. Therefore, we acquire the SMILES notations in the PubChem database by recognizing the names of compounds, and finally, a total of 233 compounds are assigned with correct SMILES notations and they are used to evaluate the performance of the predictive model. Hereinto, 89 compounds contain no less than 10 carbon atoms (C10–C17) and 144 compounds contain less than 10 carbon atoms (C3–C9). The predictive model in Scheme 4 is used to perform predictions on these compounds, and the predictive performance is presented in Fig. 15. It is observed that the predictions on relatively complex compounds (C10–C17) present a bit larger deviations than those on relatively simple compounds (C3–C9). Overall, the predictive model reveals decent and acceptable accuracy on these compounds outside the employed dataset in the model development, demonstrating the satisfactory predictability of the developed predictive model.

Another point is worth mentioning that the proposed neural-network-assisted predictive model is developed relying on a large dataset of 2566 organic compounds with a R^2 of 0.9856. In the available predictive models, the largest dataset for model development contains 1954 pure compounds and the model exhibits an R^2 of 0.9828 [16]. Therefore, the developed predictive model is considered to be the most comprehensive model for predicting logHLC. Accordingly, using the neural network and the proposed algorithm for extracting molecular features, the developed predictive model is able to provide accurate and reliable prediction for logHLC of organic compounds.

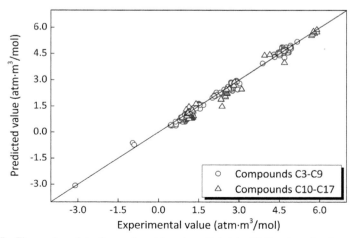

FIG. 15 The scatter plot of experimental and predicted logHLC values for the collected compounds.

4   Conclusions

This research proposes an unambiguous feature extraction strategy to avoid different feasible strategies in the characterization of molecular structures. It therefore can overcome some shortcomings of GC-based methods, such as the multiple predicted values for compounds. A four-layer neural network is then constructed to correlate the molecule structures with target property values for organic compounds. With the frequencies of molecular features as inputs, the neural network is trained with the acquired experimental data and evaluated with a test set which is not involved in the training process. During the training process, the numbers of neurons in the neural network are optimized to achieve a robust model using the fivefold cross-validation and grid search. As such, a hybrid predictive model is obtained with the combination of the proposed strategy and machine learning algorithm.

With respect to the logHLC values of pure organic compounds in water, the predictive model is built based on the experimental values of 2566 organic compounds in water. Moreover, the introduction of the PBF descriptor and two dataset dividing methods are investigated in regard to the model performance. As it turns out, four predictive models are characterized by good predictability and predictive accuracy. The statistical analysis indicates that the predictive model developed with feature vector supplemented with PBF under cluster sampling shows significantly better predictive ability. It proves that the introduction of the PBF descriptor and adopting k-means clustering in sampling enhanced the model performance.

In contrast with the reported predictive models in the literature, the developed predictive model demonstrates higher predictive accuracy although fewer molecular descriptors were used in its development. Moreover, it exhibits enhanced generality and covers more diverse organic compounds than reported models with respect to the employed comprehensive database. Therefore, the proposed strategy and model development methods can serve as a promising and effective approach to develop property predictive models, directing the reduction of pollutants in environment and the development of greener solvents. The established machine learning based models can be further used in the computer-aided molecular design framework. In addition, we can reasonably expect them to be further popularized to use for some other important environmental properties such as water solubility and the bioconcentration factor, which reveals their vital potential in the development of green chemistry and engineering.

Acknowledgments

The chapter is reproduced from Green Chemistry, 2020, 22, 3867-3876, Zihao Wang, Yang Su, Saimeng Jin, Weifeng Shen, Jingzheng Ren, Xiangping Zhang, James H. Clark, A novel unambiguous strategy of molecular feature extraction in machine learning assisted predictive models for environmental properties, by permission of The Royal Society of Chemistry.

References

[1] J.H. Clark, Green chemistry: challenges and opportunities, Green Chem. 1 (1999) 1–8.

[2] J.H. Clark, Green chemistry: today (and tomorrow), Green Chem. 8 (2006) 17–21.

[3] F.P. Byrne, S. Jin, G. Paggiola, T.H.M. Petchey, J.H. Clark, T.J. Farmer, A.J. Hunt, C.R. McElroy, J. Sherwood, Tools and techniques for solvent selection: green solvent selection guides, Sustain. Chem. Process. 4 (2016) 7.

[4] N.G. Chemmangattuvalappil, C.C. Solvason, S. Bommareddy, M.R. Eden, Combined property clustering and GC+ techniques for process and product design, Comput. Chem. Eng. 34 (2010) 582–591.

[5] T. Zhou, K. McBride, S. Linke, Z. Song, K. Sundmacher, Computer-aided solvent selection and design for efficient chemical processes, Curr. Opin. Chem. Eng. 27 (2020) 35–44.

[6] J. Sedlbauer, G. Bergin, V. Majer, Group contribution method for Henry's law constant of aqueous hydrocarbons, AICHE J. 48 (2002) 2936–2959.

[7] S.H. Hilal, S.N. Ayyampalayam, L.A. Carreira, Air-liquid partition coefficient for a diverse set of organic compounds: Henry's law constant in water and hexadecane, Environ. Sci. Technol. 42 (2008) 9231–9236.

[8] W. Shen, L. Dong, S. Wei, J. Li, H. Benyounes, X. You, V. Gerbaud, Systematic design of an extractive distillation for maximum-boiling azeotropes with heavy entrainers, AICHE J. 61 (2015) 3898–3910.

[9] D. Prat, A. Wells, J. Hayler, H. Sneddon, C.R. McElroy, S. Abou-Shehada, P.J. Dunne, CHEM21 selection guide of classical- and less classical-solvents, Green Chem. 18 (2016) 288–296.

[10] M. Tobiszewski, S. Tsakovski, V. Simeonov, J. Namieśnik, F. Pena-Pereira, A solvent selection guide based on chemometrics and multicriteria decision analysis, Green Chem. 17 (2015) 4773–4785.

[11] S. Jin, A.J. Hunt, J.H. Clark, C.R. McElroy, Acid-catalysed carboxymethylation, methylation and dehydration of alcohols and phenols with dimethyl carbonate under mild conditions, Green Chem. 18 (2016) 5839–5844.

[12] N.K. Razdan, D.M. Koshy, J.M. Prausnitz, Henry's constants of persistent organic pollutants by a group-contribution method based on scaled-particle theory, Environ. Sci. Technol. 51 (2017) 12466–12472.

[13] F. Gharagheizi, A. Eslamimanesh, A.H. Mohammadi, D. Richon, Empirical method for representing the flash-point temperature of pure compounds, Ind. Eng. Chem. Res. 50 (2011) 5877–5880.

[14] F. Gharagheizi, A. Eslamimanesh, A.H. Mohammadi, D. Richon, Empirical method for estimation of Henry's law constant of non-electrolyte organic compounds in water, J. Chem. Thermodyn. 47 (2012) 295–299.

[15] T. Puzyn, P. Rostkowski, A. Świeczkowski, A. Jędrusiak, J. Falandysz, Prediction of environmental partition coefficients and the Henry's law constants for 135 congeners of chlorodibenzothiophene, Chemosphere 62 (2006) 1817–1828.

[16] F. Gharagheizi, P. Ilani-Kashkouli, S.A. Mirkhani, N. Farahani, A.H. Mohammadi, QSPR molecular approach for estimating Henry's law constants of pure compounds in water at ambient conditions, Ind. Eng. Chem. Res. 51 (2012) 4764–4767.

[17] Y. Su, Z. Wang, S. Jin, W. Shen, J. Ren, M.R. Eden, An architecture of deep learning in QSPR modeling for the prediction of critical properties using molecular signatures, AICHE J. 65 (2019) e16678.

[18] Z. Wang, Y. Su, W. Shen, S. Jin, J.H. Clark, J. Ren, X. Zhang, Predictive deep learning models for environmental properties: the direct calculation of octanol-water partition coefficients from molecular graphs, Green Chem. 21 (2019) 4555–4565.

[19] S. Datta, V.A. Dev, M.R. Eden, Developing non-linear rate constant QSPR using decision trees and multi-gene genetic programming, Comput. Chem. Eng. 127 (2019) 150–157.

[20] M. Barycki, A. Sosnowska, T. Puzyn, AquaBoxIL—a computational tool for determining the environmental distribution profile of ionic liquids, Green Chem. 20 (2018) 3359–3370.

[21] J.I. García, H. García-Marín, J.A. Mayoral, P. Pérez, Quantitative structure-property relationships prediction of some physico-chemical properties of glycerol based solvents, Green Chem. 15 (2013) 2283–2293.

[22] S.T. Lin, S.I. Sandler, Henry's law constant of organic compounds in water from a group contribution model with multipole corrections, Chem. Eng. Sci. 57 (2002) 2727–2733.

[23] Y. Huang, H. Dong, X. Zhang, C. Li, S. Zhang, A new fragment contribution-corresponding states method for physicochemical properties prediction of ionic liquids, AICHE J. 59 (2013) 1348–1359.

[24] J. Marrero, R. Gani, Group-contribution based estimation of pure component properties, Fluid Phase Equilib. 183 (2001) 183–208.

[25] J. Marrero, R. Gani, Group-contribution-based estimation of octanol/water partition coefficient and aqueous solubility, Ind. Eng. Chem. Res. 41 (2002) 6623–6633.

[26] S. Jhamb, X. Liang, R. Gani, A.S. Hukkerikar, Estimation of physical properties of amino acids by group-contribution method, Chem. Eng. Sci. 175 (2018) 148–161.

[27] T. Zhou, S. Jhamb, X. Liang, K. Sundmacher, R. Gani, Prediction of acid dissociation constants of organic compounds using group contribution methods, Chem. Eng. Sci. 183 (2018) 95–105.

[28] X. Yao, M. Liu, X. Zhang, Z. Hu, B. Fan, Radial basis function network-based quantitative structure-property relationship for the prediction of Henry's law constant, Anal. Chim. Acta 462 (2002) 101–117.

[29] A. Eslamimanesh, F. Gharagheizi, A.H. Mohammadi, D. Richon, Artificial neural network modeling of solubility of supercritical carbon dioxide in 24 commonly used ionic liquids, Chem. Eng. Sci. 66 (2011) 3039–3044.

[30] F. Gharagheizi, A. Eslamimanesh, A.H. Mohammadi, D. Richon, Artificial neural network modeling of solubilities of 21 commonly used industrial solid compounds in supercritical carbon dioxide, Ind. Eng. Chem. Res. 50 (2010) 221–226.

[31] Y. Pan, J. Jiang, Z. Wang, Quantitative structure-property relationship studies for predicting flash points of alkanes using group bond contribution method with back-propagation neural network, J. Hazard. Mater. 147 (2007) 424–430.

[32] M. Safamirzaei, H. Modarress, M. Mohsen-Nia, Modeling and predicting the Henry's law constants of methyl ketones in aqueous sodium sulfate solutions with artificial neural network, Fluid Phase Equilib. 266 (2008) 187–194.

[33] N.J. English, D.G. Carroll, Prediction of Henry's law constants by a quantitative structure property relationship and neural networks, J. Chem. Inf. Comput. Sci. 41 (2001) 1150–1161.

[34] M. Safamirzaei, H. Modarress, Application of neural network molecular modeling for correlating and predicting Henry's law constants of gases in [bmim][PF6] at low pressures, Fluid Phase Equilib. 332 (2012) 165–172.

[35] D.R. O'Loughlin, N.J. English, Prediction of Henry's law constants via group-specific quantitative structure property relationships, Chemosphere 127 (2015) 1–9.

[36] F. Gharagheizi, R. Abbasi, B. Tirandazi, Prediction of Henry's law constant of organic compounds in water from a new group-contribution-based model, Ind. Eng. Chem. Res. 49 (2010) 10149–10152.

[37] F. Gharagheizi, A. Eslamimanesh, A.H. Mohammadi, D. Richon, Determination of parachor of various compounds using an artificial neural network-group contribution method, Ind. Eng. Chem. Res. 50 (2011) 5815–5823.

[38] F. Gharagheizi, R. Abbasi, A new neural network group contribution method for estimation of upper flash point of pure chemicals, Ind. Eng. Chem. Res. 49 (2010) 12685–12695.

[39] D. Yaffe, Y. Cohen, G. Espinosa, A. Arenas, F. Giralt, A fuzzy ARTMAP-based quantitative structure-property relationship (QSPR) for the Henry's law constant of organic compounds, J. Chem. Inf. Comput. Sci. 43 (2003) 85–112.

[40] H.P. Chao, J.F. Lee, C.T. Chiou, Determination of the Henry's law constants of low-volatility compounds via the measured air-phase transfer coefficients, Water Res. 120 (2017) 238–244.

[41] C.L. Yaws, Yaws' Critical Property Data for Chemical Engineers and Chemists, 2019. https://app.knovel.com/hotlink/toc/id:kpYCPDCECD/yaws-critical-property/yaws-critical-property. (Accessed 3 April 2019).

[42] Simplified Molecular-Input Line-Entry System, 2019. https://en.wikipedia.org/wiki/Simplified_molecular-input_line-entry_system. (Accessed 9 May 2019).

[43] D. Weininger, SMILES, a chemical language and information system. 1. Introduction to methodology and encoding rules, J. Chem. Inf. Comput. Sci. 28 (1988) 31–36.

[44] S. Kim, J. Chen, T. Cheng, A. Gindulyte, J. He, S. He, Q. Li, B.A. Shoemaker, P.A. Thiessen, B. Yu, L. Zaslavsky, J. Zhang, E.E. Bolton, PubChem 2019 update: improved access to chemical data, Nucleic Acids Res. 47 (2019) D1102–D1109.

[45] Formal Charge, 2020. https://en.wikipedia.org/wiki/Formal_charge. (Accessed 6 March 2020).

[46] N.C. Firth, N. Brown, J. Blagg, Plane of best fit: a novel method to characterize the three-dimensionality of molecules, J. Chem. Inf. Model. 52 (2012) 2516–2525.

[47] N. Ketkar, Deep Learning With Python, Apress, Berkeley, 2017.

[48] D.P. Kingma, J. Ba, Adam: A Method for Stochastic Optimization, 2014. arXiv preprint:1412.6980.

[49] T. Caliński, J. Harabasz, A dendrite method for cluster analysis, Commun. Stat. Theory Methods 3 (1974) 1–27.

Chapter 5

Intelligent approaches to forecast the chemical property: Case study in papermaking process

Yang Zhang[a], Jigeng Li[a], Mengna Hong[a], and Yi Man[b]

[a]*State Key Laboratory of Pulp and Paper Engineering, School of Light Industry and Engineering, South China University of Technology, Guangzhou, People's Republic of China,* [b]*Department of Industrial and Systems Engineering, The Hong Kong Polytechnic University, Hong Kong SAR, People's Republic of China*

1 Introduction

Traditional process control technology relies on indirect quality control of process measurable variables related to product quality. It has been difficult to meet the requirements of modern production processes. Instead, advanced control technologies such as predictive control and fuzzy control that take product quality as the direct control objective have received more and more attention [1]. Real-time online measurement of key physical and chemical parameters has become the main difficulty in process control and optimization. These parameters have a profound impact on the efficiency of the production processes and the quality of the products, such as the Kappa value in the pulping and cooking process; the quality of oil in the refining process, the concentration of bacteria in biological fermentation tanks, the crystal size in the crystallization process, etc. [2–5]. Accurate and real-time measurement of these key physical and chemical parameters is the basis for process control and optimization. However, due to technical limitations or economic reasons, they cannot be detected online by conventional sensors. Generally, the method used in the production process is an offline manual analysis method of these process variables. Because the manual laboratory analysis makes the sampling period of process variables longer, online real-time detection of process variables cannot be achieved [6].

Using soft measurement technology is one of the ways to solve this kind of variable prediction problem. Soft-sensing technology is to use auxiliary

Applications of Artificial Intelligence in Process Systems Engineering
https://doi.org/10.1016/B978-0-12-821092-5.00001-2

variables that are closely related to the dominant variable and easy to measure and realize the prediction of the dominant variable by constructing a certain mathematical relationship [7, 8]. Soft measurement technology can obtain some very important process parameters that cannot be directly measured by traditional instruments, which is of great significance to the monitoring of operating conditions and the optimization of the production process. The soft sensor modeling methods commonly used in industrial processes are mainly divided into two categories: mechanism modeling method and data-driven modeling method [9]. The mechanism model is based on the mass and energy conservation law, dynamics, and other basic laws involved in the production process. Compared with other models, the mechanism model has good explanatory properties and is the most ideal soft sensor model. For example, the continuous stirred tank reactor model proposed by Meng and Li based on the reaction mechanism, and the simplified model of 160 MW oil drum boiler proposed by Balko and Rosinová are all very classic mechanism models [10, 11].

However, mechanism modeling methods has the following shortcomings: (1) It requires a more comprehensive and in-depth understanding of the mechanism of the process and an accurate theoretical basis. The model is only suitable for the description of simple and stable process objects. In fact, the actual industrial process objects are often very complicated, the mechanism of the process is not very clear, and the production conditions are usually affected by the environment and the controller and change dynamically, so it is difficult to accurately describe the mechanism; (2) When the model is complex, a large number of differential equations are needed to characterize the mechanism process, and it is very difficult to obtain accurate numerical solutions; (3) When using the mechanism method to model the production process, it is generally necessary to simplify the object, and based on many assumptions, so that there is a big error between the model and the actual process. Therefore, it takes a lot of effort to establish an accurate mechanism model. After the model is built, it needs to be continuously improved to adapt to process changes. The above factors greatly limit the applicable scope of mechanism model. For processes with unclear mechanisms, data-driven method based on machine learning can be used to establish soft sensor model, also known as black box models [12]. This type of method treats the entire system as a black box and constructs the mathematical relationship between auxiliary variables and dominant variables [13]. Since this method does not require detailed knowledge of the internal mechanism of the process, it has become a commonly used soft sensor modeling method. Statistical methods such as Principal Component Analysis (PCA) and Partial Least Squares (PLS) and machine learning such as Artificial Neural Network (ANN) and Support Vector Machine (SVM) have provided an important theoretical basis for data-driven modeling method [14].

Papermaking is a resource-intensive basic raw material industry, accounting for 7% of global industrial energy consumption [15]. As the world's largest paper and paperboard producer, China's output in 2019 reached

107.65 million tons, accounting for about a quarter of the world. Papermaking is a complex process involving multiple variables. Beating degree is a key parameter of the pulp, which represents the drainage performance of the pulp [16]. The higher the beating degree, the lower the filtration rate, which reflects the tighter bonding between the fibers. Beating degree should not be too low or too high. If it is too low, the paper strength is not good, and if it is too high, it will be more difficult to dehydrate, that is, drying energy consumption will increase. Therefore, besting degree is closely related to the quality of paper and production energy consumption. In actual production, the measurement of beating degree mainly includes offline manual measurement and online equipment measurement methods. Offline manual measurement method takes a long time, generally more than 2 h. The measurement accuracy and timeliness cannot be guaranteed. The delayed measurement results cannot be fed back to the production process and the real-time optimization of the production process cannot be realized. The DRT-5500 beating degree meter developed by BTG can realize online beating degree measurement, but its equipment investment and maintenance are expensive, and frequent maintenance requirements have become the biggest obstacle to real-time control of the refining process [17].

Tissue paper is one of the most important products in the papermaking industry [18]. This work takes the online detection of the beating degree of tissue paper as an example. Taking the production line of a tissue papermaking mill as the research object, the data of pulp board characteristics, and the historical operation data of the refining process are collected and preprocessed. Based on the gradient boost decision tree (GBDT) algorithm, a prediction model of the beating degree of tissue paper is established. The model is verified by industrial operation data and compared with the support vector machine model. The prediction model proposed in this work can detect the beating degree in real time and accurately, which provides a basis for real-time control of the refining process.

2 Literature review

Mechanism model is an accurate mathematical model based on the internal mechanism of object process. Babu et al. proposed a model of water loss in fresh leaves for cooking and medicinal purposes based on the principle of drying kinetics, and the application of the model provided guidance for the preservation methods of fresh leaves [19]. Kim et al. proposed an atomic-level solvent extraction desalination model based on the principle of molecular dynamics. It was found that a higher surface polarity facilitates more water absorption. Partial recovery of the salt ions was attributed to the surface polarity of the organic solvent clusters [20]. Zhang et al. proposed a numerical computational framework that combines a phase field model and cohesive element for modeling progressive failure in composite material. This model can capture cracks in individual materials and interfaces separately [21]. Niu et al. proposed a tree structure method to characterize the structure of the mechanism, and two generalized models are proposed for planar compliant mechanisms. These two models can be used for dynamic analysis and parameter optimization of flexible mechanisms [22]. Liu et al. proposed a model based on the ratio of minimum necessary time for uniformly dispersing carbon nanotube to flattening time of composite powders. Using the model, the effect of milling speed on the distribution of carbon nanotube was analyzed, showing that both low milling speed and high milling speed were not conducive to carbon nanotube [23].

Compared with the mechanism model, the data-driven model is established based on statistical methods or machine learning methods and does not require the modeler to have a deep understanding of the internal process mechanism. Zafra et al. used autoregressive integrated moving average (ARIMA) model to predict the PM10 concentration in the air in Bogotá. Through model analysis, it was found that the persistence of PM10 pollution is higher when vegetation increases [24]. Yao et al. applied the random forest (RF) algorithm to the hail forecast in Shandong Peninsula. After cross-validation, the model can effectively identify the hail area and the time of hail disaster. [25]. Alameer et al. established a gold price fluctuation prediction model based on whale optimization algorithm (WOA) and neural network algorithm [26]. Ma et al. used the gray model (GM) to accurately predict China's carbon emissions in the next 5 years and found that energy consumption, economic growth, industrial structure, foreign direct investment, and urbanization are the five main factors that lead to increased carbon emissions [27]. Wu et al. combined the variational modal decomposition (VMD) algorithm, sample entropy (SE) algorithm, and long-term short-term memory network (LSTM) established a VMD-SE-LSTM hybrid model for air index prediction in Beijing and Baoding the comparison shows that the prediction accuracy of the mixed model is higher [28]. Man et al. proposed a hybrid model based on autoregressive moving average (ARMA) and vector autoregressive (VAR) algorithm to achieve accurate prediction of urban sewage influent chemical oxygen demand (COD) indicators. The application of the model can reduce the COD treatment cost of sewage treatment plants [29]. Hu et al. established a hybrid forecasting model of power load in the papermaking process based on particle swarm optimization (PSO) algorithm, genetic algorithm (GA), and feedforward neural network (BPNN) algorithm. In this research, the auxiliary algorithm is used to optimize the connection weight of the BPNN algorithm [30].

The hybrid model based on mechanism and data driven modeling method combines the advantages of both methods and has become a hot research. Sun et al. proposed a mechanism and data-driven prediction method. In this novel approach, blast furnace operational events are considered when predicting blast furnace gas generation, thus making predictions more accurate by integrating a priori mechanism knowledge associated with the blast furnace ironmaking process [31, 32]. By combining the mechanism with a data-driven model, Meng et al. accurately predicted the crystal content and major crystal size distribution parameters during the sucrose crystallization process. Based on the hybrid model, an intelligent integrated measurement and control system for the sucrose crystallization process has been developed and successfully applied in sugar production [33]. Sun et al. proposed a hybrid cooling control strategy based on a simplified nonlinear mechanical model and a data-driven model. The simulation results show that, compared with the traditional proportional integral (PI) method and active disturbance rejection method, the hybrid method has stronger robustness and lower noise sensitivity [31, 32].

Based on the above researches these conclusions can be drawn: (1) the application fields of mechanism models are limited, and most of them are targeted at simple process objects, while model prediction methods based on intelligent algorithms are widely used in the prediction of key indicators in various fields. (2) The mixed prediction model based on mechanism and data-drive has better performance than the single model. However, in some traditional process industries, such as papermaking, cement manufacturing, and glass manufacturing, relevant researches are rare. The production process of these industries is very rough. The reason is that the production process is not transparent enough, which is unfavorable to the control of product quality and the reduction of production energy consumption. Establishing the prediction model of key parameters in the production process can avoid the expensive purchase and maintenance of sensor equipment to achieve accurate prediction of key parameters.

3 Intelligent prediction method

Data-driven predictive modeling is a multiple function regression problem, so most regression methods can be used in modeling. In recent years, machine learning techniques such as artificial neural network (ANN), SVM, and ensemble learning have been widely used because of their good nonlinear approximation ability [34].

3.1 ANN

Artificial neural network is one of the main methods of machine learning. It is a complex network computing system composed of many highly interconnected neurons and realizes the storage and processing of network information through the interaction between neurons. Rumelhart et al. proposed Back Propagation (BP) algorithm, which triggered the climax of ANN research [35]. Many scholars proposed various improved algorithms based on BP algorithm to improve the performance of the model. ANN has strong nonlinear function approximation ability, which provides a new way for the modeling of nonlinear problems and has been widely used in industrial processes.

When using ANN for modeling, it is necessary to determine the structure of the model in advance, but there is no effective criterion to determine the optimal number of hidden layers and the number of nodes at each layer. Furthermore,

ANN is based on the principle of experiential risk minimization, and the model is established on the assumption that the training samples tend to be infinite. However, the actual training samples are limited, so the model tends to be too complex during training and overfitting phenomenon occurs. Especially in the case of small samples, the performance deterioration will be more obvious.

3.2 SVM

Support vector machine (SVM), proposed by Cherkassky, is a new learning method based on statistical learning theory [36]. Its basic idea is to transform the input space into a high-dimensional feature space through nonlinear kernel functions, and to find a linear relationship between input and output in the feature space. Different from ANN, SVM take structural risk minimization as the principle, not only considering the approximation accuracy of training samples, but also considering the complexity of the model. Therefore, it can effectively suppress the overfitting phenomenon, and has better generalization performance than the traditional machine learning methods such as ANN [37].

However, traditional optimization algorithms eventually reduce the problem solved to a problem of Quadratic Programming (QP), and the complexity of the algorithm depends on the number of training samples, so the process of solving could be time consuming as training samples increase. Van Gestel et al. proposed the least squares support vector machine (LSSVM) algorithm and used equality constraints to replace inequality constraints in traditional SVM [38]. At the same time, the sum of squares of errors is used as the empirical loss of training samples, so the QP problem is transformed into linear equations to solve, which reduces the complexity of calculation.

3.3 Ensemble learning

Ensemble learning mainly consists of three parts: generating different individuals, individual training, and individual synthesis. It originally originated from the neural network ensemble idea proposed by Hansen and Salamon, which significantly improves the generalization ability of the model by training multiple neural networks and integrating all the results [39]. Ensemble learning constructs multiple learners and then synthesizes them according to a certain strategy to obtain the final model output. These learners are called individuals or submodels, and can be implemented by any regression method, such as traditional multiple linear regression or intelligent algorithms such as decision trees, ANN, and SVM. Ensemble learning obtains a composite global model with higher prediction accuracy and greater universality through the combination of multiple models, so that the prediction of unknown samples is more accurate [40]. In addition, the use of integrated learning "divide and conquer" idea to divide large-scale data into multiple subdata sets can greatly reduce the time for model training. Because ensemble learning can effectively improve the generalization ability of a single model, it has become a research hotspot in machine learning. Ensemble Learning is considered by authoritative experts to be the top research direction of machine learning.

4 Beating degree prediction model of tissue paper

This work takes the production line of a tissue paper enterprise as the research object. The process flow chart of the production line is shown in Fig. 1. The production line mainly produces low-gram tissue paper, including napkins, soft draws, toilet rolls, facial tissues, and other different types of paper 17 kinds of products. The production line is mainly divided into two stages: pulping and papermaking. In the pulping stage, the original pulp board is mainly subjected to the processes of pulping, deslagging, and refining to make a fiber suspension. The main equipment of the pulping system includes the following: medium-consistency hydraulic pulper, high-concentration desander, and double-disc refiner. In the papermaking stage, after pressing, drying, and other processes, the paper is finally scraped off the cylinder by the scraper.

The technical route of this research is shown in Fig. 2, including the following aspects: (1) Collect historical data of the refining process. Including: the feed weight of each pulp board and its ratio, pulp board fiber shape data, and refining process parameters. The fiber shape data of pulp board includes, average fiber length, fiber number, twisted fiber content, fine fiber content, average fiber thickness, average fiber width, fiber content of broken tail, and pulp beating degree before refining. The parameters of the refining process include, (1) Specific energy consumption and pulp flow. (2) Data preprocessing. Refining is a mixing process of various pulp board, so it is necessary to calculate the fiber shape data of the mixed pulp. Because the collection frequency of each

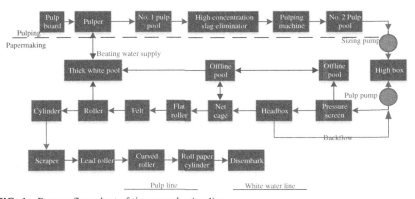

FIG. 1 Process flow chart of tissue production line.

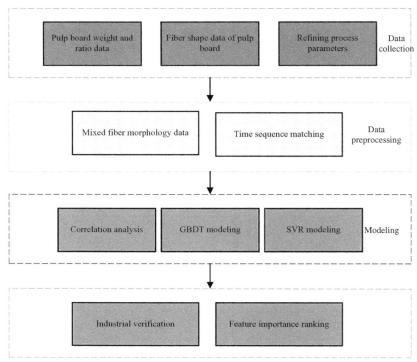

FIG. 2   Overall technology roadmap.

type of data is different, the data needs to be matched in time sequence. (3) Modeling and comparison. Correlation analysis is carried out on the preprocessed pulp fiber shape data, and the appropriate data-driven algorithm is selected according to the correlation analysis result to establish the beating degree prediction model. The GBDT and SVM algorithms are selected in this study. Use preprocessed historical data as a training set for model training and parameter tuning and compare the performance of GBDT and SVM algorithm models. (4) Industrial verification and importance ranking. Use industrial operating data to verify the accuracy of the established beating degree prediction model and analyze the GBDT model to get the relative importance of each modeling variable.

4.1   Data collection and preprocessing

4.1.1   Data collection

The production line uses commercial pulp boards as raw materials. According to the production plan, the pulp boards that need to be used are put in proportion. The quality of the raw materials is relatively stable for, which can eliminate the interference caused by raw material fluctuations. The collected data includes: the feed weight of each pulp board and its ratio, pulp board fiber shape data (11 kinds of short fiber pulp board, 7 kinds of long fiber pulp board) and refining process parameters. The fiber shape data of pulp board includes, average fiber length, fiber number, twisted fiber content, fine fiber content, average fiber thickness, average fiber width, fiber content of broken tail, and pulp beating degree before refining. The characteristic variables and definitions of fiber shape are shown in Table 1. The parameters of the refining process include, specific energy consumption of refining and pulp flow. The location of sensors is shown in Fig. 3.

4.1.2   Data preprocessing

Pulping is the process of pulp mixing, does not change the pulp properties. The properties of the pulp before refining are determined by the properties of various pulp boards and feeding ratio. By collecting the initial beating degree and fiber shape data and combining with the feeding ratio, the beating degree and fiber shape of the mixed slurry before refining can be calculated. The calculation method of initial beating degree of mixed pulp is shown in Eq. (1) [41]:

$$\ln(SR_{mix}) = \sum_{i=1}^{n} q_i \, \ln(SR_i) \tag{1}$$

TABLE 1 Characteristic variables and definitions of fiber shape.

Name	Unit	description
Fiber number	million/g	The total number of fibers per gram of absolute dry fiber
Average fiber length	mm	The sum of fiber length divided by the total number of fibers
Average fiber width	μm	The total fiber width divided by the total number of fibers
Average fiber thickness	mg/m	The mass of the absolute dry fiber per 1 m length
Twisted fiber content	%	Two or more fibers are kinked together
Fiber content of broken tail	%	The number of cut fibers accounted for the target proportion of the total number of fibers
Fine fiber content	%	The total length of fine fibers less than 0.2 mm in length divided by the total length of all fibers

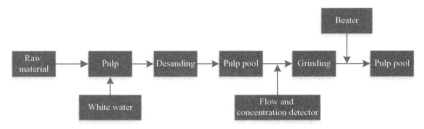

FIG. 3 Flow chart of pulp preparation unit.

where q_i represents the feeding weight of the ith type of pulp board; SR_i represents the initial beating degree of the ith type of pulp board, SR_{mix} represents the beating degree of mixed pulp before refining. The fiber shape of the mixed pulp can be calculated by Eq. (2) [42]:

$$
\begin{cases}
L_m = \dfrac{L}{N} = \sum_{i=1}^{n} q_i L_i / \sum_{i=1}^{n} q_i N_i = \sum_{i=1}^{n} q_i l_i N_i / \sum_{i=1}^{n} q_i N_i \\[2mm]
d_m = \dfrac{D}{N} = \sum_{i=1}^{n} q_i D_i / \sum_{i=1}^{n} q_i N_i = \sum_{i=1}^{n} q_i d_i N_i / \sum_{i=1}^{n} q_i N_i \\[2mm]
c_m = \dfrac{L^c}{L} = \sum_{i=1}^{n} q_i c_i L_i / \sum_{i=1}^{n} q_i L_i = \sum_{i=1}^{n} q_i c_i N_i l_i / \sum_{i=1}^{n} q_i N_i l_i \\[2mm]
r_m = \sum_{i=1}^{n} q_i M_i / \sum_{i=1}^{n} q_i L_i = \sum_{i=1}^{n} q_i r_i L_i / \sum_{i=1}^{n} q_i L_i = \sum_{i=1}^{n} q_i r_i N_i l_i / \sum_{i=1}^{n} q_i N_i l_i \\[2mm]
w_m = \dfrac{W}{N} = \sum_{i=1}^{n} q_i W_i / \sum_{i=1}^{n} q_i N_i = \sum_{i=1}^{n} q_i w_i N_i / \sum_{i=1}^{n} q_i N_i \\[2mm]
b_m = \dfrac{B}{N} = \sum_{i=1}^{n} q_i B_i / \sum_{i=1}^{n} q_i N_i = \sum_{i=1}^{n} q_i b_i N_i / \sum_{i=1}^{n} q_i N_i
\end{cases}
\tag{2}
$$

where L_m represents the average fiber length of the mixed pulp; d_m represents the twisted fiber content of the mixed pulp; c_m represents the fine fiber content of the mixed pulp; r_m represents the average fiber thickness of the mixed pulp; w_m represents the average fiber width of the mixed pulp; b_m represents the fiber content of broken tails of the mixed pulp; L represents the total fiber length of the mixed pulp; N represents the fiber number of the mixed pulp; D represents the twisted fiber number of the mixed pulp; L^c represents the total length of fibers with a length of less than 0.2 mm in the mixed pulp; W represents the total width of the fiber of the mixed pulp, B represents the number of broken tail fibers of the mixed pulp. i represents pulp board category, $i = \{1, 2, \ldots, n\}$. There are n kinds of pulp boards. q_i represents the feeding weight of the type i pulp board; L_i represents the total length of fiber of the type i pulp board; l_i represents the average length of fiber of the type i pulp board; N_i represents the number of fiber of the type i pulp board; D_i represents the number of twisted fibers of the type i pulp board; d_i represents the twisted fiber content of the type i pulp board; c_i represents the fine fiber content of the type i pulp board; M_i represents the dry fiber mass of the type i pulp board; r_i represents the average fiber thickness of the type i pulp board; W_i represents the sum of the fiber width of the type i pulp board; w_i represents the average fiber width of the fiber width of the type i pulp board; B_i represents the number of broken tail fibers of the type i pulp board; b_i is the broken tail fiber content of the type i pulp board. Brooming rate is also an important fiber shape parameter, which indicates the ratio of the degree of fiber brooming to the number of fiber ends measured. The calculation of the brooming rate is very complicated. It is necessary to know the unblooming fiber rate and the 1, 2, and 3 fiber brooming rates, and calculate the weighted average with the number of fiber breaks. It cannot be calculated directly according to the mathematical definition. This study uses the method shown in Eq. (3) to calculate brooming rate [43]:

$$\ln f_m = \sum_{i=1}^{n} q_i \ln f_i / \sum_{i=1}^{n} q_i \tag{3}$$

where f_i represents the brooming rate of the type i pulp board; q_i represents the feeding weight of the type i pulp board; f_m represents the brooming rate of the mixed pulp.

In the pulp preparation unit, due to limitations such as the liquid level of the pulp pool and the speed of the papermaking machine, the refiner works intermittently, and only the data on the period of time when the refiner is on is retained. Therefore, it is necessary to take a one-to-one correspondence between the refining process parameters and the beating degree data in a period of 5 min. In addition, the collection of pulp board weight and ratio data and pulp board fiber shape data are all performed offline. Therefore, it is also necessary to match these data with the refining process parameters and beating degree data. As shown in Fig. 4, it is the unprocessed raw data during the start-stop cycle of a certain refiner, and Fig. 5 is the processed data.

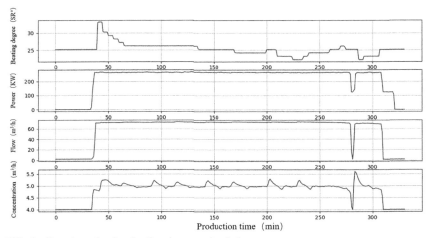

FIG. 4 Raw data of online beating degree and refining process.

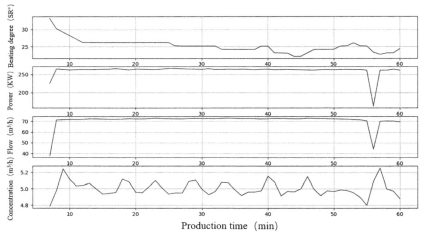

FIG. 5 Online beating degree and refining process data after processing.

4.2 Modeling the beating degree of tissue paper

4.2.1 Correlation analysis

The choice of algorithm is the most important work to establish a predictive model, and the understanding of data characteristics is the basis for choosing a suitable algorithm for modeling. Therefore, this work first analyzed the linear correlation of the pulp fiber shape data before modeling. The Pearson correlation coefficient is a quantity that studies the degree of linear correlation between variables, and the calculation method is shown in Eq. (4) [44]:

$$\rho_{X,Y} = \frac{\text{cov}(X, Y)}{\sigma_X \sigma_Y} = \frac{E((X - \mu_X)(Y - \mu_Y))}{\sigma_X \sigma_Y} = \frac{E(XY) - E(X)E(Y)}{\sqrt{E(X^2)}\sqrt{E(Y^2)}} \qquad (4)$$

where $cov(X, Y)$ represents the covariance between random variables X and Y; σ_X represents the standard deviation of X; σ_Y represents the standard deviation of Y; E represents the expectation, μ_X is the mean value of X, μ_Y is the mean value of Y. The Pearson correlation coefficient matrix between fiber shape parameters is shown in Fig. 6, where the serial numbers 0–8 indicate in turn: mixing beating degree, long fiber content, average fiber length, brooming rate, twisted fiber content, fine fiber content, average fiber thickness, average fiber width, fiber content of broken tail.

It can be seen from Fig. 6 that the correlation coefficients between the brooming rate and all other parameters are very small (the absolute value is less than 0.2). The correlation coefficient between the long fiber content and the brooming rate is very small, and the linear correlation with the twisted fiber content and the beating degree is medium (the absolute value is about 0.6). The long fiber content has a strong linear correlation with the average fiber length, fine

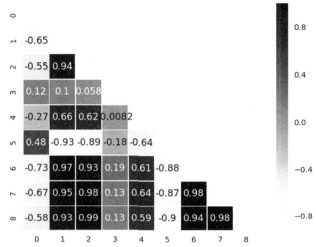

FIG. 6 Pearson correlation coefficient between pulp shape parameters.

fiber content, average fiber thickness, average fiber width, and fiber content of broken tail (absolute value higher than 0.9). The average fiber length has a high correlation with the average fiber thickness, the average fiber width, and the fiber content of the broken tail (absolute value is higher than 0.9). The average fiber thickness, average fiber width, fiber content of broken tail, and fine fiber content are highly correlated. It shows that there is a multicollinear relationship between the pulp shape parameters. Such data samples are not suitable for constructing a multiple linear regression model, and a nonlinear model needs to be established. In this work, GBDT algorithm and SVM algorithm are selected to establish the prediction model of beating degree. The variables used for modeling and the mean and minimum-maximum values of each variable are shown in Table 2.

4.2.2 Algorithm description

GBDT algorithm. The basic idea of GBDT is that the decision tree is taken as a weak learner, and each learner learns on the basis of the prediction bias of the previous learner, and the weighted sum of the simple model is established to replace the single strong learner [45]. The research shows that GBDT provides good prediction performance and robustness for the prediction space, and the modification of tree structure can have lower model complexity and better generalization ability. GBDT is well adapted and interpreted for both linear and nonlinear data sets and can be used for regression or classification problems as well as dealing with different types of response variables (numeric and nonnumeric). The prediction problem of beating degree in this work belongs to supervisory regression problem. GBDT uses classification and regression tree

TABLE 2 Modeling characteristics of beating degree.

Sign	Name	Mean, Min-max	Unit	Category
f	Pulp flow	78.28, 21.97–101.17	m^3/h	Refining process parameters
SEC	Specific energy consumption	0.3503, 0.0725–0.8035	kW/T	
SR_{mix}	Beating degree before refining	15.23, 13.4–16.0	°	Fiber shape before refining
r_{SW}	High/low long fiber content	0.44, 0–1	–	
L_m	Average fiber length	0.8003, 0.6892–1.0210	mm	
d_m	Twisted fiber content	0.3716, 0.3539–0.3925	%	
c_m	Fine fiber content	24.67, 21.34–16.83	%	
r_m	Average fiber thickness	0.0806, 0.06861–0.1045	mg/m	
w_m	Average fiber width	20.43, 18.62–24.30	µm	
b_m	Broken tail fiber content	28.45, 25.85–33.60	%	
f_m	Brooming rate	0.3716, 0.3539–0.3925	%	
Δy	Increment of beating degree	25.13, 20–32.23	°	Output

(CART) as a weak learner superposition to replace a single strong learner, as shown in Eq. (5) [46]:

$$T(x; c, R) = \sum_{m=1}^{M} c_m I(x \in R_m) \qquad (5)$$

where $R = \{R_1, R_2, \ldots, R_M\}$ represents the subspace finally formed by the input space through the segmentation rules, also known as the leaf; M is the number of leaves; $c = \{c_1, c_2, \ldots, c_m\}$, $c_m = \text{mean}(y_i | x_i \in R_m)$ is a constant, that is, the average value of the output features in each leaf space; $I = \begin{cases} 0, & x \notin R_m \\ 1, & x \in R_m \end{cases}$ is a discriminant function. When the combination $\{c, R\}$ is determined, the equation of CART is as follows:

$$F(x) = \sum_{i=0}^{N} T_i(x; c_i, R_i) \tag{6}$$

where T_i represents the ith tree ($i = 1, 2, \ldots, N$). When GBDT is constructed, one CART T_m is added for each iteration, and the T_m added for each iteration is based on the minimum loss function of the sum of the former CART trees and the actual value, as shown in Eq. (7) [47]:

$$T_m = \arg \min_{T} \sum_{k=1}^{n} \mathcal{L}\left(y_k, \sum_{j=1}^{m-1} T_j(x_k) + T(x_k)\right) \tag{7}$$

CART uses least squares loss as the loss function, and each iteration builds the mth tree on the residual of the sum of CART generated by the previous $m - 1$ times. The GBDT generated at the mth time is shown in Eq. (8) [47]:

$$F_m(x) = F_{m-1}(x) + \arg \min_{T} \sum_{i=1}^{n} \mathcal{L}(y_i, F_{m-1}(x_i) + T(x_i)) \tag{8}$$

GBDT uses the gradient descent method to find the minimum value. The maximum descent gradient direction is the negative gradient direction of the loss function under the current model F_{m-1}, as shown in Eq. (9) [48]:

$$F_m(x) = F_{m-1}(x) + \gamma_m \sum_{i=1}^{n} \nabla_{F_{m-1}} \mathcal{L}(y_i, F_{m-1}(x_i)) \tag{9}$$

where the calculation method of γ_m is shown in Eq. (10) [49]:

$$\gamma_m = \arg \min_{\gamma} \sum_{i=1}^{n} \mathcal{L}\left(y_i, F_{m-1}(x_i) - \gamma \frac{\partial \mathcal{L}(y_i, F_{m-1}(x_i))}{\partial F_{m-1}(x_i)}\right) \tag{10}$$

In GBDT modeling, to obtain good model performance and prevent overfitting, the usual approach is to limit the improvement effect of each tree on the accuracy of the algorithm, and set the shrinkage coefficient, also known as the learning rate n, as shown in Eq. (11):

$$F_m(x) = F_{m-1}(x) + \nu \gamma_m T_m(x) \tag{11}$$

where $0 < \nu < 1$, choosing a small learning rate will help improve the accuracy and fitness of the model, but it requires more iterations. To improve the efficiency of the solution, some samples are randomly selected for training during

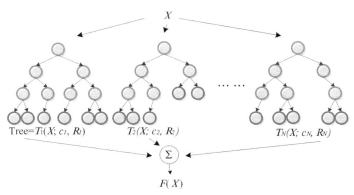

FIG. 7 GBDT structure diagram.

each iteration. The proportion of the selected samples in the training set is called the subsampling rate [50]. The structure of the GBDT model is shown in Fig. 7.

The implementation process of GBDT algorithm can be described as:

(1) Model initialization $F_0 = \arg\min_\gamma \mathcal{L}(y_i, \gamma)$. Set model parameters, including, iteration number M, learning rate ν, subsampling rate σ, etc.

(2) Iteration process.

    (a) Calculate the current residual $r_{im} =$

$$-\left[\frac{\partial \mathcal{L}(y_i, F(x_i))}{\partial F(x_i)}\right]_{F=F_{m-1}}, i = 1, 2, \ldots, N', \text{ where, } N' = \sigma \bullet N;$$

    (b) Train the CART tree T_m with the current residual set $\{(x_i, r_{im})\}_{i=1}^{N'}$;

    (c) Calculation step $\gamma_m = \arg\min_\gamma \sum_{i=1}^{n} \mathcal{L}(y_i, F_{m-1}(x_i) - \gamma T_m)$;

    (d) Update model $F_m(x) = F_{m-1}(x) + \nu \gamma_m T_m(x)$;

(3) Output F_m.

SVM algorithm. SVM can transform nonlinear problems into linear problems by mapping nonlinear problems to high-dimensional space in the form of kernel functions and has good nonlinear processing capabilities. SVM is suitable for small sample research, while retaining the generalization ability of the algorithm. The algorithm steps are as follows [51]:

(1) Determine the input variable $X = [x_1, \ldots x_p]_{n \times p}$ and output variable Y. To eliminate the difference in dimensions of each variable and improve the efficiency of the algorithm, the data needs to be standardized, as shown in Eq. (12):

$$\begin{cases} x_{ij}^* = \dfrac{x_{ij} - \overline{x_j}}{S_j} \\ S_j^2 = \dfrac{1}{n}\sum_{i=1}^{n}(x_{ij} - \overline{x_j})^2 \end{cases} \quad (12)$$

where X represents the independent variable, $X = [x_1, \ldots x_p]_{n \times p}$; Y represents the dependent variable, $Y = [y_1, \ldots y_q]_{n \times q}$; i represents the number of samples, $i = 1$, $2, \ldots, n$; j represents the sample dimension, $j = 1, 2, \ldots, p$. x_j represents the mean value of the sample in the j dimension; x_{ij} represents the j-dimensional value of the ith sample; x_{ij}^* represents the j-dimensional standardized value of the ith sample; S_j represents the standard deviation of the sample in the j dimension; S_j^2 represents the variance of the sample in dimension j.

(2) Based on the principle of structural risk minimization, the slack variables $\{\xi_i\}_{i=1}^l$ and $\{\xi_i^*\}_{i=1}^l$ are introduced, and the Lagrangian multipliers α_i, α_i^*, η_i, η_i^* are introduced to obtain the SVM model, as shown in Eq. (13):

$$\hat{y} = \omega^T x + b = \sum_{i=1}^{l} \left(\alpha_i^* - \alpha_i \right) K \left(x_i, x_j \right) + b \tag{13}$$

where ω represents the weight coefficient, b represents the bias term, and $K(x_i, x_j)$ represents the kernel function.

The kernel function can map the linear inseparable low-dimensional feature data to the high-dimensional space, thereby converting the nonlinear problem into a linear problem. Commonly used kernel functions include linear kernel function, polynomial kernel function, radial basis kernel function, and sigmoid kernel function, as shown in Eqs. (14)–(17), respectively [52]:

$$K(x_i, x) = x_i x^T \tag{14}$$

$$K(x_i, x) = \left(\gamma x_i x^T + r \right)^p, \gamma > 0 \tag{15}$$

$$K(x_i, x) = \exp \left(-\gamma \|x_i - x\|^2 \right), \gamma > 0 \tag{16}$$

$$K(x_i, x) = \tanh \left(\gamma x_i x^T + r \right), \gamma > 0 \tag{17}$$

where γ, r, and p are the parameters of each kernel function.

4.2.3 Parameter adjustment

Adjust appropriate model parameters to make the model achieve good accuracy while preventing overfitting. Overlearning on the training set will lead to an overcomplex model structure, which is the main cause of overfitting. Overfitting is manifested in insufficient interpretation of new samples and poor accuracy, which is what data-driven modeling needs to avoid. The GBDT model has four main parameters that need to be adjusted: learning rate η, tree depth d, number of iterations N, and subsampling ratio σ. The adjustment range of each parameter set in this study is shown in Table 3. Cross-validation is an effective method to prevent abnormal accuracy due to unbalanced training set segmentation [53]. The method is to evenly divide the data set into k subsets, select one of the subsets as the test set each time, and use the remaining $k - 1$

TABLE 3 GBDT model parameter adjustment table.

Parameter name	Parameter name	Parameter range
Learning rate	η	$\{0.01, 0.02, 0.05, 0.1, 0.2, 0.4, 0.9\}$
Tree depth	d	$\{3, 4, 5, 7, 9, 10\}$
Number of iterations	N	$\{10, 20, \dots, 400\}$
Subsampling ratio	σ	$\{0.001, 0.005, 0.01, 0.05, 0.1, 0.15, 0.5\}$

subsets as the training set, and repeat the verification k times. In this work, a set of parameters with the smallest cross-validation $RMSE^k$ is selected as the optimal combination. The calculation method of $RMSE^k$ is shown in Eqs. (18), (19) [54]:

$$RMSE^k = \frac{1}{k} \sum_{j=1}^{k} \sqrt{MSE_j} \qquad (18)$$

$$MSE = \frac{1}{N} \sum_{i=1}^{N} (y_i - \hat{y}_i)^2 \qquad (19)$$

The depth of the tree determines the complexity of the model, that is, the deeper the tree, the more complex the model, but if the model is too complex, overfitting will occur. Choose a larger number of iterations $N = 400$ and set $\eta = 0.1$ to observe the changing trend of the residual curve at different depths d. As shown in Fig. 8, when $d = 7$, 9, 10, the error first drops rapidly, and as the number of iterations increases, the error increases instead, which indicates that the model has overfitting. Therefore, $d = 3$, 4, and 5 are more appropriate. When d is 4 or 5, the error curve is basically the same, so choose 3 and 4 with small depth. When $d = 3$ or 4, examine the influence of different learning rates on the residual curve. As shown in Fig. 9, comparing the left and right subgraphs, it can be found that compared with $d = 3$, the error curve under all learning rates drops faster and the convergence error is smaller when $d = 4$. Therefore, choose $d = 4$. When $\eta = 0.2$, 0.4, 0.9, the error curve will first decline and then rise, which indicates that a smaller η needs to be selected. When $\eta = 0.01$, the error of each iteration decreases very little, and the error curve still has a downward trend after 400 iterations. At this time, the training efficiency of the model is low, and more iterations are required. When $\eta = 0.02$, the convergence error is higher than $a = 0.05$. In summary, choose $d = 4$ and $\eta = 0.05$.

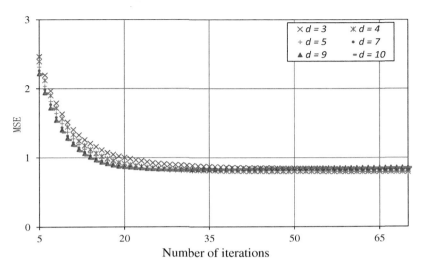

FIG. 8 Tree depth and iterative residual curve.

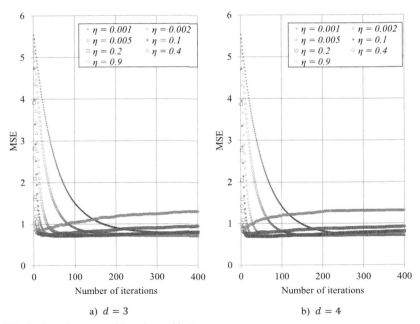

FIG. 9 Learning rate and iterative residual curve.

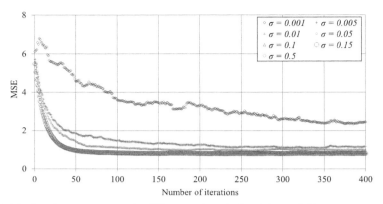

FIG. 10 Iterative error curve under different σ when $d=4$ and the $\eta=0.05$.

Observe the influence of the size of σ on the iteration error. When the variance of the data set is very small, choosing a smaller subsampling rate can ensure the randomness of the training data subset in each iteration. As shown in Fig. 10, when $\sigma=0.001$, the error is relatively large due to insufficient sample size for model learning. When σ increases to 0.005, the error decreases by half and the error curve fluctuates slightly during the iteration. When $\sigma=0.05$, 0.1, 0.15, 0.5, the error curve is stable and basically coincides, increasing a convergence error does not decrease, but increases the training time of each iteration. So, choose $\sigma=0.05$.

After d, η and σ are determined, an appropriate N needs to be selected. When the slope s of the error curve fluctuates within a small range of $0 < s < \varepsilon$, it is considered that the residual error will no longer decrease, and the number of iterations $N=60$. After adjustment, a set of parameters $\{\eta=0.05, d=4, N=60, \sigma=0.005\}$ is finally selected.

4.3 Industrial verification and feature importance ranking

In order to evaluate the accuracy of the model, the mean squared error (MSE), the mean absolute error (MAE), and the explained variance (EV) are selected as evaluation indicators. The calculation method is shown in Eqs. (19)–(21) [55]:

$$MAE = \frac{1}{N} \sum_{i=1}^{N} |y_i - \hat{y}_i| \tag{20}$$

$$EV = 1 - \frac{Var\{y - \hat{y}\}}{Var\{y\}} \tag{21}$$

where y_i represents the actual value and $\hat{y}_i$ represents the predicted value.

Using $k = 5$ as the number of folds for cross-validation, 80% of the data in each cross-validation is used as the training set and 20% of the data as the test set. The result of five-time cross validation of GBDT model is $RMSE^k = 0.9948$. Adjust the parameters of the SVM model so that the $RMSE^k$ errors of SVM and GBDT are similar. The test results are shown in Fig. 11, and the cross-check results are shown in Table 4. The $RMSE^k$ errors of SVM and GBDT are similar, indicating that the accuracy of the two algorithms is similar. The data set is randomly divided into the test set by 20%. The accuracy and time complexity of the two algorithms are shown in Table 5. At approximately the same level of accuracy, the time complexity of the GBDT algorithm is about one-fifth of that of the SVM.

The tree structure of GBDT algorithm has good interpretability. When obtaining a precise quantitative relationship between the independent variable and the dependent variable, the qualitative relationship with the dependent variable can also be determined. The structure of the GBDT model established in this study is a superposition of 60 regression trees, and the depth of each tree is 3. In the regression tree model, the relative importance of a variable is

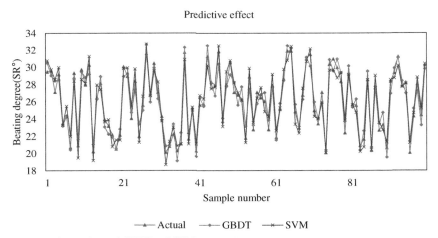

FIG. 11   Comparison of GBDT and SVM test results.

TABLE 4 Comparison of cross-check results.

	1	2	3	4	5	$RMSE^k$
GBDT	0.9039	1.0560	1.0173	1.0463	0.8404	0.9848
SVM	0.8756	1.3589	0.8683	0.9773	0.8280	0.9864

TABLE 5 Comparison of accuracy and time complexity.

	EV	MSE	MAE	Time (ms)
GBDT	0.8188	0.9637	0.7627	1.494
SVM	0.8181	0.9659	0.7442	85.337

determined based on all the split points of the tree. Specifically, the more a variable is used as a split point, the greater the impact of this variable on the result, that is, the greater the relative importance of this variable. In the GBDT model, the mean value of the relative importance of all regression tree variables can be used to express the relative importance of the GBDT model variables. The calculation method of the global importance of feature j is shown in Eqs. (22), (23) [56]:

$$\hat{J}_g^2 = \frac{1}{M} \sum_{m=1}^{M} \hat{J}_s^2 (T_m) \tag{22}$$

$$\hat{J}_s^2 (T) = \sum_{t=1}^{L-1} \hat{i}_t^2 \cdot I(v_t = j) \tag{23}$$

where M represents the number of trees; $\hat{J}_s^2 (T)$ represents the importance of a feature j in a single tree; $L - 1$ is the number of nonleaf nodes of the tree; v_t is the feature associated with node t, $\hat{i}_t^2$ represents the reduction value of the square loss after node t is split.

It can be seen from Fig. 12 that specific energy consumption and pulp flow are the variables that have higher influence on the beating degree. The twisted

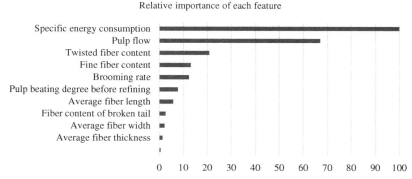

FIG. 12   Histogram of relative importance of input features of GBDT model.

fiber content, fine fiber content, and brooming rate have medium influence, and the remaining variables matter less.

5   Conclusion

With the continuous development of Internet of things, the data of process system is more and more easy to obtain. The data-driven artificial intelligence technology has gradually attracted the attention of researchers in various fields. At present, the main popular data-driven methods include ANN, SVM, and ensemble learning. The prediction method based on data-driven model is widely used in the prediction of key variables of process system. On the basis of avoiding expensive sensor equipment and maintenance investment, the disadvantages of mechanism model modeling, such as difficulty and narrow application range, are further overcome. In the process of industrial production, accurate and real-time measurement of key physical and chemical parameters is the basis of process control and optimization. However, due to the influence of sensing technology and economy and other factors, many key physical and chemical parameters cannot be accurately and real-time measured, which leads to the opacity of the production process and becomes the biggest obstacle to the transformation of industrial production to information and intelligence.

In this work, based on historical production data and intelligent algorithm, a prediction model of the beating degree of tissue paper is established, which can realize real-time and accurate prediction of the beating degree in the pulping unit. The model overcomes the problems of long manual measuring cycle of beating degree, expensive on-line monitoring equipment, and easy damage. The model uses cross validation to evaluate the accuracy and compare the performance of GBDT model and SVM model. The verification results show that the MSE of the GBDT model is 0.9637 SR° and the MAE is 0.7627 SR°. The operation time is one-fiftieth of that of SVM model. The GBDT model has good adaptability to high autocorrelation data sets and high precision for small and medium-sized data samples, which makes GBDT model have a better application prospect in the prediction of beating degree. Using the characteristics of GBDT model tree structure, this work studies the relative importance of 11 model input variables and finds that specific energy consumption and pulp flow are the variables that have higher influence on the beating degree. The twisted fiber content, fine fiber content, and brooming rate have medium influence, and the remaining variables matter less.

Acknowledgments

This work is supported by the Departmental General Research Funds (UAFT) of Department of Industrial and Systems Engineering, The Hong Kong Polytechnic University, entitled "Green Manufacturing and Better Sustainability for Papermaking Industry: Decision-Making and Optimization in Life Cycle Perspective" (G-UAFT), and the Postdoctoral Fellowships

Scheme, entitled "Research on Advanced Control Integration Framework for Greenhouse Gas Emission Reduction in Papermaking Wastewater Treatment Process" (G-YW4Y).

References

[1] M. Stoller, R. Serrão Mendes, Advanced control system for membrane processes based on the boundary flux model, Sep. Purif. Technol. 175 (2017) 527–535.

[2] P. Musekiwa, L.B. Moyo, T.A. Mamvura, G. Danha, G.S. Simate, N. Hlabangana, Optimization of pulp production from groundnut shells using chemical pulping at low temperatures, Heliyon 6 (2020) e4184.

[3] H.D. Ouattara, H.G. Ouattara, M. Droux, S. Reverchon, W. Nasser, S.L. Niamke, Lactic acid bacteria involved in cocoa beans fermentation from Ivory Coast: species diversity and citrate lyase production, Int. J. Food Microbiol. 256 (2017) 11–19.

[4] D.I. Sánchez-Machado, J. López-Cervantes, J.A. Núñez-Gastélum, G. Servín De La Mora-López, J. López-Hernández, P. Paseiro-Losada, Effect of the refining process on *Moringa oleifera* seed oil quality, Food Chem. 187 (2015) 53–57.

[5] H. Yang, P. Peczulis, P. Inguva, X. Li, J.Y.Y. Heng, Continuous protein crystallisation platform and process: case of lysozyme, Chem. Eng. Res. Des. 136 (2018) 529–535.

[6] X. Yuan, S. Qi, Y. Wang, H. Xia, A dynamic CNN for nonlinear dynamic feature learning in soft sensor modeling of industrial process data, Control. Eng. Pract. 104 (2020) 104614.

[7] P. Cao, X. Luo, Modeling for soft sensor systems and parameters updating online, J. Process Control 24 (2014) 975–990.

[8] P. Cao, X. Luo, Soft sensor model derived from wiener model structure: modeling and identification, Chin. J. Chem. Eng. 22 (2014) 538–548.

[9] B. Bidar, J. Sadeghi, F. Shahraki, M.M. Khalilipour, Data-driven soft sensor approach for online quality prediction using state dependent parameter models, Chemom. Intell. Lab. 162 (2017) 130–141.

[10] P. Balko, D. Rosinová, Nonlinear boiler-turbine unit: modelling and robust decentralized control, IFAC-PapersOnLine 49 (2016) 49–54.

[11] W. Meng, J. Li, B. Chen, H. Li, Modeling and simulation of ethylene polymerization in industrial slurry reactor series, Chin. J. Chem. Eng. 21 (2013) 850–859.

[12] A. Afram, A.S. Fung, F. Janabi-Sharifi, K. Raahemifar, Development and performance comparison of low-order black-box models for a residential HVAC system, J. Build. Eng. 15 (2018) 137–155.

[13] R. Kicsiny, Black-box model for solar storage tanks based on multiple linear regression, Renew. Energy 125 (2018) 857–865.

[14] Z. Ge, Review on data-driven modeling and monitoring for plant-wide industrial processes, Chemom. Intell. Lab. Syst. 171 (2017) 16–25.

[15] Y. Man, Y. Han, Y. Wang, J. Li, L. Chen, Y. Qian, M. Hong, Woods to goods: water consumption analysis for papermaking industry in China, J. Clean. Prod. 195 (2018) 1377–1388.

[16] D. Danielewicz, B. Surma-Ślusarska, Miscanthus × giganteus stalks as a potential non-wood raw material for the pulp and paper industry. Influence of pulping and beating conditions on the fibre and paper properties, Ind. Crop. Prod. 141 (2019) 111744.

[17] A. Panghal, R. Kaur, S. Janghu, P. Sharma, P. Sharma, N. Chhikara, Nutritional, phytochemical, functional and sensorial attributes of *Syzygium cumini* L. pulp incorporated pasta, Food Chem. 289 (2019) 723–728.

[18] Z. Zeng, M. Hong, J. Li, Y. Man, H. Liu, Z. Li, H. Zhang, Integrating process optimization with energy-efficiency scheduling to save energy for paper mills, Appl. Energy 225 (2018) 542–558.

[19] A.K. Babu, G. Kumaresan, V.A.A. Raj, R. Velraj, Review of leaf drying: mechanism and influencing parameters, drying methods, nutrient preservation, and mathematical models, Renew. Sustain. Energy Rev. 90 (2018) 536–556.

[20] M. Kim, O.K. Choi, Y. Cho, J.W. Lee, A.E. Cho, Elucidation of the desalination mechanism of solvent extraction method through molecular modeling studies, Desalination 496 (2020) 114704.

[21] P. Zhang, Y. Feng, T.Q. Bui, X. Hu, W. Yao, Modelling distinct failure mechanisms in composite materials by a combined phase field method, Compos. Struct. 232 (2020) 111551.

[22] M. Niu, B. Yang, Y. Yang, G. Meng, Two generalized models for planar compliant mechanisms based on tree structure method, Precis. Eng. 51 (2018) 137–144.

[23] Z.Y. Liu, B.L. Xiao, W.G. Wang, Z.Y. Ma, Modelling of carbon nanotube dispersion and strengthening mechanisms in Al matrix composites prepared by high energy ball milling-powder metallurgy method, Compos. A: Appl. Sci. 94 (2017) 189–198.

[24] C. Zafra, Y. Ángel, E. Torres, ARIMA analysis of the effect of land surface coverage on PM10 concentrations in a high-altitude megacity, Atmos. Pollut. Res. 8 (2017) 660–668.

[25] H. Yao, X. Li, H. Pang, L. Sheng, W. Wang, Application of random forest algorithm in hail forecasting over Shandong Peninsula, Atmos. Res. 244 (2020) 105093.

[26] Z. Alameer, M.A. Elaziz, A.A. Ewees, H. Ye, Z. Jianhua, Forecasting gold price fluctuations using improved multilayer perceptron neural network and whale optimization algorithm, Resour. Policy 61 (2019) 250–260.

[27] X. Ma, P. Jiang, Q. Jiang, Research and application of association rule algorithm and an optimized grey model in carbon emissions forecasting, Technol. Forecast. Soc. 158 (2020) 120159.

[28] Q. Wu, H. Lin, Daily urban air quality index forecasting based on variational mode decomposition, sample entropy and LSTM neural network, Sustain. Cities Soc. 50 (2019) 101657.

[29] Y. Man, Y. Hu, J. Ren, Forecasting COD load in municipal sewage based on ARMA and VAR algorithms, Resour. Conserv. Recycl. 144 (2019) 56–64.

[30] Y. Hu, J. Li, M. Hong, J. Ren, R. Lin, Y. Liu, M. Liu, Y. Man, Short term electric load forecasting model and its verification for process industrial enterprises based on hybrid GA-PSO-BPNN algorithm—a case study of papermaking process, Energy 170 (2019) 1215–1227.

[31] L. Sun, G. Li, Q.S. Hua, Y. Jin, A hybrid paradigm combining model-based and data-driven methods for fuel cell stack cooling control, Renew. Energy 147 (2020) 1642–1652.

[32] W. Sun, Z. Wang, Q. Wang, Hybrid event-, mechanism- and data-driven prediction of blast furnace gas generation, Energy 199 (2020) 117497.

[33] Y. Meng, S. Yu, J. Zhang, J. Qin, Z. Dong, G. Lu, H. Pang, Hybrid modeling based on mechanistic and data-driven approaches for cane sugar crystallization, J. Food Eng. 257 (2019) 44–55.

[34] N. Pitropakis, E. Panaousis, T. Giannetsos, E. Anastasiadis, G. Loukas, A taxonomy and survey of attacks against machine learning, Comput. Sci. Rev. 34 (2019) 100199.

[35] D.E. Rumelhart, G.E. Hinton, R.J. Williams, Learning representations by back-propagating errors, Nature (London) 323 (1986) 533–536.

[36] V. Cherkassky, The nature of statistical learning theory, IEEE Trans. Neural Netw. 8 (1997) 1564.

[37] S. Hosseini, B.M.H. Zade, New hybrid method for attack detection using combination of evolutionary algorithms, SVM, and ANN, Comput. Netw. 173 (2020) 107168.

[38] T. Van Gestel, J. Suykens, G. Lanckriet, A. Lambrechts, B. De Moor, J. Vandewalle, Bayesian framework for least-squares support vector machine classifiers, Gaussian processes, and kernel Fisher discriminant analysis, Neural Comput. 14 (2002) 1115–1147.

[39] L.K. Hansen, P. Salamon, Neural network ensembles, IEEE Trans. Pattern Anal. 12 (1990) 993–1001.

[40] K. Hiromasa, Automatic outlier sample detection based on regression analysis and repeated ensemble learning, Chemometr. Intell. Lab. 177 (2018) 47–82.

[41] L.T. Mulyantara, H. Harsono, R. Maryana, G. Jin, A.K. Das, H. Ohi, Properties of thermomechanical pulps derived from sugarcane bagasse and oil palm empty fruit bunches, Ind. Crop. Prod. 98 (2017) 139–145.

[42] H. Shi, M. Zhou, C. Li, X. Sheng, Q. Yang, N. Li, M. Niu, Surface sediments formation during auto-hydrolysis and its effects on the benzene-alcohol extractive, absorbability and chemical pulping properties of hydrolyzed acacia wood chips, Bioresour. Technol. 289 (2019).

[43] X. Lin, Z. Wu, C. Zhang, S. Liu, S. Nie, Enzymatic pulping of lignocellulosic biomass, Ind. Crop. Prod. 120 (2018) 16–24.

[44] F.H. Mohamed Salleh, S.M. Arif, S. Zainudin, M. Firdaus-Raih, Reconstructing gene regulatory networks from knock-out data using Gaussian noise model and Pearson correlation coefficient, Comput. Biol. Chem. 59 (2015) 3–14.

[45] N. Nasyrov, M. Komarov, P. Tartynskikh, N. Gorlushkina, Automated formatting verification technique of paperwork based on the gradient boosting on decision trees, Proc. Comput. Sci. 178 (2020) 365–374.

[46] A. Mollalo, B. Vahedi, S. Bhattarai, L.C. Hopkins, S. Banik, B. Vahedi, Predicting the hotspots of age-adjusted mortality rates of lower respiratory infection across the continental United States: integration of GIS, spatial statistics and machine learning algorithms, Int. J. Med. Inform. 142 (2020) 104248.

[47] G.I. Nagy, G. Barta, S. Kazi, G. Borbely, G. Simon, GEFCom2014: probabilistic solar and wind power forecasting using a generalized additive tree ensemble approach, Int. J. Forecast. 32 (2016) 1087–1093.

[48] T. Yang, W. Chen, G. Cao, Automated classification of neonatal amplitude-integrated EEG based on gradient boosting method, Biomed. Signal Process. 28 (2016) 50–57.

[49] C. Persson, P. Bacher, T. Shiga, H. Madsen, Multi-site solar power forecasting using gradient boosted regression trees, Sol. Energy 150 (2017) 423–436.

[50] D.C. Carslaw, P.J. Taylor, Analysis of air pollution data at a mixed source location using boosted regression trees, Atmos. Environ. 43 (2009) 3563–3570.

[51] G.E. Güraksın, H. Haklı, H. Uğuz, Support vector machines classification based on particle swarm optimization for bone age determination, Appl. Soft Comput. 24 (2014) 597–602.

[52] L. Sun, B. Zou, S. Fu, J. Chen, F. Wang, Speech emotion recognition based on DNN-decision tree SVM model, Speech Commun. 115 (2019) 29–37.

[53] T.I. Dhamecha, A. Noore, R. Singh, M. Vatsa, Between-subclass piece-wise linear solutions in large scale kernel SVM learning, Pattern Recogn. 95 (2019) 173–190.

[54] J. Jatnieks, M. De Lucia, D. Dransch, M. Sips, Data-driven surrogate model approach for improving the performance of reactive transport simulations, Energy Procedia 97 (2016) 447–453.

[55] V. De Martinis, F. Corman, Data-driven perspectives for energy efficient operations in railway systems: current practices and future opportunities, Transp. Res. C: Emerg. Technol. 95 (2018) 679–697.

[56] A. Zhao, L. Qi, J. Li, J. Dong, H. Yu, A hybrid spatio-temporal model for detection and severity rating of Parkinson's disease from gait data, Neurocomputing 315 (2018) 1–8.

Chapter 6

Machine learning-based energy consumption forecasting model for process industry—Hybrid PSO-LSSVM algorithm electricity consumption forecasting model for papermaking process

Yi Man, Yusha Hu, Jigeng Li, and Mengna Hong
State Key Laboratory of Pulp and Paper Engineering, School of Light Industry and Engineering, South China University of Technology, Guangzhou, People's Republic of China

1 Introduction

The total energy consumption of China increased from 2.31×10^3 Mtce in 2006 to 4.36×10^3 Mtce in 2016. The energy consumption goes up by nearly 90% in the past decade [1]. Meanwhile, the production capacity of the paper mills has increased drastically in China. From 2011 to 2017, the paper and paperboard output has increased from 32.33 to 111.3 Mt. Its energy consumption has risen from 33.6 Mtce in 2000 to 52.1 Mtce in 2017 [2]. The exponential growth of the paper industry has given rise to great pressure on power reservation and emission reduction goals. However, nowadays, the paper mills insist on the traditional operation mode due to the technology gap. There is no suitable intelligent method to optimize the process and manage energy consumption, resulting in a great amount of energy wastage.

The papermaking process devours a huge amount of electricity by about 15%–20% of the total energy usage [3]. It is necessary to cut down the power usage of the production process for energy conservation in paper mills.

Forecasting future electric loads could help the papermaking process optimize the electricity consumption structure and which could enhance energy

Applications of Artificial Intelligence in Process Systems Engineering
https://doi.org/10.1016/B978-0-12-821092-5.00017-6

efficiency and cut down the production expense [3]. Thus, developing a STELF model is the tide of future development for paper mills.

The electricity load for paper mills is influenced by many different types of elements, such as policy, economy, external environment, and many random elements in the production process (such as the disturbance between diverse electrical equipment) [4]. Hence, there is a complex nonlinear relationship between the electricity load and the real production process. It is hard to develop a STELF model for paper mills without the help of artificial intelligence methods. In the last few years, the study on electric load forecasting has achieved important progress. The STELF models have been applied to many different industries. For instance, forecasting the electricity load for steel processes in the mining and metallurgical industry [5], the electricity production forecasting for wind and photovoltaic power [6], and also the electricity load for electricity grids [7], etc. Papermaking process includes the preprocessing, postprocessing, and other related utilities such as cogeneration and wastewater treatment. The preprocessing has characteristics of process industries, which the production process is sequential and steady. The postprocessing has characteristics of discrete manufacturing industry, which is not sequential and contains a large amount of processing shifting. There are a lot of intermittent electric devices such as refiner and pulper in the preprocessing. The startup and shutdown of this intermittent electric devices are almost randomly scheduled conforming to the engineers' experience. Additionally, the unplanned paper-machine downtime often takes place because of paper break. Thus, the features of electricity load in the papermaking process shows aperiodicity, instability, and high fluctuation frequency. To forecast the future electric load for paper mills is more complicated compared with the above-mentioned works. To develop an accurate electric load forecasting is a difficult and urgent task for the papermaking process.

1.1 Literature review

In recent years, there are four main research directions for STELF: (a) linear approaches; (b) nonlinear approaches; (c) hybrid intelligent approaches; (d) deep learning.

(1) Linear approaches

Linear approaches mainly include linear regression (LR) method, exponential smoothing method, auto regression moving average (ARMA), and improved ARMA method, etc. [8]. The forecasting model based on linear approaches has the advantages of strong adaptability, stability, simple mathematical structure, and can deal with seasonal and nonstationary problems. However, according to the nonlinear characteristics of power demand time series, those mature forecasting model based on linear approaches cannot fully consider the multidimensional critical factors affecting electricity loads, which leads to low precision and cannot express the nonlinear characteristics of electricity load.

(2) Intelligent algorithms

To the above problems, forecasting models based on intelligent algorithms have been proposed. Based on intelligent algorithms, the forecasting model could learn the complexity of load sequences more accurately. The SVM algorithm and the ANN algorithm are two extensively used algorithms in the field of STELF [9].

(3) Hybrid intelligent algorithms

There are some disadvantages in intelligent algorithms, such as the results obtained from BPNN are closer to local optimized results rather than global optimum results [3]. The SVM algorithm increases the computational complexity of the forecasting model when the sample size is too large, and decrease the accuracy of the forecasting model when the sample size is too small [10]. Hence, the optimization algorithms are applied in the single algorithms, which are used to overcome disadvantages in the single approaches. The classic optimization algorithms include genetic algorithm (GA) [11], particle swarm optimization (PSO) [12], fruit fly optimization algorithm (FOA) [13], etc. Those above optimization algorithms not only alleviate the limitations of the intelligent algorithms, but also further improve the accuracy of the forecasting results.

(4) Deep learning algorithms

The deep learning algorithms are widely used in electric load forecasting because they have the ability to capture nonlinear time series features. In recent years, the most widely studied deep neural networks in electric load forecasting are recurrent neural network (RNN) [14], convolutional neural network (CNN) [15], and long short-term memory (LSTM) [16]. Although the deep learning approaches have the advantages of high precision and fast convergence, they are complicated to construct and slow in calculation [17]. The current research on deep learning for STELF is mainly aimed at process industries with relatively stable electricity consumption [18, 19]. There is still a lack of research for the process industries like the paper mills with large fluctuation. Hence, this study uses the mature intelligence hybrid algorithm to develop the STELF model.

1.2   Algorithm selection

Compared with the ANN algorithm, the SVM algorithm can avoid problems of data overfitting and local minimum. Moreover, SVM can efficiently address the practical problems of high dimension [20]. However, when addressing large sample problems, the SVM algorithm is prone to the quadratic programming problem [10]. In order to design a highly accurate forecasting model for the papermaking process, huge amounts of data are used in the forecasting model. As an alternative to the SVM algorithm, the LSSVM algorithm is superior to the SVM algorithm in the convenient calculation by transforming quadratic optimization problems into linear equations [21].

The LSSVM algorithm is one of the most widely used algorithms for solving forecasting problems. Zhu et al. [22] proposed the EMD-LSSVM algorithm to forecast carbon price. Zhang et al. [23] proposed an LSSVM-DBN-SSA-LSH-based combination forecasting model to forecast the wind power. All these state-of-the-art works achieved very good performance by the LSSVM algorithm-based forecasting models.

However, single LSSVM-based forecasting model fails to get better forecasting results for paper mills, because it is fairly random and uncertain for the choosing of the parameters in an LSSVM algorithm. There still is a lack of systemic approach for the parameters selected for the algorithm optimization. Therefore, many algorithms are utilized to find best-fit parameters of LSSVM [24]. Among the optimization algorithms of LSSVM parameters selection, PSO can fast convergence and find best-fit parameters of the LSSVM model very efficiently [25].

In the previous work, our team has designed a STELF model based on GA-PSO-BPNN algorithm for paper mills [3]. Based on the model and the results, the input variables that would highly influence the precision of the forecasting model are further discussed to improve the forecasting performance. For the purpose of facilitating the precision of the forecasting model and avoiding the possibility of the data overfitting problem, this study develops a new data-driven STELF model based on the hybrid PSO-LSSVM algorithm. The model adopts the global search advantage of the PSO algorithm to avoid the disadvantage of the LSSVM algorithm. The lag autocorrelation function is used for input variables selection. The performance of the developed model is tested by the practical electricity loads from two diverse paper mills. Since the electricity loads saving energy has the features of tremendous fluctuation ranges and frequency and irregular variation, the developed STELF model will have wider applications.

2 Methodology

The technical route for this study is shown in Fig. 1. The description of the modelling procedure is as below: (1) Input variable selection: select the input variables for the forecasting model using correlation function and lag autocorrelation function respectively. (2) Model verification: Test the performance of the proposed STELF model. More specifically, there are four sections in the modelling process:

> **Step 1**: Data acquisition and preprocessing. Collecting the electric load data from the database. During the data acquisition of the production process, data outliers often take place because of the disturbance of transmission in the dynamic acquisition, the disturbance between the electric devices, etc. Establishing the forecasting model with the data including the outliers could cause serious consequences, such as increasing the uncertainty of

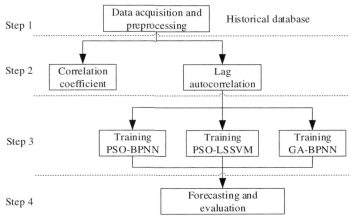

FIG. 1   Design and modelling of forecast framework.

system analysis results, leading to unreliable outputs, and reducing the precision of the STELF model. Hence, it is necessary to preprocess the abnormal data before modelling.

Step 2: Input variables selection. The input variables are chosen by correlation function and lag autocorrelation function. The electricity load of the electric devices is regarded as input variables if the absolute correlation coefficient goes beyond 0.6. The historical total electricity loads are also regarded as input variables when the absolute lag autocorrelation coefficient goes beyond 0.9. For the purpose of simplifying the computational complexity of the STELF model, this study chooses the historical total electricity loads with the top 10 highest lag autocorrelation coefficient as the input variable.

Step 3: Kernel algorithm structuring. The hybrid PSO-LSSVM algorithm is structured as the kernel algorithm of the STELF model. The PSO algorithm is utilized to find best-fit the parameters of the LSSVM algorithm. Besides, the GA-BPNN-based forecasting model and PSO-BPNN-based forecasting model are also analyzed as contrastive cases to test the precision and convergence speed of the developed PSO-LSSVM-based model.

Step 4: Forecasting performance evaluation. The forecasting results of the STELF models based on different kernel algorithms are verified by the real-time data from two diverse paper mills.

2.1   Data preprocessing

Two diverse paper mills are chosen as the test case study in this research. All the electricity load data are acquired from two paper mills. Due to the disturbance of transmission, the disturbance between the electric devices, and other disturbance, the acquired data could occasionally go beyond the practical range. To increase the precision of the model, the low-quality data needs to be identified and removed by data preprocessing.

Data preprocessing generally includes removing outliers, missing values imputation, and noising filtering. The details of data processing are shown in Appendix A.

2.2 Input variables selection methods

2.2.1 Correlation function

In order to find the influence of the electrical equipment on the total effective electric power, a statistical indicator, the correlation function, is used to analyze the correlation between the electric equipment and the total effective electric power. The correlation between the two sequences is stronger when the absolute correlation coefficient gets closer to 1. The electricity load of electric devices is regarded as the input variables if absolute correlation coefficients go beyond 0.6. The correlation function is defined by Eq. (1) [3]:

$$r_{x,y} = \frac{\sum_{i=1}^{n} (x_i - \bar{x})(y_i - \bar{y})}{\sqrt{\sum_{i=1}^{n} (x_i - \bar{x})^2} \times \sqrt{\sum_{i=1}^{n} (y_i - \bar{y})^2}} \tag{1}$$

where x and y are two diverse data series, $\bar{x}$ and $\bar{y}$ are the average of two diverse data sequences, and n is the length of the sequence.

2.2.2 Lag autocorrelation function

The papermaking process is a continuous process. The present effective power is affected by past effective electric load. In order to find the influence of past effective electric load on the present effective power, the autocorrelation function (ACF) is used to guide the selection of information feature subsets. The lag autocorrelation function is used to choose the input variables. The historical total electricity loads are regarded as the input variables if the absolute lag autocorrelation coefficients go beyond 0.9. The lag k autocorrelation coefficient r_k is defined by Eq. (2) [26]:

$$r_k = r(X_t, X_{t-k}) = \frac{\sum_{t=k+1}^{n} (x_t - \bar{x}) \times (x_{t-k} - \bar{x})}{\sum_{t=1}^{n} (x_t - \bar{x})^2} \tag{2}$$

where X is a data set based on time series, $X = \{X_t : t \in T\}$.

2.3 The algorithm descriptions

2.3.1 LSSVM algorithm

The least squares linear system is used by the LSSVM algorithm as a loss function instead of the quadratic programming method used by the conventional SVM algorithm. The basic principle of LSSVM is to construct an optimal decision function in the selected nonlinear mapping space. The theory of structural risk minimization will be used when constructing an optimal decision function. The kernel function of the original scope is used instead of the algorithm of point multiplication in the high dimensional feature space. The detail of LSSVM algorithm is shown in Appendix B.

The performance of LSSVM model largely hinges on the input variables and the parameters. The determination of these two parameters, regularization parameter c and kernel parameter σ are usually selected by experience, which readily decreases the accuracy of LSSVM. Therefore, the PSO is used to find the best-fit parameters of the LSSVM.

2.3.2 PSO algorithm

The elementary theory of the PSO algorithm has been generated from studying the social life of the fishes and birds that live in groups [25]. Instead of performing function operations on individuals, it treats each individual as a particle (no volume) in the search space (N-dimensional), flying at a certain speed (the speed is controlled by its own experience and social experience) in the search field. Where the position of the ith particle is donated as s_i, and s_i is substituted into the fitness function $F(s_i)$ to obtain the fitness value, the best place that each individual has experienced is denoted as $pbest_i$. The best place that all particles in the group have experienced is regarded as $gbest_i$. The velocity of the particle i is regarded as V_i. Generally, the range of positional variation in the nth ($1 \leq n \leq N$) dimension is limited to $[S_{min,n}, S_{max,n}]$, and the range of speed variation is limited to $[-V_{max,n}, V_{max,n}]$. For each generation, its velocity and positional variation of the nth dimension ($1 \leq n \leq N$) is updated using Eqs. (3), (4) [25]:

$$v_{in}^k = \omega \times v_{in}^{k-1} + c_1 \times \text{rand}() \times \left(pbest_{in} - s_{in}^{k-1}\right) + c_2 \times \text{rand}() \times \left(gbest_n - s_{in}^{k-1}\right)$$
(3)

$$s_{in}^k = s_{in}^{k-1} + v_{in}^{k-1}$$
(4)

where s_{in}^k is the nth segment of the position vector of the particle i at the kth iteration, v_{in}^k is the nth component of the velocity of the granule i at the kth iteration, c_1, c_2 are personal learning element and social learning element respectively, $rand()$ is random function, ω is the inertia factor.

2.3.3 The hybrid PSO-LSSVM algorithm

The flowchart of the PSO-LSSVM is shown in Fig. 2. The specific description of the PSO-BPNN algorithm is shown as below.

(1) Initialize the parameters for the PSO, containing population size (M), the location and velocity of the particles.
(2) Compute the fitness value of each particle.
(3) Comparing the fitness value obtained by each particle with the fitness value in the best local location it has experienced, the location corresponding to the smallest fitness value is selected as *pbest*. Comparing the fitness value of each particle and fitness value in the best location that the group has experienced, the location corresponding to the smallest fitness value is selected as *gbest*.
(4) Update the speed and location of each particle based on Eqs. (3), (4);
(5) Evaluate whether the end condition is met (the end condition judgment is that whether the number of iterations exceeds 200), and If the answer is no, the number of iterations will plus 1 and return to Step (2). Otherwise, performing the next steps.

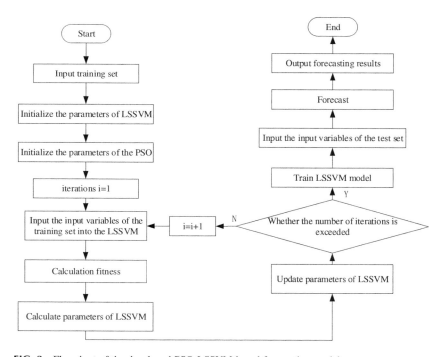

FIG. 2 Flowchart of the developed PSO-LSSVM-based forecasting model.

(6) Input the best-fit parameters (regularization parameter c and kernel coefficient σ) into the LSSVM model.

(7) Train the LSSVM model and forecast.

3 Results and discussion

The difference of input variables could influence the precision of the model. For the lack of research on the selection methods for input variables of the STELF model, the lag autocorrelation function is used to choose input variables for the forecasting model, and the correlation function is used as the comparative method to analyze the influence of different input variables on the precision of the forecasting model. To avoid the particularity of the experiment and to broaden the usage of the developed model, the study uses the data from two diverse paper mills producing different products to establish and test the same forecasting model. The GA-BPNN-based forecasting model and the PSO-BPNN-based forecasting model are applied as comparing analyze cases to validate the performances of the developed STELF model.

The mean absolute percent error (MAPE), root mean square error (RMSE), and relative error percentage (REP) are applied to assess the performance of all those forecasting models, the detail of those evaluation indexes are shown in Appendix C.

3.1 Input variables selection

3.1.1 Case 1

The electricity load data is derived from a practical papermaking enterprise in Guangdong, China in Case 1, and reserved for 60 days. The acquisition frequency is every 30 min. The acquired data is processed according to the approaches given in Appendix A. Fig. 3 shows the comparison of the processed total electric load data and the collected original data. According to Fig. 3, the features of the electricity consumption in the paper mill are aperiodicity and

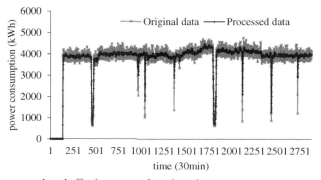

FIG. 3 Preprocessed total effective power of trend graph.

instability. The correlation function and the lag autocorrelation function described in Section 2.2 are used to choose input variables. The details of input variables selection are shown in Appendix D. The input variables chosen by the two different approaches are presented in Tables D.1 and D.2, respectively.

The initial parameters of PSO-LSSVM are set as follows: acceleration constants (c_1, c_2), that is, (2,2); maximum particle velocity (v_{max}), that is, 0.5; population scale (N), that is, 30; maximum number of iteration, that is, 100. The forecasting results are presented in Fig. 4. Fig. 4A is the contrast of the

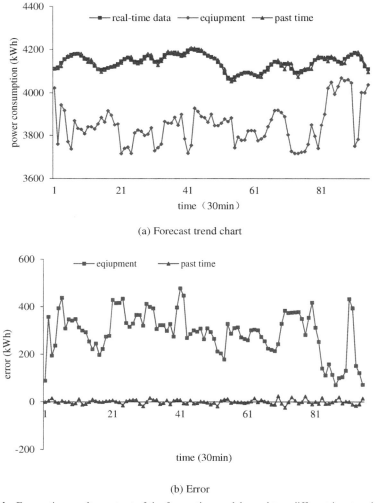

(a) Forecast trend chart

(b) Error

FIG. 4 Forecasting results contrast of the forecasting models used two different input variables selection function for Case 1.

forecasting capability between the forecasting model with two diverse input variable sets, and Fig. 4B is the error. In Fig. 4, the "equipment" represents the forecasting results of choosing the input variables using the correlation function. The "past time" represents the forecasting results of choosing the input variables using lag autocorrelation function.

To evaluate which input section method outperforms the other models, RMSE, and MAPE are applied to analyze the results of the forecasting models with diverse input variables in Case 1. The forecasting results are shown in Table 1.

The forecasting result shows that using correlation function to choose input variables for the STELF model is not suitable for Case 1. On the contrary, using the lag autocorrelation function to select input variables for the STELF model shows the great precision of the forecasting result. The errors of the forecasting model using the input variables chosen by lag autocorrelation function are between $[-100, 100]$, and the REP is between $[-2\%, 2\%]$, which satisfies the industrial requirements between $[-5\%, 5\%]$. The MAPE of the forecasting model using the input variables chosen by lag autocorrelation function is 35 times less than that of the forecasting model using the input variables chosen by correlation function.

3.1.2   Case 2

The real-time load data is derived from a practical papermaking enterprise in Hubei, China, and reserved for 60 days. The acquisition frequency is every 30 min. Data preprocessing functions described in Appendix A are used to process the original data. The processed electric load data is illustrated in Fig. 5. The input variables chosen by the same methods as Case 1 are shown in Tables D.3 and D.4, respectively.

The initial parameters for PSO-LSSVM are set as the same as that of Case 1. The forecasting results are illustrated in Fig. 6. Fig. 6A is the contrast of the forecasting performance between the forecasting model with two diverse input variable sets, and Fig. 6B is the error. In Fig. 6, the "equipment" and "past time" are described as the same meaning as in Case 1.

To evaluate which input section method outperforms the other models, RMSE and MAPE are used to analyze the results of the forecasting models with diverse input variables in Case 2. The results are presented in Table 2.

TABLE 1 Evaluation indicator of diverse situations.

Evaluate index	RMSE (kWh)	MAPE (%)
Equipment	303.87	6.95
Past time	9.252	0.17

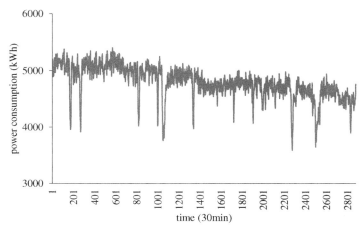

FIG. 5 Total effective power trend graph after data preprocessing.

The forecasting result shows that selecting input variables for the electric load forecasting model by using correlation function and the lag autocorrelation function both show the high precision of the forecasting result for Case 2. The errors of the two forecasting model are both between $[-100, 100]$, the REP is between $[-2\%, 2\%]$, which satisfies the industrial requirements between $[-5\%, 5\%]$. However, the MAPE of the forecasting model using the input variables chosen by lag autocorrelation function is 1 time less than that of the forecasting model using the input variables chosen by correlation function.

According to the comparative analysis of the two cases, using the lag autocorrelation function to select input variables for the electric load forecasting model achieves wider application and higher accuracy. Therefore, this study uses the lag autocorrelation function to select input variables.

3.2 Model verification

The forecasting results of the three different STELF model for the two cases are illustrated in Figs. 7 and 8. The contrast of the forecasting performance between three different models are shown in Figs. 7A and 8A. The REP between the forecasting data and the original data are illustrated in Figs. 7B and 8B.

In Figs. 7B and 8B, a benchmark of $[-0.5\%, 0.5\%]$ of the error range is set to show the accuracy of the three forecasting models more intuitively. In Case 1, the forecasting models based on PSO-LSSVM, PSO-BPNN, GA-BPNN has 93, 81, and 66 times points from the total 96 times points respectively, which lie within the forecasting error $[-0.5\%, 0.5\%]$. In Case 2, the forecasting models based on PSO-LSSVM, PSO-BPNN, GA-BPNN has 87, 80, and 30 times points from the total 96 times points respectively, which lie within the forecasting error

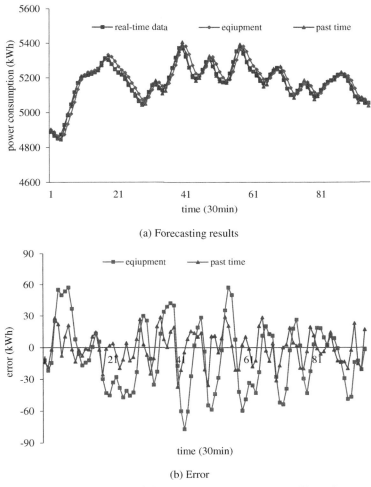

(a) Forecasting results

(b) Error

FIG 6   Forecasting result contrast of the forecasting model used two different input variables selection function for Case 2.

[−0.5%, 0.5%]. This explicitly shows that the PSO-LSSVM-based forecasting model has the best constant capability among all the adopted models.

Table 3 is the evaluation results of the three models in the two cases. The MAPE of the PSO-LSSVM-based forecasting model in Case 1 is reduced by 182% lower than that of the GA- BPNN-based forecasting model and 70% lesser than that of the PSO-BPNN-based forecasting model. The MAPE of the PSO-LSSVM-based forecasting model in Case 1 is reduced by 287% lower than that of the GA-BPNN-based forecasting model and 22% lesser than that of the PSO-BPNN-based forecasting model. The test results imply that the

TABLE 2 Evaluation indicator of diverse situations.

Evaluate index	RMSE	MAPE
Equipment	33.27	0.54
Past	15.14	0.23

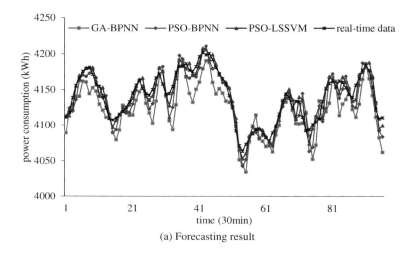

(a) Forecasting result

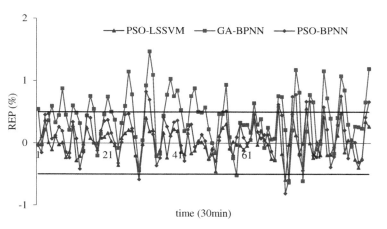

(b) Relative error percentage

FIG. 7 Forecasting result contrast for Case 1.

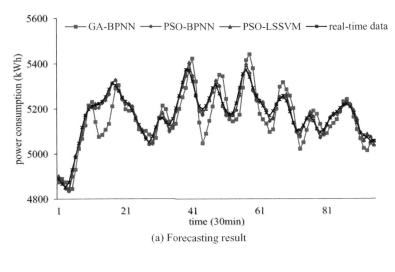

(a) Forecasting result

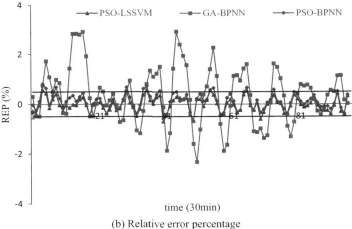

(b) Relative error percentage

FIG. 8 Forecasting result contrast in Case 2.

developed PSO-LSSVM-based model has higher precision than the contrast models in all cases.

Moreover, as shown in Figs. 7B and 8B, when three models use input variables chosen by the lag autocorrelation function, the MAPE of the three models are within $[-3\%, 3\%]$, which fully satisfies the accuracy requirements of the papermaking process industry. Thus, selecting the input variables by the lag autocorrelation function for the STELF model has universal applicability.

For the purpose of demonstrating the advantages of the developed model in recent studies, this study makes a comparison of the developed model with

TABLE 3 The forecasting performance analysis.

Study cases	PSO-LSSVM		GA-BPNN		PSO-BPNN	
	MAPE (%)	RMSE (kWh)	MAPE (%)	RMSE (kWh)	MAPE (%)	RMSE (kWh)
Case 1	0.17	9.252	0.48	23.76	0.29	14.63
Case 2	0.23	15.14	0.98	63.25	0.28	17.85

respect to the precision of the forecasted targets in different input variables selections. The results presented in Table 4. Table 4 reveals that contrasted with recent studies, the developed model in this study has a wider application range and higher precision.

4 Conclusions

STELF is of great importance to the energy management system (EMS). It is significant for developing an efficient STELF model to cut down the production expense and energy consumption of the papermaking process.

TABLE 4 Synopsis of researches on load forecasting for diverse utilization cases via AI-based.

Core algorithms	Input variables	MAPE (%)	References
PSO-LSSVM	3	0.17	This study
WLSSVM	3	3.14	[18]
AS-GCLSSVM	3	0.77	[22]
DMD	2	1.19	[27]
EEMD-ARIMA+CPSO-ENN	2	0.998	[28]
SARIMAX	1 and 2	0.9	[29]
GOA-SVM	1 and 2	1.39	[30]
GA-PSO-BPNN	1	0.77	[3]
VMD-LSTM	1	0.61	[31]

Note: "1" represents using the factors that have a great influence on the electric loads as input variables of the forecasting model, "2" represents using the periodicity of time series to select electric loads at the corresponding points in the past few cycles as the input variables, "3" represents using the historical electric loads as the input variables.

This paper develops a data-driven STELF model based on the PSO-LSSVM algorithm for the paper mills. The lag autocorrelation function is used to choose input variables. To widen the utilization of the developed model, the real-time data from two diverse paper mills are chosen to test the performance of the forecasting model.

Compared with the correlation function, the forecasting performance shows that the lag autocorrelation function to select input variables for the STELF model has higher precision. The MAPE of the forecasting model using the input variables chosen by lag autocorrelation function is 135% lesser than that of the forecasting model using the correlation function.

To testify the forecasting performance of the developed PSO-LSSVM-based model, the GA-BPNN-based forecasting model and the PSO-BPNN-based forecasting model are employed as the contrast study cases. The forecasting performance implies that the developed PSO-LSSVM-based forecasting model possesses the highest precision. The MAPE of the PSO-LSSVM-based model is merely 0.17%. Compared with recent studies, the developed forecasting model also demonstrates excellent reliability and high precision. According to the forecasting performance of the three hybrid models, using the lag autocorrelation function to choose input variables for the STELF model achieves better precise forecasting results, all the range of REP are between [−5%, 5%]. Selecting the input variables by the lag autocorrelation function for the STELF model has universal applicability.

Appendix A

Two diverse paper mills are chosen as the test case study in this research. All the electricity load data are acquired from two paper mills. Due to the disturbance of transmission, the disturbance between the electric devices, and other disturbance, the acquired data could occasionally go beyond the practical range. To increase the precision of the model, the low-quality data needs to be identified and removed by data preprocessing.

Data preprocessing generally includes removing outliers, missing values imputation, and noising filtering. There are a lot of elements that lead to huge fluctuation of electricity consumption in the papermaking process, such as the unscheduled downs, and the switching of paper products, etc. These types of data reveal the practical electricity loads and cannot be considered as an abnormal value. Hence, for the production process, the operational environment should be separated before the data preprocessing. After then, data preprocessing is used to distinguish invalid data from the real data.

The approach of removing outliers adopted extensively is the 3σ criterion. The principle of the 3σ criterion is as follows: when the distance between a value in a sequence and its average value exceeds three times the standard deviation, the probability is less than 0.3% [32]. So the value out of the distance is considered as an abnormal value and need to be removed.

Missing values in time series come from abnormal data removed. The precision of the STELF model will be diminished by the incomplete data sets. The missing data raise the uncertainty of the system analysis results, which directly leads to misjudgment of the STELF model, and then making the forecasting result deviate from the expectation and reducing the precision of the STELF model. The linear interpolation approach is applied to fill the missing data. The principle of the linear interpolation is as follows: assuming that the missing value exists in a straight line, its formula is defined by Eq. (A.1):

$$f(x) = a \times x + b \tag{A.1}$$

where x is the location of the missing data in a sequence, a, b are parameters of the linear function.

The two points closest to the missing data are used to establish the linear function, and then calculating the corresponding value of the missing data.

A large amount of data, extra data, and gauged noise are inevitable to be generated in the production process when transforming electrical signals of energy expenditure device into digital signals. These types of data also have a powerful impact on the precision of the forecasting models. Hence, after filling the missing data, the time series need to be filtered.

The moving average filtering method filters the data which have been preprocessed. It is determined by Eq. (A.2) [33]:

$$f(k) = \left. \sum_{i=k}^{t+n} y(i) \middle/ n \right. \tag{A.2}$$

where, $f(k)$ is the filtering output when the time is k, $y(i)$ is a sequence that needs to be filtered, n is the length of the sequence.

Appendix B

First, the sample ($\Psi(X)$) is mapped from the initial space (R_n) to the feature space ($\Psi(X) = (\varphi(x_1), \varphi(x_2), \ldots, \varphi(x_i))$) by the nonlinear map. Constructing the optimal decision function in the process of the nonlinear mapping, as shown in Eq. (B.1) [34]:

$$y(x) = w^T \times \varphi(x) + b \tag{B.1}$$

where w is the weight coefficients of the samples in the feature space, and w^T is the transposed matrix of w, $\varphi(x)$ is a nonlinear function and b is the bias.

On the basis of structural risk minimization theory, constrained optimization is defined by Eq. (B.2) [34]:

$$R = \frac{1}{2} \times \|w\|^2 + c \times R_{emp} \tag{B.2}$$

where $\|w\|^2$ dominates the complexity of the function, and c is the penalty coefficient. R_{emp} is the error control function, also called the insensitive loss function.

Widely used loss functions contain linear loss functions, quadratic loss functions, and hinge loss functions. The variation in the loss function makes the form of the SVM different. As the optimization objective, the loss function of the least squares linear system is the quadratic term of the error (ε_i). The resulting optimization problem of LSSVM can be formulated by Eq. (B.3) [34]:

$$\min_{w,b,\varepsilon} J(w,\varepsilon) = \frac{1}{2} \times w^T \times w + c \sum_{i=1}^{l} \varepsilon_i^2 \qquad (B.3)$$

where, $y_i = \varphi(x_i) \times w^T + b + \varepsilon_i$, $i = 1, \ldots, l$.

The Lagrangian function is represented by Eq. (B.4) [34]:

$$L(w,b,\varepsilon;a) = \frac{1}{2} \times w^T \times w + c \times \sum_{i=1}^{l} \varepsilon_i^2 - \sum_{i=1}^{l} a_i \times \left(w^T \times \varphi(x_i) + b + \varepsilon_i - y_i \right) \qquad (B.4)$$

where a_i, $i = 1, \ldots, l$, is a Lagrangian multiplier.

According to the Lagrangian function, the optimization conditions are by Eq. (B.5):

$$\frac{\partial L}{\partial w} = 0, \quad \frac{\partial L}{\partial b} = 0, \quad \frac{\partial L}{\partial a} = 0 \qquad (B.5)$$

Based on Eq. (B.5), the solution is found by dealing with the system of linear equations expressed in matrix form by Eq. (B.6) [34]:

$$\begin{bmatrix} 0 & I^T \\ I & \Omega + I/y \end{bmatrix} \begin{bmatrix} b \\ a \end{bmatrix} = \begin{bmatrix} 0 \\ y \end{bmatrix} \qquad (B.6)$$

where, $\Omega_{ij} = \varphi(x_i)^T \times \varphi(x_j) = K(x_i, x_j)$, $a = [a_1, a_2, \ldots, a_i]^T$, $y = [y_1, y_2, \ldots, y_i]^T$, I, $j = 1, 2, \ldots, l$; I; is the identity matrix, the kernel function can be described as $K(x_i, x_j) = \varphi(x_i) \times \varphi(x_j)$, $K(x_i, x_j)$ is a symmetric function conforming to the Mercer's condition.

The regression coefficients a_i and b are gained by the least squares method. The LSSVM regression model is defined by Eq. (B.7) [34]:

$$f(x) = \sum_{i=1}^{l} a_i \times K\left(x_i, x_j \right) + b \qquad (B.7)$$

Since the RBF is more widely applicable kernel function, and it does not require a priori knowledge of the data set. Therefore, this paper uses the

RBF as the kernel function of LSSVM algorithm. The RBF is defined by
Eq. (B.8) [34]:

$$K(x_i, x_j) = \exp\left(-\frac{\|x_i - x_j\|^2}{2 \times \sigma^2}\right) \tag{B.8}$$

where, σ is the kernel parameter. If σ is large, it is easy to classify all sample
points into the same class; otherwise, it will cause the overfitting problem.

Appendix C

The mean absolute percent error (MAPE), root mean square error (RMSE), and
relative error percentage (REP) are applied to assess the performance of the
developed STELF model [35–37]. RMSE, REP, and MAPE are separately
defined by Eqs. (C.1), (C.2), and (C.3):

$$\text{RMSE} = \sqrt{\left.\sum_{i=1}^{n} \left(y_{pi} - y_{oi}\right)^2 \middle/ n\right.} \tag{C.1}$$

$$\text{MAPE} = \left.\sum_{i=i}^{n} \left|\frac{y_{pi} - y_{oi}}{y_{oi}}\right| \middle/ n\right. \times 100\% \tag{C.2}$$

$$\text{REP}_i = \left.\left(y_{pi} - \hat{y}_{pi}\right) \middle/ y_{oi}\right. \times 100\% \tag{C.3}$$

where y_{pi} is the forecasting value, $\hat{y}_{pi}$ is the average of forecasting value, y_{oi} is
the real-time value, and n is the step size.

Appendix D

(1) Case 1

The correlation function and the lag autocorrelation function described in
Section 2.2 are used to choose input variables, where the electric loads of electric devices are regarded as the input variables when their absolute correlation
coefficient goes beyond 0.6, and the top 10 historical electric loads are regarded
as the input variables when their absolute lag autocorrelation coefficient goes
beyond 0.9. The input variables chosen by the two different approaches are presented in Tables D.1 and D.2, respectively.

(2) Case 2

The input variables of Case 2 chosen by the same methods as Case 1 are shown
in Tables D.3 and D.4, respectively.

TABLE D.1 Input Variables selected by the correlation function.

Order	Input variables	Unit
1	Refining power	kWh
2	Slurry power	kWh
3	Fan power	kWh
4	Transmission power	kWh
5	Vacuum pump power	kWh
6	Production Plan	t

TABLE D.2 Input variables selected by the lag autocorrelation function.

Case	Input variables	Unit
Case 1	$I_{t-1,1}$, $I_{t-2,1}$, $I_{t-3,1}$, $I_{t-4,1}$, $I_{t-5,1}$, $I_{t-6,1}$, $I_{t-7,1}$, $I_{t-8,1}$, $I_{t-9,1}$, $I_{t-10,1}$	kWh

Note: $I_{t-i,j}$ is the electricity load of the past ith sample point at the tth sample point of the jth reconstruction sequence.

TABLE D.3 Input variables chosen by the correlation function.

No.	Input variables	Unit
1	Refining power	kWh
2	pulping power	kWh
3	Vacuum pump power	kWh
4	drive side power	kWh
5	Lightning power	kWh
6	Total pulp production power	kWh
7	Total paper production power	kWh
8	Production planning	T

TABLE D.4 Input variables selected by the lag autocorrelation function.

Case	Input variables	Unit
Case 2	$I_{t-1,1}$, $I_{t-2,1}$, $I_{t-3,1}$, $I_{t-4,1}$, $I_{t-5,1}$, $I_{t-6,1}$, $I_{t-7,1}$, $I_{t-8,1}$, $I_{t-9,1}$, $I_{t-10,1}$	kWh

Note: $I_{t-i,j}$ is the electricity load of the past ith sample point at the tth sample point of the jth reconstruction sequence.

References

[1] National Bureau of Statistics of the People's Republic of China, 2017 China Energy Statistical Yearbook, China Statistical Press, Beijing, 2018. (Online), Available: http://www.stats.gov. cn/tjsj/ndsj/2018/indexch.htm.

[2] Y. Man, Y. Han, J. Li, M. Hong, Review of energy consumption research for papermaking industry based on life cycle analysis, Chin. J. Chem. Eng. 27 (2019) 1543–1553, https://doi.org/10.1016/j.cjche.2018.08.017.

[3] Y. Hu, J. Li, M. Hong, J. Ren, R. Lin, Y. Liu, M. Liu, Y. Man, Short term electric load forecasting model and its verification for process industrial enterprises based on hybrid GA-PSO-BPNN algorithm—a case study of papermaking process, Energy 170 (2019) 1215–1227, https://doi.org/10.1016/j.energy.2018.12.208.

[4] L. Na, X. Luo, Forecast and scenario simulation analysis of power demand considering a variety of external environmental factors, in: China International Conference on Electricity Distribution, CICED, 2016, https://doi.org/10.1109/CICED.2016.7576115.

[5] Y. Xuan, Q. Yue, Forecast of steel demand and the availability of depreciated steel scrap in China, Resour. Conserv. Recycl. 109 (2016) 1–12. 11, 1836–1845 https://doi.org/10.1016/j.resconrec.2016.02.003.

[6] J. Wang, T. Niu, H. Lu, W. Yang, P. Du, A novel framework of reservoir computing for deterministic and probabilistic wind power forecasting, IEEE Trans. Sustain. Energy 11 (2020) 337–349, https://doi.org/10.1109/TSTE.2019.2890875.

[7] W. Sulandari, Subanar, M.H. Lee, P.C. Rodrigues, Indonesian electricity load forecasting using singular spectrum analysis, fuzzy systems and neural networks, Energy 190 (2020) 116408, https://doi.org/10.1016/j.energy.2019.116408.

[8] Y. Weng, X. Wang, J. Hua, H. Wang, M. Kang, F.Y. Wang, Forecasting horticultural products price using ARIMA model and neural network based on a large-scale data set collected by web crawler, IEEE Trans. Comput. Soc. Syst. 6 (2019) 547–553, https://doi.org/10.1109/TCSS.2019.2914499.

[9] M.A. Mat Daut, M.Y. Hassan, H. Abdullah, H.A. Rahman, M.P. Abdullah, F. Hussin, Building electrical energy consumption forecasting analysis using conventional and artificial intelligence methods: a review, Renew. Sustain. Energy Rev. 70 (2017) 1108–1118, https://doi.org/10.1016/j.rser.2016.12.015.

[10] M. Barman, N.B. Dev Choudhury, Season specific approach for short-term load forecasting based on hybrid FA-SVM and similarity concept, Energy 174 (2019) 886–896, https://doi.org/10.1016/j.energy.2019.03.010.

[11] L.G. Chen, H.D. Chiang, N. Dong, R.P. Liu, Group-based chaos genetic algorithm and non-linear ensemble of neural networks for short-term load forecasting, IET Gener. Transm. Distrib. 10 (2016) 1440–1447, https://doi.org/10.1049/iet-gtd.2015.1068.

[12] Y. Wen, D. AlHakeem, P. Mandal, S. Chakraborty, Y.K. Wu, T. Senjyu, S. Paudyal, T.-L. Tseng, Performance evaluation of probabilistic methods based on bootstrap and quantile regression to quantify PV power point forecast uncertainty, IEEE Trans. Neural Netw. Learn. Syst. 99 (2019) 1–11, https://doi.org/10.1109/tnnls.2019.2918795.

[13] X. Xu, D. Niu, Q. Wang, P. Wang, D.D. Wu, Intelligent forecasting model for regional power grid with distributed generation, IEEE Syst. J. 11 (2017) 1836–1845, https://doi.org/10.1109/JSYST.2015.2438315.

[14] J. Kim, J. Moon, E. Hwang, P. Kang, Recurrent inception convolution neural network for multi short-term load forecasting, Energy Build. 194 (2019) 328–341, https://doi.org/10.1016/j.enbuild.2019.04.034.

[15] G. Sideratos, A. Ikonomopoulos, N.D. Hatziargyriou, A novel fuzzy-based ensemble model for load forecasting using hybrid deep neural networks, Electr. Power Syst. Res. 178 (2020) 106025, https://doi.org/10.1016/j.epsr.2019.106025.

[16] W. Kong, Z.Y. Dong, Y. Jia, D.J. Hill, Y. Xu, Y. Zhang, Short-term residential load forecasting based on LSTM recurrent neural network, IEEE Trans. Smart Grid 10 (2019) 841–851, https://doi.org/10.1109/TSG.2017.2753802.

[17] X. Wei, Z. Yang, Y. Liu, D. Wei, L. Jia, Y. Li, Railway track fastener defect detection based on image processing and deep learning techniques: a comparative study, Eng. Appl. Artif. Intell. 80 (2019) 66–81, https://doi.org/10.1016/j.engappai.2019.01.008.

[18] K. Chen, K. Chen, Q. Wang, Z. He, J. Hu, J. He, Short-term load forecasting with deep residual networks, IEEE Trans. Smart Grid 10 (2019) 943–3952, https://doi.org/10.1109/TSG.2018.2844307.

[19] C. Wang, J.H. Wang, S.S. Gu, X. Wang, Y.X. Zhang, Elongation prediction of steel-strips in annealing furnace with deep learning via improved incremental extreme learning machine, Int. J. Control Autom. Syst. 15 (2017) 1–12, https://doi.org/10.1007/s12555-015-0463-7.

[20] X. Zhang, J. Wang, K. Zhang, Short-term electric load forecasting based on singular spectrum analysis and support vector machine optimized by Cuckoo search algorithm, Electr. Power Syst. Res. 146 (2017) 270–275, https://doi.org/10.1016/j.epsr.2017.01.035.

[21] W. Li, Z. Xu, D. Xu, D. Dai, L. Van Gool, Domain generalization and adaptation using low rank exemplar SVMs, IEEE Trans. Pattern Anal. Mach. Intell. 40 (2018) 1114–1127, https://doi.org/10.1109/TPAMI.2017.2704624.

[22] B. Zhu, S. Ye, P. Wang, K. He, T. Zhang, Y.M. Wei, A novel multiscale nonlinear ensemble leaning paradigm for carbon price forecasting, Energy Econ. 70 (2018) 143–157, https://doi.org/10.1016/j.eneco.2017.12.030.

[23] Y. Zhang, J. Le, X. Liao, F. Zheng, Y. Li, A novel combination forecasting model for wind power integrating least square support vector machine, deep belief network, singular spectrum analysis and locality-sensitive hashing, Energy 168 (2019) 558–572, https://doi.org/10.1016/j.energy.2018.11.128.

[24] H. Liu, C. Chen, X. Lv, X. Wu, M. Liu, Deterministic wind energy forecasting: a review of intelligent predictors and auxiliary methods, Energy Convers. Manage. 190 (2019) 328–345, https://doi.org/10.1016/j.enconman.2019.05.020.

[25] Y. Chen, Y. Yang, C. Liu, C. Li, L. Li, A hybrid application algorithm based on the support vector machine and artificial intelligence: an example of electric load forecasting, Appl. Math. Model. 39 (2015) 2617–2632, https://doi.org/10.1016/j.apm.2014.10.065.

[26] Y. Yu, J. Li, Residuals-based deep least square support vector machine with redundancy test based model selection to predict time series, Tsinghua Sci. Technol. 24 (2019) 706–715, https://doi.org/10.26599/TST.2018.9010092.

[27] N. Mohan, K.P. Soman, S. Sachin Kumar, A data-driven strategy for short-term electric load forecasting using dynamic mode decomposition model, Appl. Energy 232 (2018) 229–244, https://doi.org/10.1016/j.apenergy.2018.09.190.

[28] W.Q. Li, L. Chang, A combination model with variable weight optimization for short-term electrical load forecasting, Energy 164 (2018) 575–593, https://doi.org/10.1016/j.energy.2018.09.027.

[29] N. Elamin, M. Fukushige, Modeling and forecasting hourly electricity demand by SARIMAX with interactions, Energy 165 (2018) 257–268, https://doi.org/10.1016/j.energy.2018.09.157.

[30] M. Barman, N.B. Dev Choudhury, S. Sutradhar, A regional hybrid GOA-SVM model based on similar day approach for short-term load forecasting in Assam, India, Energy 145 (2018) 710–720, https://doi.org/10.1016/j.energy.2017.12.156.

[31] F. He, J. Zhou, Z.k. Feng, G. Liu, Y. Yang, A hybrid short-term load forecasting model based on variational mode decomposition and long short-term memory networks considering relevant factors with Bayesian optimization algorithm, Appl. Energy 237 (2019) 103–116, https://doi.org/10.1016/j.apenergy.2019.01.055.

[32] J. Xu, Q. Xiao, Y. Fei, S. Wang, J. Huang, Accurate estimation of mixing time in a direct contact boiling heat transfer process using statistical methods, Int. Commun. Heat Mass Transf. 75 (2016) 162–168, https://doi.org/10.1016/j.icheatmasstransfer.2016.04.012.

[33] K. Mivule, C. Turner, Applying moving average filtering for non-interactive differential privacy settings, Procedia Comput. Sci. 36 (2014) 409–415, https://doi.org/10.1016/j.procs.2014.09.013.

[34] X. Xue, M. Xiao, Deformation evaluation on surrounding rocks of underground caverns based on PSO-LSSVM, Tunn. Undergr. Space Technol. 69 (2017) 171–181, https://doi.org/10.1016/j.tust.2017.06.019.

[35] F. Ferracuti, A. Fonti, L. Ciabattoni, S. Pizzuti, A. Arteconi, L. Helsen, G. Comodi, Data-driven models for short-term thermal behaviour prediction in real buildings, Appl. Energy 204 (2017) 1375–1387, https://doi.org/10.1016/j.apenergy.2017.05.015.

[36] N. Ghadimi, A. Akbarimajd, H. Shayeghi, O. Abedinia, Two stage forecast engine with feature selection technique and improved meta-heuristic algorithm for electricity load forecasting, Energy 161 (2018) 130–142, https://doi.org/10.1016/j.energy.2018.07.088.

[37] L. Xu, S. Wang, R. Tang, Probabilistic load forecasting for buildings considering weather forecasting uncertainty and uncertain peak load, Appl. Energy 237 (2019) 180–195, https://doi.org/10.1016/j.apenergy.2019.01.022.

Chapter 7

Artificial intelligence algorithm application in wastewater treatment plants: Case study for COD load prediction

Zifei Wang and Yi Man
State Key Laboratory of Pulp and Paper Engineering, School of Light Industry and Engineering, South China University of Technology, Guangzhou, People's Republic of China

1 Introduction

Urban sewage is a mixture of domestic and industrial sewage that also contains a portion of natural precipitation [1]. Due to the variety of industrial products and the different production technologies, there are a variety of pollutants in industrial sewage, and their concentrations fluctuate greatly. Although the number of pollutants in domestic sewage is not as high as that in industrial sewage, the concentration of pollutants in domestic sewage also varies. Therefore, the sources of urban sewage are wide and uncertain, which causes fluctuations in the inflow and quality of urban sewage [2]. The goal of sewage treatment has evolved over the years. It started with the need for sanitation and then moved on to environmental protection [3]. Currently, most mainstream urban sewage treatment plants adopt an activated sludge process, including secondary treatment. The A2O process (also known as the AAO process) is short for "anaerobic-anoxic-oxic process" and is a commonly used sewage treatment process. The synchronous removal of phosphorus and nitrogen by the A2O process consists of two parts. First, phosphorus is removed. The phosphorus-accumulating bacteria in the sewage are released in an anaerobic state and then absorbed in the aerobic state and discharged with the remaining sludge. Second, denitrification occurs. Dissolved oxygen, $<0.5\,mg/L$, facultative oxygen, and nitrogen-removing bacteria use the biochemical oxygen demand (BOD) in the water as a hydrogen (organic carbon source) supply, and the nitrate and nitrite in the mixture reflux of the aerobic pool are reduced to nitrogen and released into the atmosphere, so as to carry out denitrification. The energy

Applications of Artificial Intelligence in Process Systems Engineering
https://doi.org/10.1016/B978-0-12-821092-5.00009-7

143

consumption of the A2O process occurs mainly in the inlet pump room, which lifts sewage, and the blower room, which provides oxygen for the secondary biological treatment. These areas account for approximately 70% of the plant's energy consumption. The power consumption of the aeration system accounts for approximately 40%–50% of the power consumption for the entire plant. This unit is the largest power consumer in the secondary sewage treatment system and the key to energy savings in sewage treatment. Because the water quality fluctuates greatly and the discharge standard is increasingly strict, to ensure that the sewage quality after treatment meets the standard, many sewage treatment plants will increase the chemical input and aeration. Although such a practice can ensure that the effluent quality reaches the standard, it will also cause energy waste and, potentially, chemical pollution. With the increasingly strict discharge standards and the increase in sewage treatment capacity, such a practice will obviously bring a substantial cost pressure to the sewage treatment plant. For example, China is both a populous and industrial country, discharging a huge amount of domestic sewage and industrial wastewater every year. In addition, with the acceleration of China's urbanization and industrialization, cities are becoming larger, and the urban population is increasing rapidly. At the same time, large industrial projects are increasing every year, resulting in an annual increase of 5% in the discharge of urban sewage [4]. In 2017, China's urban sewage discharge was $6.99 \times 10^9 \, m^3$, its treatment capacity was $6.21 \times 10^9 \, m^3$ [5], and the average power consumption by sewage treatment plants was $0.29 \, kW \, h/m^3$. Sewage treatment has a significant impact on China's energy consumption and economic growth. Other countries also face the same problems as does China. Some scholars have conducted studies aimed at improving the sewage treatment efficiency. For example, Pan et al. [6] studied and analyzed the treatment efficiency of urban sewage in the Yangtze River Basin cities of China. They found that the average sewage treatment efficiency of 0.51 is still far from a perfect efficiency of 1. Analysis of sewage treatment efficiency is certainly helpful in reducing the high treatment cost resulting from excessive aeration in sewage treatment plants. However, to truly solve this problem and achieve an effective management, it is necessary to obtain accurate data on sewage parameters to accurately control the addition of chemicals and aeration. The water inflow, inlet temperature, inlet pH, inlet ammonia nitrogen, inlet COD, inlet BOD, and total inlet solid suspended matter are commonly used in sewage treatment. The water inflow and inlet temperature are relatively easy to measure and can be measured online. Although the water inlet pH value can also be measured online, the result is not very stable. If a laboratory measurement method is adopted, more accurate results can be obtained; however, due to hysteresis, laboratory results cannot be used as the basis for controlling the aeration. Currently, indexes such as water inlet COD, water inlet BOD, water inlet ammonia nitrogen, and total water inlet solid suspended matter can only be measured in the laboratory and cannot be measured online. Therefore, one way to accurately measure these parameters is to develop more advanced measurement sensors that meet the requirements; another way is to study the mechanism of the sewage treatment process, establish the mechanism model, and derive the corresponding parameter values. The corresponding parameter values could also be predicted through an intelligent algorithm. Manufacturing the corresponding sensor is the most direct way; however, the research on and development of a sensor cannot be done quickly and requires the investment of a great amount of manpower, material resources, and time. Therefore, it is not practical wait for sensors to be developed. It is also not feasible to calculate the corresponding index values through mechanism derivations in sewage treatment. For example, although the COD can be calculated using a formula, it is based on the premise that the exact chemical formula is known. In the process of sewage treatment, there are many kinds of pollutants, and the concentration of the pollutants fluctuates greatly; thus, it is impossible to calculate the COD value of sewage. An AI algorithm can predict the required parameters based on historical data and the potential relationships among the data, which can solve the problem of hysteresis quality and guarantee prediction accuracy. In this study, the application of artificial intelligence algorithms in sewage treatment plants is reviewed, and it is found that they play an important role in the prediction of key water quality parameters in sewage treatment plants. Therefore, this study takes a sewage treatment plant in Guangzhou, China as a case and uses an artificial intelligence algorithm to predict the water inlet COD load.

2 Literature review

If we know the inlet water quality of the sewage treatment plant in advance, we can adjust the chemical dosage and aeration time to avoid energy wastage. To help wastewater treatment plants reduce energy consumption by controlling aeration and chemical use more accurately while ensuring effluent water quality, many prediction models based on artificial intelligence algorithms have been applied to predict water quality parameters. Two types of prediction models are used for sewage water quality parameters: intelligent algorithm-based and neural network-based models.

Prediction models based on intelligent algorithms: Niu et al. [7] proposed a prediction model based on a genetic algorithm-deep belief network (GA-DBN) for two important water quality indexes of papermaking wastewater COD and suspended solids (SS). The results show that the GA-DBN model can be used to describe the complex papermaking wastewater treatment process with fewer variables or sample construction models, and better model performance and a higher prediction accuracy can be achieved. Hvala et al. [8] predicted two water quality indicators (total nitrogen and phosphorus) by combining mechanism and data-driven models, namely, an activated sludge model, and a machine learning method based on a Gaussian process (GP) model. This method maintains physical transparency and achieves a high prediction accuracy. Lotfi et al. [9] used a linear stochastic model (ARIMA) and nonlinear outlier robust limit learning machine (ORELM) to model sewage quality parameters. The model successfully predicted the BOD, COD, total dissolved solids (TDS), TSS, and other indexes. To realize online monitoring of the COD and total phosphorus (TP), Wang et al. [10] performed principal component analysis (PCA) to determine the important variables for predicting COD and TP. The multiple regression method uses variables recommended by the principal component analysis to predict COD and TP. Chen et al. [11] proposed a gray dynamic model combined with a GA to establish a prediction model for the effluent quality at an industrial wastewater treatment plant in southern Taiwan. Nadiri et al. [12] proposed a fuzzy logic (FL) and supervised committee fuzzy logic (SCFL) model to predict water quality parameters. The BOD, COD, and TSS prediction results show that the SCFL model performs better than does the FL model. Zeng et al. [13] proposed an explicit model predictive control (EMPC) method to predict and control the water quality and economic costs for sewage treatment plants. Lee et al. [14] proposed a robust adaptive partial least squares regression (PLS) model to detect and predict the sewage treatment process, and the results showed that the model performed well on the prediction and detection of complex sewage treatment processes. Vasilaki et al. [15] used dense-based clustering, support vector machine (SVM), and support vector regression (SVR) models to estimate the dissolved N_2O concentration at different stages of the SBR system. This method accurately predicts the N_2O emission as a potential parameter. Tomperi et al. [16] used a genetic algorithm to select the optimal subsets of input variables such as SS, BOD, COD, total nitrogen, and total phosphorus in sewage and then predicted these water quality indicators using a method that combines linear regression with optical monitoring. Lin et al. [17] established a prediction model based on a cluster extension using the concepts of matter element and correlation functions and applied the model to the prediction of wastewater discharge. The application results show that the method is scientific and reliable. Verma et al. [18] used a multilayer perceptron, k-nearest neighbor, multivariable adaptive regression spline, SVM, and random forest to construct a time series prediction model of TSS. Historical TSS values are used as input parameters to predict the current and future values of the TSS. The sliding window method is adopted to improve the prediction results. The results show that TSS can be predicted up to 7 days in advance. Ilter et al. [19] established a MIMO (multiinput multioutput) fuzzy logic model. The volume organic load rate (OLR), volume chemical oxygen demand (TCOD) removal rate (RV), inlet alkalinity, inlet pH value, and outlet pH value were used as inputs to predict the methane production rate of a 90-L medium-temperature upflow anaerobic sludge bed reactor (UASB) for molasses wastewater treatment. The results show that the model has a good prediction effect, and the determination coefficient reaches 0.98. Liu et al. [20] proposed a dynamic nuclear extremum learning machine (DKELM) method to predict the key quality indexes of sewage COD. A delay coefficient is introduced, and the kernel function is embedded in the extremum learning machine

(ELM) to extract dynamic information and obtain a better prediction accuracy. Chan [21] used multiple regression, first-order dynamic, and mass balance analyses to predict the removal rate of ammonia nitrogen from sewage treatment systems and established a prediction model. Chen et al. [22] proposed a least-squares SVM model to predict the COD in effluent. The multiscale chaotic search algorithm is applied to optimize the model parameters, and the genetic algorithm is included to accelerate the search. The simulation results show that this method has a higher precision and stronger generalization ability and requires less computation. The RMSE is reduced from 21.43 to 6.83. Huang et al. [23] proposed an improved least-squares support vector machine (LS-SVR). A benchmark simulation model 1 (BSM1) was used to generate input and output data, and prediction models for effluent parameters COD, BOD, TN, NH3, and TSS were established. The parameters of the LS-SVR were optimized via particle swarm optimization (PSO), and a more accurate model was obtained. Yang et al. [24] proposed a COD prediction model of sewage effluent based on a LS-SVR. The model converts inequality constraints into equality constraints and quadratic programming into solving a set of linear equations, which simplifies the learning process and improves the computational efficiency. Compared with the BP neural network, the experimental results verify the effectiveness of the LS-SVM method in the prediction of the COD in effluent.

Prediction models based on neural networks: Zhao et al. [25] used a cascade forward back propagation neural network and feedforward back propagation neural network to predict the total organic carbon concentration (TOC) in sewage, and the experimental results showed that the performance of the FFBPNN model was significantly better than that of the CFBPNN model. Schmitt et al. [26] established an artificial neural network (ANN) to predict the transmembrane pressure of an anoxic membrane bioreactor for domestic sewage treatment. The results show that the ANN is an effective method for predicting the transmembrane pressure of domestic sewage systems. Luo et al. [27] artificially described the process of activated sludge circulation and established a fuzzy neural network model to predict the circulation flow of activated sludge. The simulation results show that the activated sludge circulation model based on this network has a strong adaptive capacity, simple network structure, and fast learning speed. The effectiveness of this method is proven by the successful prediction of effluent sludge circulation flow based on input. Han et al. [28] introduced a system consisting of an online sensor and a sludge volume index (SVI) prediction device. This system uses the hierarchical radial basis function neural network to predict the SVI of a WWTP. Based on the hierarchical structure, sequential information can be learned online by adjusting the output weight connecting the hidden layer and the output layer. An extended ultimate learning machine (EELM) method is proposed to train the HRBF weight. In this EEL-HRBF implementation, the EELM method provides a better generalization performance during the learning process. Ran et al. [29] proposed a wastewater quality prediction method based on a PCA time-delay neural network to obtain real-time data and achieve real-time closed-loop control. This method involves a principal component analysis, time-delay neural network, and model updating, and the offline model is trained using the GABP algorithm. The model has the advantages of capable real-time performance, high stability, high precision, and easy updating and can be used for online prediction of the sewage BOD. Revollar et al. [30] used neural networks and fuzzy models to predict sewage water quality and energy consumption, thus providing guidance for cost reduction. Pai et al. [31] used an ANN and three adaptive neuro fuzzy reasoning systems (ANFIS) to predict the effluent suspended matter (SS), COD, and pH of a sewage treatment plant in an industrial park. The results show that the adaptive neural fuzzy inference system can be used to predict the effluent water quality. Samli et al. [32] predicted COD and SS parameters in sewage using an ANN. In addition, an ANN was used to calculate the influence of various parameters on toxicity. The results showed that the COD was the parameter with the greatest influence on toxicity except for the color parameter, while the SS has the least influence. Liang et al. [33] combined the fuzzy rough set method with an RBF neural network to predict the biochemical oxygen demand, COD, and other important sewage outputs. Fu et al. [34] proposed an improved wavelet denoising technique combined with an adaptive neuro-fuzzy inference system (ANFIS) model to predict water quality parameters, such as total dissolved solids (TDSS) and electrical conductivity (EC). The results show that compared with the experimental results of other wavelet denoising AI models, the wavelet-ANFIS model has the best predictive performance. Ahmed et al. [35] proposed an ANN model for the short-term prediction of the inflow rate of wastewater into the largest gold bar wastewater treatment plant (GBWWTP) in Edmonton, Alberta, Canada. Alain et al. [36] used the AI model of an ANN-GA to predict the effectiveness of the electrooxidation (EO) process for treating sulfate with wastewater bromophenol blue dye, with electrolysis time, flow rate, current density, pH, and dye concentration as input variables and color change efficiency as the output variable. Mohammad et al. [37] established a multiple nonlinear autoregressive exogenous neural network based on five input parameters—total dissolved sulfide, biochemical oxygen demand (BOD_5), temperature, flow rate, and pH—and used it predict H_2S concentration in sewage treatment plants.

Through this review of previous studies, it was found that artificial intelligence algorithms have already been applied in sewage treatment processes. These algorithms are mainly used to predict water quality parameters that are difficult to measure either in a laboratory or online. The application of artificial intelligence algorithms solves the problems of difficult-to-measure water quality parameters and measurement hysteresis and results in the high-precision prediction of water quality parameters.

3 Artificial intelligence algorithm model

Although the geometric average method, arithmetic average method, weighted average method, and other algorithms can be used for prediction and are easy to implement, the prediction effect of these algorithms for complex nonlinear problems is not satisfactory. Artificial intelligence algorithms can be used for linear/nonlinear classification as well as for regression. Artificial intelligence algorithms generally have the following characteristics: low generalization error, easily explainable, and low computational complexity. Because most practical problems are nonlinear and multivariable, artificial intelligence algorithms are needed to solve such complex forecasting problems. According to different model training methods, artificial intelligence algorithms can be classified into four types: supervised learning, unsupervised learning, semisupervised learning, and reinforcement learning. Common algorithms are the Boltzmann machine, convolutional neural network, recurrent neural network, Bayesian network, classification and regression tree, random forest, SVM, and cluster analysis. With the current rapid development of science and technology, artificial intelligence algorithms are being applied not only in sewage treatment plants, but also in various other industries and fields. Hopkins et al. [38] used AI to predict postoperative infection with an accuracy rate of 98.45%. Roca et al. [39] used convolutional neural network and random forest algorithms to predict EDSS scores of patients with multiple sclerosis and obtained an $MSE = 2.2$ for the verification data set and $MSE = 3$ for the test data set (mean EDSS error $= 1.7$). Nguyen et al. [40] established prediction models based on genetic algorithm-spline smoothing, artificial neural networks, and SVMs and predicted mine explosion pressures. The results showed that the prediction based on the genetic algorithm-spline smoothing was the best. Based on GA, ANN, and data mining (DM) methods of time series (TS) analyses, Saud [41] developed a new artificial intelligence and data-driven prediction model called GANNATS to analyze and predict the energy market. Diptendu et al. [42] used a time series prediction method to conduct experiments on large (28-year) daily economic time series datasets to verify the success of automatic prediction of structures, with an average prediction accuracy of 88.05%, an average prediction accuracy of 91.24%, and an average recall rate of 93.42%. Yin et al. [43] established a plant population photosynthetic rate prediction model based on a SVM by taking greenhouse environmental parameters as input parameters and the photosynthetic rate as the output parameter. PSO and a GA were used to optimize the SVM parameters, and the results showed that the model had a high accuracy. The correlation coefficients of the PSO- and GA-based photosynthesis prediction models were 0.9883 and 0.9878, respectively. Daniyan et al. [44] used an artificial neural network to predict the state and potential failure of a trolley wheel bearing. Yan et al. [45] proposed a hybrid artificial intelligence model combining a back propagation neural network (BPNN), GA, and adaptive enhancement algorithm (AdaBoost) to predict the strength change in coal in a CO_2 geologic sequestration more accurately. Zhang et al. [46] used a mixed artificial intelligence model to predict the uniaxial compressive strength of concrete with oil palm shells. Cotrufo et al. [47] established a model for the predictive control of institutional architecture based on decision tree, random forest, and SVM algorithms. Artificial intelligence algorithms can be widely used in various fields because they have the characteristics of strong generalization and high prediction accuracy and can solve complex problems. In addition, the application of artificial intelligence algorithms can reduce the required investment in manpower and equipment, and the general application of artificial intelligence algorithms is the future development trend in various industries. A gradient boosting decision tree (GBDT) is an iterative decision tree algorithm that is composed of multiple decision trees, and the results from all the trees determine the final outcome. GBDT is thought to have a natural advantage in finding a variety of distinguishing features and feature combinations and is effective when used for regression predictions. Therefore, the GBDT algorithm is used in this study to establish a prediction model for the COD load in sewage treatment plants.

The GBDT algorithm is based on the idea that the decision tree is a weak learner, and each learner learns based on the prediction bias of the previous learner, establishing the weighted sum of the simple model and replacing the single strong learner [48–50].

The GBDT algorithm uses the classification and regression tree (CART) as a weak learner superposition instead of a single strong learner, as shown in Eq. (1) [48]:

$$T(x; c, R) = \sum_{m=1}^{M} c_m I(x \in R_m) \tag{1}$$

where $R = \{R_1, R_2, \ldots, R_M\}$ is the input space and the final subspace formed by the partition rule, which is also called the leaf. M is the number of leaves. $c = \{c_1, c_2, \ldots, c_m\}$, and $c_m = \text{mean}(y_i | x_i \in R_m)$ is the mean of the output features in each leaf space. I is a discriminant function $I = \begin{cases} 0, & x \notin R_m \\ 1, & x \in R_m \end{cases}$. When the unique $\{c, R\}$ combination is determined, the following CART formula is determined:

$$F(x) = \sum_{i=0}^{N} T_i(x; c_i, R_i) \tag{2}$$

where T_i, $i = 1, 2, \ldots, N$, and there are N trees. Thus, the GBDT is constructed. A tree is added for each iteration. Each CART tree T_m added is based on the minimum loss function of the sum of the former $m - 1$ CART trees and the actual value, as shown in Eq. (3):

$$T_m = \arg\min_T \sum_{k=1}^{n} \mathcal{L}\left(y_k, \sum_{j=1}^{m-1} T_j(x_k) + T(x_k)\right) \tag{3}$$

With the least squares loss as the loss function, each iteration builds the mth tree on the residual of the CART sum generated by the first m-1 trees. The GBDT generated by the mth iteration is shown in Eq. (4):

$$F_m(x) = F_{m-1}(x) + \arg\min_{T} \sum_{i=1}^{n} \mathcal{L}(y_i, F_{m-1}(x_i) + T(x_i)) \tag{4}$$

Gradient boosting adopts the gradient descent method to obtain the minimum objective value, for which the maximum gradient direction is the negative gradient direction of the loss function under the current model F_{m-1}, as shown in Eq. (5):

$$F_m(x) = F_{m-1}(x) + \gamma_m \sum_{i=1}^{n} \nabla_{F_{m-1}} \mathcal{L}(y_i, F_{m-1}(x_i)) \tag{5}$$

where γ_m is obtained from Eq. (6):

$$\gamma_m = \arg\min_{\gamma} \sum_{i=1}^{n} \mathcal{L}\left(y_i, F_{m-1}(x_i) - \gamma \frac{\partial \mathcal{L}(y_i, F_{m-1}(x_i))}{\partial F_{m-1}(x_i)}\right) \tag{6}$$

To achieve satisfactory model performance and prevent overfitting, each tree is limited to improve the accuracy of the algorithm, and the contraction factor, also known as the learning rate ν, is set as shown in Eq. (7):

$$F_m(x) = F_{m-1}(x) + \nu \gamma_m T_m(x) \tag{7}$$

where $0 < \nu < 1$. Choosing a small learning rate is conducive to the accuracy and fitness of the model; however, more iterations are required [51, 52]. Similarly, to improve the accuracy and efficiency of the approximation from the iteration process, part of the training sample is randomly selected in each iteration, and the proportion of selected samples in the training set is called the sub-sampling rate [48]. The structure of the GBDT model is shown in Fig. 1.

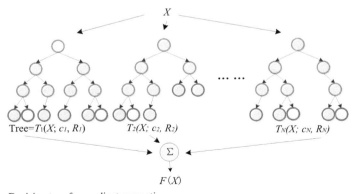

FIG. 1   Decision tree for gradient promotion.

The implementation process of GBDT algorithm can be described as follows [48]:

(1) Model initialization, $F_0 = \arg \min_\gamma \mathcal{L}(y_i, \gamma)$. Set the model parameters, iteration number M, learning rate v, and sub-sampling rate.

(2) For $m = 1$ to M do:

    a. Calculate the current residual. $r_{im} = -\left[\dfrac{\partial \mathcal{L}(y_i, F(x_i))}{\partial F(x_i)}\right]_{F=F_{m-1}}$, $i = 1$, $2,\ldots,N'$, where $N' = \sigma \cdot N$, the product of the subsampling rate, and the sample tree;

    b. Train CART tree Tm with current residual set $\{(x_i, r_{im})\}_{i=1}^{N'}$;

    c. Calculate the step size, $\gamma_m = \arg \min_\gamma \sum_{i=1}^{n} \mathcal{L}(y_i, F_{m-1}(x_i) - \gamma T_m)$;

    d. Update the model, $F_m(x) = F_{m-1}(x) + v\gamma_m T_m(x)$;

(3) Output F_m.

Although GBDT has the advantages of high prediction accuracy and can deal with nonlinear problems, it also has disadvantages. For example, GBDT is not suitable for processing high-dimensional data. GBDT cannot carry out parallel training of data. Therefore, it is necessary to reduce the dimensionality of high-dimensional data in order to better apply GBDT algorithm. The improvement of GBDT algorithm is also a way to achieve better results.

4 Case study: COD prediction model based on the GBDT algorithm

4.1 Data source

The municipal sewage data used in this study are from a sewage treatment plant in Guangzhou that can treat 5 million tons of sewage annually. The temperature of the wastewater varies from 15°C to 30°C, with an average annual temperature of 20°C. This study is based on data from the A2O process wastewater treatment plant, and the required data are obtained from the historical database of the plant. The sampling frequency was 1 min, and data for 20 days were collected, for a total of 28,800 data points. The data obtained include water inflow, water inlet temperature, water inlet pH, water inlet ammonia nitrogen, water inlet COD, and inlet COD load. The GBDT algorithm was used to establish a prediction model for sewage water quality. The inputs for the model were the water inflow, water inlet temperature, water inlet pH, water inlet ammonia nitrogen, and water inlet COD, and the output was the water inlet COD load. A total of 75% of the data was used as a training set, and 25% of the data was used as a test set to predict 7200 data points. The data analysis and modeling work were performed in Python3.6.

4.2 Model evaluation index

In this study, the mean absolute percentage error (MAPE) and root mean square error (RMSE), which are commonly used evaluation indexes, were used to determine the performance of the prediction model. The MAPE is used to represent the accuracy of the model; the smaller the MAPE, the higher is the accuracy of the model. The RMSE is used to represent the stability of the model; the smaller the RMSE, the higher the stability. The determination coefficient is R^2, which has a maximum value 1. The closer R^2 is to 1, the smaller the prediction error. MAPE, RMSE, and R^2 are calculated as follows:

$$MAPE = \sum_{t=1}^{n} \left| \frac{true - predict}{true} \right| \times \frac{100}{n} \tag{8}$$

$$RMSE = \sqrt{\frac{1}{n}\sum_{t=1}^{n}(true - predict)^2} \tag{9}$$

$$R^2 = 1 - \frac{\sum (true - p\hat{r}edict)^2}{\sum (true - \overline{predict})^2} \tag{10}$$

4.3 Model parameter selection

Four parameters are used to adjust this model: learning_rate, N_estimators, max_DEPTH, and min_samples_split.

4.3.1 Learning_rate

The learning_rate is used to reduce the contribution of each tree and is one of the key parameters of the model; therefore, it is necessary to select a reasonable value to improve the prediction accuracy of the model.

As can be seen from Figs. 2–4, with the learning rate increasing from 0.1 to 0.9, the three evaluation indices of the model fluctuated; however, the overall trend was that the prediction results of the model became worse. When the learning rate was equal to 0.1, all three evaluation indexes obtained the optimal value; the RMSE was 165.99, MAPE was 0.12, and R^2 was 0.99. Therefore, the learning rate of this model was selected to be 0.1.

4.3.2 N_estimators

N_estimators is the number of gradient enhancements to perform. Gradient enhancement has a strong robustness to overfitting; therefore, a large number of gradient enhancements can usually achieve better results.

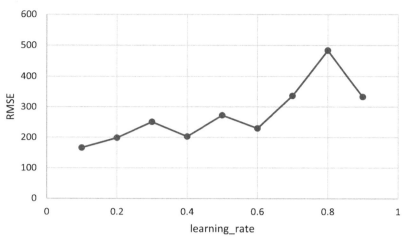

FIG. 2 RMSE for the model under different learning_rates.

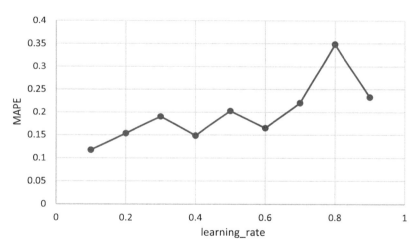

FIG. 3 MAPE for the model under different learning_rates.

With the increase in n_estimators, in general, the three indicators of the model became better; however, after the n_estimators increased to 2500, the accuracy of the model was not significantly improved. Thus, the results obtained at 2500 were the best (Figs. 5–7).

4.3.3 Max_depth

Max_depth is the maximum depth of a single regression estimator, which limits the number of nodes in the tree.

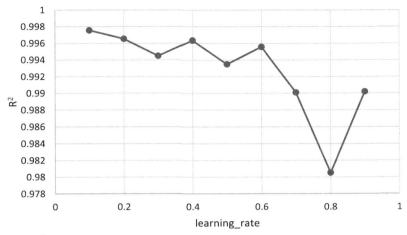

FIG. 4 R^2 for the model under different learning_rates.

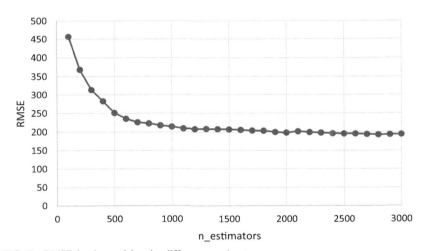

FIG. 5   RMSE for the model under different n_estimators.

As shown in Figs. 8–10, when the n_estimators increased to 2000, the RMSE and R^2 of the model were slightly better when the max_depth was equal to 4. However, when the max_depth was equal to 3, the MAPE was significantly better than when the max_depth was equal to 4; therefore, it was considered that the model was most accurate when the max_depth was equal to 3.

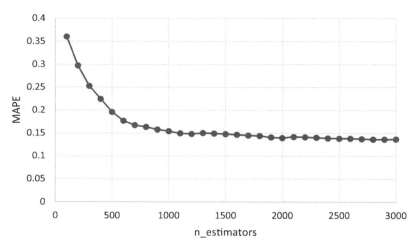

FIG. 6 MAPE for the model under different n_estimators.

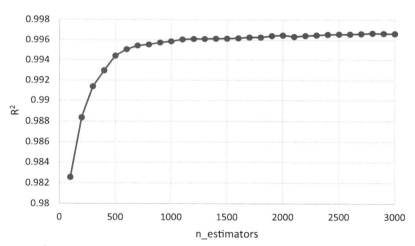

FIG. 7 R^2 for the model under different n_estimators.

4.3.4 Min_samples_split

The min_samples_split is the minimum number of samples required to split an internal node.

From Figs. 11–13, it can be seen that after the n_estimators reached 2000, the min_samples_split had little impact on the model accuracy; thus, the min_samples_split was chosen to be 2.

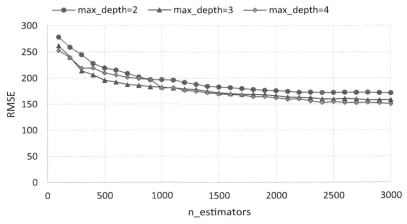

FIG. 8   RMSE for the model under different values of max_depth.

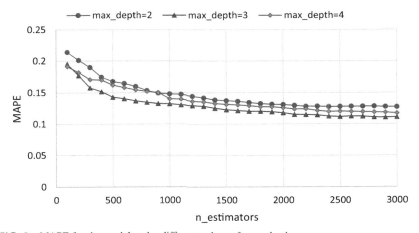

FIG. 9   MAPE for the model under different values of max_depth.

5   Discussion of results

Based on the model parameters mentioned in Section 4.3, 75% of the data were used to train the model, and an inlet COD load of 7200 points was predicted. The results are shown in Fig. 14. As shown in Fig. 15. From Fig. 14, we can intuitively find that the predicted result is very close to the actual value. Even in places where water quality fluctuates, the model's prediction is still good. This indicates that the model has strong adaptability to data. This is very

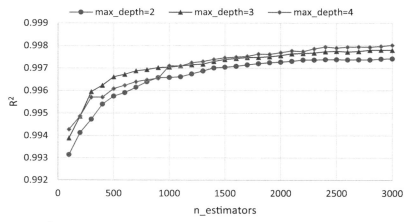

FIG. 10 R^2 for the model under different values of max_depth.

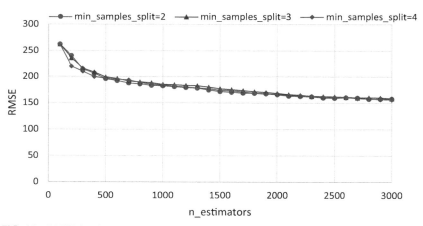

FIG. 11 RMSE for the model under different min_samples_splits.

important, because the sewage data is not smooth and orderly usually. The maximum relative error does not reach -1.5%, and the overall relative error range is $[-1.4\%, 0.8\%]$. The values for the MSE, MAPE, and R^2 are 161.12, 0.116, and 0.998, respectively. This prediction effect was excellent. The results were compared with other sewage load forecasting cases and were shown in Table 1. Although facing different objects, MAPE value can be used to compare the prediction effect of different models. It can be seen from Table 1 that the MAPE value of the prediction results obtained by this model is much lower than that of

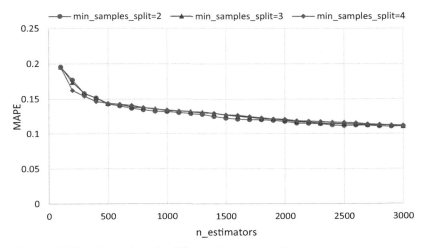

FIG. 12   MAPE for the model under different min_samples_splits.

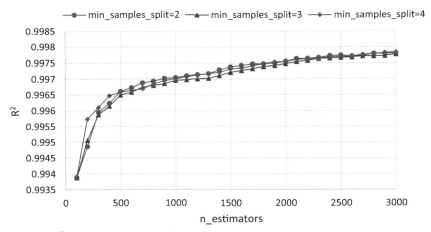

FIG. 13 R^2 for the model under different min_samples_splits.

other models. This indicates that the prediction effect of this model is better than VAR+ARMA model, LSSVM model, HANN model, ANN model, SCFL model.

6   Conclusion

To achieve the goal of reducing energy consumption in sewage treatment plants under the premise of reaching the effluent quality standard, it is necessary to

Result

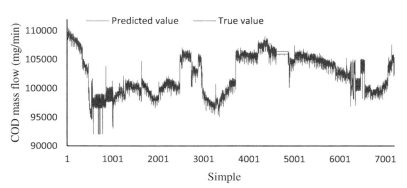

FIG. 14 Prediction results.

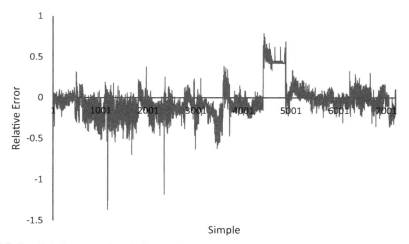

FIG. 15 Relative error of prediction results.

apply an artificial intelligence algorithm. It is feasible to apply artificial intelligence algorithms in sewage treatment plants, and this can achieve a good parameter prediction effect, which can help sewage treatment plants to realize effective management without adding sensors.

This study takes the prediction of COD load as a case, and the results show that the prediction effect is excellent. Specifically, the value with the largest relative error does not reach -1.5%, while the overall forecast relative error

TABLE 1 The results of different researches.

Core algorithms	Application scenario	MAPE (%)	References
GBDT	Wastewater	0.116	This study
VAR+ARMA	Wastewater	1.08	[4]
LSSVM		1.54	
HANN	Drinking water	20.3	[53]
ANN	Petroleum refinery wastewater	15.82	[54]
SCFL	Wastewater	3.66	[12]

range is $[-1.4\%, 0.8\%]$, the RMSE is 161.12, MAPE is 0.116, and R^2 is 0.998. In this study, a COD load prediction was taken as a case, and highly satisfactory results were obtained. Similarly, water quality parameters such as BOD, TSS, TN, and TP could also be predicted. Artificial intelligence algorithms can predict water quality parameters that are difficult to measure in real time. Using artificial intelligence algorithms to solve problems in sewage treatment plants will continue to be a developing trend. The development of sewage treatment plants is inseparable from the application of artificial intelligence algorithms.

References

[1] K. Murashko, M. Nikku, E. Sermyagina, et al., Techno-economic analysis of a decentralized wastewater treatment plant operating in closed-loop. A finnish case study, J. Water Process Eng. 25 (2018) 278–294.
[2] Z. Wang, Y. Man, Y. Hu, J. Li, M. Hong, P. Cui, A deep learning based dynamic COD prediction model for urban sewage, Environ. Sci.: Water Res. Technol. 5 (2019) 2210–2218.
[3] K. Solon, E.I.P. Volcke, M. Spérandio, M.C.M. van Loosdrecht, Resource recovery and wastewater treatment modelling, Environ. Sci.: Water Res. Technol. 5 (2019) 631–642.
[4] Y. Man, Y. Hu, J. Ren, Forecasting COD load in municipal sewage based on ARMA and VAR algorithms, Resour. Conserv. Recycl. 144 (2019) 56–64.
[5] Republic of China, China Energy Statistical Yearbook (2018), https://doi.org/10.1111/j.1462-5822.2009.01366.x.
[6] D. Pan, W. Hong, F. Kong, Efficiency evaluation of urban wastewater treatment: evidence from 113 cities in the Yangtze River Economic Belt of China, J. Environ. Manage. 110940 (2020).
[7] G. Niu, X. Yi, C. Chen, X. Li, D. Han, B. Yan, M. Huang, G. Ying, A novel effluent quality predicting model based on genetic-deep belief network algorithm for cleaner production in a full-scale paper-making wastewater treatment, J. Clean. Prod. 265 (2020) 121787.
[8] N. Hvala, J. Kocijan, Design of a hybrid mechanistic/Gaussian process model to predict full-scale wastewater treatment plant effluent, Comput. Chem. Eng. 140 (2020) 106934.

[9] K. Lotfi, H. Bonakdari, I. Ebtehaj, F.S. Mjalli, M. Zeynoddin, R. Delatolla, B. Gharabaghi, Predicting wastewater treatment plant quality parameters using a novel hybrid linear-nonlinear methodology, J. Environ. Manage. 240 (2019) 463–474.

[10] X. Wang, H. Ratnaweera, J.A. Holm, V. Olsbu, Statistical monitoring and dynamic simulation of a wastewater treatment plant: a combined approach to achieve model predictive control, J. Environ. Manage. 193 (2017) 1–7.

[11] H.W. Chen, R.F. Yu, S.K. Ning, et al., Forecasting effluent quality of an industry wastewater treatment plant by evolutionary Grey dynamic model, Resour. Conserv. Recycl. 54 (2010) 235–241.

[12] A.A. Nadiri, S. Shokri, F.T.C. Tsai, et al., Prediction of effluent quality parameters of a wastewater treatment plant using a supervised committee fuzzy logic model, J. Clean. Prod. 180 (2018) 539–549.

[13] J. Zeng, J.F. Liu, Economic model predictive control of wastewater treatment processes, Ind. Eng. Chem. Res. 54 (21) (2015) 5710–5721.

[14] H.W. Lee, M.W. Lee, J.M. Park, Robust adaptive partial least squares modeling of a full-scale industrial wastewater treatment process, Ind. Eng. Chem. Res. 46 (3) (2007) 955–964.

[15] V. Vasilaki, V. Conca, N. Frison, A.L. Eusebi, F. Fatone, E. Katsou, A knowledge discovery framework to predict the N_2O emissions in the wastewater sector, Water Res. 178 (2020) 115799.

[16] J. Tomperi, E. Koivuranta, K. Leiviskä, Predicting the effluent quality of an industrial wastewater treatment plant by way of optical monitoring, J. Water Process Eng. 16 (2017) 283–289.

[17] L. Liangfang, Z. Tao, Y. Yongquan, L. Shangfang, Extension cluster prediction model used in predicting wastewater emissions, in: 2008 International Symposium on Computer Science and Computational Technology, Shanghai, 2008, pp. 72–75.

[18] A. Verma, X. Wei, A. Kusiak, Predicting the total suspended solids in wastewater: a data-mining approach, Eng. Appl. Artif. Intell. 26 (2013) 1366–1372.

[19] F.I. Turkdogan-Aydınol, K. Yetilmezsoy, A fuzzy-logic-based model to predict biogas and methane production rates in a pilot-scale mesophilic UASB reactor treating molasses wastewater, J. Hazard. Mater. 182 (2010) 460–471.

[20] H. Liu, Y. Zhang, H. Zhang, Prediction of effluent quality in papermaking wastewater treatment processes using dynamic kernel-based extreme learning machine, Process Biochem. 97 (2020) 72–79.

[21] S.Y. Chan, Y.F. Tsang, L.H. Cui, H. Chua, Domestic wastewater treatment using batch-fed constructed wetland and predictive model development for NH_3-N removal, Process Biochem. 43 (2008) 297–305.

[22] C. Zhi-ming, H. Jue, Wastewater treatment prediction based on chaos-GA optimized LS-SVM, in: 2011 Chinese Control and Decision Conference (CCDC), Mianyang, 2011, pp. 4013–4016.

[23] Z. Huang, J. Luo, X. Li, Y. Zhou, Prediction of effluent parameters of wastewater treatment plant based on improved least square support vector machine with PSO, in: 2009 First International Conference on Information Science and Engineering, Nanjing, 2009, pp. 4058–4061.

[24] Y. Bao-lei, Z. De-an, Z. Jun, Prediction system of sewage outflow COD based on LS-SVM, in: 2011 2nd International Conference on Intelligent Control and Information Processing, Harbin, 2011, pp. 399–402.

[25] G. Zhao, N. Li, B. Li, W. Li, Y. Liu, T. Chai, ANN model for predicting acrylonitrile wastewater degradation in supercritical water oxidation, Sci. Total Environ. 704 (2020) 135336.

[26] F. Schmitt, R. Banu, I.-T. Yeom, K.-U. Do, Development of artificial neural networks to predict membrane fouling in an anoxic-aerobic membrane bioreactor treating domestic wastewater, Biochem. Eng. J. 133 (2018) 47–58.

[27] L. Luolong, Y. Zhouliyou, Xuyuge, Predicting wastewater sludge recycle performance based on fuzzy neural network, in: 2011 International Conference on Networking, Sensing and Control, Delft, 2011, pp. 266–269.

[28] H. Han, J. Qiao, Hierarchical neural network modeling approach to predict sludge volume index of wastewater treatment process, IEEE Trans. Control Syst. Technol. 21 (6) (2013) 2423–2431.

[29] W. Ran, J. Qiao, X. Ye, Soft-measuring approach to on-line predict BOD based on PCA time-delay neural network, in: Fifth World Congress on Intelligent Control and Automation (IEEE Cat. No. 04EX788), Hangzhou, China, vol. 4, 2004, pp. 3483–3487.

[30] S. Revollar, H. Álvarez, R. Lamanna, P. Vega, A. Goldar, Economic model predictive control of a wastewater treatment plant using neural and fuzzy models, Comput. Aided Chem. Eng. 43 (2018) 1237–1242.

[31] T.Y. Pai, P.Y. Yang, S.C. Wang, M.H. Lo, C.F. Chiang, J.L. Kuo, H.H. Chu, H.C. Su, L.F. Yu, H.C. Hu, Y.H. Chang, Predicting effluent from the wastewater treatment plant of industrial park based on fuzzy network and influent quality, Appl. Math. Model. 35 (2011) 3674–3684.

[32] R. Samli, V.Z. Sonmez, N. Sivri, Modeling the toxicity of textile industry wastewater using artificial neural networks, in: 2017 Electric Electronics, Computer Science, Biomedical Engineerings' Meeting (EBBT), Istanbul, 2017, pp. 1–5.

[33] J. Liang, F. Luo, R. Yu, Y. Xu, Wastewater effluent prediction based on fuzzy-rough sets RBF neural networks, in: 2010 International Conference on Networking, Sensing and Control (ICNSC), Chicago, IL, 2010, pp. 393–397.

[34] Z. Fu, J. Cheng, M. Yang, J. Batista, Wastewater discharge quality prediction using stratified sampling and wavelet de-noising ANFIS model, Comput. Electr. Eng. 85 (2020) 106701.

[35] A.G. El-Din, D.W. Smith, A neural network model to predict the wastewater inflow incorporating rainfall events, Water Res. 36 (2002) 1115–1126.

[36] A.R. Picos-Benítez, B.L. Martínez-Vargas, S.M. Duron-Torres, E. Brillas, J.M. Peralta-Hernández, The use of artificial intelligence models in the prediction of optimum operational conditions for the treatment of dye wastewaters with similar structural characteristics, Process Saf. Environ. Prot. 143 (2020) 36–44.

[37] M. Zounemat-Kermani, D. Stephan, R. Hinkelmann, Multivariate NARX neural network in prediction gaseous emissions within the influent chamber of wastewater treatment plants, Atmos. Pollut. Res. 10 (2019) 1812–1822.

[38] B.S. Hopkins, A. Mazmudar, C. Driscoll, M. Svet, J. Goergen, M. Kelsten, N.A. Shlobin, K. Kesavabhotla, Z.A. Smith, N.S. Dahdaleh, Using artificial intelligence (AI) to predict postoperative surgical site infection: a retrospective cohort of 4046 posterior spinal fusions, Clin. Neurol. Neurosurg. 192 (2020) 105718.

[39] P. Roca, A. Attye, L. Colas, A. Tucholka, Artificial intelligence to predict clinical disability in patients with multiple sclerosis using FLAIR MRI, Diagn. Interv. Imaging 101 (2020) 795–802.

[40] H. Nguyen, X.-N. Bui, Soft computing models for predicting blast-induced air over-pressure: a novel artificial intelligence approach, Appl. Soft Comput. 92 (2020) 106292.

[41] M. Saud, Al-Fattah, A new artificial intelligence GANNATS model predicts gasoline demand of Saudi Arabia, J. Pet. Sci. Eng. 194 (2020) 107528.

[42] D. Bhattacharya, J. Mukhoti, A. Konar, Learning regularity in an economic time-series for structure prediction, Appl. Soft Comput. 76 (2019) 31–44.

[43] Y. Jian, L. Xinying, Z. Man, L. Han, Photosynthetic rate prediction of tomato plant population based on PSO and GA, IFAC-PapersOnLine 51 (2018) 61–66.

[44] I. Daniyan, K. Mpofu, M. Oyesola, B. Ramatsetse, A. Adeodu, Artificial intelligence for predictive maintenance in the railcar learning factories, Procedia Manuf. 45 (2020) 13–18.

[45] H. Yan, J. Zhang, N. Zhou, M. Li, Application of hybrid artificial intelligence model to predict coal strength alteration during CO_2 geological sequestration in coal seams, Sci. Total Environ. 71 (2020) 135029.

[46] J. Zhang, L. Dong, Y. Wang, Predicting uniaxial compressive strength of oil palm shell concrete using a hybrid artificial intelligence model, J. Build. Eng. 30 (2020) 101282.

[47] N. Cotrufo, E. Saloux, J.M. Hardy, J.A. Candanedo, R. Platon, A practical artificial intelligence-based approach for predictive control in commercial and institutional buildings, Energy Build. 206 (2020) 109563.

[48] J.H. Friedman, Stochastic gradient boosting, Comput. Stat. Data Anal. 38 (4) (2002) 367–378.

[49] J. Elith, J.R. Leathwick, T. Hastie, A working guide to boosted regression trees, J. Anim. Ecol. 77 (4) (2008) 802–813.

[50] J. Ye, J.H. Chow, J. Chen, et al., Stochastic gradient boosted distributed decision trees, in: Proceedings of the 18th ACM Conference on Information and Knowledge Management, ACM, 2009, pp. 2061–2064.

[51] S. Touzani, J. Granderson, S. Fernandes, Gradient boosting machine for modeling the energy consumption of commercial buildings, Energy Build. 158 (2018) 1533–1543.

[52] C. Persson, P. Bacher, T. Shiga, et al., Multi-site solar power forecasting using gradient boosted regression trees, Sol. Energy 150 (2017) 423–436.

[53] Y.Y. Zhang, X. Gao, K. Smith, G. Inial, S.M. Liu, L.B. Conil, B.C. Pan, Integrating water quality and operation into prediction of water production in drinking water treatment plants by genetic algorithm enhanced artificial neural network, Water Res. 164 (2019) 114888.

[54] G. Hayder, M.Z. Ramli, M.A. Malek, A. Khamis, N.M. Hilmin, Prediction model development for petroleum refinery wastewater treatment, J. Water Process Eng. 4 (2014) 1–5.

Chapter 8

Application of machine learning algorithms to predict the performance of coal gasification process

Zeynep Ceylan[a] and Selim Ceylan[b]

[a]*Samsun University, Faculty of Engineering, Industrial Engineering Department, Samsun, Turkey,* [b]*Ondokuz Mayıs University, Faculty of Engineering, Chemical Engineering Department, Samsun, Turkey*

1 Introduction

Gasification technology is a popular strategy for power, energy, or chemical production due to its product range and positive effect on the environment compared to conventional combustion technologies. The gasification process can be defined as the thermal oxidation of carbon-rich raw materials, such as coal or biomass to produce valuable gas products. The coal gasification process is a widespread technology due to the abundance and availability of coal reserves on a global scale. Coal gasification provides the conversion of solid coal to gas by simultaneously occurring the complex heterogeneous gas–solid reactions involved in the formation of synthesis gas. The synthesis gas (SynGas) mainly consists of carbon dioxide (CO_2), hydrogen (H_2), carbon monoxide (CO), water vapor (H_2O), methane (CH_4), and other gaseous hydrocarbons that can be used as fuel or chemical sources. Also, several products are formed during the process, such as a small amount of ash and condensable tar [1]. The main process is the reaction of the carbon of the coal or biomass with air and steam at pressures below 10 MPa and at temperatures above 750 °C to form synthesis gas with a high proportion of CO and H_2 and less CO_2 and CH_4 [2].

Gasification performance depends on feeding type, process conditions, and gasifier configuration. An in-depth understanding of the fuel properties that affect gasification behavior is required to ensure optimum process efficiency. Thus, the characteristics of the feed determined based on the proximate analysis (fixed carbon, volatile matter, ash, and moisture) and ultimate analysis (carbon,

Applications of Artificial Intelligence in Process Systems Engineering
https://doi.org/10.1016/B978-0-12-821092-5.00003-6

hydrogen, oxygen, nitrogen, and sulfur) provide a pre-evaluation of the fuel quality. The volatile matter content is proportional to the gasification reactivity of the solid fuel. In contrast, the ash content negatively affects the gasification process due to slag and agglomeration, which limits the transfer of heat and mass in the gasification reactor. The carbon, hydrogen, and fixed carbon content promotes the quantity and quality of the gas product. The moisture in the composition of fuel affects the gasifier efficiency due to the endothermic nature of water vaporization and the water-gas shift reactions that occur during gasification. The temperature to be selected in the gasification process has an important effect on the gas production rate and gas composition. The high amount of heat required to perform this high endothermic reaction at elevated temperatures can be achieved by burning the coal outside or by burning some of the coal in the reactor using oxygen. At higher temperatures, the formation of CO_2 and its conversion into CO increases, resulting in an increase in syngas quality and generation rate. However, as with any chemical process, it is appropriate to choose a temperature by considering economy and efficiency together. Coal or biomass feeding rates affect the residence time of the fuel particles in the gasification reactor and thus the product composition. The ratio of air to fuel is another important parameter in a gasification system. Selecting a higher air/fuel ratio than the optimum causes the existence of excess air in the gas product and thus obtaining a gas composition result in lower thermal value. Therefore, the performance of the gasification process is adversely affected. The vapor/fuel ratio increases the H_2 production by increasing the partial pressure of H_2O in the gasifier. Thus, it supports water-gas, water-gas drift, and methane reformation reactions. In addition, the presence of high H_2 levels in the product increases the heating value of the product gas [3]. Apart from all these parameters, different gasification reactors with different structures and operating principles have an essential effect on product composition.

Considering all these parameters, many experiments are required to design a high-performance gasifier. However, gasification experiments are costly, time-consuming, and labor-intensive. Besides, insufficient and ineffective experiments or analyzes can cause problems in installing, operating, or optimizing systems. This problem can be overcome by using modeling tools. A performance assessment for the gasification system can be carried out through appropriately developed models. Modeling of thermochemical processes can be performed by examining the systems in terms of thermodynamic and kinetic events. In general, models based on process thermodynamics ignore the limitations of heat and mass transfer by considering the gasification medium as homogeneous. The models produced based on these assumptions are independent of the gasifier type. However, models derived regardless of the gasifier type are not useful to explain and simulate the effects of process conditions on gasifier efficiency and performance for each system. The kinetic models developed for the systems represent the kinetics of various complex reactions that occur during the gasification process. The kinetic model can predict process performance in terms of product composition and temperature profiles at various operating conditions. However, the development of such models requires a detailed clarification of the multiple physicochemical events affecting the process [3]. The raw material structure, processing conditions, and the nature of the gasification process create a complex system. Therefore, in order to model the system based on thermodynamic and kinetic principles, it is necessary to examine various complex endothermic and exothermic reactions that occur during gasification with comprehensive and costly experiments. Besides, ideal assumptions in the development of models create significant difficulties in the development of a successful phenomenological process model [4].

In order to overcome problems arising from over-simplifications and the assumptions made to model the systems, artificial intelligence (AI) methods are widely applied as reliable tools for the analysis and prediction of complex and nonlinear systems [5]. AI techniques provide accurate and precise information in a shorter time at low cost. Machine learning (ML) is a sub-branch of AI and has been used in a wide variety of applications where conventional models or algorithms are insufficient to perform the required tasks [6].

ML models have shown very successful results in many engineering areas that require to understand complex patterns with raw-level data [7]. ML algorithms use computational techniques to "learn" information directly from raw data without prior knowledge of an equation or model. ML algorithms investigate correlations in data to reveal associations that provide a better understanding of the systems. These algorithms help researchers make better decisions and make more accurate predictions with lower cost and shorter analysis time [8].

In recent years, especially with the development of artificial intelligence, many models have been developed to predict the gasification process outputs with high reliability. These models range from traditional methods, such as multivariate regression (MVR) to advanced techniques, such as Artificial Neural Networks (ANNs), Nonlinear Auto-Regressive Exogenous Neural Networks (NARXNNs), and Support Vector Regression (SVR). Table 1 summarizes the details of various studies available in the literature [3, 4, 6, 9–20]. As seen in Table 1, the ANNs model, which performs well especially in large datasets, is widely used for estimating the different outcomes of the gasification process. Besides, with the development of AI, different approaches based on ML have been developed to effectively predict system outcomes with limited experimental data [21].

In this chapter, we aimed to show the application potential of different ML algorithms in coal gasification systems. According to the literature, there are several studies that apply different ML algorithms in biomass gasification systems. However, the application of ML approaches for the coal gasification process is limited. Therefore, in this study, we applied various ML algorithms, such as Sequential Minimal Optimization Regression, Gaussian Process Regression, Lazy K-Star, Lazy IBk, Alternating Model Tree, Random Forest, and M5Rules to create regression models on coal gasification data. To the best of our knowledge, this is the first attempt to use and compare multiple ML methods to predict the outputs of the coal gasification process.

TABLE 1 Previous studies on the estimation of prediction ofs of the gasification process.

Reference	Feedstock	Reactor	Method(s)	Inputs	Outputs
Tiwary et al. [3]	Coal-biomass blends	Fluidized bed gasifier	– GP, – ANNs, – SVR	T, Coal-biomass feed rates, FC, VM, ash, MC, air/fuel ratio, steam/fuel ratio, C, H, O, Co-gasification reaction rate constant	– Total gas yield – LHV of gas – CCE – CGE
Puig-Arnavat et al. [4]	Biomass	– Circulating fluidized bed gasifier – Bubbling fluidized bed gasifier	– ANNs	Ash, MC, C, O, H, ER, and T	– CO, CO_2, H_2, and CH_4 – Gas yield
Elmaz et al. [6]	Biomass	Fixed bed downdraft gasifier	– PR – SVR, – DTR, – -MLP	ER, fuel flow rate, temperature distribution, C, H, O, N, MC, VM, FC, and ash	– CO, CO_2, H_2, and CH_4 – HHV
Yucel et al. [9]	Biomass	Fixed bed downdraft gasifier	– ANNs – NARX	FC, VM, ash, MC, air/fuel ratio, steam/fuel ratio, C, H, O, ER, air flow rate, temperature distribution, and reduction temperature	– CO, CO_2, H_2, and CH_4 – Calorific value
Ozbas et al. [10]	Biomass	Fixed bed updraft gasifier	– LR, – KNN, – SVR, – DTR	Time, CO, CO_2, CH_4, O_2, and heating value, T	– H_2 production
Serrano et al. [11]	Biomass	Bubbling fluidized bed gasifier	– BPNN	C, H, O, MC, ash, ER, reaction temperature, steam/biomass mass ratio, and bed material	– CO, CO_2, H_2, and CH_4 – Gas yield

Reference	Feedstock	Gasifier type	Model	Inputs	Outputs
Ozonoh et al. [12]	Coal, Biomass, and Coal-Biomass Blends	Fluidized bed gasifier	– ANNs	C, H, N, O, ash, MC, S, VM, LHV_{fuel}, ER, and T	– CO, CO_2, H_2, and CH_4 – LHV of gas – Gas yield – CGE – CCE
Elmaz and Yücel [13]	Biomass	Fixed bed downdraft gasifier	– NARXNN	C, H, O, N, FC, VM, MC, ash, and ER	– CO, CO_2, H_2, and CH_4 – HHV – T_0
Nougués et al. [14]	Coal	Fluidized bed gasifier	– ANNs	Coal feed, airflow, heating power, and water flow	– Syngas composition – Reactor temperature – Differential pressure loss through the bed
Chavan et al. [15]	Coal	Fluidized bed gasifier	– MVR – ANNs	FC, VM, MM, air feed per kg of coal, steam feed per kg of coal, and T	– Gas yield – The heating value
Patil-Shinde et al. [16]	Coal	Fluidized bed gasifier	– GP – ANNs	Fuel ratio, ash, specific surface area of coal, activation energy of gasification, coal feed rate, gasifier bed temperature, ash discharge rate, air/coal ratio	– CO+H_2 generation rate – Syngas production rate – Carbon conversion – Heating value
Pandey et al. [17]	Biomass	Fluidized bed gasifier	ANNs	C, H, N, S, O, MC, Ash, ER, and T	– LHV of product gas, – LHVp – Gas yield
Baruah et al. [18]	Biomass	Fixed bed downdraft gasifier	ANNs	C, H, O, ash, MC, reduction zone temperature	– CO, CO_2, H_2, and CH_4

Continued

TABLE 1 Previous studies on the estimation of prediction ofs of the gasification process—cont'd

Reference	Feedstock	Reactor	Method(s)	Inputs	Outputs
George et al. [19]	Biomass	Bubbling fluidized bed gasifier	ANNs	C, H, O, MC, ash, ER, and T	– CO, CO_2, H_2, and CH_4
Mutlu and Yucel [20]	Biomass	Fixed bed downdraft gasifier	– RF – LS-SVM	Temperature distribution, ER, and fuel flow rate	– CO, CO_2, H_2, and CH_4 – HHV

Abbreviations: ANNs, artificial neural networks; BPNN, back-propagation neural networks; C, carbon; CO_2, carbon dioxide; CO, carbon monoxide; CCE, carbon conversion efficiency; CGE, cold gas efficiency; DTR, decision tree regression; ER, equivalence ratio; FC, fixed carbon; GP, genetic programming; H_2, hydrogen; H_2O, water vapor; H_4, methane; HHV, higher heating value; KNN, K-nearest neighbors; LHV, lower heating value; LHVp, lower heating value of gasification products including tars and entrained char; LR, linear regression; LS-SVM, least-squares support vector machine; MC, moisture content; MM, mineral matter; MLP, multilayer perceptron; MVR, multivariate regression; N, nitrogen; NARX, non-linear autoregressive exogenous; NARXNN, nonlinear auto-regressive exogenous neural network; O, oxygen; PR, polynomial regression; RF, random forest; S, sulfur; SVR, support vector regression; T, temperature; VM, volatile matter.

2 Materials and methods

2.1 Dataset description

In this study, 106 coal gasification data obtained from 18 fluidized bed gasifiers were used. The gasification data was taken from a previously published paper by Chavan et al. [15]. The provided data consist of 6 inputs (i.e., independent factors) and two outputs (i.e., response factors). The proximate analysis and gasification process parameters were used to predict the product gas yield (GY) and the product gas heating value (HV). The proximate analysis includes the mineral matter (MM), fixed carbon (FC), and volatile matter (VM) contents. These are the main parameters of the coal and are evaluated during any coal gasification run as they provide a primary assessment of the coal quality. On the other hand, bed temperature (T), air feed per kg of coal (AF), and steam feed per kg of coal (SF) are used as gasification process parameters. The summary of the statistical evaluation of the dataset is shown in Table 2.

TABLE 2 Descriptive statistics of the variables.

			Statistics			
Inputs	Symbol	Unit	Min	Max	Mean	St. dev[a]
Mineral matter[b]	MM	(wt%)	9.05	49.98	27.32	10.14
Fixed carbon	FC	(wt%)	25.83	61.00	39.20	11.41
Volatile matter	VM	(wt%)	21.29	49.56	33.81	6.47
Air feeding	AF	(Nm^3/kg coal)	1.34	5.84	2.35	0.82
Steam feeding	SF	(kg/kg coal)	0.10	0.63	0.35	0.11
Temperature	T	(°C)	720.00	980.00	858.80	62.14
Outputs						
Gas yield	GY	(Nm^3/kg coal)	1.90	5.55	2.90	0.74
Heating value	HV	(MJ/Nm^3)	1.42	5.50	3.91	0.94

[a]Standard deviation.
[b]Mineral matter (wt%) = 1.1 × Ash (wt%).

2.2 Data preparation

The analysis was performed using WEKA (Waikato Environment for Knowledge Analysis) software. WEKA has data pre-processing, regression, classification, association rules, clustering, and visualization tools. It includes different open-source ML algorithms that are used to solve real-world data mining problems. Although many ML algorithms have been developed, their performance largely depends on the quality of the data. However, raw data is often irrelevant, messy, or incomplete and needs to be repaired or removed. Therefore, data preparation (also called "data pre-processing") is an important step that helps make the dataset suitable for ML.

The outliers and extreme values can negatively impact the prediction performance of the ML algorithms. Therefore, they must be removed as they do not represent the behavior of the underlying system. In WEKA, the outlier values fall outside the boundaries by three times interquartile ranges, while extreme values fall outside by six times the boundaries. In this study, outliers and extreme values were determined using the "*InterquartileRange*" filter and then discarded from the dataset.

Data normalization is significant to create a better learning experience under different data structures. Thus, all inputs and outputs data were normalized using the min-max normalization technique to provide constant variability and minimize the effects of high variance variables on the prediction results. The min-max normalization formula is as in Eq. (1):

$$Z_{norm} = \frac{Z_i - Z_{min}}{Z_{max} - Z_{min}} \tag{1}$$

where Z_i and Z_{norm} represent the experimental value of the i_{th} data point and normalized data, respectively. Z_{max} and Z_{min} show the maximum and minimum values of the data, respectively.

In order to check the robustness and prediction ability of the ML models, the normalized data were randomly divided into two sets, namely training and testing sets. In this regard, 85% of the dataset was used for training, and the remaining 15% for testing the models.

2.3 Evaluation criteria

The ML algorithms were evaluated based on five well-known statistical indices, including the coefficient of determination (R^2), mean absolute error (MAE), root mean square error (RMSE), relative absolute error (RAE in %), and root relative squared error (RRSE in %). In this study, the model with the lowest MAE, RMSE, RAE, and RRSE and the highest R^2 values is considered the best candidate model. The formulation of these criteria is presented in Eqs. (2–6) [8, 22].

$$R^2 = 1 - \frac{\sum_{i=1}^{n}\left(Yi_{predicted} - Yi_{observed}\right)^2}{\sum_{i=1}^{n}\left(Yi_{observed} - \overline{Y}_{observed}\right)^2} \qquad (2)$$

$$MAE = \frac{1}{n}\sum_{i=1}^{n}\left|Yi_{observed} - Yi_{predicted}\right| \qquad (3)$$

$$RMSE = \sqrt{\frac{1}{n}\sum_{i=1}^{n}\left(Yi_{observed} - Yi_{predicted}\right)^2} \qquad (4)$$

$$RAE = \frac{\sum_{i=1}^{n}\left|Yi_{predicted} - Yi_{observed}\right|}{\sum_{i=1}^{n}\left|Yi_{observed} - \overline{Y}_{observed}\right|} \qquad (5)$$

$$RRSE = \sqrt{\frac{\sum_{i=1}^{n}\left(Yi_{predicted} - Yi_{observed}\right)^2}{\sum_{i=1}^{n}\left(Yi_{observed} - \overline{Y}_{observed}\right)^2}} \qquad (6)$$

where $Yi_{predicted}$, $Yi_{observed}$, and $\overline{Y}_{observed}$ represent the predicted, observed, and the average of the observed values of coal gasification performance parameters, respectively. Also, n denotes the number of samples.

2.4 Modeling

In the literature, ANNs are frequently used for the estimation of gasification process outputs [3, 4, 9, 12, 14–18]. However, there are many powerful algorithms developed in ML to predict system outputs by learning system characteristics effectively. Thus, ML techniques can contribute to the discovery of the rules and characteristics underlying the coal gasification process and predict process outcomes.

In this study, seven ML algorithms belonging to different classes in WEKA are presented; function-based, lazy, tree-based, and rule-based techniques. The methods selected from such classes are *sequential minimal optimization regression (SMOreg)* and *Gaussian process regression (GPR)* from function-based models; *K-Star* (K^*) and *IBk* from lazy models; *Alternating Model Tree (AMT)* and *Random Forest (RF)* from tree-based models; *M5Rules* from rule-based models. In this study, the hyperparameter values used in the ML methods were set as default in WEKA software and are presented in Table 3. The mentioned ML algorithms are briefly explained in the subsections from 2.4.1 to 2.4.7. The general process used in this study is shown in Fig. 1.

2.4.1 Sequential minimal optimization-based support vector regression

Sequential minimal optimization-based regression (SMOreg) is an algorithm in WEKA which implements support vector machines (SVM) for the regression problems [23]. Support Vector Regression (SVR) was first proposed by Cortes and Vapnik in 1995 [24]. It is a widely popular ML algorithm in classification

TABLE 3 Hyperparameters of the ML methods.

Method	Parameter	Value
GPR	Kernel type	PUK[a]
	Kernel parameter, ω	1
	Kernel parameter, σ	1
	Noise	1
SMOreg	Kernel type	PUK[a]
	Kernel parameter, ω	1
	Kernel parameter, σ	1
	The complexity parameter, C	1
	Regoptimizer	RegSMOImproved
Lazy K-star	Global blend	20
	Missing mode	Average column entropy curves
Lazy IBk	Number of neighbors	1
	Nearest neighbor search algorithm	LinearNNSearch
AMT	Number of iterations	10
	Shrinkage	1
RF	Maximum depth of the tree	0
	Number of execution slots	1
	Number of trees	100
M5Rules	Minimum number of instances	4

[a]PUK kernel was selected because it gives the best performance compared to other kernel functions.

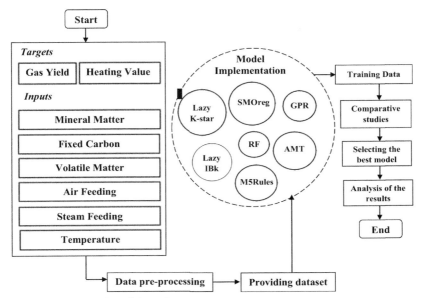

FIG. 1 A schematic view of the study.

and regression problems [25, 26]. It provides good generalization performance even in small datasets. SVR uses a nonlinear mapping procedure to map input data to a higher dimensional feature space. For the data transformation, a kernel function is used. The SVR function is detailed in Eq. (7) [27]:

$$f(x) = \sum_i^k (\alpha_i - \alpha_i^*) k(x_i, x_j) + b \qquad (7)$$

where x_i and x_j are the i_{th} and j_{th} input vectors of dimension D, $k(x_i, x_j)$ is the kernel function, α_i and α_i^* are Lagrangian multipliers, b is the intercept.

In the literature, there are different kernel functions in different forms. The most commonly used kernel functions are linear, Pearson VII kernel function (PUK), polynomial, Gaussian (Radial basis function, RBF), and sigmoid [28]. In this study, PUK kernel function was used with SMOreg for the training of the model as it gives the best performance compared to others. Also, the popular algorithm RegSMOImproved was used for parameter settings due to its short implementation time [29]. The PUK function can be expressed as [30]:

$$k(x_i x_j) = \frac{1}{\left[1 + \left(\frac{\sqrt[2]{\|x_i - x_j\|^2} \sqrt{2^{(1/\omega)} - 1}}{\sigma} \right)^2 \right]^\omega} \qquad (8)$$

where x_i and x_j are the i_{th} and j_{th} input vectors of dimension D, respectively, $k(x_i, x_j)$ is the kernel function, ω and σ are the actual shape and width parameters, respectively.

2.4.2 Gaussian process regression

Gaussian process regression (GPR) was first introduced by Rasmussen and Williams [31]. GPR is used effectively for both regression and probabilistic classification problems. It is a nonparametric kernel-based probabilistic algorithm. It can handle complex relationships between inputs and outputs using the Gaussian process [32]. Gaussian process (GP) is a set of random variables, such that any finite number of which has a joint Gaussian distribution. GP is defined by its mean function $\mu(x)$ and a kernel (covariance) function $k(x, x')$ [31].

$$y = g(x) \sim GP(\mu(x), k(x, x')) \tag{9}$$

$$\mu(x) = E[f(x)] \tag{10}$$

$$k(x, x') = E[(f(x) - \mu(x))(f(x') - \mu(x'))] \tag{11}$$

where y is the target values, $E[.]$ denotes expectation. Typically, the mean function is considered to be zero while the kernel function is restricted to generate positive matrices [33]. The generated matrix by applying the kernel function to an input variables vector can then be expressed as $K(x, x')$. The Eq. (9) can be shown as follows [31]:

$$K(x, x') = \begin{bmatrix} k(x_1, x_1) \, k(x_1, x_2) \ldots k(x_1, x_n) \\ k(x_2, x_1) \, k(x_2, x_2) \ldots k(x_2, x_n) \\ \vdots \ddots \vdots \\ k(x_n, x_1) \, k(x_n, x_2) \ldots k(x_n, x_n) \end{bmatrix} \tag{12}$$

$$\begin{bmatrix} g(x_1) \\ \vdots \\ g(x_n) \end{bmatrix} \sim \left(\begin{bmatrix} \mu(x_1) \\ \vdots \\ \mu(x_n) \end{bmatrix} \begin{bmatrix} k(x_1, x_1) & \ldots & k(x_1, x_n) \\ \vdots & \ddots & \vdots \\ k(x_n, x_1) & \ldots & k(x_n, x_n) \end{bmatrix} \right) \tag{13}$$

WEKA software includes polynomial, RBF, and PUK kernel functions to solve problems. In this study, the PUK kernel function was used due to its high performance.

2.4.3 Lazy K-star

K-star (K^*) is one of the lazy learner algorithms used for various classification problems [34]. Lazy learning is a local learning approach that stores training instances, and these instances do not operate until classification time [35]. K^* is an instance-based classifier that applies the k-Nearest Neighbor ($k-NN$) method. In this algorithm, n instances are partitioned into k clusters according to the nearest mean procedure. The entropy distance measure is used for placing a new instance to the nearest class. New instances, x, are assigned to the closest class, y_j, where $j = 1, \ldots, k$. The K^* formulation is defined as follows [22]:

$$K^*(y_i, x) = - \ln P^*(y_i, x) \tag{14}$$

where P^* is the probability that x will reach y through a random path. The use of entropy as a measure distance allows the K^* algorithm to take a consistent approach to handle real-valued, symbolic, and missing attributes [36]. In many classification problems in the literature, K^* provided satisfactory results. However, choosing an appropriate k value can be difficult, and the method can give different results if the size and density of the clusters change [37]. Detailed information about K^* algorithm can be referenced in the literature [36].

2.4.4 Lazy IBk

IBk algorithm is instance-based learning that implements the k-NN classifier [38]. IBk algorithm is a nonparametric method used in both classification and regression problems [39, 40]. The IBk algorithm is a lazy learner that does not build a model from the training dataset; instead, it generates a prediction for a test sample [41]. The IBk algorithm relies on a distance measure, such as Euclidean, Minkowski, Manhattan, or Chebyshev to locate k "close" instances in the training data for each test sample. Then, it uses the selected cases to make predictions. In this algorithm, "k" is the number of nearest neighbors and can be determined precisely in the object editor or adjusted automatically based on cross-validation. Various search algorithms in WEKA can be used to find nearest neighbors, such as ball tree, cover tree, linear search, etc.

2.4.5 Alternating model tree

An alternating model tree (AMT) is a new ensemble learning algorithm proposed by Frank et al. [42]. AMT is an alternative decision tree with a single tree structure [43]. In recent years, it has been used in many different studies due to its high accuracy in classification and regression problems [22, 44]. There are two different nodes in AMT implementation: a *splitter node* and a *prediction node* [45]. The splitter node is the same as an internal node in a standard decision tree. This node aims to split numerical attributes by the median value of the attribute to reduce computational complexity and data fragmentation. The predictor node estimates the numeric output at that node using simple linear regression functions. In AMT, the tree is grown by forward stage-wise additive modeling instead of a boosting algorithm to minimize squared error. Also, the size of the tree, which depends on the number of iterations of the forward additive modeling, is determined using cross-validation [42].

2.4.6 Random forest

Random forest (RF) is an ensemble learning algorithm, which was first proposed by Breiman [46]. RF algorithm trains a large number of individual decision trees and combines them to make a more accurate prediction. It uses a random sampling technique and a subset of the training data to create the final model. In this method, many decision trees are created, and a bootstrap sample technique is used to train each tree from the set of training data. Finally, the RF algorithm selects the class with the highest number of votes.

The RF algorithm is widely used in the literature due to its satisfactory performance compared to other ML methods [47]. It has advantages, such as fast training and overcoming the over-fitting problems associated with decision tree methods. Besides, the RF method is easy to develop as it does not require too much hyperparameter settings or feature scaling [48]. In this study, the RF algorithm consisting of 100 decision trees was used to estimate the gasification outputs of coal (Table 3).

2.4.7 M5Rules

The M5Rules model is a nonparametric decision tree learner used for various classification and prediction problems [49]. Especially in regression problems, it is among the most effective rule-based learning techniques. It has a straightforward working method that extracts if-then rules from model trees. Due to its fast working, it is used in many applications of ML [50]. This technique generates a decision list for regression problems using a divide-and-conquer strategy [51]. M5Rules follows the following steps [52]:

- A model tree is built using the M5 model tree to the full training dataset.
- The tree is pruned in different ways, the best leaf selected, and the tree is discarded.
- The best leaf is turned into an IF-THEN rule. All instances covered by the rule from the dataset are removed.
- Until all instances are included in the rules, the previous procedure is repeated.

3   Results and discussion

3.1   Comparison of ML algorithms

Total gas yield is a crucial performance parameter that measures the rate of conversion of solid coal to syngas. On the other hand, the heating value of the product gas is important for energy applications and represents the energy content of the syngas. Thus, both outputs need to be estimated in order to reliably evaluate the gasification performance of coal. As mentioned earlier, the primary motivation of this study is to provide a comprehensive ML analysis for the prediction of product gas generation and product gas heating value using coal properties and gasification process parameters. All models were created for predicting single output, and all methods were applied using an 85/15% splitting strategy. The reliability of each model was measured by five accuracy criteria, namely, R^2,

MAE, RMSE, RAE%, and RRSE%. The calculated values are given in Tables 4 and 5 for heating value and gas yield outputs, respectively.

From the results in Tables 4 and 5, it is seen that most of the ML algorithms have good correlations, achieving high values ($R^2 > 0.90$) in both training and testing stages. However, the GPR and M5Rules models performed less than other ML algorithms to estimate both outputs.

As can be seen in Table 4, RF and lazy K-star performed better than other ML algorithms to predict HV output. In the training stage, the statistical indexes (R^2, MAE, RMSE, RAE, and RRSE) results showed that the best correlations and the lowest errors were obtained with the RF (0.9851, 0.017, 0.0312, 7.8621%, and 12.4322%) and K-star (0.9722, 0.0333, 0.0449, 15.3684%, and 17.8912%) models, respectively. Similarly, R^2, MAE, RMSE, RAE (%), and RRSE (%) values for the RF and lazy K-star models in the test dataset were

TABLE 4 Prediction performances of ML models for estimating the product gas heating value.

Dataset	Method	Statistics				
		R^2	MAE	RMSE	RAE	RRSE
Train set	K-star	0.9722	0.0333	0.0449	15.3684	17.8912
	GPR	0.8580	0.0824	0.1022	38.0578	40.7031
	SMOreg	0.9295	0.0423	0.0669	19.5179	26.6482
	AMT	0.9270	0.0521	0.0679	24.0652	27.0464
	RF	0.9851	0.0170	0.0312	7.8621	12.4322
	M5Rules	0.8339	0.0756	0.1025	34.8952	40.8475
	IBk	0.9655	0.0327	0.0468	15.1065	18.6372
Test set	K-star	0.9594	0.0421	0.0576	19.5772	24.4067
	GPR	0.8122	0.0709	0.1017	32.9253	43.0827
	SMOreg	0.9157	0.0560	0.0748	26.0216	31.6753
	AMT	0.9099	0.0614	0.0753	28.5462	31.8839
	RF	0.9730	0.0338	0.0451	15.7148	19.1181
	M5Rules	0.8010	0.0788	0.1102	36.6381	46.6895
	IBk	0.9430	0.0459	0.0599	21.3483	25.3491

The RAE and RRSE values are in (%).

TABLE 5 Prediction performances of ML models for estimating the product gas generation.

Dataset	Method	R^2	MAE	RMSE	RAE	RRSE
				Statistics		
Train set	K-star	0.9833	0.0212	0.0308	12.3452	13.5939
	GPR	0.9090	0.0435	0.0684	25.2969	30.1745
	SMOreg	0.9635	0.0317	0.0434	18.4633	19.1553
	AMT	0.9604	0.0328	0.0452	19.0657	19.9503
	RF	0.9998	0.0015	0.0028	0.8872	1.2385
	M5Rules	0.8699	0.0606	0.0817	35.2436	36.0542
	IBk	0.9722	0.0282	0.0380	16.4058	16.7653
Test set	K-star	0.9763	0.0363	0.0475	14.9424	16.8627
	GPR	0.8894	0.0632	0.0965	26.0295	34.3041
	SMOreg	0.9430	0.0521	0.0726	21.4569	25.7808
	AMT	0.9424	0.0544	0.0768	22.3995	27.2960
	RF	0.9928	0.0214	0.0258	8.8001	9.1592
	M5Rules	0.8391	0.0818	0.1156	33.7033	41.0620
	IBk	0.9590	0.0445	0.0594	18.3532	21.1062

The RAE and RRSE values are in (%).

(0.9730, 0.0338, 0.0451, 15.7148%, and 19.1181%) and (0.9594, 0.0421, 0.0576, 19.5772%, and 24.4067%), respectively.

It was observed that for the GY output, the RF algorithm has the highest R^2 value (0.9998) and the lowest MAE (0.0015), RMSE (0.0028), RAE (0.8872%), and RRSE (1.2385%) values in the training set. These values show the superiority of the RF model compared to other methods. In this regard, the best results were obtained with the RF model where the R^2, MAE, RMSE, RAE, and RRSE values of the testing dataset were 0.9928, 0.0214, 0.0258, 8.8001%, and 9.1592%, respectively.

According to the above results, it can be said that the RF technique is the more accurate and consistent compared to the other six models, which has also been found in many previous studies [53–56]. To better understand the predictive success of the RF model, the actual values were plotted against the predicted values in Fig. 2.

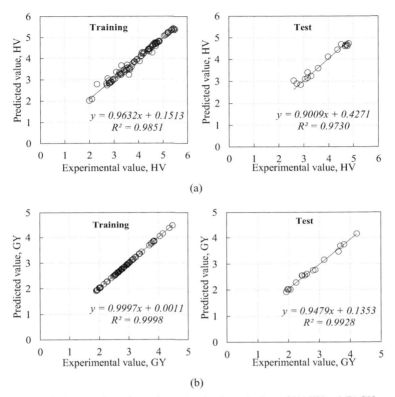

FIG. 2   Graphical comparison of experimental and estimated values of (A) HV and (B) GY outputs with the RF model.

3.2   Comparison with previous studies

In this section, a review of existing models in the literature was conducted to analyze the performance of the RF model, which performs well for both outputs of the coal gasification process. Table 6 gives the performance comparison of the models based on R^2 and RMSE values.

As can be seen in Table 6, there are several studies that apply different mathematical models to predict the outputs of various gasification system, such as biomass and coal. However, unlike biomass gasification systems, very few researchers have focused on estimating the outputs of the coal gasification system.

As mentioned before, the ANNs model is widely used in gasification studies [3, 6, 11, 12, 15]. Although the ANNs model obtains relatively satisfactory results for predicting gasification products, it needs a large number of experimental samples in the training dataset to achieve high performance. Also, it has several issues, such as choosing the appropriate network type and parameters

TABLE 6 Performance comparison of the developed RF model with the literature.

Output	Reference	Method	Accuracy	
			R^2	RMSE
Heating value	Tiwary et al. [3]	GP	0.951	0.256
		MLP	0.938	0.281
		SVR	0.951	0.252
	Elmaz et al. [6]	SVR	0.886	0.552
		DTR	0.921	0.286
		PR (Quadratic)	0.713	0.866
		PR (Cubic)	0.858	0.610
		MLP	0.931	0.221
	Chavan et al. [15]	MVR	0.677	0.435
		ANNs	0.893	0.250
	This study	RF	0.973	0.034
Gas yield	Puig-Arnavat et al. [4]	ANNs	0.992	0.075
	Serrano et al. [11]	FFBP	0.987	0.017
	Chavan et al. [15]	MVR	0.901	0.216
		ANNs	0.943	0.162
	This study	RF	0.993	0.026

Abbreviations: ANNs, artificial neural networks; DTR, decision tree regression; FFBP, feedforward backpropagation; GP, genetic programming; LR, linear regression; MLP, multilayer perceptron; MVR, multivariate regression; PR, polynomial regression; RF, random forest; SVR, support vector regression.

and overfitting of the data [57]. To overcome the shortcoming of ANNs, many studies use powerful ML tools that can reveal nonlinear relationships with high accuracy even in small dataset.

As seen in Table 6, the RF model showed the best performance among the other prediction models. The main reason why RF model is superior to the other models is that RF has an ensemble learning structure. RF uses the bootstrap sampling method to create Ntotal sub-data for each decision tree, and then estimates are obtained by averaging each tree prediction. In addition, it may be due to the characteristic of the RF model that can handle unbalanced samples [58]. All these features make the learning process more stable and robust, reducing the impact of noise data and the possibility of over-fitting.

4 Conclusions and future perspectives

In this study, various ML models were investigated to predict gas yield and heating value from coal gasification in a fluidized bed gasifier. Various physicochemical properties of coal and process parameters were used as inputs. It was seen that the performance of the examined models was in good agreement with the experimental data sets. Thus, it can be said that ML algorithms can be applied as an alternative method for modeling nonlinear systems, such as thermochemical processes with complex reaction mechanisms. Analysis results showed that the RF model performed significantly better than the rest of the methods used in this study with R^2, MAE, RMSE, RAE (%), and RRSE (%) values for gas yield output (0.9928, 0.0214, 0.0258, 8.8001%, and 9.1592%) and for heating value output (0.9730, 0.0338, 0.0451, 15.7148%, and 19.1181%), respectively.

In conclusion, ML models can be used to model gasification products without further adaptation. According to analysis results, the RF method is appropriate for understanding, exploring, and making accurate predictions of the gasification process. The findings of this study may enable researchers in academia and industry to learn more about gasifier systems that can be used in both energy and chemical production. The authors believe that tuning the hyperparameters of ML algorithms with optimization algorithms to achieve higher performance could be a good topic for future studies. Besides, sensitivity analysis can be applied to obtain the best structure of the used methods.

References

[1] S. Ding, Z. Li, H. Zhou, Y. Hu, L. Liu, Municipal solid waste gasification for environmental management: a modeling study, energy sources, Part A Recover. Util. Environ. Eff. 41 (2019) 1640–1648, https://doi.org/10.1080/15567036.2018.1549139.

[2] M.J. Chadwick, N.H. Highton, N. Lindman (Eds.), 6—Coal conversion technologies, in: Environmental Impacts of Coal Mining & Utilization, 1st, Amsterdam, Pergamon, 1987, pp. 105–155.

[3] S. Tiwary, S.B. Ghugare, P.D. Chavan, S. Saha, S. Datta, G. Sahu, S.S. Tambe, Co-gasification of high ash coal–biomass blends in a fluidized bed gasifier: experimental study and computational intelligence-based modeling, Waste Biomass Valoriz. 11 (2020) 323–341, https://doi.org/10.1007/s12649-018-0378-7.

[4] M. Puig-Arnavat, J.A. Hernández, J.C. Bruno, A. Coronas, Artificial neural network models for biomass gasification in fluidized bed gasifiers, Biomass Bioenergy 49 (2013) 279–289.

[5] Z. Ceylan, E. Pekel, S. Ceylan, S. Bulkan, Biomass higher heating value prediction analysis by ANFIS, PSO- ANFIS and GA-ANFIS, Global Nest J. 20 (2018) 589–597.

[6] F. Elmaz, Ö. Yücel, A.Y. Mutlu, Predictive modeling of biomass gasification with machine learning-based regression methods, Energy 191 (2020) 116541, https://doi.org/10.1016/j.energy.2019.116541.

[7] G. Kakavas, N. Malliaropoulos, R. Pruna, N. Maffulli, Artificial intelligence. A tool for sports trauma prediction, Injury (2019), https://doi.org/10.1016/j.injury.2019.08.033.

[8] Z. Ceylan, Estimation of coal elemental composition from proximate analysis using machine learning techniques, energy sources, Part A Recover. Util. Environ. Eff. 42 (2020) 2576–2592, https://doi.org/10.1080/15567036.2020.1790696.

[9] O. Yucel, E.S. Aydin, H. Sadikoglu, Comparison of the different artificial neural networks in prediction of biomass gasification products, Int. J. Energy Res. 43 (2019) 5992–6003, https://doi.org/10.1002/er.4682.

[10] E.E. Ozbas, D. Aksu, A. Ongen, M.A. Aydin, H.K. Ozcan, Hydrogen production via biomass gasification, and modeling by supervised machine learning algorithms, Int. J. Hydrog. Energy 44 (2019) 17260–17268, https://doi.org/10.1016/j.ijhydene.2019.02.108.

[11] D. Serrano, I. Golpour, S. Sánchez-Delgado, Predicting the effect of bed materials in bubbling fluidized bed gasification using artificial neural networks (ANNs) modeling approach, Fuel 266 (2020) 117021, https://doi.org/10.1016/j.fuel.2020.117021.

[12] M. Ozonoh, B.O. Oboirien, A. Higginson, M.O. Daramola, Performance evaluation of gasification system efficiency using artificial neural network, Renew. Energy 145 (2020) 2253–2270, https://doi.org/10.1016/j.renene.2019.07.136.

[13] F. Elmaz, Ö. Yücel, Data-driven identification and model predictive control of biomass gasification process for maximum energy production, Energy 195 (2020) 117037, https://doi.org/10.1016/j.energy.2020.117037.

[14] J.M. Nougués, Y.G. Pan, E. Velo, L. Puigjaner, Identification of a pilot scale fluidised-bed coal gasification unit by using neural networks, Appl. Therm. Eng. 20 (2000) 1561–1575, https://doi.org/10.1016/S1359-4311(00)00023-5.

[15] P.D. Chavan, T. Sharma, B.K. Mall, B.D. Rajurkar, S.S. Tambe, B.K. Sharma, B.D. Kulkarni, Development of data-driven models for fluidized-bed coal gasification process, Fuel 93 (2012) 44–51, https://doi.org/10.1016/j.fuel.2011.11.039.

[16] V. Patil-Shinde, T. Kulkarni, R. Kulkarni, P.D. Chavan, T. Sharma, B.K. Sharma, S.S. Tambe, B.D. Kulkarni, Artificial intelligence-based modeling of high ash coal gasification in a pilot plant scale fluidized bed gasifier, Ind. Eng. Chem. Res. 53 (2014) 18678–18689, https://doi.org/10.1021/ie500593j.

[17] D.S. Pandey, S. Das, I. Pan, J.J. Leahy, W. Kwapinski, Artificial neural network based modelling approach for municipal solid waste gasification in a fluidized bed reactor, Waste Manag. 58 (2016) 202–213, https://doi.org/10.1016/j.wasman.2016.08.023.

[18] D. Baruah, D.C. Baruah, M.K. Hazarika, Artificial neural network based modeling of biomass gasification in fixed bed downdraft gasifiers, Biomass Bioenergy 98 (2017) 264–271, https://doi.org/10.1016/j.biombioe.2017.01.029.

[19] J. George, P. Arun, C. Muraleedharan, Assessment of producer gas composition in air gasification of biomass using artificial neural network model, Int. J. Hydrog. Energy 43 (2018) 9558–9568, https://doi.org/10.1016/j.ijhydene.2018.04.007.

[20] A.Y. Mutlu, O. Yucel, An artificial intelligence based approach to predicting syngas composition for downdraft biomass gasification, Energy 165 (2018) 895–901, https://doi.org/10.1016/j.energy.2018.09.131.

[21] D.S. Pandey, I. Pan, S. Das, J.J. Leahy, W. Kwapinski, Multi-gene genetic programming based predictive models for municipal solid waste gasification in a fluidized bed gasifier, Bioresour. Technol. 179 (2015) 524–533, https://doi.org/10.1016/j.biortech.2014.12.048.

[22] W. Gao, J. Alsarraf, H. Moayedi, A. Shahsavar, H. Nguyen, Comprehensive preference learning and feature validity for designing energy-efficient residential buildings using machine learning paradigms, Appl. Soft Comput. J. 84 (2019) 105748, https://doi.org/10.1016/j.asoc.2019.105748.

[23] J. Platt, Fast training of support vector machines using sequential minimal optimization, Advances in Kernel Method: Support Vector Learning, MIT Press, Cambridge, MA, United States, 1999, pp. 185–208.

[24] C. Cortes, V. Vapnik, Support-vector networks, Mach. Learn. 20 (1995) 273–297.

[25] Z. Ceylan, S. Bulkan, S. Elevli, Prediction of medical waste generation using SVR, GM (1,1) and ARIMA models: a case study for megacity Istanbul, J. Environ. Heal. Sci. Eng. 18 (2) (2020) 687–697, https://doi.org/10.1007/s40201-020-00495-8.

[26] Z. Ceylan, Assessment of agricultural energy consumption of Turkey by MLR and Bayesian optimized SVR and GPR models, J. Forecast. 39 (6) (2020) 944–956, https://doi.org/10.1002/for.2673.

[27] A.J. Smola, B.S.C.H. Olkopf, A Tutorial on Support Vector Regression *, 2004, pp. 199–222.

[28] R. Gujar, V. Vakharia, Prediction and validation of alternative fillers used in micro surfacing mix-design using machine learning techniques, Constr. Build. Mater. 207 (2019) 519–527, https://doi.org/10.1016/j.conbuildmat.2019.02.136.

[29] S.K. Shevade, S.S. Keerthi, C. Bhattacharyya, K.R.K. Murthy, Improvements to the SMO algorithm for SVM regression, IEEE Trans. Neural Netw. 11 (2000) 1188–1193, https://doi.org/10.1109/72.870050.

[30] K. Pearson, X. Contributions to the mathematical theory of evolution.—II. Skew variation in homogeneous material, Philos. Trans. R. Soc. Lond. A (1895) 343–414.

[31] C.K.I. Williams, C.E. Rasmussen, Gaussian Processes for Machine Learning, MIT Press Cambridge, MA, 2006.

[32] Z. Ceylan, Estimation of municipal waste generation of Turkey using socio-economic indicators by Bayesian optimization tuned Gaussian process regression, Waste Manag. Res. 38 (8) (2020) 840–850, https://doi.org/10.1177/0734242X20906877.

[33] N. Lawrence, Probabilistic non-linear principal component analysis with Gaussian process latent variable models, J. Mach. Learn. Res. 6 (2005) 1783–1816.

[34] S. Ravikumar, H. Kanagasabapathy, V. Muralidharan, Fault diagnosis of self-aligning troughing rollers in belt conveyor system using k-star algorithm, Meas. J. Int. Meas. Confed. 133 (2019) 341–349, https://doi.org/10.1016/j.measurement.2018.10.001.

[35] N.S. Altman, An introduction to kernel and nearest-neighbor nonparametric regression, Am. Stat. 46 (1992) 175–185.

[36] J.G. Cleary, L.E. Trigg, K*: An instance-based learner using an entropic distance measure, in: A. Prieditis (Ed.), Machine Learning Proceedings 1995, Morgan Kaufmann, San Francisco, CA, 1995, pp. 108–114, https://doi.org/10.1016/B978-1-55860-377-6.50022-0.

[37] S.S. Roy, D. Kaul, R. Roy, C. Barna, S. Mehta, A. Misra, Prediction of customer satisfaction using naive Bayes, MultiClass classifier, K-star and IBK, in: V.E. Balas, L.C. Jain, M.M. Balas (Eds.), Soft Computing Applications, Springer International Publishing, Cham, 2018, pp. 153–161.

[38] D.W. Aha, D. Kibler, M.K. Albert, Instance-based learning algorithms, Mach. Learn. 6 (1991) 37–66, https://doi.org/10.1007/BF00153759.

[39] B. Raza, A. Aslam, A. Sher, A.K. Malik, M. Faheem, Autonomic performance prediction framework for data warehouse queries using lazy learning approach, Appl. Soft Comput. J. 91 (2020) 106216, https://doi.org/10.1016/j.asoc.2020.106216.

[40] A. Chellam, L. Ramanathan, S. Ramani, Intrusion detection in computer networks using lazy learning algorithm, Proc. Comput. Sci. 132 (2018) 928–936, https://doi.org/10.1016/j.procs.2018.05.108.

[41] F.A. Ozbay, B. Alatas, Fake news detection within online social media using supervised artificial intelligence algorithms, Phys. A Stat. Mech. Appl. 540 (2020) 123174, https://doi.org/10.1016/j.physa.2019.123174.

[42] E. Frank, M. Mayo, S. Kramer, Alternating model trees, in: Proceedings of the 30th Annual ACM Symposium on Applied Computing 13–17-April, 2015, pp. 871–878, https://doi.org/10.1145/2695664.2695848.

[43] W.L. Ruzzo, Tree-size bounded alternation, J. Comput. Syst. Sci. 21 (1980) 218–235, https://doi.org/10.1016/0022-0000(80)90036-7.

[44] H. Moayedi, B. Aghel, L.K. Foong, D.T. Bui, Feature validity during machine learning paradigms for predicting biodiesel purity, Fuel 262 (2020) 116498, https://doi.org/10.1016/j.fuel.2019.116498.

[45] Y. Freund, L. Mason, The alternating decision tree learning algorithm, in: Icml, 1999, pp. 124–133.

[46] L. Breiman, Random forests, Mach. Learn. 45 (2001) 5–32, https://doi.org/10.1023/A:1010933404324.

[47] T. Niu, Y. Chen, Y. Yuan, Measuring urban poverty using multi-source data and a random forest algorithm: a case study in Guangzhou, Sustain. Cities Soc. 54 (2020) 102014, https://doi.org/10.1016/j.scs.2020.102014.

[48] S. Misra, H. Li, Chapter 9 - Noninvasive fracture characterization based on the classification of sonic wave travel times, in: S. Misra, H. Li, J. He (Eds.), Machine Learning for Subsurface Characterization, Gulf Professional Publishing, 2020, pp. 243–287.

[49] G. Holmes, M. Hall, E. Prank, Generating rule sets from model trees, in: N. Foo (Ed.), Advanced Topics in Artificial Intelligence, Springer, Berlin Heidelberg, Berlin, Heidelberg, 1999, pp. 1–12.

[50] I.H. Witten, E. Frank, Data mining: practical machine learning tools and techniques with Java implementations, ACM SIGMOD Rec. 31 (2002) 76–77.

[51] J.R. Quinlan, Learning with continuous classes, Aust. Jt. Conf. Artif. Intell. 92 (1992) 343–348.

[52] A.E. Charalampakis, G.C. Tsiatas, S.B. Kotsiantis, Machine learning and nonlinear models for the estimation of fundamental period of vibration of masonry infilled RC frame structures, Eng. Struct. 216 (2020) 110765, https://doi.org/10.1016/j.engstruct.2020.110765.

[53] H. You, Z. Ma, Y. Tang, Y. Wang, J. Yan, M. Ni, K. Cen, Q. Huang, Comparison of ANN (MLP), ANFIS, SVM, and RF models for the online classification of heating value of burning municipal solid waste in circulating fluidized bed incinerators, Waste Manag. 68 (2017) 186–197, https://doi.org/10.1016/j.wasman.2017.03.044.

[54] C. Lei, J. Deng, K. Cao, Y. Xiao, L. Ma, W. Wang, T. Ma, C. Shu, A comparison of random forest and support vector machine approaches to predict coal spontaneous combustion in gob, Fuel 239 (2019) 297–311, https://doi.org/10.1016/j.fuel.2018.11.006.

[55] J. Xing, K. Luo, H. Wang, Z. Gao, J. Fan, A comprehensive study on estimating higher heating value of biomass from proximate and ultimate analysis with machine learning approaches, Energy 188 (2019) 116077, https://doi.org/10.1016/j.energy.2019.116077.

[56] J. Xing, H. Wang, K. Luo, S. Wang, Y. Bai, J. Fan, Predictive single-step kinetic model of biomass devolatilization for CFD applications: a comparison study of empirical correlations (EC), artificial neural networks (ANN) and random forest (RF), Renew. Energy 136 (2019) 104–114, https://doi.org/10.1016/j.renene.2018.12.088.

[57] A.P. Piotrowski, J.J. Napiorkowski, A comparison of methods to avoid overfitting in neural networks training in the case of catchment runoff modelling, J. Hydrol. 476 (2013) 97–111, https://doi.org/10.1016/j.jhydrol.2012.10.019.

[58] A.T. Azar, H.I. Elshazly, A.E. Hassanien, A.M. Elkorany, A random forest classifier for lymph diseases, Comput. Methods Prog. Biomed. 113 (2014) 465–473, https://doi.org/10.1016/j.cmpb.2013.11.004.

Chapter 9

Artificial neural network and its applications: Unraveling the efficiency for hydrogen production

Sushreeta Paul[a], Vijay Kumar[b,c], and Priyanka Jha[a]
[a]*Amity Institute of Biotechnology, Amity University, Kolkata, West Bengal, India,* [b]*Plant Biotechnology Lab, Division of Research and Development, Lovely Professional University, Phagwara, Punjab, India,* [c]*Department of Biotechnology, Lovely Faculty of Technology and Sciences, Lovely Professional University, Phagwara, Punjab, India*

1 Introduction

Data models are specific models or arrangements that organize data and standardizes them based on the relation to each another and also in relation to the real world. Data models are designed for conceptual representation of data and also the relation between various data objects [1]. The main purpose behind any data model is designing of the data architecture in conceptual, physical, and logical levels. It helps to clear out redundant data and create a base data which would minimize labor as well as expenditure of time, in the long run.

A predictive model commonly uses statistical techniques that use data and statistics to predict data outcomes with data models. It predicts future events or outcomes by analyzing data patterns that would forecast future results [2]. It is also referred as predictive analysis or machine learning. It is a form of data mining technology that utilizes historical data to predict and generate future outcomes. The predictive analysis can learn eventually how various data points can interconnect with each other. The predictive models are of various types such as:

> *One-variable-at-a-time (OVAT)*: Also known as Monothetic analysis in which experiments are designed based on testing different factors, one at a time as an alternative to employing multiple factors simultaneously.
> *Response surface methodology (RSM)*: It is a model which uses mathematical and statistical method for analyzing a process in which the response or

Applications of Artificial Intelligence in Process Systems Engineering
https://doi.org/10.1016/B978-0-12-821092-5.00014-0
187

output of the process is affected by various variables to optimize the response.

Fuzzy logic theory: It is directed toward optimization of data "based on truths" rather than true or false which are mostly employed for decision making, route planning, and controlling systems.

Artificial neural network (ANN): The model is based on networking inspired by the neural networks of the nervous system inside the human brain along with its information storage property. The neural networks motivated by the human brain working mechanism, are formed with a web of organized interconnected nodes in hierarchy, hence building the base for artificial intelligence.

2 Artificial neural network

In biological terms, a human brain can perform various tasks such as pattern recognition, perception and motor control, estimation, and optimization. The ANNs are inspired by its close similarity to the working process of nervous systems which consists of simple but plentiful nerve cells working simultaneously with learning capability [3]. The ANNs collect information by identifying different data patterns and relationships and are trained via experience (feeding various data). An ANN is comprised of numerous single functional units known as artificial neurons or processing elements (PE). The processing elements are weighted links representing the neural structure arranged in layers.

However, before studying ANN in detail, it is essential to understand the functioning outline of the human brain as ANN is all about mimicking the functioning of the human brain [4]. A human brain consists of an estimated 100 billion nerve cells. Neurons are the primary essential functional units of the nervous system in a human body and are accountable for receiving sensory signals, which are short-lived impulses and cause spikes in membrane. Neurons communicate within each other through electrochemical impulses known as synapses. The synapses are produced between axon terminals on the signal dispatching neurons and the dendrites on the signal receiving neurons.

The classification of the nervous system is as follows:

Peripheral Nervous System (PNS): The PNS comprises of group of neurons and axons that branch out of the brain and spinal cord. The PNS mainly consists of spinal and cranial nerves and is predominantly distributed throughout the body.

Central Nervous System (CNS): The CNS is primarily made up of brain and spinal cord. It mainly functions by receiving sensory information from PNS and hence, controls the motor response.

Each neuron keeps on receiving a multitude of incoming signals which ultimately reach the cell body (Fig. 1). Then it is transmitted to other neurons via a fiber that gets eventually branched and is known as axon. An impulse

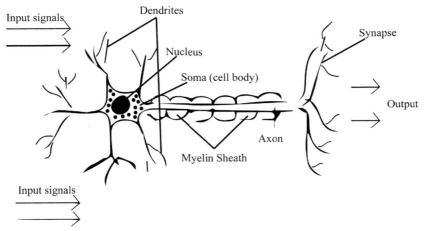

FIG. 1 Structure of a typical neuron. *(Modified after S. Mohammed, E.R. Abdessamad, R. Saadane, K.A. Hatim, Performance evaluation of spectrum sensing implementation using artificial neural networks and energy detection method, in: International Conference on Electronics, Control, Optimization and Computer Science (ICECOCS), Kenitra, 2018, pp. 1–6, https://doi. org/10.1109/ICECOCS.2018.8610506.)*

has both inhibitory and excitatory effects depending on the type of incoming signals which ultimately lead to impulse generation depending on its synaptic connections with other neurons. The neurons in turn, regulate the membrane potential for an active processing of the information.

Thus, we learn the concept of biological nervous systems that are linked to one another and are stimulated by electrical signals sent via axons. The impulse undergoes a series of biochemical changes after crossing the synaptic cleft to a postsynaptic nerve cell. Depending on these input signals, the output gets linearized or nonlinearized. This entire concept led to the framing of the ANN model. The ANN is a computational structure that mimics the structural arrangements and attributes of biological neural system. Signals or information passing through the neural network impacts the ANN structure as such networks learn to rely on the input provided and the output obtained [5]. The functioning capabilities of the ANNs maintain similarity to the neurons working mechanism in our nervous system and this term was coined by Warren S McCulloch and Walter Pitts in the early 1970s.

The basic mechanism in ANN starts with neurons with variable inputs providing outputs for various other neurons (Fig. 2). The neuron collates and combines the inputs and generates a single output that is compared to a training set also known as a set of known input and output values. The output value is either 0 or 1 or between 0 and 1. In case, if the output received from the neuron do not match the known value in the training set, then, the structure rearranges

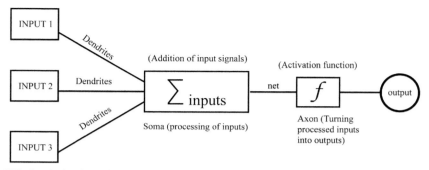

FIG. 2 A simple artificial neuron. *(Modified from M. Shariati, M.S. Mafipour, P. Mehrabi, A. Bahadori, Y. Zandi, M.N.A. Salih, H. Nguyen, J. Dou, X. Song, S. Poi-Ngian, Application of a hybrid artificial neural network-particle swarm optimization (ANN-PSO) model in behavior prediction of channel shear connectors embedded in normal and high-strength concrete, Appl. Sci. 9 (2019) 5534.)*

by adjusting the weights. The process repeats until the error lies within the threshold range while the weight fixes itself to existing position to be tested again against the dataset not used in training dataset.

3   Principle of ANN

A neural network principally consists of three layers such as input layer, hidden layer, and output layer. The input layer is the foremost layer in the ANN structure which primarily functions by receiving the information. The hidden layers are arranged as either a single layer or multiple layers in between the ANN structure (Fig. 3). The key function performed by hidden layers is to carry out different types of mathematical computations on the input and also recognizing the patterns through computation. The output layer then receives the analysis done by hidden layers after a thorough mathematical analysis [5].

Let us consider two input signals X_1 and X_2, and the weights assigned to them as W_1 and W_2, respectively.

Now each of these neurons are associated with a numerical value called the bias (b) which is added up to the input sum, that is

$$Y = (W_1X_1 + W_2X_2 + b) \tag{1}$$

The obtained sum is passed through a threshold function called "activation function." Artificial neural networks utilize not only activation function but also sigmoid function. The (activation) sigmoid function is expressed as:

$$\sigma(x) = e^x/(1 + e^x) \tag{2}$$

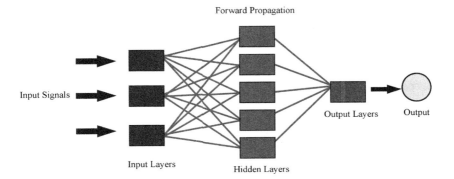

FIG. 3   Architecture of artificial neural network model. *(Modified from V. Laxmi, H. Rohil, License plate recognition system using back propagation neural network, Int. J. Comput. Appl. 99 (2014) 29.)*

Now the result of the activation function determines whether a particular neuron will get activated or not. The triggered signal can both be 0/1, or any real value between 0 and 1 depending upon the binary or real valued artificial neurons we are dealing with. The activated neuron transmits data to the next layer over the channels. This process by which the data is propagated to the next layer is called "forward propagation." In the output layer the neuron with the highest value eventually determines the ultimate output or result. The function which calculates the output is primarily composed of two parts. The first part evaluates the input signals and determines the total input/net input, whereas the second part propagates the net input signal, nonlinearly to obtain the output value, that is.

$$f : X \rightarrow Y \tag{3}$$

All these assigned values considered are basically only probabilities. When the assumed value turns out to be incorrect that is, when the magnitude of error of predicted output in comparison to the real output is more than anticipated, the information is channelized back to the network. This is termed as "backward propagation."

The main purpose behind backward propagation is changing the required weights so as to minimize error as much as possible.

$$dE/dW_1 = (dE/dA) \, X(dA/dS) \, X \, (dS/dW_1); \tag{4}$$

$$dE/dW_2 = (dE/dA) \, X \, (dA/dS) \, X \, (dS/dW_2) \tag{5}$$

[Here, dE = Loss function].

Once the old weights are adjusted, using gradient descent procedure, new weights are updated as:

$$W_1' = W_1 - \eta dE/dW_1 \tag{6}$$

$W_2' = W_2 - \eta dE/dW_2$, where W_1' and W_2' are new weights assigned, respectively.

The weights are then adjusted which eventually trains the network and minimizes the error rate, making the model more reliable. This cycle of forward propagation and backward propagation is repeatedly performed with multiple inputs until the predicted output matches with the actual output. The error rate is minimized so that the output close to its true value is obtained and this marks as the end of the training procedure.

The ANN method has been witnessed to work precisely when nonlinear dependence is considered between inputs and outputs. That does not restrict ANN for possible usage in finding the linear relationship, however, with less preciseness.

4   Methodology of ANN model

The key criteria determining success of the ANN model is the choice of ANN training and architectural parameters [6]. Back propagation algorithms are used for development and training of ANN models. Statistically, the developed ANN models are tested to check their prediction accuracy, and eventually the optimal ANN model is selected (Fig. 4) [6].

The steps followed to develop an ANN model consist of compilation and data preparation for training and validation of ANN model, subsequent selection of structural parameters followed by training of ANN model and lastly testing the model and result analysis.

4.1   Collection of data and its preparation

The initial step to develop an ANN model is to collect prepare the data. Various data forms are converted in the step of data preparation and converted to an ANN friendly format. Data are prepared via normalization process in binary order, that is, 0 to 1, or −1 to 1.

The following equation of normalization [7] is used:

$$x_{norm} = 2.\{(x - x_{min})/(x_{max} - x_{min})\} - 1 \tag{7}$$

where x is the data to be normalized and x_{min} and x_{max} are the lowest and highest values of the raw data and every input and output factors are normalized within the range of ± 1.

4.2   Selection of structural parameters

The training of the ANN model and the structural parameters are important factors that are carefully considered for optimal selection of ANN model.

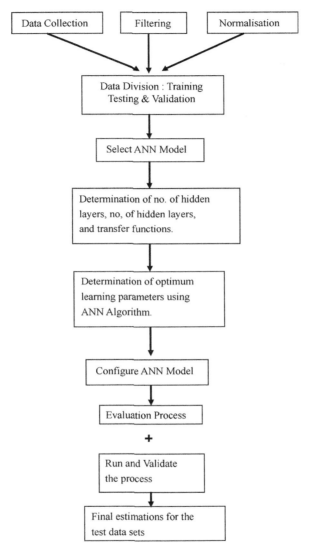

FIG. 4 Flowchart of ANN model methodology. *(Modified from H. Li, Z. Zhang, Z. Liu, Application of artificial neural networks for catalysis: a review, Catalysts 7 (2017) 306.)*

The trial/error method is mostly used for selection of parameter settings. The essential parameters under consideration are learning efficiency, momentum, number of neurons present in the hidden layers along with the exact number of hidden layers.

The learning efficiency or pace and momentum determines speed and effectiveness of the process, during which ANN adjusts its weight and hence, the training rate. Higher learning rate would result in increased convergence, but this process can be quite unstable and some distillations might occur. Identifying the exact number of hidden layers and neurons involved can be a tedious task. Any rise in the number of neurons leads to change the subsequent complexity, hence, affecting the training pace. However, an increase in neuron numbers in the hidden layers can solve complex datasets.

4.3   ANN model training

The ANN model training represents the input-output information according to BP algorithm using to modify the weights. Randomizations within specific intervals are done to establish initial values of weights. The initial values can also be established as per one of the widely used techniques known as Nguyen-Widrow method. Subsequently, the data obtained from the training data set are reiterated in random order and the weights are updated with every repetition of data as per the BP algorithm [8]. The iteration of the complete training dataset refers to one epoch and an ANN training requires such hundreds to thousands of epochs. However, in such cases, overtraining is a growing concern that can be avoided by early stopping and regularization methods [8]. The early stopping method is mostly used due to its simple understanding and better application over regularization methods. However, to use the early stopping method, the data requires to be separated into three sets such as training set, validation set, and test set.

The training set is employed to find the ANN weights whereas the validation set is a prerequisite for assessing ANN model performance and also to choose the precise time to stop the training procedure. The test set is employed to check ANN model efficiency.

4.4   Model testing and result analysis

All the ANN outputs are renormalized before assessing the performance of the ANN model by using the following equation [7]:

$$x_{denorm} = 1/2\,(o+1)\,(x_{\max} - x_{\min}) + x_{\min} \tag{8}$$

where o is the force obtained from ANN model, and $x_{\min}$ and $x_{\max}$ are the lowest and highest values of the raw data.

Training time is an important criterion for selection of ANN model and usually gets reduced with fewer free parameters [6]. The best performance obtained from the test results does not correlate to the best predictions and therefore, developed models are employed to obtain more precise analysis. Usually, smaller models are used for such analysis to bypass complexity issues in the hidden layers and assure better optimization results.

5   Applications of ANN model

5.1   Forecasting of river flow

Forecasting of thunderstorms is a very difficult task in weather prediction, due to their very specific and small spatial extension along with nonlinearity in their dynamics [9]. Studies from 2013 had shown that by application of ANN models, it had developed few algorithms which successfully predicted thunderstorm time series. Hence, this model turned out to be very beneficial especially to the meteorologists for forecasting real-time thunderstorm [9]. The ANN models have shown promising results for analyzing the simulation of river flow. The key characteristics of the models include quick completion and development of the model, data parsimony over conventional models. Application of ANN models has been seen to work well in river flow forecast. In a study by Cigozoglu [9], river water flow forecast and estimation of daily water flow for the rivers in East Mediterranean sea in Turkey were analyzed. Multilayer perceptron network was employed as ANN structure and compared to traditional statistical and scholastic model [9]. The results suggested better efficiency of the ANN model over conventional ones.

Study by Pramanik et al. [10] employed wavelet-ANN hybrid model with distinct wavelet transformation functions for processing the flow data with corresponding time into the wavelet coefficients of various frequency groups. The neural network models were developed for 1–3 days of river flow forecast of Brahmani River, India [10]. In another study by scientist Kişi [11], suggested the river flow data on monthly basis for Canakdere river and Goksudere river in eastern black sea zone of Turkey. Three different ANN models were used to evaluate the flow data including feed-forward neural network, generalized regression neural network, and radial basis ANN. Out of the three various models, generalized regression neural network model revealed better results for the monthly forecast of river flow [11]. Several related studies have been mentioned in Table 1.

5.2   ANN model for prediction of depression in aged population

Depression is the main cause of aging, especially in geriatric population which immensely adds to mortality and morbidity rate. ANN Model serves as an important tool for prediction of the various socio-demographic variables and co-morbid conditions. It is also useful as machine learning classifier within the predictive modeling technique. This sophisticated technology is able to identify relation between input (predictor) and output (predicted) variables, which is a concept inspired by the model of the human brain and its functioning mechanisms. This model helps to identify and screen the elderly and aged ones from the general population who are undergoing depression and refers to a

TABLE 1 Overview of neural networks used to detect river flow forecast.

Sl. no.	Modeling technique	Location	Application	Reference
1	Multilayer perceptron network (MLP), radial basis function network (RBF)	Apure River Basin, Venezuela	River flow forecast	[12]
2	Artificial neural network (ANN) and support vector regression (SVR)	Kharjegil and Ponel stations, Iran	River flow forecast	[13]
3	Feed forward back propagation neural network (FFBPNN), generalized regression neural network (GRNN), radial basis function neural network (RBFNN) and statistical autoregressive (AR) model	Jinsha river basin in the upper part of the Yangtze River, China	Flood prediction	[14]

psychiatric consultant. The ANN modeling technique is primarily data-driven for predictions. ANN modeling approach has also been utilized to predict autism in general population and depression in aged group (Table 2). Aging brain remains highly vulnerable to depression and hence, depression is one of the primary reasons for death in elderly groups. Study by Sau and Bhakta [20] has predicted depression in the elderly group via ANN models in Kolkata, India. The multilayer perceptron network was employed to develop the ANN model for a total sample size of 126 people.

A recent study by Jabłońska and Zajdel suggested prediction of behavior of females using the Instagram social networking site. The study predicted mood swings, anxiety level, self-esteem, depression, and similar other factors for such users. Total sample sizes of 974 females with an average age of 18–49 years were considered for the study. Multilayer perceptron and radial basis function models were used to predict the behavior. The multilayer perceptron model was found to have better efficiency than radial basis function model [21].

A recent study by Allahyari [22] indicated the utilization of ANN model to predict elderly depression. Various factors such as age variation, marital status, income, employment status, gender was taken into consideration for the

TABLE 2 Various types of neural networks used in developmental disorder.

Sl. no.	Modeling technique	Key characteristics	Reference
1	Spiking neural networks	Atypical neural functions	[15]
2	Recurrent neural networks	Stereotypic behavior	[16]
3	Convolutional neural networks	Local processing bias	[17]
4	Feed forward neural networks and recurrent neural networks	Extreme attention to every details	[18]
5	Recurrent neural networks	Decreased generalization capacity	[19]

analysis. It was found that the most suitable and efficient ANN model had 33 neurons present in hidden layers.

5.3   Application in various sectors

A stock market widely functions in dealing with company's stock at reasonable price. The supply and demand for stocks and shares allow flexibility to run the stock industry [23]. ANN plays a key role in forecasting the stock market. Stock Market is one of the most emerging sectors in any country for trading of a company's stocks and derivatives at an agreed price [24]. Intelligent Trading Systems rather than fundamental analysis used for prediction of the stock prices, basically utilizing the ANN model, so that immediate investment decisions can be made. The accuracy of these algorithms is directly proportional to the rate of gain, hence serves as an important tool of prediction. ARIMA and MLP (multilayer perceptron) were used for a better valuation of performances by Khashei and Rahimi [24].

Selvamuthu et al. [23] employed various algorithms including Levenberg-Marquardt, Scaled Conjugate Gradient, and Bayesian Regularization to predict the stock market. The scaled conjugate gradient algorithm provided the validation of the predicted data. Qiu et al. [25], recently employed long short-term memory (LSTM) neural network to predict the opening prices of the share market. The LSTM is one of the most frequently used networks of recurrent neural networks with wide application outlook.

The multilayer perceptron, dynamic artificial neural network, and hybrid neural network approach were used by Guresen et al. [26], to forecast the stock prices of NASDAQ stock exchange index. Results suggest that MLP approach yielded best output for the prediction. In another case from China [27], the stock exchange forecast was done using Elman neural network and BP neural network. Results suggested higher efficiency of Elman neural network over BP neural network for predictions for the Chinese stock market. In the same way, ANN has been employed in many essential activities for proper and precise predictions (Table 3). ANN modeling finds most important in the field of health care organizations, globally [37–39]. Artificial intelligence utilizes ANN modeling in the background is thought to be a potential strategy for better development of health care management all around the world, making it more cost-effective. Value-based decision making turned out to be an important key for better management of health care organizations. ANN model has found a wide-ranged application in the analysis of financial and operational data, improving services, and reducing data costs [40]. AI has been used in major diseases like cancers or cardiology departments, and ANN had been repeatedly used as a common machine learning technique. Clinical diagnosis,

TABLE 3 Application of ANN in various sectors.

Sl. no.	Study set	Model employed	Reference
1	Climate forecast	General circulation model followed by deep neural network	[28]
2	Climate forecast	Artificial neural network	[29]
3	Total knee arthroplasty	Artificial neural network	[30]
4	Ecotoxicity assessment	Artificial neural network and ecological structure-activity relationship (ECOSAR) model	[31]
5	Ecotoxicity assessment	Ecological structure-activity relationship (ECOSAR) model	[32]
6	Water pollution	Linear regression analysis and artificial neural network	[33]
7	Water pollution	Artificial neural network	[34]
8	Soil pollution	Artificial neural network, response surface methodology, genetic algorithm	[35]
9	Soil pollution	Back propagation artificial neural network	[36]

speech recognition, image analysis, and important interpretation of data along with drug development are the most.

River salinity of River Murray at Murray Bridge is determined and estimated using ANN modeling. High salinity was recorded previously due to repeated pumping but had been difficult in keeping salinity levels in check. But ANN modeling has been relatively accurate and quite specific in determination of the water quality in the region. Construction management (CM) is highly unpredictable as it has to deal with uncertainties related to time, quality, cost, and safety [41]. ANN model comes very beneficial in predictions related to cost, risk, and safety, as well as in the productivity of labor and equipment. Moreover, feed-forward back propagation networking type had been used extensively. It became quite evident that ANN can serve as a good model for analysis and optimization of data and produce accurate results henceforth.

6 ANN and hydrogen production

Hydrogen as a fuel is one of the most widely used clean fuel. It gains predominance over other biomass-based fuels due to its quick-burning capacity with no byproduct release. ANN has been used extensively by scientists in the fields of science and engineering; it is popular as the most capable and successful technique for the modeling of hydrogen production. Subsequent modeling and optimization of biohydrogen production provide better insight into the role of various dependent and independent variables involved in fermentation process [42]. There are various studies on the control of bioprocesses in environmental and industrial applications. ANN has a network of two layers that utilizes a sigmoid and linear transfer function which can be utilized for modeling nonlinear relations. The substrate concentration which remains the limiting factor, decides the biomass concentration in hydrogen production. The Monod equation is widely used for unstructured and nonsegregated microbial growth model in which growth depends on substrate utilization. Hence, microbial growth can also be modeled by the Monod equation [43], expressed as:

$$\mu = (\mu_{\max} \cdot S)/K_s + S \tag{9}$$

where μ is the specific growth rate, $\mu_{\max}$ is the maximum specific growth rate, K_s is the saturation constant and S is the limiting substrate concentration [42].

Hydrogen can be used as an efficient fuel both for transportation as well as for production of energy. The use of anaerobic microorganism has always been suspected to be promising in the field of biotechnology and applying ANN prediction will help in investigating the effects of factors that affect the production (Table 4).

A study by Rosales-Colunga et al. [51] predicted biohydrogen production utilizing back propagation approach in ANN model. A total of 102 experimental data were generated using 7 different experiments. The input parameters considered for neural networking were dissolved CO_2, pH, and redox potential

TABLE 4 Overview on hydrogen production using various optimization approaches.

Sl. no.	Modeling technique	Input factors	Reference
1	ANN	Substrate concentration, pH and temperature	[44]
2	ANN	Catalyst type, catalyst concentration, biomass variation, temperature	[45]
3	ANN and RSM	Molasses concentration, inoculums concentration, temperature	[46]
4	ANN and fuzzy logic model	Organic waste ratio, COD, acidification time, pH, temperature	[47]
5	RSM and ANN	pH, substrate concentration, temperature	[48]
6	ANN	Time, influent COD, effluent pH and VFAs	[49]
7	ANN	Pressure, temperature, cryogenic, compressor.	[50]

(ORP) monitored during cheese whey fermentation process. The input variable and the output variable data were scales in the range of $(-1, +1)$ and $(0, +1)$ respectively. The outer layer of the BPNN structure consisted a node which predicted for hydrogen production whereas more three nodes were present in the input layer for pH, DCO_2 and ORP. The neurons present in the hidden layer were nonlinear with sigmoid activation function and output layer neurons had linear activation functions. The training was made in accordance to minimal square methodology with respect to error function. The training of the network was accomplished utilizing conjugated gradient algorithm and 25 full cycles of the algorithm was employed to reach convergence and a minimum error of 0.0016. The values for BPNN parameter W was assigned between $(-0.5, +0.5)$. The W_1 values were denoted for weights between input layer and hidden layer whereas, W_2 values were denoted for weights between hidden layer and output layer. The observed and predicted data had a good fit trend for the microbial cultures at pH 5.5 and 6. The R^2 value for hydrogen production by the ANN model was found to be 0.955 [51]. This work was considered first of its own kind to report hydrogen production using genetically modified organisms by BPNN approach [51].

In another study by Nasr et al. [52] back propagation ANN was employed to predict hydrogen production using 313 data points collected from 26 various experiments. The input factors considered were pH, temperature, hydraulic retention time, and substrate and biomass concentration in batch reactions using various substrates. The range of input variables considered were: pH (5.5–7.5), substrate concentration (0.3–58.6 g COD/L), biomass concentration (0.9–17.6 g COD/L), temperature (20–55°C), and maximum fermentation time for batch cultures (97 h). The input and output variables were normalized in the range of (−1, +1) to avoid errors and minimization of training time. The 60%, 20%, and 20% of data was used to train, validate, and test the ANN model, respectively. The input and output layer consisted of five and one neurons, respectively. A double layer hidden layer configuration was selected over a single layer configuration as high error percentage occurred in the later. The neurons on the hidden layer were nonlinear with sigmoid transfer function. The feed forward ANN structure was utilized in the study which comprised of back propagation algorithm. During the ANN training, the error between the predicted and experimental data MSE was analyzed and back propagated through the network in all cycles. The connection weights were denoted as W_1, W_2, and W_3 for matrix representing weights between input layer and first hidden layer, first hidden layer to second hidden layer and second hidden layer to output layer, respectively. The ANN model was predicted utilizing correlation coefficient and mean absolute error. Correlation coefficient of 0.988, 0.987, and 0.996 was obtained for training, validating, and testing data points, respectively for hydrogen production [52]. The proposed model obtained R^2 value of 0.976 using a new data set.

Zamaniyan et al. [53] employed back propagation feed-forward ANN for modeling of industrial hydrogen plant. The essential parameters considered for the modeling were feed temperature, reformer pressure, steam to carbon ratio and carbon dioxide to methane ratio. The ANN training data for the study was generated utilizing mathematical modeling and simulation of plant. A three-layer feed forward BPNN was employed to model the industrial hydrogen plant unit. The number of hidden networks used in the study was 2–5 and the least MSE of 0.00045 was obtained with 5 hidden layers. The dataset generated was divided in three sets of 50%, 25%, and 25% for training, validating, and testing the model. The average R^2 obtained was 0.986 suggesting good fit of hydrogen production by ANN modeling [53]. The study indicated that with change of effective variables, the model predicted presence of mole fraction of hydrogen and carbon monoxide, temperature, pressure in the product. The study suggested that ANN is capable of modeling nonlinear process such as hydrogen production plant [53].

Influence of input factors such as pH, catalyst loading, Pt% and glycerol% have been considered for hydrogen production via neural network modeling [54]. Response surface methodology approach and genetic algorithm-based ANN models were compared for hydrogen production. The Box-Behnken design was employed to generate 29 experiments including five center points. A three-layer feed forward ANN structure was employed with back propagation gradient descendent algorithm. The data were divided in the sets of 23, 3, and 3 for training, validating, and testing. The transfer functions used for neurons in hidden layer was hyperbolic tangent sigmoid function and for input and output layers two linear functions were used. The genetic algorithm optimization technique was employed to optimize the model. To examine the level of influence of variables on response, Garson's method was opted [54]. The optimum value of each variable used was: 50% of glycerol, 3.9 g/L catalyst loading, pH 4.5 and 3.1% of Pt. The R^2 value for RSM and ANN model obtained was 0.9748 and 0.9913, hence suggesting higher accuracy for ANN model [54].

Similarly, Jha et al. [55], suggested ANN outperforming and precisely predicting the hydrogen production along with COD removal from anaerobic sludge in uplift anaerobic sludge blanket reactor. The input parameters considered for prediction were hydraulic retention time, immobilized cell volume and temperature. The ANN structure was constructed using MLP architecture. The back propagation algorithm was employed to train the data. The model training was done using 80% of the observed data whereas the rest 20% data were employed for cross validation. The root mean square error was reduced by training the committee for 12,000 epochs. The connecting influence of individual input variables were analyzed by neural network-based sensitivity analysis. The RSM model for the experiments suggested R^2 value of 0.90 for hydrogen production and 0.99 for COD removal whereas ANN model provided R^2 value of 0.99 for both the output factors [55]. The result also suggested the optimum hydrogen production and COD removal at HRT of 23.4 h, ICV of 60% and temperature of 37.5°C. It is very important to predict hydrogen production using a comprehensive model for the design, monitoring, and management of bio-hydrogen producing bioreactors. Inside any ANN, the neurons in the hidden layer unit contribute to a certain connection to the input variables relating to the basis vector of that unit. The output layer comprises linear neurons and the exact number of neurons in input and output layers is equivalent to number of input and output systems.

7 Conclusion

Artificial Neural Network is thus simulations which are derived from the biological functions of neurons, which consists of densely interconnected processing units that use computational algorithms. Such properties often allow ANN to be known as natural intelligent system and due to this nature ANN is also referred to as natural intelligent systems and machine learning algorithms. The literature in this chapter focuses on ANN training and architectural parameters that are impacting significant effects in various spheres of mankind. This study has included various case studies on prediction of fields mainly focusing on hydrogen production. Apart from that employment of ANN in other fields like weather forecasting, stock market, construction management, health care organization, and environmental ecotoxicity has been observed. It has been implemented across all health care organizations, based on decision making, especially by taking advantage of hybrid models. ANN has proved to be beneficial in determination of prediction of weather forecasting results including prediction of meteorological parameters during premonsoon thunderstorms. The algorithm has proved to be very advantageous especially during demonetization in the stock market in India. It has also played a key role in predicting levels of ecotoxicity in the environment beforehand, and thus helping in prevention of pollution to a great extent. The ANN also plays a significant role in predicting and analyzing the hydrogen production with a vast variety of biomass. A number of input parameters influencing the fermentation remain involved while analyzing the H_2 production. ANN doesn't involve simplification of any problem or incorporation of any assumptions; instead, it can be trained to obtain better results as newer data gets available. It has allowed solving problems like clustering, classification, pattern recognition, determining outliers, prediction, and has turned out to be an extensively useful tool for predictive modeling.

Acknowledgments

The authors would like to thank Amity Institute of Biotechnology, Amity University, Kolkata campus for the infrastructural facilities provided. VK would like to acknowledge the Department of Science and Technology (DST), SERB, Govt. of India, under Startup Research Scheme (File No. SRG/2019/001279).

References

[1] A. El-Shahat, Artificial Neural Network (ANN): Smart and Energy Systems Applications, Scholar Press Publishing, Germany, 2014, ISBN: 978-3-639-71114-1.

[2] C. Stergiou, D. Siganos, Neural Networks, Imperial College, 1996.

[3] D. Hassabis, D. Kumaran, C. Summerfield, M. Botvinick, Neuroscience-inspired artificial intelligence, Neuron 95 (2017) 245–258.

[4] F. Rosenblatt, The perceptron: a probabilistic model for information storage and organization in the brain, Psychol. Rev. 65 (1958) 386–408.

[5] T.J. Huang, Imitating the brain with neurocomputer a "new" way towards artificial general intelligence, Int. J. Autom. Comput. 14 (2017) 520–531.

[6] S.A. Kalogirou, Artificial neural networks in renewable energy systems applications: a review, Renew. Sust. Energ. Rev. 5 (2001) 373–401.

[7] R. Mohammadzadeh Kakhki, M. Mohammadpoor, R. Faridi, M. Bahadori, The development of an artificial neural network—genetic algorithm model (ANN-GA) for the adsorption and photocatalysis of methylene blue on a novel sulfur–nitrogen co-doped Fe_2O_3 nanostructure surface, RSC Adv. 10 (2020) 5951.

[8] D.E. Rumelhart, G.E. Hinton, R.J. Williams, Learning internal representations by error propagation, in: Parallel Distributed Processing: Explorations in the Microstructure of Cognition, vol. 1, MIT Press, Cambridge, MA, 1986 (Chapter 8).

[9] H.K. Cigozoglu, Estimation, forecasting and extrapolation of river flows by artificial neural networks, Hydrol. Sci. J. 48 (2003) 349–361.

[10] N. Pramanik, R.K. Panda, A. Singh, Daily river flow forecasting using wavelet ANN hybrid models, J. Hydroinf. 13 (2010) 49–63.

[11] Ö. Kişi, River flow forecasting and estimation using different artificial neural network techniques, Hydrol. Res. 39 (2008) 27–40.

[12] Y.B. Dibike, D.P. Solomatine, River flow forecasting using artificial neural networks, Phys. Chem. Earth Part B: Hydrol. Oceans Atmos. 26 (2001) 1–7.

[13] A.M. Kalteh, Monthly river flow forecasting using artificial neural network and support vector regression models coupled with wavelet transform, Comput. Geosci. 54 (2013) 1–8.

[14] M. Tayyab, J. Zhou, X. Zeng, R. Adnan, Discharge forecasting by applying artificial neural networks at the Jinsha river basin, China, Eur. Sci. J. 12 (2016) 108.

[15] J. Park, K. Ichinose, Y. Kawai, J. Suzuki, M. Asada, H. Mori, Macroscopic cluster organizations change the complexity of neural activity, Entropy 21 (2019) 214.

[16] H. Idei, S. Murata, Y. Chen, Y. Yamashita, J. Tani, T. Ogata, Reduced behavioral flexibility by aberrant sensory precision in autism spectrum disorder: a neurorobotics experiment, in: 2017 Joint IEEE International Conference on Development and Learning and Epigenetic Robotics (ICDL-EpiRob), 2017.

[17] Y. Nagai, T. Moriwaki, M. Asada, Influence of excitation/inhibition imbalance on local processing bias in autism spectrum disorder, in: Proceedings of the 37th Annual Meeting of the Cognitive Science Society, 2015, pp. 1685–1690.

[18] I.L. Cohen, Neural network analysis of learning in autism, Neural Networks Psychopathol. (1998) 274–315.

[19] A. Philippsen, Y. Nagai, Understanding the cognitive mechanisms underlying autistic behavior: a recurrent neural network study, in: 2018 Joint IEEE International Conference on Development and learning and epigenetic robotics (ICDL-EpiRob), IEEE, 2018, pp. 84–90.

[20] A. Sau, I. Bhakta, Artificial neural network (ANN) model to predict depression among geriatric population at a slum in Kolkata, India, J. Clin. Diagn. Res. 11 (2017) VC01–VC04.

[21] M.R. Jabłońska, R. Zajdel, Artificial neural networks for predicting social comparison effects among female Instagram users, PLoS One 15 (2020) e0229354.

[22] E. Allahyari, Predicting elderly depression: an artificial neural network model, Iran. J. Psychiatry Behav. Sci. 13 (2019) e98497.

[23] D. Selvamuthu, V. Kumar, A. Mishra, Indian stock market prediction using artificial neural networks on tick data, Financ. Innovation 5 (2019) 16.

[24] M. Khashei, Z.H. Rahimi, Performance evaluation of series and parallel strategies for financial time series forecasting, Financ. Innov. 3 (2017) 24.

[25] J. Qiu, B. Wang, C. Zhou, Forecasting stock prices with long-short term memory neural network based on attention mechanism, PLoS One 15 (2020), e0227222.

[26] E. Guresen, G. Kayakutlu, T.U. Daim, Using artificial neural network models in stock market index prediction, Expert Syst. Appl. 38 (2011) 10389–10397.

[27] B. Wu, T. Duan, A performance comparison of neural networks in forecasting stock price trend, Int. J. Comput. Intell. Syst. 10 (2017) 336–346.

[28] S. Scher, G. Messori, Weather and climate forecasting with neural networks: using general circulation models (GCMs) with different complexity as a study ground, Geosci. Model Dev. 12 (2019) 2797–2809.

[29] L.B. Nguyen, M.H. Le, Application of artificial neural network and climate indices to drought forecasting in south-Central Vietnam, Pol. J. Environ. Stud. 29 (2020) 1–11.

[30] P.N. Ramkumar, J.M. Karnuta, S.M. Navarro, H.S. Haeberle, G.R. Scuderi, M.A. Mont, V.E. Krebs, B.M. Patterson, Deep learning preoperatively predicts value metrics for primary total knee arthroplasty: development and validation of an artificial neural network model, J. Arthroplast. 34 (2019) 2220–2227.

[31] P. Hou, O. Jolliet, J. Zhu, M. Xu, Estimate ecotoxicity characterization factors for chemicals in life cycle assessment using machine learning models, Environ. Int. 135 (2020) 105393.

[32] M. Takata, B.-L. Lin, M. Xue, Y. Zushi, A. Terada, M. Hosomi, Predicting the acute ecotoxicity of chemical substances by machine learning using graph theory, Chemosphere (2019) 124604.

[33] K.U. Ahamad, P. Raj, N.H. Barbhuiya, A. Deep, Surface water quality modeling by regression analysis and artificial neural network, Adv. Waste Manage. (2018) 215–230.

[34] S. Bansal, G. Ganesan, Advanced evaluation methodology for water quality assessment using artificial neural network approach, Water Resour. Manage. 33 (2019) 3127–3141.

[35] F. Mohammadi, M.R. Samaei, A. Azhdarpoor, H. Teiri, A. Badeenezhad, S. Rostami, Modeling and optimizing pyrene removal from the soil by phytoremediation using response surface methodology, artificial neural networks, and genetic algorithm, Chemosphere 237 (2019) 124486.

[36] H. Bao, J. Wang, J. Li, H. Zhang, F. Wu, Effects of corn straw on dissipation of polycyclic aromatic hydrocarbons and potential application of backpropagation artificial neural network prediction model for PAHs bioremediation, Ecotoxicol. Environ. Saf. 186 (2019) 109745.

[37] E.P. Goss, G.S. Vozikis, Improving health care organizational management through neural network learning, Health Care Manage. Sci. 5 (2002) 221–227.

[38] H. Kaur, S.K. Wasan, Empirical study on applications of data mining techniques in healthcare, J. Comput. Sci. 2 (2006) 194–200.

[39] J. Nolting, Developing a neural network model for health care, AMIA Ann. Symp. Proc. 2006 (2006) 1049.

[40] M.A. Boyacioglu, Y. Kara, Ö.K. Baykan, Predicting bank financial failures using neural networks, support vector machines and multivariate statistical methods: a comparative analysis in the sample of savings deposit insurance fund (SDIF) transferred banks in Turkey, Expert Syst. Appl. 36 (2009) 3355–3366.

[41] A.H. Boussabaine, The use of artificial neural networks in construction management: a review, Constr. Manage. Econ. 14 (1996) 427–436.

[42] K. Nath, D. Das, Modeling and optimization of fermentative hydrogen production, Bioresour. Technol. 102 (2011) 8569–8581.

[43] N. Panikov, Microbial Growth Kinetics, Springer Netherlands, 1995, p. 378.

[44] A. El-Shafie, Neural network nonlinear modeling for hydrogen production using anaerobic fermentation, Neural Comput. Applic. 24 (2012) 539–547.

[45] A. Karaci, A. Caglar, B. Aydinli, S. Pekol, The pyrolysis process verification of hydrogen rich gas (H-rG) production by artificial neural network (ANN), Int. J. Hydrog. Energy 41 (2016) 4570–4578.

[46] J.K. Whiteman, E.B. Gueguim Kana, Comparative assessment of the artificial neural network and response surface modelling efficiencies for biohydrogen production on sugar cane molasses, BioEnergy Res. 7 (2013) 295–305.

[47] E.L. Moreno Cárdenas, A.D. Zapata-Zapata, D. Kim, Modeling dark fermentation of coffee mucilage wastes for hydrogen production: artificial neural network model vs. fuzzy logic model, Energies 13 (2020) 1663.

[48] C. Mahata, S. Ray, D. Das, Optimization of dark fermentative hydrogen production from organic wastes using acidogenic mixed consortia, Energy Convers. Manage. 219 (2020) 113047.

[49] M.K. Yogeswari, K. Dharmalingam, P. Mullai, Implementation of artificial neural network model for continuous hydrogen production using confectionery wastewater, J. Environ. Manage. 252 (2019) 109684.

[50] N.D. Vo, D.H. Oh, J.-H. Kang, M. Oh, C.H. Lee, Dynamic-model-based artificial neural network for H_2 recovery and CO_2 capture from hydrogen tail gas, Appl. Energy 273 (2020) 115263.

[51] L.M. Rosales-Colunga, R.G. García, A. De León Rodríguez, Estimation of hydrogen production in genetically modified *E. coli* fermentations using an artificial neural network, Int. J. Hydrog. Energy 35 (2010) 13186–13192.

[52] N. Nasr, H. Hafez, M.H. El Naggar, G. Nakhla, Application of artificial neural networks for modeling of biohydrogen production, Int. J. Hydrog. Energy 38 (2013) 3189–3195.

[53] A. Zamaniyan, F. Joda, A. Behroozsarand, H. Ebrahimi, Application of artificial neural networks (ANN) for modeling of industrial hydrogen plant, Int. J. Hydrog. Energy 38 (2013) 6289–6297.

[54] M.R.K. Estahbanati, M. Feilizadeh, M.C. Iliuta, Photocatalytic valorization of glycerol to hydrogen: optimization of operating parameters by artificial neural network, Appl. Catal. B: Environ. 209 (2017) 483–492.

[55] P. Jha, E.B.G. Kana, S. Schmidt, Can artificial neural network and response surface methodology reliably predict hydrogen production and COD removal in an UASB bioreactor? Int. J. Hydrog. Energy 42 (2017) 18875–18883.

Chapter 10

Fault diagnosis in industrial processes based on predictive and descriptive machine learning methods

Ahmed Ragab, Mohamed El Koujok, Hakim Ghezzaz, and Mouloud Amazouz
CanmetENERGY-Natural Resources Canada (NRCan), Varennes, Canada

1 Introduction

Process monitoring and fault diagnosis are of great importance for safety, productivity, and the sustainability of industrial processes. The availability of huge volumes of industrial data (referred to as Massive Data or Big Data) represents an important asset to fulfill the needs for efficient fault diagnosis. The complexity of the industrial processes makes fault diagnosis very difficult for engineers and operators; thus decision support tools that automatically diagnose faults are looked for.

Most of the current fault detection and diagnosis (FDD) methods are subject to specific assumptions and approximations, some of which are mathematical, while others relate to practical implementation issues such as the quality and quantity of data required to validate a proposed diagnostic method. For that reason, many diagnostic methods are not tractable or easily accessible to practitioners. The developers in the industrial community are willing to avoid the incurred risk before trialing a particular diagnostic method. To properly select a diagnostic method for practical implementation, it is important not only to have the mathematical understanding of each method but also to know how a particular industrial application intends to use the method and its outputs.

The challenges faced when developing a diagnostic method with useful and friendly outputs to end-users are dependent on the applied algorithm and its computing infrastructure. Therefore, it is essential to test, compare, and validate different diagnostic methods. Due to the complexity of fault diagnosis problems in the industry, it is sometimes necessary to exploit an ensemble of different

Applications of Artificial Intelligence in Process Systems Engineering
https://doi.org/10.1016/B978-0-12-821092-5.00002-4
207

methods to enhance the efficiency of fault diagnosis tasks. The big challenge is to develop a decision agent that combines those methods in order to elicit a unified decision that provides the correct diagnosis.

Broadly, the FDD methods can be divided into model-based and data-driven methods. The model-based methods require in-depth understanding and analysis of physical phenomena in the targeted processes in order to derive accurate mathematical models. Model-based FDD methods may not be the most practical tools since faults can vary from one component to another in the plant. Moreover, obtaining an effective detailed model can be very difficult, time consuming, and expensive, particularly for large-scale processes with many variables. Furthermore, such methods may not be able to capture the detailed dynamics which are difficult to describe, thus generating less accurate models. Even if an accurate mathematical model is built for a certain process, the experimental evaluation of such model is a tedious task. Consequently, it is not advantageous to build a model-based fault diagnosis method for complex industrial processes such as pulp mills, in which important quality parameters (e.g., the key performance indicator of the biomass recovery boiler) need to be modeled properly. Recently, many practical cases are utilizing data-driven FDD methods since it is easier to gather data rather than to build accurate mathematical models. The availability of huge volumes of industrial data represents an important asset to fulfill the needs for efficient data-driven FDD methods in the industry. The development and implementation of a sophisticated data-driven FDD method are carried out by performing analysis, knowledge discovery and visualization of multivariable data based on advanced machine learning (ML) and pattern recognition techniques.

This chapter surveys the widely applied and novel data-driven methodologies suggested in the literature for process monitoring and fault diagnosis from the application point of view. It discusses practical issues that need to be considered when selecting an appropriate data-driven fault diagnosis method for industrial applications. The chapter also presents the main characteristics of FDD methods and categorization diagrams to assist developers in selecting the appropriate methods for diagnosing faults within their specific operational environment. Easiness-to-use by process operators, applicability to complex processes, and interpretation facilities are among the common desirable characteristics of a fault diagnosis system.

This chapter provides a state-of-the-art review of the data-driven FDD methods that have been developed for complex industrial systems focusing on ML-based methods. Among common ML techniques, the top fault diagnosis algorithms are discussed in this chapter according to their efficiencies and widespread popularities. A number of common predictive and descriptive ML techniques have been discussed according to their pros and cons. A literature review was conducted on the characteristics of these methods, according to a multitude of papers and recent reviews. The chapter also presents a number of methodologies applied to real case studies in industrial plants located in Canada. Each methodology combines diversified predictive and descriptive methods integrated together. Finally, the chapter concludes the results and briefly lists some of the lessons learned through these case studies.

2 FDD in large-scale processes

FDD is at the heart of any process monitoring system. As a wide area of research for more than three decades, FDD has been attractive to researchers and engineers in different disciplines and fields of engineering such as process monitoring, control, manufacturing, automotive industry, process industry, etc. Several approaches have been proposed for fault detection and diagnosis depending on system properties and information availability. Fault detection is defined as the task of determining when a system is experiencing malfunctioning operations. While fault diagnosis is the task of locating and analyzing the source of a fault once it is detected. A detailed schematic diagram and explanation for the FDD process are found in [1]. FDD is an ongoing significant research field and an open-ended problem for researchers as well as for practitioners, due to the increasing need for maintainability, reliability, and safety of complex chemical processes. The complexity of the interacted mechanisms in such processes can result in huge losses even due to a slight error. Those complex processes are classified as nonlinear dynamical systems with a huge number of observed variables. As such, fault diagnosis in such processes becomes a big challenge and problematic for engineers and operators. It is very difficult for operators to detect and diagnose the system state by visual inspection of all data variables' values, and hence an automated efficient tool that performs the two tasks of FDD is looked for.

2.1 Existing FDD methods: Limitation and needs

Generally, the FDD methods broadly fall into two major categories: model-based and data-driven approaches [2]. The model-based approach is highly dependent on mathematical models of the system. The more precise a model is the more reliable the results obtained. The model-based methods can be further classified as qualitative and quantitative. A quantitative model is a functional representation of the relationships between the system's inputs and outputs based on the fundamental understanding of the system's physics. The physical and damage propagation models are examples of quantitative models used to detect the faults and predict future failures as well [3]. These models are capable of even detecting unanticipated faults because they rely on inconsistencies or residuals between the actual system and expected behavior of a system (the model output). The most commonly used techniques for generating and analyzing residuals are parameter estimation, parity relations, and observers [4]. The qualitative models on the other hand represent the relationships between the system's inputs and outputs in terms of qualitative functions centered on different units in the system. Examples of qualitative models are causal models, abstraction hierarchies, directed graphs (digraphs) [5–7], Petri Nets (PNs) and others [8–10]. However, in many industrial processes, it is not possible to obtain a precise model due to process high dimensionality as well as the complexity of the system that results in large modeling costs and time required to obtain an accurate model. Moreover, model-based FDD approaches may not be the most practical approach since the faults are highly interacting in complex plants. As a result, it is not advantageous to build model-based FDD methods for complex industrial processes (e.g., biomass recovery boiler, heat exchangers in pulp mills).

In contrast to model-based, the data-driven FDD methods require transforming a sufficient amount of historical data into a priori knowledge to build a diagnostic model [4]. They are based on the concept of learning and testing, which keeps improving as the knowledge is accumulated [11]. It is noticed from the overall trend of the reviewed literature that there is a consistent growth in data-driven methods specifically in the fault diagnosis domain. Developments in data acquisition allow for more data collection and storage, which leads to the availability of a large amount of process historical data, making the data-driven approach a significantly attractive choice for monitoring and FDD, rather than building physical models from domain knowledge as in the model-based approach. However, an important remark is that the efficiency of data-driven approaches is highly dependent on the quantity and quality of the acquired data [1].

Although there is a wide repertoire of online fault diagnostic tools available in the market, only a few of them are generally used in the industry, particularly for large-scale chemical plants. Moreover, state-of-the-art diagnostic tools are unable to properly cope with the identification of the root causes of faults in complex equipment. Most of the knowledge discovery techniques that are used in FDD are not tractable or accessible to practitioners, since they have been undertaken based on theory rather than practical problem application. In addition, the majority of these methods are subject to specific assumptions and approximations, some of which are mathematical, while others relate to practical implementation issues. Consequently, a commercialized-based online diagnosis tool is often needed to provide the decision-maker with comprehensive knowledge about the process operational state. The potential benefits offered by such a tool are enough to support an investment decision.

2.2   Selection criteria for FDD methods in process industries

Given the limitations of existing FDD methods, it is therefore necessary for the developers in the industrial community to avoid the incurred risk before trialing a particular diagnostic method, since an immature understanding of the proposed solutions can hinder the project to deliver its promised outcomes in a timely fashion. To properly select a diagnostic method for a certain practical implementation, it is important not only to understand the computing infrastructure of each method, but also how a particular industrial application intends to utilize the model and its outputs. Thus, there is a need to develop an efficient and tractable diagnostic approach with useful outputs to end-users. The applied diagnostic scheme in large-scale problems should provide the decision-making system with predictive and descriptive information about a given fault in an initial phase in order to reduce the losses caused by that fault. The selection characteristics considered in this chapter to select an efficient diagnostic method are as follows: robustness to noise and uncertainty, explainability, adaptability to process changes, handling of process complexity, minimal tuning by the user, and generalization and flexibility. These characteristics are shown in Fig. 1 and are briefly explained as follows.

- *Handling processes complexity.* FDD methods should have the ability to cope with high dimensional, nonlinear, and dynamic industrial processes.
- *Robustness to noise and uncertainties.* Provide rapid and reliable detection and isolation of system faults, while taking into account the system's exogenous disturbances and uncertainties.
- *Explanation facility.* Provide interpretations on how faults have originated and propagated through the system. This includes the ability to find the root cause of the problem.
- *Adaptability to process changes.* Ability to identify novel faults although it is difficult to achieve due to the interacting nature of most faults and the system's fast dynamics.

FIG. 1   Characteristics of efficient FDD system.

- *Minimal human efforts.* Avoid tedious settings and parameter tuning by machine learning nonexpert users.
- *Generic and flexible.* An FDD method should be as generic and flexible as possible, and easy to apply to a broad category of different applications and processes.

3 Data-driven FDD methods in industrial systems: A review

This section presents a literature review for the widely applied data-driven FDD methods suggested for process monitoring and fault diagnosis, from the industrial applications point of view. An abundance of researches on FDD methods ranging from analytical and statistical methods to AI methods was reported in the most cited review papers [12–19]. Although those papers are very comprehensive and detailed, they did not focus on the FDD in complex industrial processes. The review papers published by [20, 21] surveyed the data-driven fault diagnosis methods that were applied in industrial processes.

Other papers published by [13, 22, 23] are discussing the trends in fault diagnosis of chemical processes. According to the reviews found in all of those papers, it is necessary to summarize and categorize the FDD methods that are used in the industry. However, the categorization of FDD methods is a tedious task since there are no standards to categorize them and some overlaps between the categories are unavoidable. As shown in Fig. 2, data-driven FDD methods are mainly categorized into three major groups; ML-based FDD methods, multivariate statistical process monitoring (MVSPM) methods, hybrid and integrated methods. ML-based FDD approaches rely on the availability of condition monitoring data and draw on learning techniques from the area of computational intelligence, and statistical learning where artificial neural networks (ANNs), support vector machines (SVM), decision tree (DT), and others are employed to map the system measurements into fault growth models. ML-based FDD methods are further divided into predictive and descriptive methods.

It is worth mentioning that the ML-based FDD methods are categorized as supervised and unsupervised. In supervised learning, the data are labeled and the classes (normal condition and faults) are defined using the a priori knowledge provided by means of expert or a person who has a sufficient experience about the process. This is the case of industrial systems where a database containing recorded observations of previous faults of the system is available. In unsupervised learning, the task is to identify and describe the hidden structures in unlabeled data. Clustering algorithms are employed to group the data into classes with similar characteristics. In statistical MVSPM diagnostics, the statistical learning algorithms are employed to process the acquired historical data in order to discover the hidden patterns in such data. Linear feature reduction methods including principal component analysis (PCA), partial least squares (PLS), independent component analysis (ICA), and Fisher discriminant analysis

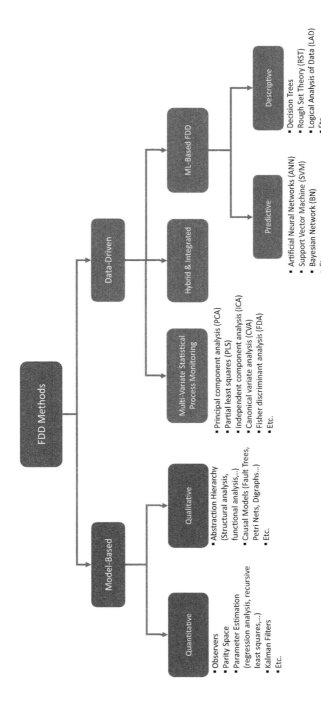

FIG. 2 Possible categorization of FDD methods.

(FDA), and their nonlinear versions such as Kernel PCA (KPCA) and Kernel FDA (KFDA) are also demonstrated as common MVSPM methods.

In this chapter, we focus on the predictive and descriptive ML-based FDD methods along with their pros and cons discussed in details. However, the chapter gives a brief explanation of MVSPM and integrated methods. It presents three case studies applying hybrid methods.

3.1 Machine learning-based FDD methods

Machine learning and pattern recognition methods have been introduced to the field of FDD to deal with the huge amount of data and to extract useful diagnostic information. The ML-based FDD methods exploit machine learning as a multidisciplinary field that uses various methods and algorithms from statistical learning, artificial intelligence (AI), and combinatorial optimization.

Supervised machine learning methods are mainly categorized into predictive and descriptive approaches [24, 25]. Predictive ML involves using some variables in the data to predict unknown or future values of other variables of interest. Descriptive ML focuses on finding human interpretable patterns that describe the hidden phenomena in the data. The relative importance of predictive and descriptive ML approaches can vary considerably for particular industrial applications. In what follows, a review for common machine learning FDD methods in each approach that are found in industrial applications along with their pros and cons is discussed.

3.1.1 Predictive FDD methods

The general concept of predictive machine learning methods is to learn certain properties from a training dataset to build a model that is capable of making predictions [26]. Predictive ML methods can be divided into two groups: regression and pattern classification. ML regression methods are based on the analysis of relationships between variables and trends in the data to make predictions about numerical variables, for example, the prediction of a key performance indicator (KPI) in industrial processes. In contrast to ML regression, the task of pattern classification is to assign discrete class labels to particular observations as outcomes of a prediction. To go back to the above example: a pattern classification task could be the prediction of a *high* or *low* KPI. The most common predictive methods are the SVM and the ANN [27].

3.1.1.1 Artificial neural network (ANN)

The ANN was inspired by biological neurons [28]. The ANN is proven to be a powerful tool for learning and constructing a nonlinear mapping when a set of input and output data is provided [27]. Compared with the traditional analytical methods, the ANN approach has the advantage of an excellent nonlinear approximation ability [29]. The ANN was used commonly in the fault diagnosis of industrial systems, due to its inherent pattern recognition capabilities and its ability to handle noisy data [2]. The ANN is the most widely used to generate residuals that can be monitored through simple thresholds [30].

The feedforward neural network consists of different types of layers; input layer, hidden layer, and output layer, as depicted in Fig. 3. The input data variables need to be normalized before they are fed into the input layer. One common way is to apply a variable scaling so that all the values are in the range between 0 and 1, using the following form:

$$x' = \frac{x - x_{min}}{x_{max} - x_{min}} \tag{1}$$

The information contained in the input data is transformed in the hidden layers into higher representations through a nonlinear transformations using an activation function, as shown in Fig. 4. The activation at hidden unit j is given as:

$$net_j = \sum_{i=1}^{n} x_j w_{ji} + b \tag{2}$$

The activation function at the hidden unit j is given as:

$$z_j = \sigma(net_j) \tag{3}$$

In the context of monitoring processes, the paper [31] presents a hybrid multivariate method based on ANN classifier and PCA improved by GA. The method determines the main principal components used to detect faults during the operation of the industrial process. The method is applied to simulated data collected from the TEP. Versions of ANN were used to deal with the dynamic monitoring processes; recurrent neural networks (RNN) and hamming neural network (HNN). The HNN is specially designed for two-class classification problems in industrial systems [32]. Kim et al. have presented a pattern recognition-based diagnosis approach as an application of the HNN to diagnose state-of-health (SOH) for a polymer electrolyte membrane (PEM) fuel cell [33]. This paper [34] proposes a fault diagnosis method based on the long short-term memory (LSTM) network. The LSTM parameters are optimized to get a higher fault diagnosis accuracy for the targeted chemical process. The developed method is validated based on the simulation experiment of the TEP chemical process. The results show that the optimized LSTM model gives better performance than comparable neural networks such as multi-layer perceptron method and others.

The self-organizing map (SOM) is a specific type of unsupervised neural network that is used as a dimensionality reduction and classification tool of high dimensional data. It was originated by [35]. The SOM were used as an important tool for fault diagnosis and monitoring with intuitive visualization, especially for state transitions in continuous processes. The SOM was applied by Francesco Corona et al. to visualize process measurements, by extracting the

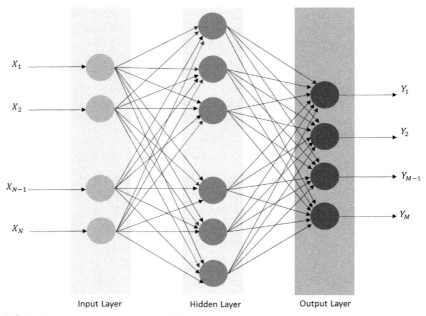

FIG. 3   The conceptual structure of ANN.

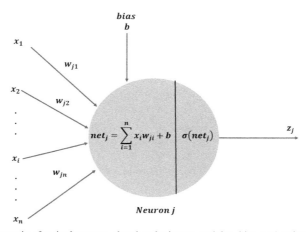

FIG. 4   Schematic of a single neuron showing the inputs, weights, bias, and activation function.

relevant information and exploiting the topological structure of the data points [36]. In that chapter, it is confirmed that the proposed SOM method is capable of providing valuable information for direct application to process monitoring.

A distinguishing characteristic of the SOM is that it can preserve the topology of the input data from a high-dimensional space onto the output map in such a way that relative distances between input data are more or less preserved. The input data observations, located close to each other in the input space, are mapped to the nearby neurons on the output map. As a result, the SOM can be used as a clustering tool of high-dimensional data with valuable data visualization ability [37]. After training, the output neurons of the SOM are clustered and labeled. When a new observation is collected from the monitored process, the signatures of normal and abnormal transitions are visually inspected in order to monitor and detect faults. This is an important advantage of the SOM where the response of new observations can be tracked using a dynamic trajectory graph on the map. This is very helpful to monitor the whole process, and if a fault occurs, then the response neuron will leave the normal area to a fault area.

An FDD method was proposed by Hongyang Yu et al. based on SOM to detect and diagnose faults in non-Gaussian processes [38]. The effectiveness of the proposed method has been verified using the Tennessee Eastman chemical process. The results demonstrated that the proposed method was able to detect faults at an early stage, and the root-cause variables were correctly identified. Chen and Yan proposed an improved fault diagnosis method by combining correlative component analysis with SOM (CCA–SOM) [37]. The CCA was used to extract fault classification information, and then the identified correlative components are fed to the SOM to distinguish and visualize the various types of faults. That method was proven suitable for complex industrial processes. The TEP process is employed as a case study to illustrate the effectiveness of the proposed method. P. Juntunen et al. applied self-organizing map (SOM) techniques combined with K-means clustering to model water quality [39]. The proposed method comprises two phases. In the first phase, the online and laboratory data of the treatment process are exploited to construct a SOM. In the second one, the reference vectors of the SOM are classified by K-means algorithm into clusters to present different states of the process. Finally, the results are interpreted by analyzing the reference vectors in such states.

Recently, deep learning (DL) has emerged as a powerful FDD approach [16]. The deep neural networks (DNNs) are at the heart of DL used as a promising method used to enhance the detection, classification and prediction accuracy in industrial processes [40,41]. These networks address the high nonlinearities and strong correlations in data. DNN architectures, instead of shallow ones, are developed to learn the different feature representations from the raw data through a layer-by-layer successive learning process.

The paper [42] proposes a deep neural network (DNN) with active learning for inducing chemical fault diagnosis. The method exploits a large amount of chemical sensor data, which is a combination of deep learning and active learning criterion to target the difficulty of consecutive fault diagnosis. This method uses a stacked sparse autoencoder to train the deep learners based on fault data to minimize the loss of information. This chapter [43] proposes an extended deep belief network (EDBN) for fault diagnosis in industrial processes. The proposed method is used to fully exploit the raw data combined with the hidden features as inputs to each extended restricted Boltzmann machine (ERBM) during the pre-training phase. Then, a dynamic EDBN-based fault classifier is trained to take the dynamic characteristics of process data into consideration. The proposed method is applied to the TEP for fault classification and the results show that EDBN has better performance than traditional deep belief networks. The results show that the proposed method improves the divisibility between faults and normal process, and exhibits a better performance on the accuracy of fault classification for the TEP data. This paper [44] proposed a deep neural network for fault detection and diagnosis, and compared the oversampling by a generative adversarial network (GAN) to standard oversampling techniques. The objective is to address the problem data imbalance in which the amount of fault data is much less than that of normal data.

The work in [45] proposes a fault diagnosis framework based on a deep transfer network (DTN), which generalizes the deep learning model to domain adaptation scenarios. The proposed method exploits the discrimination structures in the labeled data in the source domain to adapt the conditional distribution of unlabeled target data, and thus guarantee a more accurate distribution matching. This chapter [46] presents an approach for automated fault diagnosis based on deep learning to locate multiple classes of faults under real-time working conditions.

The limitations of the ANN in fault diagnosis are related to its training. As data variables and fault patterns (classes) increase, a more complex network has to be designed to diagnose the faults properly. Although these networks have many different learning algorithms, they are not so great when it comes to knowledge presentation. We cannot just present knowledge by saying a particular weight would give us the best result; the knowledge is distributed throughout the whole network. This makes the neural network as a black box for the user (the architecture of the network could not be characterized). Another drawback is the tedious parameter tuning of the network structure. Moreover, the ANN algorithm often does not show the desired performance, in the case of a limited number of samples. This is attributed to the fact that the ANN is based on the principle of empirical risk minimization, but not expected risk minimization, and this can lead to poor minimization performance.

The ANN is combined with fuzzy logic in many fault diagnosis structures, in order to exploit the adaptive capability of the former and the transparency of the latter [47]. The disadvantage of fuzzy logic is the potential exponential explosion in the number of rules as the number of variables increases. Another drawback is its dependency on human expertise. Although this expertise is valuable, it is subject to some errors that often come from imprecise knowledge and inaccurate reasoning. The combined neuro-fuzzy structures have been proven to have superior recognition accuracy and better generalization capability compared with a single ANN [48].

Lau et al. purposed an online fault diagnosis framework for a dynamical process incorporating multi-scale principal component analysis (MSPCA) for feature extraction and adaptive neuro-fuzzy inference system (ANFIS) for learning the fault-symptom correlation from the process historical data [49].

3.1.1.2   Support vector machines (SVM)

SVM is a machine learning technique that is based on statistical learning theory (SLT) [50]. SVMs were originally designed as binary classifiers (dichotomizers). The main concept is to find the optimal hyperplane by minimizing the upper bound of the generalization error by maximizing the margin between the separating hyperplane and the two classes of positive and negative data [51,52]. The optimal hyperplane separating the two classes can be solved as an optimization problem found in, as shown in Fig. 5 [53]. In most practical situations, the data points of one class cannot be separated from another class by a straight line, in their original space. Consequently, the data points in the original are lifted to another feature space where a straight hyperplane that separates them can be found. This is done by using the concept of kernels that map input space, to a higher dimensional feature space [54].

SVM is based on the structural risk minimization (SRM) principle whereas in the ANNs, traditional empirical risk minimization is used for minimization of the estimation error for the training data [23]. The hyperplanes of SVMs are obtained by solving constrained optimization problems, and the output is determined by maximizing the following function [53]:

$$L(\alpha) = \sum_{i=1}^{N} \alpha_i - \frac{1}{2} \sum_{i=1}^{N} \sum_{j=1}^{N} \alpha_i \alpha_j y_i y_j Z_i^t Z_j \tag{4}$$

Constrained to

$$\alpha_i \geq 0, \forall i \text{ and } \sum_{i=1}^{n} \alpha_i y_i = 0 \tag{5}$$

where N is the number of data points, α is a Lagrange multiplier obtained from the solved optimization problem, and $Z = \varphi(x, x')$ is a kernel function that performs the nonlinear mapping of the original data points x into the feature space. The resolution principle of the SVM is a quadratic programming tool whose local optimal value is just its global value.

Multi-classifiers SVM can be obtained by the combination of binary classifiers in several methods such as one-against-one (OAO), one-against-all (OAA), and direct acyclic graph (DAG), that are found in [55]. In that work,

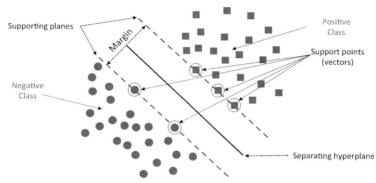

FIG. 5 The main concept of SVM learning.

a comparison of these methods is presented and pointed out that the one against one method has outstanding performance than the other methods. The details of the theory, methodology, and software of SVM are available in [56]. The algorithms for SVM are also implemented in the publicly available software package "Weka" [57].

Onel et al. presented a feature selection algorithm based on nonlinear SVM for fault detection and diagnosis in continuous processes applied to the TEP benchmark [58]. The presented feature selection algorithm is derived from the sensitivity analysis of a C-parameterized SVM objective function. The results show that the proposed feature selection algorithm improves the performance of fault diagnosis in terms of detection accuracy and latency. This chapter [59] presents a method for process monitoring in batch processes based on high dimensional process data with nonlinear SVM. The objective is to retrieve the most informative process measurements for accurate and simultaneous fault detection and diagnosis. The proposed method is applied to an extensive benchmark data set which includes process data describing 22,200 batches with 15 faults. The results show that the proposed method provides a promising decision support tool for the online fault diagnosis of batch processes.

The multi-class SVM is adopted as an FDD tool for chillers fault detection and diagnosis [60]. It has a number of desirable features: first, it is accurate and is capable of dealing with unseen samples. Second, it requires only a small amount of training samples. Finally, it needs simple computation in decision-making. All these features make it very attractive for fault diagnosis in chemical processes [61,62]. The SVM training always seeks a globally optimized solution and avoids poor generalization problems (namely the over-fitting). It is concluded from the literature that the SVM has the ability to deal with a large number of features. It also provides a unique solution and is a strongly regularized method appropriate for ill-posed problems. Support vector machine has been intensively studied in order to deal with the nonlinearity in the industrial process. Recently, Yin reviews the research and development of fault diagnosis and monitoring methods based on SVM in [23].

Some extensions of the SVM applied in fault diagnosis in industrial applications; proximal SVM (PSVM) [63], support vector data description (SVDD) [64], multi-kernel support vector machines (MK-SVMs) [65], Lagrangian SVM (LSVM) [66], and support vector clustering (SVC) [67]. The LSVM was first proposed by [68] as a modified version of SVM. It is based on an implicit Lagrangian formulation of the dual of a simple reformulation of the standard quadratic problem of SVM. The LSVM is shown to be a very fast and simple algorithm compared to the standard SVM. It is applied in the biomedical domain as a suction detection system that can precisely classify pump flow patterns. Ming Yang Sheng has applied the LSVM in the crane gear fault diagnosis [69]. The weight parameters were obtained by the particle swarm optimization algorithm. The results in those papers show that the LSVM gives superior performance in terms of classification accuracy, stability, learning speed, and good robustness compared to the original SVM algorithm.

Taqvi et al. presented a fault diagnosis approach based on an MK-SVM to classify the internal and external faults in a distillation column such as reflux failure, change in reboiler duty, column tray upsets, and change in feed composition, flow, and temperature [58]. In this work, a dataset is generated using Aspen plus dynamics simulation at normal and faulty states. The results show that the proposed model has accurate root cause isolation with a high fault detection rate of 99.77% and a very low false alarm rate of 0.23%. A diagnosis method based on MK-SVM and incremental learning is proposed in [65]. The proposed incremental MK-SVM (iMK-SVM) method uses a linear combination of single kernels to achieve accurate fault classification based on the observed data. The objective of incremental learning is to tune and update the diagnosis that carried out by the MK-SVM in an automatic manner. The iMK-SVM method thus quickly adapts to new observations and provides a more accurate fault diagnosis.

The idea of the SVDD is to identify observations of a specific class among all other observations, by learning from a training set containing only the observations of that class. The SVDD first maps the data from the original space into the feature space by a nonlinear transformation, and then a hypersphere with the minimum volume is constructed in the feature space by solving an optimization problem [70]. The SVDD is applied by Mahadevan et al. as an unsupervised classifier to diagnose faults in chemical processes [71]. Beghi et al. applied the SVDD to diagnose faults in ventilation and air conditioning (HVAC) chiller systems [72]. In that paper, it is employed as a novelty detection system to identify the unknown status and possible faults, by exploiting the PCA to accent novelties with respect to normal operations variability.

The support vector clustering (SVC)-based method has the capability of unsupervised learning with unlabeled training data. A set of support vectors are identified to characterize the probability density distribution of training data and then the minimum spheres surrounding different clusters are constructed within the high dimensional feature space [73].

The optimal solutions on normal or faulty cluster spheres are obtained by solving a minimization problem and the sphere centers are accordingly estimated. For new monitored observation, its geometric distances to each cluster sphere center are calculated and a probabilistic index is then calculated with respect to all clusters. More details about the SVC are found in [74]. A novel SVC-based probabilistic method is proposed for unsupervised chemical process monitoring and fault classification in [67]. The proposed SVC method is applied to two test scenarios in the TEP and is compared to the conventional K-nearest neighbor Fisher discriminant analysis (KNN-FDA) and K-nearest neighbor support vector machine (KNN-SVM) methods.

It is concluded from the literature that the SVM is not only a useful statistical theory but also a method that can be useful in fault diagnosis in many practical engineering problems [23]. However, one of the limitations of the SVM approach is that the accuracy of the model depends on the choice of the defined kernel and penalty parameters. They play a crucial role and should be optimized to prevent over-fitting and to yield better generalization performance. Shengwei Fei et al. proposed a method called support vector machine with genetic algorithm (SVMG) to diagnose faults in power transformers [75]. In that work, the genetic algorithm was used to select appropriate free parameters of SVM. Yuan et al. proposed an artificial immunization algorithm (AIA) to optimize the parameters of SVM [76].

3.1.1.3 Bayesian networks (BN)

The Bayesian network (BN) is a probabilistic graphical model where each variable represents a node and the edges of the graph represent dependencies between nodes [77]. Fig. 6 shows a BN with four variables X_1, X_2, X_3, and Y. The network structure is determined from prior process knowledge [78]. The main concept of the BN is to enter evidence in the network (evidence is the observation of the values of a set of variables) [79]. Mathematically, the Bayesian network structure is defined as a directed acyclic graph (DAG) $G = \{V, E\}$, where V and E are sets of nodes and edges, respectively. Each node in the DAG represents a measurement variable and each directed edge connects two nodes from a parent node (representing a cause of an event) to a children node (representing the effect of an event), and no cycles are involved. More details are found in [27].

Variables with no parents are called root nodes while variables with no child nodes are called leaf nodes. The conditional independency among all connected variables is an important property for the Bayesian network. The joint probability

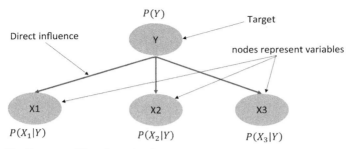

FIG. 6   The Structure of Bayesian network.

distribution of all variables X_1, X_2, ...X_n is factorized using the chain rule of Bayesian network, given as [27]:

$$p(X_1, X_2, ...X_n) = \prod_{i=1}^{n} p\left(X_i \mid X_{parent(i)}\right) \tag{6}$$

where $X_{parent(i)}$ denotes the parent variables of variable v_i graph G and $p(X_i \mid X_{parent(i)})$ is the conditional probability distribution (CPD).

The information given by the evidence is propagated in the network in order to update the knowledge and obtain a posteriori probabilities on the non-observed variables [80]. This propagation mechanism is called inference. Given a training dataset, the graph structure of the BN model is defined and a probability distribution associated with each node in the graph is calculated by maximum likelihood estimation (MLE) [81]. The MLE algorithm is used to estimate the probability of a node based on its frequency of occurrence in the dataset [82].

The Naïve Bayes Network (NBN) (also called Bayes classifier) does not take into account the correlations between variables which are frequent in industrial systems [83]. As a consequence, the classification efficiency is poor in such cases [81]. To solve this major problem generated by the correlation between the variables some extensions to the NBN have been developed [84]. It is reported in [85] that the case when the independence assumption of the variables is verified is the NBN classifier is optimal in terms of misclassification rate. The tree augmented Naïve (TAN) Bayesian classifier, and other extensions were proposed, and are performing better than the NBN classifier because they take into account some of the existing correlations between the process variables [81].

The BN allows the consideration of the temporal dimension using Dynamic Bayesian Networks (DBN) [86]. Moreover, the knowledge provided by an expert can also result in the creation of latent variables between two or more nodes [87]. This is the case for example of unsupervised problems where the class is never measured.

In the context of fault diagnosis, the structure of the BN classifier is constructed by using only key variables in the faults database [88]. Zhang et al. present a Bayesian network classifier based on the Gaussian mixture model (GMM) and a non-imputation method to deal with the FDD problems for these high-dimensional data incomplete industrial processes [89]. The TEP is used as a case study in that paper. The paper [90] has successfully implemented a fault diagnosis method for airplane engines using a TAN. The particle swarm optimization (PSO) algorithm is used for learning the network structure from a large dataset.

Zhu et al. present a distributed Bayesian network approach for fault detection and isolation in large-scale plant-wide processes [91]. The approach decomposes the entire plant-wide process into blocks and Bayesian networks are constructed for different blocks that are then fused into a global Bayesian network. The approach is used a missing data imputation method for state estimation based on which the T2 and Q statistics and uses a Bayesian decision fusion mechanism for fault detection. A Bayesian contribution index is developed for fault isolation. The approach is applied to the TEP and shows that it can be feasible for modeling large-scale processes. Don & Khan introduced a methodology for fault diagnosis based on a combined approach of data and process knowledge-driven techniques [92]. The methodology combines the Hidden Markov Model (HMM) with the BN. The HMM is used in that work to detect the abnormalities based on process history while the BN is used to diagnose the root causes of faults. The performance of the proposed methodology is tested and validated based on 10 selected faults of the TEP data. This paper proposes a Bayesian network-based probabilistic ensemble learning (PEL-BN) method that captures the mixed characteristics of multiple faults by integrating decisions derived from different diagnosis models [93]. A simulation process and a real industrial process are used to validate the performance of the proposed method, and the results show that the proposed method improves the diagnosis performance of single faults and is a feasible solution for multiple faults diagnosis. This paper proposes a fault diagnosis method that integrates BN and fault trees [94]. The proposed method is applied to qualitatively and quantitatively diagnose faults of turbo compressor that is a piece of vital equipment in petrochemical plants. It ranks the faults in descending order according to their severity.

A fault detection and isolation method in chemical processes based on the Bayesian network was proposed in [95] to isolate the contributing variables that are responsible for a fault. The paper demonstrated the impact of taking into account causality in the detection and isolation of faults. Verron et al. purposed and evaluated the performance of a fault diagnosis method for industrial processes. This method is based on the use of a Bayesian network classifier, based on the identification and selection of important variables [96]. This variable selection is made by computing the mutual information between each process variable and the class variable.

The advantage of Bayesian network is its adaptability. The network has directed edges which can model causal effects. However, the BN classifiers could not make choice easy if there are too many variables in the monitored process. The mean limitation of the predictive machine learning methods in FDD is that they are not so great when it comes to knowledge presentation. The architecture of the predictive model is considered as a black box to the user. Another limitation is the tedious parameter tuning of such models.

3.1.2 Descriptive FDD methods

Descriptive machine learning methods do exactly what the name implies they describe, or summarize raw data and make it interpretable by humans [97]. The key advantage of descriptive ML methods over the predictive ones is that these methods enable the process operators to learn from the historical data and allow them to analyze the past behaviors and to understand how they might influence future outcomes. Among the descriptive ML methods, the decision trees (DT) [98], rough set theory (RST) [99] and LAD [100] are popular ones.

3.1.2.1 Decision trees

The decision tree is one of the most widely used and practical methods for inductive inference, introduced by [98]. It is a method for approximating discrete-valued functions that is robust to noisy data and capable of learning disjunctive expressions [101]. It is a flow-chart-like structure where each node represents a test on a variable, and each branch represents an outcome of the test and the values of the class variable are placed at the leaves of the tree, as shown in Fig. 7 [11]. Therefore, learned trees are represented as sets of if-then rules that are readable (interpretable) to the human. Decision trees are quite popular and ranked #1 in the pre-eminent paper entitled "Top 10 Algorithms in Data Mining" that was published by [102]. Each branch represents an outcome of the test, and leaf nodes represent decision classes. Some measures are used to estimate the quality of tests such as the entropy, Gini's index, Sum-Minority, Max-Minority and Sum-Impurity [103,104]. The maximal-discernibility (MD) algorithm, presented in [105], uses discernibility measures to evaluate the quality of tests. The advantage of the decision trees is that they are represented as sets of if-then rules that are readable (interpretable) to the human [106].

Decision trees aim to search for rules hidden in a very large amount of data. They have already been shown to be interpretable, efficient, problem independent and able to treat largescale applications [106]. Among the decision tree methods, classification and regression trees (C&RT) and C5.0 (the successor of C4.5) are the two well-known and widely used algorithms [107]. J48 is an open-source Java implementation of the C4.5 algorithm in the Weka data mining tool [108]. The iterative dichotomizer 3 (ID3) invented by Ross Quinlan [98] is the precursor to the C4.5 algorithm, and is typically used in machine learning to generate a decision tree from a dataset. A novel hybrid classification

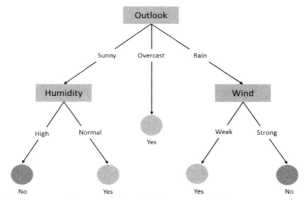

FIG. 7 The decision tree of common *Play-Tennis* example [11].

method has been proposed in [109], based on the C4.5 decision tree classifier and one-against-all (OAA) strategy, to improve the classification accuracy in the case of multi-class classification.

The C4.5 algorithm is combined with other methods and used for fault diagnosis in chemical processes for which diverse incomplete observations were produced [110]. Demetgul applied the decision trees algorithm for diagnosing faults in modular production systems (MPS) in [111]. The C5.0 algorithm is used in [112] for bivariate process online shift monitoring and fault identification is proposed under the assumption of a constant variance-covariance matrix. Two decision tree classifiers were built, one for process monitoring and the other for fault identification. A decision tree is applied in [113] to identify the causes of failures in large Internet sites. A method based on C4.5 decision tree and PCA is proposed in [114]. In that method, the PCA is used to reduce the dimensionality of data then the C4.5 is trained to generate a decision model with diagnostic knowledge.

Among the limitations is that large decision trees with many branches are difficult and time-consuming to interpret by the user. The problem of searching for the shortest decision tree has shown to be NP-hard, and one of the main challenges related to this approach is how to construct an optimal tree for a given dataset. Therefore, heuristic algorithms are used to find the tree that is very close to the optimal one. Well-known algorithms for constructing decision trees are classification and regression trees (C&RT), C4.5, C5.0, and the iterative dichotomizer 3 (ID3) [107]. Those algorithms are mainly based on the top-down recursive strategy. The training of decision trees requires optimization tasks for determining the best split of each node and selecting optimal combining weights to prune the decision tree. To avoid the problem of overfitting, the majority of decision tree algorithms use a post-pruning strategy that involves constructing the tree from the data until all possible leaf nodes have been reached then pruning [107]. Nguyen proposed the RSABR approach for the construction of a decision tree based on managing the discernible objects. The method uses a discernibility measure in the induction of the maximal-discernibility decision tree [105]. It considers a problem called MD partition for searching for optimal binary partitions with respect to discernibility. The proposed method aims at avoiding overfitting, searching for the best tests, and creating accurate decision trees.

This paper [115] presents an ML method based on an ensemble of pattern trees to predict hot metal temperature (HMT), which is related to the product quality and thermal state, to maintain the high productivity of the blast furnace. The efficiency of the proposed method was verified through its application to an industrial blast furnace ironmaking process. The application results have demonstrated that the proposed method performs better than the conventional PLS, decision trees, random forest, and artificial neural networks.

3.1.2.2   Logical analysis of data

Logical analysis of data (LAD) is a machine learning classification technique that is based on the theories of Boolean algebra and combinatorial optimization, introduced in 1988 by [100]. It is a rule-based machine learning classification method where the fundamental concept is the extraction of human-interpretable patterns from the data to discover hidden knowledge from a set of training observations [116].

LAD was used on a great variety of classification problems and it reacts well to noisy and erroneous measurements, and the results obtained in [116,117] indicate that the LAD's classification accuracy is often superior to the other classification techniques. It has been used for the first time in condition-based maintenance (CBM) as an effective diagnostic technique that relies on extracting patterns from the training datasets that contain two classes of observations; positive and negative [118]. Each extracted pattern represents the logical conjunction between variables for either the set of positive or the set of negative observations [116]. The patterns are then translated into a decision model that is used as a classifier for the new observations.

In addition to its good accuracy, LAD is not based on any statistical analysis, which makes it capable of dealing with highly correlated or time-varying covariates, without the need to satisfy any statistical assumptions. Another advantage is its transparency in describing the cause-effect relation for the degradation and failure processes [119]. The discovered knowledge in terms of the patterns has physical interpretations that describe the physical events of degradation or failure.

LAD mainly consists of three main steps: data binarization, pattern extraction (generation), and selection, and theory formation [116]. The binarization step involves the transformation of each nonbinary variable, whether numerical, categorical, or ordinal, into a set of binary attributes. This is because LAD is based on certain concepts from the Boolean functions. The pattern extraction is the cornerstone of the LAD approach. The pattern is defined as a logical conjunction of some variables in the data. A pattern is said to cover a certain observation if it is true for that particular observation, that is if its logical conjunction is valid [116]. The pattern generation step is terminated when every observation in the training data is covered by at least one pattern.

Fig. 8 shows how the distinction is made between observations that are classified to *Class 0* and those are classified in the opposite class (*Class 1*). The green pattern $(\overline{X_8}X_9)$ that is, $X_8 = 0$ *AND* $X_9 = 1$ appears in observations 1 through 10 which are belonging to *Class 0*, and does not appear in any observation in *Class 1*. Similarly, for *Class 1* observations, the red pattern $(X_3\overline{X_7})$ appears in observations 11 and 12, and the yellow pattern $(X_1\overline{X_5})$ appears in observations 12, 13 and 14. The two patterns in *Class 1* do not appear in any observation in *Class 0*.

There are three common methods for pattern extraction: (1) enumeration-based methods [116], (2) heuristic and meta-heuristic methods [120–122], and (3) mixed-integer and linear programming (MILP) optimization-based methods [117,123]. More details about the steps of LAD through an illustrative example are found in [124].

It should be noted that not all the patterns are taken into consideration in the final LAD model. A pattern selection procedure is therefore required in order to find the set of strongest patterns that guarantee the coverage of all observations while avoiding the overfitting of the LAD model. This is a key challenge to achieve the most generalized explanatory power. In the last step of LAD, the theory formation, the extracted patterns are finally represented as sets of if-then rules that are readable (interpretable) to the human [124]. An extension to multi-class applications, that involves modifying the architecture of two class LAD, is proposed in [119]. The method aims at modifying the architecture of LAD as a dichotomy to a multi-class decision model (polychotomizer). The proposed method has the advantage that it generates a less complex decision model, in a better execution time. In that work, a unified multi-classifier is obtained by the combination of binary classifiers.

Observation	X_1	X_2	X_3	X_4	X_5	X_6	X_7	X_8	X_9	Class
1	1	0	0	0	1	1	0	0	1	0
2	0	0	0	0	0	1	1	0	1	0
3	1	1	0	0		1	0	0	1	0
4	1	1	0	0		1	1	0	1	0
5	0	0	1	1		1	0	0	1	0
6	1	1	1	1		1	1	0	1	0
7	1	0	1	1		1	0	0	1	0
8	0	0	1	1		1	1	0	1	0
9	1	0	1	0		1	1	0	1	0
10	1	1	1	1		1	0	0	1	0
11	0	0	1	0		1	0	0	0	1
12	1	0	1	1	0	1	0	1	1	1
13	1	0	1	1	0	1	1	1	1	1
14	1	1	1	1	0	0	0	0	0	1

FIG. 8 Representation of the positive and negative patterns and their coverage.

LAD was used in [125] to detect and isolate rogue components. By monitoring certain performance indicators and expert system knowledge, patterns that are unique to rogue components are discovered. In [117] an approach for automatic diagnosis of faults in rolling bearings by using a modified pattern generation method with LAD is presented. The vibration signals in that paper were used for the detection of bearing faults at an earlier stage of the crack propagation. Ragab et al. developed two reliability-based prognostic methodologies to predict the health states of equipment, based on the lifetime and condition monitoring (CM) data [126,127]. The methodology exploits the condition monitoring data collected just before the occurrence of complete failure in equipment. Ragab et al. proposed a methodology for multiple failure modes prognostics in rotating machinery in [128]. The methodology merges multiclass LAD with a set of nonparametric cumulative incidence functions (CIFs). It considers the condition monitoring data that is collected from a system that experiences several competing failure modes over its life span. Jocelyn et al. applied LAD to machinery-related occupational accidents. In that work, LAD was used to characterize different types of machinery-related accidents and to relate them to the root causes of faults [129].

Other pattern classification methods have some similarities with LAD. Such methods are rule-induction methods that are based on the extraction of descriptive rules that permit the user to understand the most interesting relationships in data. The most common examples are the rough set theory (RST) such as the ENDER [130], SLIPPER [131], MLRules [132], RuleFit [133], and LRI [134].

3.1.2.3 Rough set theory

Rough set theory (RST) was introduced by Pawlak as an extension of set theory [135]. It is used for the study of data that are characterized by incomplete values to classify imprecise or incomplete knowledge [136]. The RST is an effective tool to handle fuzzy and uncertain problems, and it can process the knowledge by obtaining the minimal representation of information [137]. The RST deals with the automatic generation of the rules for the classification of observations. The fundamental concept of rough set theory is indiscernibility, which is mainly used to define the equivalence classes for the observations [99]. The distinction is made between observations that may definitely be classified to a certain category and that may possibly be classified [138]. Considering specific variables, observations are indiscernible (similar) if they are characterized by the same variable.

Rough set structures allow the formulation of the notions of lower and upper approximations of each class in the data. The objective is to search for the optimal rough approximation with the minimal boundary region separating between classes of observations [139]. The RST approximations are determined for the training observations only. Therefore, as in the problem of searching for an extension theory in the LAD approach, a big challenge in the RST inductive learning approach is to construct optimal approximations for the extension of the rough membership function. The RST is not only a useful tool for rule extraction and classification but it can be treated as a feature selection method [139]. In rough set theory, the feature selection problem is defined in terms of reducts that is, subsets of the most informative variables in a given dataset.

Researches have been conducted to modify and improve the classical RST approach in order to construct rough classifiers [140]. The RST has the same advantage as LAD in data analysis that provides efficient algorithms for finding hidden patterns in data without any preliminary or additional information about data. In addition to its interpretability, one of the advantages of RST is that it can handle uncertain problems, and it can process the knowledge by obtaining the minimal representation of any type of information. More details about RST are found in [99]. Chikalov et al. presented a comprehensive discussion on the similarities and differences between LAD and RST in [106].

Because the rough set theory has a superior ability to extract the uncertainty information, it has been widely applied in fault diagnosis, especially it has made great achievements in extracting interpretable rules [141]. Feature reduction is one of the main steps of rough set theory [142]. This paper [143] proposes an approach based on rough set theory to decrease the cost of fault diagnosis in industrial systems. The proposed approach is implemented to extract the hidden relations and implement rule-based control for the sustainability of the rotary clinker kiln.

Konar et al. propose a rough set-based method for multi-class fault diagnosis of induction motor [144]. Hua et al. proposed the rough set theory as a diagnostic method for transformer fault diagnosis [145]. Zhou et al. proposed an integrated rough set and neural network algorithm-based fault diagnosis for power transformers, using dissolved gas analysis (DGA) [146]. This approach takes advantage of the feature reduction ability of the rough set and the good adaptability of ANN.

Rough set theory is applied to generate the rules from the vibration signals of different faults collected from monoblock centrifugal pump in [147]. The vibration signals are preprocessed then rules are generated using rough set theory and classified using fuzzy logic. The rough set theory is used in [136] to locate and diagnose the distribution feeder faults in order to reduce the duration of the outage. The RST is combined with the Bayesian network in [148] to deal with uncertain problems in transformer fault diagnosis.

3.2 Multi-variate statistical process monitoring methods

Multi-variate statistical process monitoring (MVSPM) methods are among the most fundamental tools used widely in chemical and biochemical processes. These methods include PCA, projection to latent structures (PLS), independent component analysis (ICA), Fisher discriminant analysis (FDA), canonical variate analysis (CVA), and others.

This paper [21] illustrates the different aspects of multivariate statistical process monitoring methods and SVM as common statistical FDD methods in fault detection and diagnosis. It clarifies how these methods can deal with a certain spectrum of process issues, such as dynamic and non-linearity issues. In what follows, such methods are discussed.

3.2.1 Principal component analysis (PCA)

The PCA was used as of deterministic variance-preserving method because of its simple form and its ability to handle a large number of process data [149]. PCA preserves the significant variability information extracted from the process measurements and proposed originally for the purpose of dimensionality reduction of the huge amount of correlated data [150]. It has been widely and successfully applied as a fault detection and diagnosis methods in industrial processes [151]. Fault isolation based on PCA relies on score distance and residual distance decomposition contribution plots, as reported in [149].

PCA transforms correlated and uncorrelated variables into two independent spaces. Assume that a recorded data X represents m process sensor measurements with N observations collected and represents the normal process operation. The data is normalized to zero mean and unit variance. The covariance matrix is calculated as follows:

$$S = \frac{1}{N-1} X^T X \tag{7}$$

To get the most variability from the preprocessed normal operation data set, the singular value decomposition is performed on the covariance matrix S, as follows:

$$S = \frac{1}{N-1} X^T X = P \Lambda P^T \tag{8}$$

where $\Lambda = diag\,(\lambda_1, \cdots, \lambda_m)$ is a matrix with the eigenvalues of S, ranked in a descending order and P is a matrix of the corresponding eigenvectors. The normalized data matrix X is divided into the two subspaces $(\hat{X} + \tilde{X})$ and the eigenvectors of P is also separated into:

$$P = \begin{bmatrix} \hat{P} & \tilde{P} \end{bmatrix} \tag{9}$$

where $\hat{P}$ and $\tilde{P}$ are the principal and residual subspaces, respectively. Suppose that n is the number of principal components (PCs), then $\hat{P}$ has the eigenvectors corresponding to the first n largest eigenvalues in Λ, and $\tilde{P}$ contains the eigenvectors corresponding to the remaining $m - n$ smallest eigenvalues, and as a result $\Lambda = [\Lambda_n\ \Lambda_{m-n}]$. There are several methods to select n and consequently the appropriate PCs number [152]. The cross-validation is an example of these approaches [153].

When projecting X onto the abovementioned two orthogonal subspaces, a score matrix T can be obtained by the following formula:

$$T = XP = X \begin{bmatrix} \hat{P} & \tilde{P} \end{bmatrix} = X\hat{P} + X\tilde{P} \tag{10}$$

The original data X can be reconstructed by the following formula:

$$X = \hat{X} + \tilde{X} = \hat{T}\hat{P} + \tilde{T}\tilde{P} \tag{11}$$

The PCA approach has a major limitation that is in addition to its linear nature, there is a lack of exploitation of autocorrelation [18]. Moreover, the PCA-based process monitoring method claims that the assumptions on the multivariate Gaussian distribution of the measurement variables have to be fulfilled, which is not always valid in chemical processes; this is due to time-varying behavior and possible high sampling frequencies [154]. Moreover, the minor principal components would normally represent insignificant variance in the data for the linear case, but this cannot be verified certainly for nonlinear data [155]. More principal components have to be retained in order to represent nonlinear data properly. It is also difficult to discern which minor components capture nonlinearity and which represent insignificant variation [156].

This paper [157] proposes a procedure to determine the principal components to be retained using the discriminant information contained in them. The proposed method is based on the use of statistical tests to select the components with greater separability power between classes.

3.2.2 Projection to latent structures (PLS)

Projection to latent structures (PLS) is another basic multivariate statistical method that is used extensively for model building, fault detection, and diagnosis [149]. It extracts the correlation model from the process inputs and outputs [158]. Unlike PCA, by the use of the off-line trained correlation model and online process measurements, the PLS model could provide more beneficial information for the purpose of fault diagnosis purposes, and for the online prediction of key performance indicators (KPIs) of an industrial process [159]. In other words, the PLS method inclines to discover the faults that occurred in process inputs, which might influence the KPI. As one can guess, one of the major uses of PLS is for predicting variables, but this is not its only purpose as a model. It is a good tool for process understanding and troubleshooting [152]. The first step in PLS is to extend latent variable methods to using more than one block of data [160]. The data variables (columns) are divided into two blocks: called X and Y. Accordingly, the PLS extracts two sets of scores, one set for X and another set for Y.

$$T = XW \text{ for the X space} \tag{12}$$

$$U = YC \text{ for the Y space} \tag{13}$$

The scores T and U are found subject to the constraints that $W^TW = 1$ and $C^TC = 1$. More details are found in [152,160].

The process dynamics and nonlinearity are very important aspects in industrial processes. In order to design a robust FDD system that is able to cope with the real industrial environment, more and more attention has been paid to the PCA- and PLS-based process monitoring methods. It is evident that, similar to the PCA-based schemes, for successful application of the PLS-based approaches, the input and output variables should also follow the multivariate Gaussian distribution [161]. To successfully apply the standard such methods, a linear process that runs under stationary operating conditions is a primary assumption [162].

The work in [6] proposes a hybrid fault diagnosis method that is based on the signed digraph (SDG) and the PLS to diagnose faults in the benchmark pulp mill process. The process produces pulp from wood chips, and is considered as one of the biggest processes in the fault diagnosis area [163].

As an extension to the PLS method, a method called total projection to latent structure (TPLS) is proposed in [51]. The main concept in the TPLS is to acquire a further decomposition on input space, thus allowing the properties of each subspace to be analyzed in detail, which is more suitable for process monitoring. The so-called concurrent projection to latent structures (CPLS) method is proposed in [164]. In the CPLS method, the input and output data are concurrently projected to five subspaces, thus fault detection indices are developed based on the five subspaces for various types of fault detection alarms. It is concluded from the results obtained in that paper that the proposed method correctly detects the various faulty cases based on simulated data in the Tennessee Eastman process monitoring the results show that the CPLS-based monitoring effectively detects faults thus avoiding false alarms due to disturbances that are effectively eliminated by the control strategy.

3.2.3   Independent component analysis

ICA can be viewed as a useful extension of PCA, although it has a different objective. Whereas PCA can only impose independence up to second-order statistical information (mean and variance) while constraining the direction vectors to be orthogonal [165]. The ICA method not only decorrelates the data but also reduces higher-order statistical dependencies, thus the ICA method reveals more useful information from observed data than the PCA method [166].

ICA has been recently extended and applied for the purpose of fault detection, particularly for the processes whose measurements are non-Gaussian distributed [165]. Although the calculation involved in ICA is more complicated than the standard PCA and PLS approaches, it is still worthy to investigate and deliver more extensions for process monitoring [167].

3.2.4 Fisher discriminant analysis

Fisher discriminant analysis (FDA) was originally proposed for dimensionality reduction and has been recently widely applied in the multivariate statistic fields and pattern classification [168]. For process monitoring, the FDA method is an efficient tool for the purpose of fault classification [169–171]. With consideration of an additional class of normal operating conditions, the FDA was applied on industrial processes for fault detection to isolate normal process data from a mixture of normal and abnormal historical data [170]. A contribution plot based on the fault directions in pairwise FDA is proposed to enhance the ability of fault diagnosis in multivariate statistical monitoring. The method shows superior capability for fault diagnosis to the contribution plots method based on PCA. The FDA was used with SVM in multiple fault diagnosis in [169]. The objective of the proposed method is to mask the datasets with irrelevant information in order to decrease the effectiveness of the classification techniques. The TEP simulator was used to generate overlapping datasets to evaluate the classification performance.

3.2.5 Kernel methods—kernel PCA (KPCA), kernel PLS (KPLS), and kernel ICA (KICA)

To tackle the nonlinearity issues of the industrial process, variants like kernel PCA (KPCA) [155,157] and kernel PLS (KPLS) [172], and kernel ICA (KICA) [173] were proposed and have been intensively studied in the fault diagnosis. The main concept of using the kernel is to facilitate analysis by mapping the input space into a higher-dimensional kernel feature space, where the data distributions become approximately linear [174].

Gan Liang et al. have studied the use of sparse kernel principal angles for online process monitoring in order to surmount the problem of overlearning that sometimes occurs when kernel methods are used [175]. Simulations on the Tennessee Eastman process indicated that this approach could be used to effectively capture nonlinear relationships in the process variables, but in a significantly simpler way than with kernel PCA. This paper [176] presents an adaptive KPCA (AKPCA) fault diagnosis method in complex chemical processes. The method is based on a moving window integrating the threshold method to adaptively extract the kernel principal. It is applied to diagnose faults of the TEP. The results verify that this proposed method can improve the cumulative contribution rate in comparison with the KPCA that is based on thresholds.

With KPCA, A careful selection of the kernels should be considered. The kernel functions are usually selected empirically, and this may have an adverse effect on the performance of the diagnostic method, if the kernels are not suited to the structure of the data [177]. A maximum variance unfolding method was proposed by [178] to optimize the kernel function over a family of kernels. Simulation with the Tennessee Eastman problem suggested that this promoted more precise monitoring in the feature space. To address the problem of kernel selection, Shao et al. have proposed a method to select a kernel function learning method to fit the kernel function to the training data [179].

Recently, an attractive research field could be observed in the literature, in which process monitoring is achieved by extending the applications of PCA and PLS to the nonlinear dynamic processes. To extend the PCA method to dynamic systems, the paper [180] presented a study of PCA and PLS on lagged variables to develop dynamic models (this so-called dynamic PCA) for dynamic processes. Komulainen et al. used the dynamic PLS (DPLS) in the online monitoring of an industrial dearomatization process, in a similar way to the DPCA [181]. The canonical variate analysis (CVA) is proposed as a fault detection method by using the pairs of variables selected from the input–output data set that maximizes the correlation, hence is more suitable for dynamic monitoring than DPCA and DPLS [182].

Odiowei et al. proposed a monitoring technique using the CVA with the upper control limit (UCL) derived from the estimated probability density function through kernel density estimations and applied to the simulated nonlinear TEP [183]. It is concluded in that paper that the CVA model outperforms the DPCA and the DPLS approaches. It is able to reveal the difference in dynamic behavior between the normal operation and the operation with Fault 9mor efficiently than the DPCA and the DPLS. Odiowei and Cao have proposed state space independent component analysis (SSICA) to deal with nonlinear dynamic process monitoring [184]. The SSICA algorithm uses CVA to construct a state space, from which statistically independent components can be extracted for process monitoring. The SSICA could detect faults earlier than other methods, such as the traditional CVA, in the TEP simulated data. Ding et al. proposed and compared the so-called left coprime factorization (LCF) of a process with the PLS for the prediction and diagnosis of key performance indicators in complex industrial processes [185]. The proposed KPI prediction and diagnosis method is applied to an industrial hot strip mill. The obtained results demonstrate the effectiveness of the proposed scheme for dynamic processes.

3.3   Integrated and hybrid FDD methods

A hybrid data-driven FDD method merges at least two intelligent technologies. Integrating two or more FDD methods can exploit the advantages of each of them. Recently, a multitude of researches focusses on integrating a set of preprocessing and learning algorithms to develop hybrid FDD methods. It is reported in the literature that integrated methods that merge characteristics of different methods can give better results and hence satisfy the requirements for process monitoring [186].

Research has focused on integrating a set of different learning algorithms to develop hybrid methods [187,188]. Integrating different learning algorithms aims to merge their characteristics and merits to exploit the advantages of each and provide better results [189–195]. The concept of fusing several individual learning models (each with its own pattern representation) to create a single strong model that outperforms each individual model was studied by our team at CanmetENERGY Varennes, Natural Resources Canada (NRCan). We have applied the developed models to complex equipment in pulp mills in Canada.

In fault diagnosis, as an example of merging two intelligent methods, combinations of fuzzy logic and neural networks offer benefits. The black-box approach of pure neural networks does not allow utilization of qualitative knowledge of faults and their symptoms, whereas fuzzy logic-based fault diagnostics systems are often static, that is, they do not allow updating throughout the experiments. With neuro-fuzzy networks, a better understanding of the diagnosis process can be achieved, and the fault diagnosis model can be adapted to provide solutions that are more accurate. The overall structure of the Neuro-Fuzzy system is very alike neural network. The system consists of five layers; an input layer and an output layer as well as three middle layers in order to present the fuzzy system. It uses expert knowledge to present data and then it uses a neural network in order to develop if-then rules and adjust input/output membership function to improve the overall performance of the system.

Feature selection techniques and clustering algorithms are most commonly linked to the FDD methods, as common examples of merging FDD methods with preprocessing algorithms. Feature selection techniques are used to reduce the huge dimensionality of feature space, which is one of the most common problems in the fault diagnosis domain [63,196]. Not all the features are useful for the classification process and some of the irrelevant or redundant features tend to degrade the performance of FDD method. If all features are input into the FDD directly without selection, they will make the classification process slower and degrade the classification accuracy as well [197].

It is, therefore, necessary to acquire rich faulty information, thus superior features need to be selected from the original feature set to obviously and properly characterize the system conditions. The feature selection procedure further improves the diagnosis accuracy and reduces the computational burden of FDD methods. Many techniques have been proposed to perform feature selection; distance evaluation technique (DET) [196], particle swarm algorithm (PSA) [198], and genetic algorithm (GA) [199] were conducted for extracting optimal features and reduce the dimensionality of feature space. A feature selection technique employing sequential backward selection (SBS) is found in [51].

The distance evaluation technique (DET) is one of the most simple and efficient distance measure techniques in the field of fault diagnosis [3]. In DET, the idea is to select the features that make the distance within a set of classes shorter and the distance among classes longer. An evaluation factor (weights) for each feature from the original feature set is calculated. Based on these factors, all the features in the feature set are ranked in ascending order. Accordingly, the features with the largest weights can be selected and applied to the diagnosis procedure. In the paper published by Widodo et al., 2007, the features are extracted using PCA and ICA, and then the superior features are selected using DET [52].

Fuzzy logic is used as a clustering technique and combined with common machine learning techniques in unsupervised classification problems, in different domains. A kernel-induced fuzzy clustering method [200] and fuzzy clustering with multi-medoids (FCMM) [201] were proposed as unsupervised classification methods.

In the fault diagnosis domain, Liu et al. provided a solution based on fuzzy clustering [202]. They combined fuzzy c-means clustering and fuzzy integral techniques to form a two-step diagnosis strategy. A set of classifiers were first constructed based on different feature groups. The recognition rates of each classifier for existing faults were then obtained, which reflects the importance of the classifier in recognizing each fault. In the next step, both membership degrees reflecting the initial judgments of the classifiers for the current faults were applied as inputs of a fuzzy integral fusion procedure and a final decision was then made. The proposed method is evaluated by data collected from rolling element bearings with three kinds of faults. The method gives superior accuracy to that of a single classifier. This procedure is highly expected to be applied for chemical processes diagnosis of multi-faults given relevant faults (e.g., fouling, gas, and liquid leakages which are the most investigated ones).

Eslamloueyan proposed a hierarchical artificial neural network (HANN) for isolating the faults of the TEP [203]. The proposed HANN divides the fault patterns space into a few sub-spaces by using a fuzzy C-means clustering algorithm. For each sub-space of fault patterns, a neural network has been trained in order to diagnose the faults of that sub-space.

When applying fuzzy clustering for fault diagnosis, several critical points should be taken into account: determination of the optimal number of clusters, feature selection, and objective function [204]. The optimal number of clusters can be given either by prior knowledge or considering some validity indices such as partition coefficient and exponential separation (PCAES) [205], and others found in [206] and [207].

4 Fault diagnosis in pulp and paper mills: Case studies

Our knowledge in applying and adopting AI for abnormal event management and process operation optimization in the pulp and paper industry is shared through this chapter. Several of the elements that we focus on can be related to the experiences and lessons learned in different pulp and paper mills. We will discuss these challenges and share our experience in applying several AI tools successfully in real complex operational cases in different Canadian industries with the assistance of process experts. We have adopted and combined different ML and DL methods to actively monitor and optimize the operations of complex processes and to bring attention to problems before they occur. We already shared some of these use cases with the scientific community through articles that have been already published in specialized international journals [25,124,208]. Rather than propose entirely hypothetical situations, we have chosen to refer to publications that share our knowledge and give real operational cases related to FDD in a number of process industries.

We targeted several complex equipment and processes in the pulp and paper industry. It is known that this industry is one of the energy-intensive industries that remains a fundamental part of the Canadian economy. Pulp and paper mills convert predominantly woody material and agricultural residues into a wide variety of pulps, papers, paperboards, and other cellulose-based products [209]. These mills may exist separately or as integrated plants. Integrated mills use common auxiliary systems for both pulping and papermaking such as steam, electric generation, and wastewater treatment. The pulp and paper-making processes mainly use a huge amount of energy for their operations [209] with the Kraft process being the most dominating chemical pulping process worldwide [209].

The Kraft pulping process consists of several steps including wood chips cooking, pulp washing, bleaching, pulp drying, and chemical recovery. The wood chips are dissolved using the pulping chemicals, primarily NaOH and Na2S, resulting in a black liquor (BL) that is separated from the pulp in the brown stock washing step. The black liquor is concentrated in the evaporators and then sent to the black liquor recovery boiler (BLRB) to recover the inorganic chemicals for reuse. This makes the mill a closed cycle process. More detail about the Kraft process is found in [210].

Mechanical pulping is also an energy-intensive process. It is categorized as thermomechanical pulping (TMP), stone groundwood pulping (SGW), wood chips refining and others [211]. The TMP mills are used for the production of high-yield pulp. A typical TMP process consists of several stages including washing the wood chips to remove contaminants, preheating to soften the chips, and two refining stages (fiber separation and fibrillation), a latency chest for removal of fiber latency. More detail about the operation of TMP mills can be found in [211].

Process engineers and operators in pulp and paper mills frequently face energy efficiency issues related to abnormal (faulty) events, such as deteriorations in process performance, production losses, and reduction in product quality [212]. In these complex industrial processes, human expertise is heavily involved in fault diagnosis procedures to deeply understand and analyze a system's state; therefore, a reliable tool that supports executing this task is highly sought after [25]. We applied AI techniques in more than 15 case studies in different sectors in Canada (Pulp &Paper, oil refining, steel making, food, etc.). Through this section, we would like to share our experience in applying AI solutions to selected case studies. The section discusses a number of case studies involving fault diagnosis of complex equipment in different Kraft and TMP mills located in Canada. Inherently, these mills feature complex dynamics and profound interactions among process units.

4.1 Case 1: Black liquor recovery boiler (BLRB) in kraft pulp and paper mills

The BLRB is a major piece of equipment in Kraft P&P mills, made up of several units, shown in Fig. 9 [210]. It maintains the pulping process at a high level of efficiency and availability. Basically, the BLRB has three functions:

- Recovers and converts the chemicals used in the process
- Produces steam through the combustion of the BL organic components
- Eliminates undesirable by-products

One of the KPIs that reflects the efficiency of the BLRB is steam production divided by black liquor flow (SP/BLF). Several factors affect the SP/BLF such as solids percentage in the BL, combustion air in the boiler furnace, fouling rate of the tubes in the economizer, super-heater and boiler bank and inlet water temperature. A low SP/BLF value generally indicates a poor BLRB performance, often due to abnormal operation. The objective of applying LAD is to monitor the SP/BLF and to diagnose and interpret any fault that decreases its value. This paper [124] tested the LAD as a powerful descriptive machine-learning classification method to discover the hidden phenomena in the BLRB. In that work, LAD was used to analyze and diagnose the cause of low steam production by black liquor flow (SP/BLF). The revealed patterns are finally combined to build a decision model that serves to diagnose abnormal situations during the boiler operation, and to explain their potential causes to the process operator.

4.2 Case 2: Reboiler system in thermomechanical (TMP) pulp mills

In the TMP, significant quantities of electrical energy are applied in the refiners, but only a very small fraction of the energy is used in wood chip refining and most of this energy is converted into dirty steam. The dirty steam energy content is typically recovered in the reboiler system for clean steam production or recovered in the heat recovery system for hot water production. Fig. 10 shows a simplified schematic of the reboiler system.

The paper [25] contains more detail about the operation of the reboiler system. The dirty steam comes from the two TMP lines and it is used in the reboiler for producing a low-pressure (LP) clean steam. The LP steam network is further fed by another source of steam extracted from the turbine. The clean steam produced is used by two paper machines and other steam users at the mill. The undesired event in the reboiler system is explained briefly in the following. An increase in the dirty steam pressure leads to the opening of the dirty steam header vent valve. This consequently leads to energy losses to the atmosphere; moreover, it decreases the overall efficiency of the heat recovery system. The

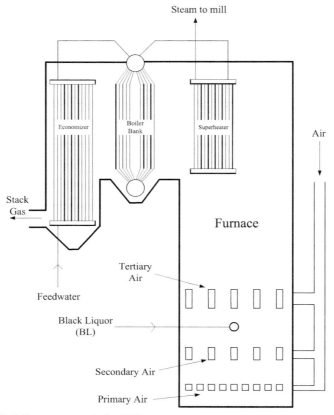

FIG. 9 Black liquor recovery boiler [210].

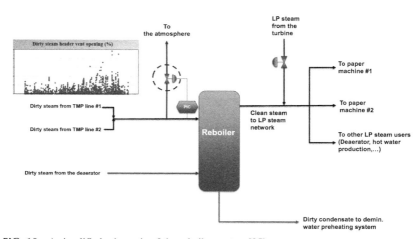

FIG. 10 A simplified schematic of the reboiler system [25].

paper [25] proposes an innovative methodology that combines human expertise, in the form of fault tree analysis (FTA), with additional knowledge extracted by LAD as a descriptive ML method. The proposed methodology was demonstrated using fault trees constructed for the reboiler system in a TMP. The fault tree for the reboiler system was updated successfully with minimal effort from process experts. This paper [208] proposes a multi-agent-based methodology for real-time fault diagnosis for a reboiler system in TMP. The proposed methodology combines both supervised and semi-supervised methods in a parallel-serial structure, exploiting their respective strengths. This method aims to build a decision support tool that helps process operators identify and better manage faulty situations whether known or novel, moreover, it provides the process expert with meaningful explanations of detected faults.

4.3 Case 3: Heat recovery system for hot water production in TMPs

The purpose of the heat recovery system (HRS) in the TMP is to recover the thermal energy from dirty steam that emerges from the refining lines of the wood chips. Fig. 11 shows a simplified diagram for the HRS. It consists of several direct contact condensers for condensing the dirty steam [25]. The hot, dirty condensate is then filtered through the screen to remove the solid particles before feeding the heat exchangers. The heat exchangers transfer the energy to the warm water, and the resulting hot water is reserved in a tank and its level is controlled. A temperature control loop is used to achieve the desired temperature for the hot water used by the paper machines in the mill. If the desired temperature is not reached, fresh steam is used in the heater to compensate. The peaks in fresh steam consumption are not only costly, but they also negatively affect the electricity production in the turbine. Therefore, these undesired peaks in fresh steam are considered to be undesired events and should be avoided. An enriched FTA for the HRS is developed in [25] with minimal human effort. The initial FTA developed by the mill expert was enriched based on the historical data collected using LAD and other ML methods.

5 Concluding remarks and lesson learned

Fault detection and diagnosis in large-scale process industries is a big challenge for engineers and operators. Despite the important work done recently by academic and private institutions, there are still practical challenges involved in designing efficient FDD systems for complex industrial processes. These processes are categorized as highly nonlinear dynamical systems with a huge number of observed variables. In large-scale processes, performance decay, instrument errors, inappropriate control tuning, and operator errors are among the sources of process operation inefficiencies and other abnormal events (faults). This is due to several factors like the complexity of the process, a lack of adequate models, incomplete knowledge and uncertain data, and the amount

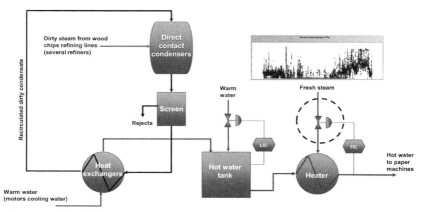

FIG. 11   A simplified schematic of the heat recovery system in the TMP [25].

of effort and expertise required to maintain the process operations in optimal conditions. Therefore, there is a need to provide the process operators with easy-to-use descriptive and prescriptive decision support systems (DSS) that help them in finding and analyzing the root causes of inefficiencies and abnormal events, as well as to advise them with the optimal corrective actions.

The use of AI has exploded over the past few years in different industrial problems, driven by advanced and relatively inexpensive computing infrastructures and the increasing availability of datasets. Machine learning (ML) and deep learning (DL) have become at the heart of AI, and have been proven as promising technologies for fault diagnosis. Process industries have started to leverage the capabilities of ML and DL techniques across various sectors to address fault diagnosis related challenges. One of the main strengths of these techniques is their capability in analyzing heterogeneous sources of data such as numbers, text, images, and videos to build predictive and descriptive diagnostic models in order to monitor, evaluate, and manage energy consumption.

This chapter addresses practical issues that need to be considered when selecting the appropriate ML models for fault diagnosis in some industrial applications. The objective is to assist engineers and researchers in selecting the proper tool for detecting and diagnosing faults within their specific business environment. The chapter also explores the strengths and weaknesses of the ML-based diagnostic methods to establish the criteria that make them better suited to certain applications than others. This provides an important starting point for researchers and practitioners in considering diagnostics options.

Our intention is to offer rational and transparent approaches for developing ML-based models from available data to achieve energy-efficient processes. We hope readers will benefit from the suggestions offered in this chapter and be able to derive a sufficient understanding to know when our recommendations may fit with their particular requirements by making the appropriate adjustments.

We briefly share our experience in adopting and applying ML tools in real operational cases related to fault diagnosis in different Canadian mills. The challenges we faced have allowed us to get some remarks and learn some lessons and insights that are shared in this chapter. We targeted several complex equipment and processes in the pulp and paper industry that do not have understandable phenomena. It is known that this industry is one of the energy-intensive industries that remains a fundamental part of the Canadian economy. Process engineers and operators in pulp and paper mills frequently face diagnostics issues related to abnormal (faulty) events that result in deteriorations in process performance, production losses, and reduction in product quality.

It is still challenging to adopt ML techniques to some process industries' equipment that have complex dynamic and nonlinear nature. Examples of these equipment are heat exchangers networks, recovery boilers, and reboiler systems in pulp and paper mills. These equipment require the intervention of human expertise to tailor the ML techniques and validate the resulting solutions to address the targeted problem. In these complex industrial processes, human expertise is heavily involved in fault diagnosis procedures to deeply understand and analyze the system's state. Therefore, a reliable tool that supports executing this task in an automated fashion is highly sought after. With all the hype around digitalization, AI, and IIoT, what is missing are solutions leveraging the knowledge of domain experts to solve complex real-world problems. Also, due to the complexity of problems in heavy process industries, it is sometimes necessary to use a combination of AI methods in addition to human expertise and first principles. This combination can exploit all possible opportunities that achieve energy-efficient industrial processes.

An important finding is that the hybridization of AI methods is reported to be effective in the advancement of prediction models, particularly for complex systems. It is well known among researchers and practitioners in different industrial applications that there is no single ML technique that perfectly works in all types of diagnostics issues. According to the literature, a method may outperform all others for a specific subset of problems but cannot achieve the optimal performance in all situations. Therefore, the pros of each method could be exploited and integrated to potentially offer complementary information about the diagnosed faults. Independent decisions of each method are then combined to finally derive an elicited optimal decision.

Our conclusion is that given the complexity of the equipment and processes in pulp and paper mills, ML can be a promising modeling framework, which has been previously used by our team in the context of optimizing the performance of some processes in these mills. The diverse ML models we have developed have made it possible to more accurately predict certain problems in pulp mills. Although this technology has made prominent contributions, its potential has not been fully tapped in this industry. However, expert opinions in this industry can be fraught with and full of bias, thus posing extra challenges. Some of these challenges are related to data privacy and others related to obtaining timely and cost-effective solutions. Domain experts in the pulp and paper industry, to a large extent, are still hesitant to adopt these tools. Particularly, this industry is still skeptical about the use of AI methods, since they are data-based solutions rather than traditional physics- and knowledge-based approaches, and operators/engineers do not yet have sufficient experience to deal with AI and data-based techniques.

Acknowledgment

This work was supported by Natural Resources Canada's OERD (Office for Energy Research and Development) program.

References

[1] J.Z. Sikorska, M. Hodkiewicz, L. Ma, Prognostic modelling options for remaining useful life estimation by industry, Mech. Syst. Signal Process. 25 (5) (2011) 1803–1836.

[2] G.Z. Vachtsevanos, Intelligent Fault Diagnosis and Prognosis for Engineering Systems, John Wiley & Sons, 2006.

[3] Y. Lei, Z. He, Y. Zi, Application of an intelligent classification method to mechanical fault diagnosis, Expert Syst. Appl. 36 (6) (2009) 9941–9948.

[4] A.K.S. Jardine, D. Lin, D. Banjevic, A review on machinery diagnostics and prognostics implementing condition-based maintenance, Mech. Syst. Signal Process. 20 (7) (2006) 1483–1510.

[5] G. Dong, W. Chongguang, B. Zhang, M. Xin, Signed directed graph and qualitative trend analysis based fault diagnosis in chemical industry, Chin. J. Chem. Eng. 18 (2) (2010) 265–276.

[6] G. Lee, T. Tosukhowong, J.H. Lee, C. Han, Fault diagnosis using the hybrid method of signed digraph and partial least squares with time delay: the pulp mill process, Ind. Eng. Chem. Res. 45 (26) (2006) 9061–9074.

[7] M.R. Maurya, R. Rengaswamy, V. Venkatasubramanian, A signed directed graph and qualitative trend analysis-based framework for incipient fault diagnosis, Chem. Eng. Res. Des. 85 (10) (2007) 1407–1422.

[8] H.-C. Liu, Q.-L. Lin, M.-L. Ren, Fault diagnosis and cause analysis using fuzzy evidential reasoning approach and dynamic adaptive fuzzy petri nets, Comput. Ind. Eng. 66 (4) (2013) 899–908.

[9] M.M. Mansour, M.A.A. Wahab, W.M. Soliman, Petri nets for fault diagnosis of large power generation station, Ain Shams Eng. J. 4 (4) (2013) 831–842.

[10] J. Sun, S.-Y. Qin, Y.-H. Song, Fault diagnosis of electric power systems based on fuzzy petri nets, IEEE Trans. Power Syst. 19 (4) (2004) 2053–2059.

[11] I. Witten, E. Frank, M. Hall, C. Pal, Data Mining: Practical Machine Learning Tools and Techniques, Elsevier Inc, 2017.

[12] V. Agrawal, B.K. Panigrahi, P.M.V. Subbarao, Review of control and fault diagnosis methods applied to coal mills, J. Process Control 32 (2015) 138–153.

[13] Z. Gao, C. Cecati, S.X. Ding, A survey of fault diagnosis and fault-tolerant techniques—Part I: fault diagnosis with model-based and signal-based approaches, IEEE Trans. Ind. Electron. 62 (6) (2015) 3757–3767.

[14] Z. Liu, L. Zhang, A review of failure modes, condition monitoring and fault diagnosis methods for large-scale wind turbine bearings, Measurement 149 (2020) 107002.

[15] W.Y. Liu, B.P. Tang, J.G. Han, X.N. Lu, N.N. Hu, Z.Z. He, The structure healthy condition monitoring and fault diagnosis methods in wind turbines: a review, Renew. Sust. Energ. Rev. 44 (2015) 466–472.

[16] R. Liu, B. Yang, E. Zio, X. Chen, Artificial intelligence for fault diagnosis of rotating machinery: a review, Mech. Syst. Signal Process. 108 (2018) 33–47.

[17] J. Ma, J. Jiang, Applications of fault detection and diagnosis methods in nuclear power plants: a review, Prog. Nucl. Energy 53 (3) (2011) 255–266.

[18] V. Venkatasubramanian, R. Rengaswamy, S.N. Kavuri, A review of process fault detection and diagnosis: part II: qualitative models and search strategies, Comput. Chem. Eng. 27 (3) (2003) 313–326.

[19] V. Venkatasubramanian, R. Rengaswamy, K. Yin, S.N. Kavuri, A review of process fault detection and diagnosis: part I: quantitative model-based methods, Comput. Chem. Eng. 27 (3) (2003) 293–311.

[20] S.J. Qin, Survey on data-driven industrial process monitoring and diagnosis, Annu. Rev. Control. 36 (2) (2012) 220–234.

[21] S. Yin, S.X. Ding, X. Xie, H. Luo, A review on basic data-driven approaches for industrial process monitoring, IEEE Trans. Ind. Electron. 61 (11) (2014) 6418–6428.

[22] C. Cecati, A Survey of Fault Diagnosis and Fault-Tolerant Techniques—Part II: Fault Diagnosis with Knowledge-Based and Hybrid/Active Approaches, 2015.

[23] Z. Yin, J. Hou, Recent advances on SVM based fault diagnosis and process monitoring in complicated industrial processes, Neurocomputing 174 (2016) 643–650.

[24] N.M. Nasrabadi, Pattern recognition and machine learning, J. Electron. Imaging 16 (4) (2007) 49901.

[25] A. Ragab, M. El Koujok, H. Ghezzaz, M. Amazouz, M.-S. Ouali, S. Yacout, Deep understanding in industrial processes by complementing human expertise with interpretable patterns of machine learning, Expert Syst. Appl. 122 (2019) 388–405.

[26] E. Alpaydin, Introduction to Machine Learning, MIT Press, 2020.

[27] C.M. Bishop, Pattern Recognition and Machine Learning, Springer, 2006.

[28] B. Yegnanarayana, Artificial Neural Networks, PHI Learning Pvt Ltd, 2009.

[29] B. Samanta, K.R. Al-Balushi, Artificial neural network based fault diagnostics of rolling element bearings using time-domain features, Mech. Syst. Signal Process. 17 (2) (2003) 317–328.

[30] J. Korbicz, J.M. Koscielny, Z. Kowalczuk, W. Cholewa, Fault Diagnosis: Models, Artificial Intelligence, Applications, Springer Science & Business Media, 2012.

[31] M.N. Nashalji, M.A. Shoorehdeli, M. Teshnehlab, Fault detection of the Tennessee Eastman process using improved PCA and neural classifier, in: Soft Computing in Industrial Applications, Springer, 2010, pp. 41–50.

[32] Z. Zheng, R. Petrone, M.-C. Péra, D. Hissel, M. Becherif, C. Pianese, M. Sorrentino, A review on non-model based diagnosis methodologies for PEM fuel cell stacks and systems, Int. J. Hydrog. Energy 38 (21) (2013) 8914–8926.

[33] J. Kim, I. Lee, Y. Tak, B.H. Cho, State-of-health diagnosis based on hamming neural network using output voltage pattern recognition for a PEM fuel cell, Int. J. Hydrog. Energy 37 (5) (2012) 4280–4289.

[34] Y. Han, N. Ding, Z. Geng, Z. Wang, C. Chu, An optimized long short-term memory network based fault diagnosis model for chemical processes, J. Process Control 92 (2020) 161–168.

[35] T. Kohonen, The self-organizing map, Proc. IEEE 78 (9) (1990) 1464–1480.

[36] F. Corona, M. Mulas, R. Baratti, J.A. Romagnoli, On the topological modeling and analysis of industrial process data using the SOM, Comput. Chem. Eng. 34 (12) (2010) 2022–2032.

[37] X. Chen, X. Yan, Using improved self-organizing map for fault diagnosis in chemical industry process, Chem. Eng. Res. Des. 90 (12) (2012) 2262–2277.

[38] H. Yu, F. Khan, V. Garaniya, A. Ahmad, Self-organizing map based fault diagnosis technique for non-Gaussian processes, Ind. Eng. Chem. Res. 53 (21) (2014) 8831–8843.

[39] P. Juntunen, M. Liukkonen, M. Lehtola, Y. Hiltunen, Cluster analysis by self-organizing maps: an application to the modelling of water quality in a treatment process, Appl. Soft Comput. 13 (7) (2013) 3191–3196.

[40] G.S. Chadha, A. Schwung, Comparison of deep neural network architectures for fault detection in Tennessee Eastman process, in: 2017 22nd IEEE International Conference on Emerging Technologies and Factory Automation (ETFA), 2017, pp. 1–8.

[41] F. Lv, C. Wen, Z. Bao, M. Liu, Fault diagnosis based on deep learning, in: 2016 American Control Conference (ACC), 2016, pp. 6851–6856.

[42] P. Jiang, Z. Hu, J. Liu, S. Yu, F. Wu, Fault diagnosis based on chemical sensor data with an active deep neural network, Sensors 16 (10) (2016) 1695.

[43] Y. Wang, Z. Pan, X. Yuan, C. Yang, W. Gui, A novel deep learning based fault diagnosis approach for chemical process with extended deep belief network, ISA Trans. 96 (2020) 457–467.

[44] Y.O. Lee, J. Jo, J. Hwang, Application of deep neural network and generative adversarial network to industrial maintenance: a case study of induction motor fault detection, in: 2017 IEEE International Conference on Big Data (Big Data), 2017, pp. 3248–3253.

[45] T. Han, C. Liu, W. Yang, D. Jiang, Deep transfer network with joint distribution adaptation: a new intelligent fault diagnosis framework for industry application, ISA Trans. 97 (2020) 269–281.

[46] R. Iqbal, T. Maniak, F. Doctor, C. Karyotis, Fault detection and isolation in industrial processes using deep learning approaches, IEEE Trans. Ind. Inf. 15 (5) (2019) 3077–3084.

[47] D. Nauck, R. Kruse, A neuro-fuzzy method to learn fuzzy classification rules from data, Fuzzy Sets Syst. 89 (3) (1997) 277–288.

[48] B.-S. Yang, M.-S. Oh, A.C.C. Tan, et al., Fault diagnosis of induction motor based on decision trees and adaptive neuro-fuzzy inference, Expert Syst. Appl. 36 (2) (2009) 1840–1849.

[49] C.K. Lau, K. Ghosh, M.A. Hussain, C.R.C. Hassan, Fault diagnosis of Tennessee Eastman process with multi-scale PCA and ANFIS, Chemom. Intell. Lab. Syst. 120 (2013) 1–14.

[50] V. Vapnik, S.E. Golowich, A.J. Smola, Support vector method for function approximation, regression estimation and signal processing, in: Advances in Neural Information Processing Systems, 1997, pp. 281–287.

[51] J. Qu, M.J. Zuo, Support vector machine based data processing algorithm for wear degree classification of slurry pump systems, Measurement 43 (6) (2010) 781–791.

[52] A. Widodo, B.-S. Yang, Support vector machine in machine condition monitoring and fault diagnosis, Mech. Syst. Signal Process. 21 (6) (2007) 2560–2574.

[53] S.R. Gunn, et al., Support vector machines for classification and regression, ISIS Tech. Rep. 14 (1) (1998) 5–16.

[54] B.-S. Yang, A. Widodo, Support vector machine for machine fault diagnosis and prognosis, J. Syst. Design Dyn. 2 (1) (2008) 12–23.

[55] C.-W. Hsu, C.-J. Lin, A comparison of methods for multiclass support vector machines, IEEE Trans. Neural Netw. 13 (2) (2002) 415–425.

[56] C.-C. Chang, C.-J. Lin, LIBSVM: a library for support vector machines, ACM Trans. Intell. Syst. Technol. 2 (3) (2011) 1–27.

[57] R.R. Bouckaert, E. Frank, M. Hall, R. Kirkby, P. Reutemann, A. Seewald, D. Scuse, WEKA Manual for Version 3-9-1, University of Waikato, Hamilton, New Zealand, 2016.

[58] M. Onel, C.A. Kieslich, E.N. Pistikopoulos, A nonlinear support vector machine-based feature selection approach for fault detection and diagnosis: application to the Tennessee Eastman process, AICHE J. 65 (3) (2019) 992–1005.

[59] M. Onel, C.A. Kieslich, Y.A. Guzman, C.A. Floudas, E.N. Pistikopoulos, Big data approach to batch process monitoring: simultaneous fault detection and diagnosis using nonlinear support vector machine-based feature selection, Comput. Chem. Eng. 115 (2018) 46–63.

[60] H. Han, B. Gu, T. Wang, Z.R. Li, Important sensors for chiller fault detection and diagnosis (FDD) from the perspective of feature selection and machine learning, Int. J. Refrig. 34 (2) (2011) 586–599.

[61] A. Kulkarni, V.K. Jayaraman, B.D. Kulkarni, Knowledge incorporated support vector machines to detect faults in Tennessee Eastman process, Comput. Chem. Eng. 29 (10) (2005) 2128–2133.

[62] S. Yin, X. Gao, H.R. Karimi, X. Zhu, Study on support vector machine-based fault detection in tennessee eastman process, in: Abstract and Applied Analysis, vol. 2014, 2014.

[63] V. Sugumaran, V. Muralidharan, K.I. Ramachandran, Feature selection using decision tree and classification through proximal support vector machine for fault diagnostics of roller bearing, Mech. Syst. Signal Process. 21 (2) (2007) 930–942.

[64] Z. Ge, F. Gao, Z. Song, Batch process monitoring based on support vector data description method, J. Process Control 21 (6) (2011) 949–959.

[65] F. Ye, Z. Zhang, K. Chakrabarty, X. Gu, Knowledge-Driven Board-Level Functional Fault Diagnosis, Springer, 2016.

[66] Y. Wang, M.A. Simaan, A suction detection system for rotary blood pumps based on the Lagrangian support vector machine algorithm, IEEE J. Biomed. Health Inform. 17 (3) (2013) 654–663.

[67] J. Yu, A support vector clustering-based probabilistic method for unsupervised fault detection and classification of complex chemical processes using unlabeled data, AICHE J. 59 (2) (2013) 407–419.

[68] O.L. Mangasarian, D.R. Musicant, Lagrangian support vector machines, J. Mach. Learn. Res. 1 (Mar) (2001) 161–177.

[69] M. Sheng, Y. Chen, Q. Dai, A novel Lagrangian support vector machine and application in the crane gear fault diagnosis system, in: Advances in Mechanical and Electronic Engineering, Springer, 2012, pp. 369–373.

[70] D.M.J. Tax, R.P.W. Duin, Support vector domain description, Pattern Recogn. Lett. 20 (11–13) (1999) 1191–1199.

[71] S. Mahadevan, S.L. Shah, Fault detection and diagnosis in process data using one-class support vector machines, J. Process Control 19 (10) (2009) 1627–1639.

[72] A. Beghi, L. Cecchinato, C. Corazzol, M. Rampazzo, F. Simmini, G.A. Susto, A one-class svm based tool for machine learning novelty detection in hvac chiller systems, IFAC Proc. Vol. 47 (3) (2014) 1953–1958.

[73] A. Ben-Hur, D. Horn, H.T. Siegelmann, V. Vapnik, Support vector clustering, J. Mach. Learn. Res. 2 (Dec) (2001) 125–137.

[74] J. Lee, D. Lee, An improved cluster labeling method for support vector clustering, IEEE Trans. Pattern Anal. Mach. Intell. 27 (3) (2005) 461–464.

[75] S. Fei, X. Zhang, Fault diagnosis of power transformer based on support vector machine with genetic algorithm, Expert Syst. Appl. 36 (8) (2009) 11352–11357.

[76] S. Yuan, F. Chu, Fault diagnosis based on support vector machines with parameter optimisation by artificial immunisation algorithm, Mech. Syst. Signal Process. 21 (3) (2007) 1318–1330.

[77] T.D. Nielsen, F.V. Jensen, Bayesian Networks and Decision Graphs, Springer Science & Business Media, 2009.

[78] D. Heckerman, D. Geiger, D.M. Chickering, Learning Bayesian networks: the combination of knowledge and statistical data, Mach. Learn. 20 (3) (1995) 197–243.

[79] F.V. Jensen, et al., An Introduction to Bayesian Networks, vol. 210, UCL Press, London, 1996.

[80] R.E. Neapolitan, et al., Learning Bayesian Networks, vol. 38, Pearson Prentice Hall, Upper Saddle River, NJ, 2004.

[81] N. Friedman, D. Geiger, M. Goldszmidt, Bayesian network classifiers, Mach. Learn. 29 (2–3) (1997) 131–163.

[82] G.H. John, P. Langley, Estimating continuous distributions in Bayesian classifiers, ArXiv Preprint ArXiv 1302 (2013) 4964.

[83] K.M. Leung, Naive Bayesian Classifier, vol. 2007, Polytechnic University Department of Computer Science/Finance and Risk Engineering, 2007, pp. 123–156.

[84] J. Cheng, R. Greiner, Comparing Bayesian network classifiers, in: Proceedings of the Fifteenth Conference on Uncertainty in Artificial Intelligence, 1999, pp. 101–108.

[85] S. Yang, K.-C. Chang, Comparison of score metrics for Bayesian network learning, IEEE Trans. Syst. Man Cybern. Part A Syst. Hum. 32 (3) (2002) 419–428.

[86] U. Lerner, R. Parr, D. Koller, G. Biswas, et al., Bayesian fault detection and diagnosis in dynamic systems, in: Aaai/iaai, 2000, pp. 531–537.

[87] K.P. Murphy, S. Russell, Dynamic Bayesian Networks: Representation, Inference and Learning [PhD Thesis], University of California, 2002.

[88] C. Romessis, K. Mathioudakis, Bayesian betwork approach for gas path fault diagnosis, ASME. J. Eng. Gas Turbines Power 128 (1) (2006) 64–72, https://doi.org/10.1115/1.1924536.

[89] Z. Zhang, J. Zhu, F. Pan, Fault detection and diagnosis for data incomplete industrial systems with new Bayesian network approach, J. Syst. Eng. Electron. 24 (3) (2013) 500–511.

[90] F. Sahin, M.Ç. Yavuz, Z. Arnavut, Ö. Uluyol, Fault diagnosis for airplane engines using Bayesian networks and distributed particle swarm optimization, Parallel Comput. 33 (2) (2007) 124–143.

[91] J. Zhu, Z. Ge, Z. Song, L. Zhou, G. Chen, Large-scale plant-wide process modeling and hierarchical monitoring: a distributed Bayesian network approach, J. Process Control 65 (2018) 91–106.

[92] M.G. Don, F. Khan, Dynamic process fault detection and diagnosis based on a combined approach of hidden Markov and Bayesian network model, Chem. Eng. Sci. 201 (2019) 82–96.

[93] W. Yu, C. Zhao, Online fault diagnosis for industrial processes with Bayesian network-based probabilistic ensemble learning strategy, IEEE Trans. Autom. Sci. Eng. 16 (4) (2019) 1922–1932.

[94] A. Lakehal, M. Nahal, R. Harouz, Development and application of a decision making tool for fault diagnosis of turbocompressor based on Bayesian network and fault tree, Manag. Prod. Eng. Rev. 10 (2019) 16–24.

[95] S. Verron, J. Li, T. Tiplica, Fault detection and isolation of faults in a multivariate process with Bayesian network, J. Process Control 20 (8) (2010) 902–911.

[96] S. Verron, T. Tiplica, A. Kobi, Fault diagnosis with bayesian networks: application to the Tennessee Eastman process, in: 2006 IEEE International Conference on Industrial Technology, 2006, pp. 98–103.

[97] P.K. Novak, N. Lavrač, G.I. Webb, Supervised descriptive rule discovery: a unifying survey of contrast set, emerging pattern and subgroup mining, J. Mach. Learn. Res. 10 (2) (2009).

[98] J.R. Quinlan, Induction of decision trees, Mach. Learn. 1 (1) (1986) 81–106.

[99] Z. Pawlak, Rough set theory and its applications to data analysis, Cybern. Syst. 29 (7) (1998) 661–688.

[100] Y. Crama, P.L. Hammer, T. Ibaraki, Cause-effect relationships and partially defined Boolean functions, Ann. Oper. Res. 16 (1) (1988) 299–325.

[101] M.I. Jordan, T.M. Mitchell, Machine learning: trends, perspectives, and prospects, Science 349 (6245) (2015) 255–260.

[102] X. Wu, V. Kumar, J.R. Quinlan, J. Ghosh, Q. Yang, H. Motoda, et al., Top 10 algorithms in data mining, Knowl. Inf. Syst. 14 (1) (2008) 1–37.

[103] L. Rokach, O.Z. Maimon, Data Mining With Decision Trees: Theory and Applications, vol. 69, World Scientific, 2008.

[104] I.H. Witten, E. Frank, M.A. Hall, C.J. Pal, Data Mining: Practical Machine Learning Tools and Techniques, Morgan Kaufmann, 2016.

[105] H.S. Nguyen, Approximate boolean reasoning: foundations and applications in data mining, in: Transactions on rough sets V, Springer, 2006, pp. 334–506.

[106] I. Chikalov, V. Lozin, I. Lozina, M. Moshkov, H.S. Nguyen, A. Skowron, B. Zielosko, Three Approaches to Data Analysis: Test Theory, Rough Sets and Logical Analysis Of data, vol. 41, Springer Science & Business Media, 2012.

[107] R. Lior, et al., Data Mining With Decision Trees: Theory and Applications, vol. 81, World Scientific, 2014.

[108] T.C. Smith, E. Frank, Introducing machine learning concepts with WEKA, in: Statistical Genomics, Springer, 2016, pp. 353–378.

[109] K. Polat, Güne\cs, S., A novel hybrid intelligent method based on C4. 5 decision tree classifier and one-against-all approach for multi-class classification problems, Expert Syst. Appl. 36 (2) (2009) 1587–1592.

[110] M. Askarian, G. Escudero, M. Graells, R. Zarghami, F. Jalali-Farahani, N. Mostoufi, Fault diagnosis of chemical processes with incomplete observations: a comparative study, Comput. Chem. Eng. 84 (2016) 104–116.

[111] M. Demetgul, Fault diagnosis on production systems with support vector machine and decision trees algorithms, Int. J. Adv. Manuf. Technol. 67 (9–12) (2013) 2183–2194.

[112] S.-G. He, Z. He, G.A. Wang, Online monitoring and fault identification of mean shifts in bivariate processes using decision tree learning techniques, J. Intell. Manuf. 24 (1) (2013) 25–34.

[113] M. Chen, A.X. Zheng, J. Lloyd, M.I. Jordan, E. Brewer, Failure diagnosis using decision trees, in: International Conference on Autonomic Computing, 2004. Proceedings, 2004, pp. 36–43.

[114] W. Sun, J. Chen, J. Li, Decision tree and PCA-based fault diagnosis of rotating machinery, Mech. Syst. Signal Process. 21 (3) (2007) 1300–1317.

[115] X. Zhang, M. Kano, S. Matsuzaki, Ensemble pattern trees for predicting hot metal temperature in blast furnace, Comput. Chem. Eng. 121 (2019) 442–449.

[116] E. Boros, P.L. Hammer, T. Ibaraki, A. Kogan, E. Mayoraz, I. Muchnik, An implementation of logical analysis of data, IEEE Trans. Knowl. Data Eng. 12 (2) (2000) 292–306.

[117] M.-A. Mortada, S. Yacout, A. Lakis, Diagnosis of rotor bearings using logical analysis of data, J. Qual. Maint. Eng. 17 (4) (2011) 371–397, https://doi.org/10.1108/13552511111180186.

[118] S. Yacout, D. Salamanca, M.-A. Mortada, Tool and Method for Fault Detection of Devices by Condition Based Maintenance, 2017 (Google Patents).

[119] M.-A. Mortada, S. Yacout, A. Lakis, Fault diagnosis in power transformers using multi-class logical analysis of data, J. Intell. Manuf. 25 (6) (2014) 1429–1439.

[120] P.L. Hammer, T.O. Bonates, Logical analysis of data—an overview: from combinatorial optimization to medical applications, Ann. Oper. Res. 148 (1) (2006) 203–225.

[121] H.H. Kim, J.Y. Choi, Pattern generation for multi-class LAD using iterative genetic algorithm with flexible chromosomes and multiple populations, Expert Syst. Appl. 42 (2) (2015) 833–843.

[122] H.H. Kim, J.Y. Choi, Hierarchical multi-class LAD based on OvA-binary tree using genetic algorithm, Expert Syst. Appl. 42 (21) (2015) 8134–8145.

[123] H.S. Ryoo, I.-Y. Jang, Milp approach to pattern generation in logical analysis of data, Discret. Appl. Math. 157 (4) (2009) 749–761.

[124] A. Ragab, M. El-Koujok, B. Poulin, M. Amazouz, S. Yacout, Fault diagnosis in industrial chemical processes using interpretable patterns based on logical analysis of data, Expert Syst. Appl. 95 (2018) 368–383.

[125] M.-A. Mortada, T. Carroll, S. Yacout, A. Lakis, Rogue components: their effect and control using logical analysis of data, J. Intell. Manuf. 23 (2) (2012) 289–302.

[126] A. Ragab, M.-S. Ouali, S. Yacout, H. Osman, Remaining useful life prediction using prognostic methodology based on logical analysis of data and Kaplan—Meier estimation, J. Intell. Manuf. 27 (5) (2016) 943–958.

[127] A. Ragab, S. Yacout, M.-S. Ouali, H. Osman, Pattern-based prognostic methodology for condition-based maintenance using selected and weighted survival curves, Qual. Reliab. Eng. Int. 33 (8) (2017) 1753–1772.

[128] A. Ragab, S. Yacout, M.-S. Ouali, H. Osman, Prognostics of multiple failure modes in rotating machinery using a pattern-based classifier and cumulative incidence functions, J. Intell. Manuf. 30 (1) (2019) 255–274.

[129] S. Jocelyn, Y. Chinniah, M.-S. Ouali, S. Yacout, Application of logical analysis of data to machinery-related accident prevention based on scarce data, Reliab. Eng. Syst. Saf. 159 (2017) 223–236.

[130] K. Dembczyński, W. Kotłowski, R. Słowiński, ENDER: a statistical framework for boosting decision rules, Data Min. Knowl. Disc. 21 (1) (2010) 52–90.

[131] W.W. Cohen, Y. Singer, A simple, fast, and effective rule learner, AAAI/IAAI 99 (335–342) (1999) 3.

[132] K. Dembczyński, W. Kotłowski, R. Słowiński, Maximum likelihood rule ensembles, in: Proceedings of the 25th International Conference on Machine Learning, 2008, pp. 224–231.

[133] J. Friedman, B. Popescu, Rulefit (Version 3)[Computer Software], 2012.

[134] S. Weiss, Lightweight Rule Induction, 2003 (Google Patents).

[135] Z. Pawlak, Rough sets, Int. J. Comput. Inform. Sci. 11 (5) (1982) 341–356.

[136] J.-T. Peng, C.F. Chien, T.L.B. Tseng, Rough set theory for data mining for fault diagnosis on distribution feeder, IEE Proc. Gener. Transm. Distrib. 151 (6) (2004) 689–697.

[137] R. Bello, R. Falcon, J.L. Verdegay, Uncertainty Management with Fuzzy and Rough Sets: Recent Advances and Applications, vol. 377, Springer, 2019.

[138] F.E.H. Tay, L. Shen, Fault diagnosis based on rough set theory, Eng. Appl. Artif. Intell. 16 (1) (2003) 39–43.

[139] R.W. Swiniarski, A. Skowron, Rough set methods in feature selection and recognition, Pattern Recogn. Lett. 24 (6) (2003) 833–849.

[140] T.Y. Lin, N. Cercone, Rough Sets and Data Mining: Analysis of Imprecise Data, Springer Science & Business Media, 2012.

[141] X. Wang, E.C.C. Tsang, S. Zhao, D. Chen, D.S. Yeung, Learning fuzzy rules from fuzzy samples based on rough set technique, Inf. Sci. 177 (20) (2007) 4493–4514.

[142] M. Inuiguchi, T. Miyajima, Rough set based rule induction from two decision tables, Eur. J. Oper. Res. 181 (3) (2007) 1540–1553.

[143] H.A. Nabwey, An approach based on rough sets theory and grey system for implementation of rule-based control for sustainability of rotary clinker kiln, Int. J. Eng. Res. Technol. 12 (12) (2019) 2604–2610.

[144] P. Konar, M. Saha, J. Sil, P. Chattopadhyay, Fault diagnosis of induction motor using CWT and rough-set theory, in: 2013 IEEE Symposium on Computational Intelligence in Control and Automation (CICA), 2013, pp. 17–23.

[145] Z. Ai-Hua, Y. Yi, S. Hong, Z. Xiao-Hui, Power transformer fault diagnosis based on integrated of rough set theory and evidence theory, in: 2013 Third International Conference on Intelligent System Design and Engineering Applications, 2013, pp. 1049–1052.

[146] A. Zhou, H. Song, H. Xiao, X. Zeng, Power transformer fault diagnosis based on integrated of rough set theory and neural network, in: 2012 Second International Conference on Intelligent System Design and Engineering Application, 2012, pp. 1463–1465.

[147] V. Muralidharan, V. Sugumaran, Rough set based rule learning and fuzzy classification of wavelet features for fault diagnosis of monoblock centrifugal pump, Measurement 46 (9) (2013) 3057–3063.

[148] Q. Xie, H. Zeng, L. Ruan, X. Chen, H. Zhang, Transformer fault diagnosis based on bayesian network and rough set reduction theory, in: IEEE 2013 Tencon-Spring, 2013, pp. 262–266.

[149] L.H. Chiang, E.L. Russell, R.D. Braatz, Fault diagnosis in chemical processes using fisher discriminant analysis, discriminant partial least squares, and principal component analysis, Chemom. Intell. Lab. Syst. 50 (2) (2000) 243–252.

[150] A. Prieto-Moreno, O. Llanes-Santiago, E. García-Moreno, Principal components selection for dimensionality reduction using discriminant information applied to fault diagnosis, J. Process Control 33 (2015) 14–24.

[151] Q. Jiang, X. Yan, W. Zhao, Fault detection and diagnosis in chemical processes using sensitive principal component analysis, Ind. Eng. Chem. Res. 52 (4) (2013) 1635–1644.

[152] K. Dunn, Process improvement using data, in: Experimentation for Improvement, vol. 4, Creative Commons Attribution-ShareAlike, Hamilton, Ontario, Canada, 2019, pp. 325–404.

[153] G. Diana, C. Tommasi, Cross-validation methods in principal component analysis: a comparison, JISS 11 (1) (2002) 71–82.

[154] W. Ku, R.H. Storer, C. Georgakis, Disturbance detection and isolation by dynamic principal component analysis, Chemom. Intell. Lab. Syst. 30 (1) (1995) 179–196.

[155] J.-M. Lee, C. Yoo, S.W. Choi, P.A. Vanrolleghem, I.-B. Lee, Nonlinear process monitoring using kernel principal component analysis, Chem. Eng. Sci. 59 (1) (2004) 223–234.

[156] D. Dong, T.J. McAvoy, Nonlinear principal component analysis—based on principal curves and neural networks, Comput. Chem. Eng. 20 (1) (1996) 65–78.

[157] J.M.B. de Lázaro, A.P. Moreno, O.L. Santiago, A.J. da Silva Neto, Optimizing kernel methods to reduce dimensionality in fault diagnosis of industrial systems, Comput. Ind. Eng. 87 (2015) 140–149.

[158] U. Kruger, G. Dimitriadis, Diagnosis of process faults in chemical systems using a local partial least squares approach, AICHE J. 54 (10) (2008) 2581–2596.

[159] S. Yin, X. Zhu, O. Kaynak, Improved PLS focused on key-performance-indicator-related fault diagnosis, IEEE Trans. Ind. Electron. 62 (3) (2014) 1651–1658.

[160] H. Abdi, Partial least squares regression and projection on latent structure regression (PLS regression), Wiley Interdiscip. Rev. Comput. Stat. 2 (1) (2010) 97–106.

[161] S. Yin, S.X. Ding, A. Haghani, H. Hao, P. Zhang, A comparison study of basic data-driven fault diagnosis and process monitoring methods on the benchmark Tennessee Eastman process, J. Process Control 22 (9) (2012) 1567–1581.

[162] J. Gertler, Fault Detection and Diagnosis in Engineering Systems, CRC Press, 1998.

[163] J.J. Castro, F.J. Doyle III, A pulp mill benchmark problem for control: problem description, J. Process Control 14 (1) (2004) 17–29.

[164] S.J. Qin, Y. Zheng, Quality-relevant and process-relevant fault monitoring with concurrent projection to latent structures, AICHE J. 59 (2) (2013) 496–504.

[165] A. Hyvärinen, E. Oja, Independent component analysis: algorithms and applications, Neural Netw. 13 (4–5) (2000) 411–430.

[166] J. Shlens, A tutorial on principal component analysis, ArXiv Preprint ArXiv 1404 (2014) 1100.

[167] J.-M. Lee, S.J. Qin, I.-B. Lee, Fault detection and diagnosis based on modified independent component analysis, AICHE J. 52 (10) (2006) 3501–3514.

[168] B. Scholkopft, K.-R. Mullert, Fisher discriminant analysis with kernels, Neural Netw. Signal Process. IX 1 (1) (1999) 1.

[169] L.H. Chiang, M.E. Kotanchek, A.K. Kordon, Fault diagnosis based on fisher discriminant analysis and support vector machines, Comput. Chem. Eng. 28 (8) (2004) 1389–1401.

[170] Q.P. He, S.J. Qin, J. Wang, A new fault diagnosis method using fault directions in fisher discriminant analysis, AICHE J. 51 (2) (2005) 555–571.

[171] J. Yu, Localized fisher discriminant analysis based complex chemical process monitoring, AICHE J. 57 (7) (2011) 1817–1828.

[172] Y. Zhang, H. Zhou, S.J. Qin, T. Chai, Decentralized fault diagnosis of large-scale processes using multiblock kernel partial least squares, IEEE Trans. Ind. Inf. 6 (1) (2009) 3–10.

[173] Y. Zhang, S.J. Qin, Fault detection of nonlinear processes using multiway kernel independent component analysis, Ind. Eng. Chem. Res. 46 (23) (2007) 7780–7787.

[174] B. Schölkopf, A. Smola, K.-R. Müller, Nonlinear component analysis as a kernel eigenvalue problem, Neural Comput. 10 (5) (1998) 1299–1319.

[175] G.A.N. Liang-zhi, Online faults detection based on sparse kernel principal angles, Microcomput. Inf. 2010 (25) (2010) 63.

[176] Z. Geng, F. Liu, Y. Han, Q. Zhu, Y. He, Fault diagnosis of chemical processes based on a novel adaptive kernel principal component analysis, in: 2019 12th Asian Control Conference (ASCC), 2019, pp. 1495–1500.

[177] J.-D. Shao, G. Rong, J.M. Lee, Learning a data-dependent kernel function for KPCA-based nonlinear process monitoring, Chem. Eng. Res. Des. 87 (11) (2009) 1471–1480.

[178] K.Q. Weinberger, L.K. Saul, An introduction to nonlinear dimensionality reduction by maximum variance unfolding, in: AAAI, vol. 6, 2006, pp. 1683–1686.

[179] R. Shao, W. Hu, Y. Wang, X. Qi, The fault feature extraction and classification of gear using principal component analysis and kernel principal component analysis based on the wavelet packet transform, Measurement 54 (2014) 118–132.

[180] J. Chen, K.-C. Liu, On-line batch process monitoring using dynamic PCA and dynamic PLS models, Chem. Eng. Sci. 57 (1) (2002) 63–75.

[181] T. Komulainen, M. Sourander, S.-L. Jämsä-Jounela, An online application of dynamic PLS to a dearomatization process, Comput. Chem. Eng. 28 (12) (2004) 2611–2619.

[182] B.C. Juricek, D.E. Seborg, W.E. Larimore, Fault detection using canonical variate analysis, Ind. Eng. Chem. Res. 43 (2) (2004) 458–474.

[183] P.-E.P. Odiowei, Y. Cao, Nonlinear dynamic process monitoring using canonical variate analysis and kernel density estimations, IEEE Trans. Ind. Inf. 6 (1) (2009) 36–45.

[184] P.P. Odiowei, Y. Cao, State-space independent component analysis for nonlinear dynamic process monitoring, Chemom. Intell. Lab. Syst. 103 (1) (2010) 59–65.

[185] S.X. Ding, S. Yin, K. Peng, H. Hao, B. Shen, A novel scheme for key performance indicator prediction and diagnosis with application to an industrial hot strip mill, IEEE Trans. Ind. Inf. 9 (4) (2012) 2239–2247.

[186] Z. Zhang, Y. Wang, K. Wang, Fault diagnosis and prognosis using wavelet packet decomposition, Fourier transform and artificial neural network, J. Intell. Manuf. 24 (6) (2013) 1213–1227.

[187] Y. Lei, F. Jia, J. Lin, S. Xing, S.X. Ding, An intelligent fault diagnosis method using unsupervised feature learning towards mechanical big data, IEEE Trans. Ind. Electron. 63 (5) (2016) 3137–3147.

[188] K. Tidriri, N. Chatti, S. Verron, T. Tiplica, Bridging data-driven and model-based approaches for process fault diagnosis and health monitoring: a review of researches and future challenges, Annu. Rev. Control. 42 (2016) 63–81.

[189] Z. Chen, C. Jiang, L. Xie, A novel ensemble ELM for human activity recognition using smartphone sensors, IEEE Trans. Ind. Inf. 15 (5) (2018) 2691–2699.

[190] Z. Hua, H. Yu, Y. Hua, Adaptive ensemble fault diagnosis based on online learning of personalized decision parameters, IEEE Trans. Ind. Electron. 65 (11) (2018) 8882–8894.

[191] H. Parvin, M. MirnabiBaboli, H. Alinejad-Rokny, Proposing a classifier ensemble framework based on classifier selection and decision tree, Eng. Appl. Artif. Intell. 37 (2015) 34–42.

[192] R. Polikar, Ensemble based systems in decision making, IEEE Circuits Syst. Mag. 6 (3) (2006) 21–45.

[193] L. Rokach, Ensemble-based classifiers, Artif. Intell. Rev. 33 (1–2) (2010) 1–39.

[194] Z. Wu, W. Lin, Y. Ji, An integrated ensemble learning model for imbalanced fault diagnostics and prognostics, IEEE Access 6 (2018) 8394–8402.

[195] X. Zhang, J. Zhou, Multi-fault diagnosis for rolling element bearings based on ensemble empirical mode decomposition and optimized support vector machines, Mech. Syst. Signal Process. 41 (1–2) (2013) 127–140.

[196] Y. Lei, Z. He, Y. Zi, X. Chen, New clustering algorithm-based fault diagnosis using compensation distance evaluation technique, Mech. Syst. Signal Process. 22 (2) (2008) 419–435.

[197] K. Zhang, Y. Li, P. Scarf, A. Ball, Feature selection for high-dimensional machinery fault diagnosis data using multiple models and radial basis function networks, Neurocomputing 74 (17) (2011) 2941–2952.

[198] S.-W. Lin, K.-C. Ying, S.-C. Chen, Z.-J. Lee, Particle swarm optimization for parameter determination and feature selection of support vector machines, Expert Syst. Appl. 35 (4) (2008) 1817–1824.

[199] S. Li, H. Wu, D. Wan, J. Zhu, An effective feature selection method for hyperspectral image classification based on genetic algorithm and support vector machine, Knowl. Based Syst. 24 (1) (2011) 40–48.

[200] S. Das, S. Sil, Kernel-induced fuzzy clustering of image pixels with an improved differential evolution algorithm, Inf. Sci. 180 (8) (2010) 1237–1256.

[201] J.-P. Mei, L. Chen, Fuzzy relational clustering around medoids: a unified view, Fuzzy Sets Syst. 183 (1) (2011) 44–56.

[202] X. Liu, L. Ma, J. Mathew, Machinery fault diagnosis based on fuzzy measure and fuzzy integral data fusion techniques, Mech. Syst. Signal Process. 23 (3) (2009) 690–700.

[203] R. Eslamloueyan, Designing a hierarchical neural network based on fuzzy clustering for fault diagnosis of the Tennessee–Eastman process, Appl. Soft Comput. 11 (1) (2011) 1407–1415.

[204] J.C. Bezdek, Pattern Recognition With Fuzzy Objective Function Algorithms, Springer Science & Business Media, 2013.

[205] K.-L. Wu, M.-S. Yang, A cluster validity index for fuzzy clustering, Pattern Recogn. Lett. 26 (9) (2005) 1275–1291.

[206] K.R. Žalik, Cluster validity index for estimation of fuzzy clusters of different sizes and densities, Pattern Recogn. 43 (10) (2010) 3374–3390.

[207] W. Wang, Y. Zhang, On fuzzy cluster validity indices, Fuzzy Sets Syst. 158 (19) (2007) 2095–2117.

[208] M. El Koujok, A. Ragab, H. Ghezzaz, M. Amazouz, A multi-agent-based methodology for known and novel faults diagnosis in industrial processes, IEEE Trans. Ind. Inf. 17 (5) (2020) 3358–3366, https://doi.org/10.1109/TII.2020.3011069.

[209] P. Bajpai, Brief description of the pulp and papermaking process, in: Biotechnology for Pulp and Paper Processing, Springer, 2018, pp. 9–26.

[210] E. Vakkilainen, et al., Kraft Recovery Boilers—Principles and Practice, Suomen Soodakattilayhdistys r.y., 2005.

[211] C.J. Biermann, Handbook of Pulping and Papermaking, Elsevier, 1996.

[212] J. Gertler, Fault Detection and Diagnosis in Engineering Systems, Routledge, 2017.

Chapter 11

Application of artificial intelligence in modeling, control, and fault diagnosis

Mohsen Hadian[a], Seyed Mohammad Ebrahimi Saryazdi[b], Ardashir Mohammadzadeh[c], and Masoud Babaei[d]

[a]Department of Mechanical Engineering, University of Saskatchewan, Saskatoon, Canada, [b]Department of Energy Engineering, Sharif University of Technology, Tehran, Iran, [c]Department of Electrical Engineering, University of Bonab, Bonab, Iran, [d]Department of Control Engineering, Tarbiat Modares University, Tehran, Iran

Advanced control systems are becoming more and more sophisticated every day. For safety-critical systems such as chemical processes, nuclear reactors, aircraft and spacecraft, the issue of reliability, acceptable performance, and environmental protection are of particular importance. If a fault occurs, the damages to finance, human, and environment could be severe. As a result, there is a growing need for online monitoring and Fault Detection, or to identify system faults to improve reliability. Accordingly, it is viable to apply the preliminary signs and to prevent the system from stopping and the occurrence of a catastrophe to a large extent.

Over the past 2 decades, research on fault detection has attracted a great deal of attention. These advances are mainly due to the tendency of systems to automate, increase demand, and ensure system security. However, other reasons, such as tremendous advances in mathematical modeling, estimation, and parameter recognition, have also contributed to the dramatic advances in computer computing. With the growth of complex systems, there has been a significant desire to develop methods for detecting and isolating faults. They are also utilized to increase the quality of products. Several engineering sciences are merged in the field of fault diagnosis science.

The impact of failure on a system can be devastating due to the increasing complexity of systems. Feedback control is an essential component of overall system monitoring. Fault diagnosis is the second component that will have significant economic, industrial, and social consequences if it is safe and cost-effective. Due to the growing need for reliability and safety of technical devices

Applications of Artificial Intelligence in Process Systems Engineering
https://doi.org/10.1016/B978-0-12-821092-5.00006-1

255

and their components, monitoring methods have been widely welcomed as part of the general control of processes. A fundamental prerequisite for further development of automated monitoring is the rapid detection of process faults. The use of computers and process microcomputers enables the use of methods that lead to faster detection of process faults than traditional processes.

Today, it is entirely impossible to fully rely on the human operator to manage abnormal events and emergencies for a variety of reasons. One of these reasons is the size and complexity of modern process equipment. In addition, the emphasis is usually on a rapid diagnosis, which in itself causes certain limitations and needs in diagnostic procedures.

Moreover, fault diagnosis becomes more complicated when one considers the fact that process measurements may be insufficient, incomplete, or unreliable for a variety of reasons, such as sensor failure or bias. Given the circumstances, it is not unreasonable for human operators to make erroneous decisions and take steps to make matters worse. Industrial statistics show that 70% of industrial accidents are caused by human error.

There is also a growing demand for fault-tolerant automated systems that can operate automatically and reliably in the presence of faults and failures of sensors, operators, and components. Because the fault detection system is an essential component of the fault tolerable automatic system, there is an urgent need to upgrade intelligent systems that can automatically detect the presence, isolate locations, and estimate the severity of faults in various components of a complex dynamic operating system.

On the other hand, correctly identifying the severity of the faults is a precious advantage for maintenance measures. Correctly assessing the severity of the faults makes it very easy to quickly identify the initial faults and identify behaviors outside the scope. In this way, it is possible to plan and execute intelligently exclusive actions to prevent system crashes, catastrophic failures, and incomplete actions by users and controllers. These methods are also used to control and improve the quality of products.

Fault diagnostic methods can be divided into two general categories of model-based and data-based methods. In this chapter, we will look at the second category, which includes fuzzy logic, neural network, and support vector machines (SVM).

1   Artificial neural network

1.1   Introduction

The phenomenological and empirical models are used to simulate the performance of a system. Parameters such as the pros and cons of each of these models, problems definition, and goal of the simulations should be considered to select one of them. The phenomenological models apply the fundamental chemical, and physical laws namely conservation laws of momentum, mass, and energy, to describe the process and performance of systems with nonlinear algebraic or differential equations [1]. The main disadvantage of these models is not only the difficulties to solve these equations (to approximate nonlinear equations), but also to develop the model and quantify the phenomenology through mathematical concept. The numerical calculations—which take a considerable computational time—are fraught with accuracy problems due to errors in the process parameters, algebraic errors, iteration errors, approximation errors, and round-off errors [2]. In addition, there can be some aspects, in some systems, with unknown or partially unknown parameters, that can be obtained through system identification procedure that consists of experimental measurements and fitting them to the mathematical model [3].

The artificial neural network (ANN) introduced by proposing the multilayer perceptron (MLP) model [4] and Hebb rule [5] to simulate the performance of neurons in the human brain in the 1940s. In 1956, the Perceptron concept [6] was developed using linear optimization. Adaptive Linear Unit network model [7] was proposed in 1959 and then had been applied in a communication application and weather forecasting. In 1974, back propagation (BP) algorithm was developed to handle the difficulties in the complex neural network related to solving nonlinear problems [8–10]. The deep learning developed the advantages of which are the usage of traditional machine learning approaches, and the inspiration of statistical learning. In 1986, Restricted Boltzmann Machine (RBM) was proposed by using the probability distribution of the Boltzmann Machine concept [11], and the hidden layers to characterize the input data. Next, Auto Encoder (AE) was developed by applying the layer-by-layer greedy learning algorithm to reduce the loss function [12]. In 1995, Recurrent Neural Network (RNN), was introduced with directed topology connections between neurons to learn feature from sequence data [13]. In 1997, Long Short-Term Memory (LSTM) which is a modified version of RNN, was put forward to overcome vanishing gradient problem and handle complex time sequence data [14]. In 1998, Convolutional Neural Network (CNN) was developed, which is suitable to deal with 2D inputs such as images [15]. Through the development of deep learning models, problems such as time-consuming model training and parameter optimization become more severe, resulting in overfitting or local optimization problems. Deep Belief Network (DBN), which allowed bidirectional connections in the top layer to tackle computational complexity, was proposed and successfully applied to several problems in 2006 [16, 17]. In DBN, layer-wise pretraining and fine-tuning were applied in the learning process of the parameters. In order to solve problems with high nonlinear inputs, Deep AE was put forwarded using more hidden layers [18]. In this network, a greedy layer-by-layer unsupervised learning algorithm was applied in the pre-train process, and eventually, the BP algorithm was used to the fine-tuning process. Meanwhile, Sparse Auto Encoder (SAE) was proposed for dimensionality reduction and sparse representations learning [19, 20]. In 2009, Deep Boltzmann Machine (DBM) using a bidirectional structure was developed to obtain robust learning process for ambiguous input data [21]. In 2010, Denoising Auto Encoder (DAE) was proposed to discover robust features of problems with the stochastically corrupted input data [22]. In 2012, a deep structure of CNN was introduced, which is named Deep Convolutional Neural Network (DCNN) [23]. In 2014, Generative Adversarial Network (GAN) was presented [24] using the discriminative model for training and classification with both real and generated random samples.

1.2 The architecture of neural networks

The basic concept of an ANN was put forward based on a mathematical model of the biological neuron consists of the internal activation function (the biological neuron cell body), output connection (the axon of the biological neuron), weights (the synapses of the biological neuron), and output signal. ANNs can be classified regarding neuron connections structure into two groups: feedforward neural networks (single-layer perceptron, MLP, and radial basis networks (RBF)) and RNNs (Hopfield network, Elman neural network (ENN), Jordan neural network, and recurrent multilayer network). Another classification can be presented for ANNs based on the input–output ratio, which categorizes them into three classes: class I (more inputs than outputs), class II (the same number of inputs and outputs), and class III (more outputs than inputs) [25]. Another classification of ANNs based on the organization is categorized into individual networks, and stakes network (ensembles) [26]. Moreover, ANNs considering the type of learning approach can be grouped into three main types: supervised (such as decision trees, regression analysis, and neural networks), unsupervised (such as cluster analysis, correlation, factor analysis (principal component analysis (PCA)), and statistical measures), and reinforcement learning (such as genetic algorithms (GA)). In supervised learning, both predictor variables and a target variable are required. Then weights are adjusted according to the deviation of the target variables from the desired outputs is negotiable. This learning approach is recommended in some cases that it is desired to predict the value of some variable. Notable examples of supervised learning algorithm are Backpropagation (BP), Levenberg–Marquardt, and gradient descent. On the contrary, there are no target variables to set the weights in unsupervised and reinforced learning. In unsupervised learning, only the input training patterns are provided to cluster into different classes. The goal of this learning method is to identify and select characteristics to classify the input data [27, 28]. Kohonen Rule and Carpenter-Grossberg Adaptive Resonance Theory (ART) algorithm are an example of unsupervised learning algorithms. Reinforcement learning employs a critic based on given output from the environment to identify the ideal behavior within a specific environment automatically. If the system of process define and model correctly, the global optimum can be obtained [27].

The optimal topology of a neural network is determined regarding the characteristics of the system being modeled. The neural networks are applied in a wide range of application areas. However, the methodology for finding the optimal neural network is still under discussion. There are several approaches to identify the optimal topology as follows:

1. *Statistical and empirical methods*: The problem and its definition have a considerable impact on finding the best type of these methods. The trial-and-error method is known as the primary method of this category, which develops several neural networks with various characteristics and compares their errors to find the best networks. However, the global optimal network may not be obtained. The trial-and-error methods devote particular attention to the overfitting, which appears during uncontrolled training, resulting in unreliable results of the neural network. Thus, it is recommended to watch the errors in the training phase carefully. Curteanu et al. have applied trial-and-error methods in the chemical engineering (CE) applications to find the most appropriate topologies for several types of neural networks such as MLP [29–32], feedforward generalized neural networks [33], modular neural networks [34], Jordan-Elman networks [35, 36], or stacked neural networks [33, 37]. The most systematic and comprehensive method of this category for the design of experiments is called the Taguchi method. Benardos and Vosniakos [38] have applied the combination of experiment design with Taguchi method using the Levenberg-Marquardt (LM) algorithm and pointed out the optimal sets of a neural network of MLP type. Sukthomya and Tannock [39] have applied the same method for identifying the optimal parameters of an MLP network, namely training algorithm, number of hidden layers, the total number of neurons, activation functions, and testing percentage of data.

2. *Constructive and/or destructive methods*: Constructively or destructively developing a too small or large initial architecture, respectively; and network performance improvement are obtained in these methods by adding or removing the layers or neuron, respectively. Using descendent gradient method in these methods can lead to a local minimum [40]. Therefore, there is no guarantee to find the optimal network topology [41]. Delogu et al. [42] applied a constructive method to select the number of neurons and their weights automatically. In another study, Islam et al. developed an approach for removing and adding hidden neurons [42], which reduces the network size by adding the correlated hidden nodes and increases the dimensionality of the network by dividing the hidden neurons. Xing and Hu [43] removed input and hidden neurons, applying an MLP network.

3. *Evolutionary algorithms*: In these methodologies, the reduction in the negative impact of the permutation [44] and the improvement in neural network optimization produce [45–47] has been subject to considerable studies. The advantages of applying evolutionary approaches to optimize ANN structure

include (i) the balanced assessment between exploitation and exploration; (ii) parallelization; (iii) scalability; (iv) avoiding the local minimum; (v) flexibility (implementing the same evolutionary method to various problem domains); and (vi) fitness flexibility [48]. The ANN-evolutionary approaches are categorized into two groups. In the first one, evolutionary algorithms are implemented to create the structure of an ANN, and then another algorithm such as BP algorithm is applied to train the network. In the second group, evolutionary algorithms are applied in both structural design and training [49]. Benardos and Vosniakos [38] developed a determination methodology to find the best architecture using GA, and of different criteria to evaluate an ANN's performance and its complexity. In this study, the number of hidden layers and hidden neurons is considered as the most critical elements of feedforward ANN architecture, which may overfit the training data and thus obtain poor generalization. Kordík et al. proposed the novel meta-learning principles for neural network topology and function optimization [50]. In this approach, supervised feedforward neural network called Group Adaptive Models Evolution is developed by combining several neuron types trained with different evolutionary optimization algorithms.

1.2.1 Feedforward neural networks

MLP, generalized regression neural network (GRNN), radial basis function (RBF) networks, CNN are the well-documented and widespread type of feedforward network. Feedforward multilayered networks are applied commonly in light of its straightforward implementation [51], easy programmability, and encouraging results [25]. MLP consists of an input layer whose number of neurons based on the input variables, hidden layer, and an output layer whose number of neurons based on the number of output variables. In simple cases, applying a single hidden layer with a large number of nodes can result in acceptable accuracy. However, in complex cases, additional layers are recommended to obtain an acceptable deviation and improved generalization capability [25]. With regard to the optimal topology, Fernandes and Lona [25] proposed several rules to find a suitable topology depending on the three mentioned different classes of neural networks (I, II, and III). Fig. 1 reveals the structure of this kind of ANN.

- Generalized feedforward networks (GFFN) are the type of MLP network with connections, which can jump over one or more layers [52]. GFFNs need less training epochs than MLP to solve the same problem.
- CNN is a multilayer feedforward ANN which is applied in applications such as 2D image processing [15], natural language processing, and speech recognition [53]. CNN is composed of stacking convolutional layers, pooling layers, and fully connected layers (see Fig. 2). In CNN, the feature learning is obtained by alternating and stacking convolutional layers convolved with

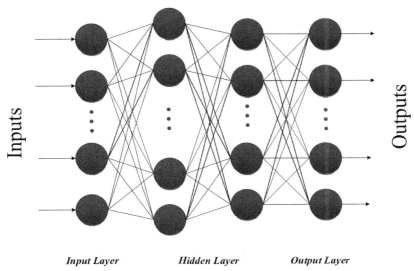

Inputs

Outputs

Input Layer   *Hidden Layer*   *Output Layer*

FIG. 1   The architecture of the MLP network.

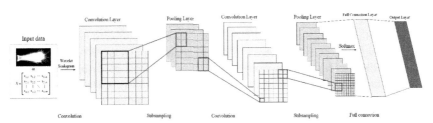

FIG. 2   Architecture of CNN [53].

input data using multiple local kernel filters, and pooling processes such as max pooling and average pooling. Then, a 2D feature map resulted from feature learning layer is converted into a 1D vector in a fully connected layer. Finally, a softmax function is applied to construct the model. Trained process of CNN is carried out using gradient-based backpropagation by minimizing the minimum mean squared error or cross-entropy loss function.

- RBM is a two-layer neural network including visible and hidden layer with a symmetric connection between these layers. However, neuron of the same layer has no connections to one another (see Fig. 3A). Data coding and dimension reduction of input data are carried out using the parameters of hidden layers to create features automatically. Then, data classification and regression are implemented using supervised learning algorithms, namely BP Neural Network, logistic regression, Naïve Bayes, and SVM.

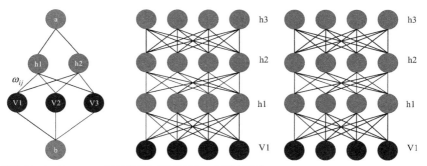

FIG. 3 Architecture of (A) RBM, (B) DBN, and (C) DBM.

Different variant models of RBM have been developed as a prime learning module, and have been received a growing number of attention [11].

DBN is developed by stacking multiple RBM networks in 2006 [16]. Regarding DBN training, a fast greedy algorithm is applied initially, and then the contractive wake-sleep algorithm is implemented to set the parameters of original deep architecture [16]. Bayesian Belief Network and RBM are employed to an area close to, and far from the visible layers, respectively. This network consists of undirected higher layer and directed lower ones, as illustrated in Fig. 3B).

DBM is a deep structured RBM network with a group of hierarchy hidden layers. In this network, there is no connection within a layer, or between non-neighboring layers, however, the full connections between two adjacent layers are allowable, as indicated in Fig. 3C. The ability of DBM to learn complex structures and provide a high-level representation of input data stems from stacking multi-RBM networks [21]. Unlike the DBN model with mixed directed/undirected connections, DBM network consists of undirected graphical ones. The trained process of DBM model is computationally demanding compared to the DBN model with a layer-wisely training process.

1.2.2 Recurrent neural networks

RNN contains both feedforward and feedback or recurrent connections between layers and nodes, as shown in Fig. 4. The feedback loops introduce a dynamic into the network [25] and lead to time-consuming and complex training process [54, 55]. Adding feedback connections gives the network ability to save information in hidden layers. Thus, RNNs are suitable to learn features of sequential data. Notable types of this network are simple recurrent, fully recurrent (such as Hopfield network and Boltzmann machine), echo state, LSTM, bidirectional, hierarchical, and stochastic neural networks. Recurrent networks consist of three main architectures, namely Internally Recurrent Network (IRN), Externally Recurrent Networks (ERN), and Internal-External Recurrent Network (IERN). The IRN or Elman network proposed employing a time-delayed

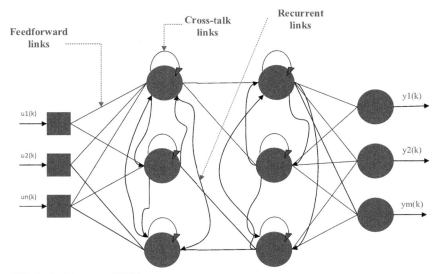

FIG. 4 Architecture of RNN.

feedback connection to the hidden nodes [56]; however, ERN implements a time-delayed feedback connection from the output layer to the hidden layer. Moreover, RNN, regarding the possible location of feedbacks, can be classified into globally recurrent networks, locally recurrent networks, and locally recurrent globally feedforward networks [55, 57]. Globally recurrent networks contain feedback between neurons of different layers or between neurons of the same layer. Fully, partially, and state-space recurrent networks are the primary types of this network. Fully recurrent networks are not suitable to apply in applications with nonlinear system identification. Partially networks such as Jordan networks [58] and Elman networks [56] implement the MLP network with an additional layer of units called context units to provide temporal information from the previous activations of the hidden neurons [52]. That is to say, this RNN is composed of input, hidden, context, and output layers. Adding this layer results in a faster training process and fewer stability process [55, 59], and therefore, partially RNN is applied to model dynamic processes successfully [54, 58, 60]. Recently, partially RNN with additional recurrent connections from the context units to themselves is developed [61, 62]. The state-space network is another type of globally RNN which contain feedback from hidden layer to the input layer through a bank of unit delays [55, 59, 63]. These feedback connections are unknown during the training process; therefore, the training process is implemented by minimizing the simulation error. The main pitfall of this models is an unstable training process and sensitive dependence on initial conditions. However, this model can be model nonlinear dynamic systems [63]. On the other hand, locally recurrent networks only feedbacks inside neuron models are implemented. This network has a similar structure to static feedforward one by replacing dynamic neurons instead of standard static ones. Compared to global RNN, this model has a simple and stable training process. Finally, locally recurrent globally feedforward networks have an architecture similar to feedforward networks and the dynamic properties of recurrent ones by integrating the best features of both feedforward and globally recurrent networks. These neural networks are recognized as the state-of-the-art network, and further studies on stability assessment, the implementation of novel learning procedures, robustness evaluation, etc., should be carried out.

RNNs cannot store long-term memory because of employing Backpropagation Through Time (BPTT) to train RNN, which lead to vanishing or exploding problems. Several studies are carried out to address this issue, which resulted in developing a modified version of RNNs, such as LSTM [14] and gated recurrent unit (GRU) [64]. LSTM can capture long-term dependencies of different time scales adaptively, proposing cell state concept. LSTM implements a processor with three gates (forget gate layer, input gate layer, and output gate layer) to set the cell state. GRU is a simplified version of LSTM, which is proposed in 2014 [64]. GRU is also used as a novel cell state to overcome the gradient vanishing problem. Compared to LSTM, reduction in the number of parameters improve the convergence speed.

1.2.3 Stacked neural networks

Stack neural network [65], ensemble neural network [66, 67], or aggregated neural network [68] are proposed by integrating a series of neural networks to improve flexibility and robustness of modeling performance. Fig. 5 reveals the structure of stacked neural networks. The method to design and implement

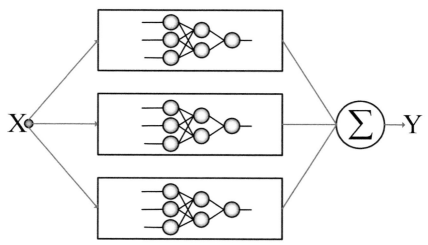

FIG. 5 The architecture of the staked neural network.

various stacked neural network are reported by Sridhar et al. [69], and Wolpert [70]. Tian et al. [71] noted that a stacked neural network perform more efficient than a single best-trained network because of inability to find an optimal network architecture, activation function, or learning algorithm in all single neural networks. Neural networks of the ensemble model have the same relation, despite this, they are capable of using different structures, weights differently initiated, and distinct training rules [72]. In this ANN model, three additional parameters, namely the combination method of parallel models, the number of randomly removed data from the initial dataset, and the number of levels. The method to integrate the outputs of ANNs included in the ensemble is one of the most critical problems of the stacked networks. Output Average, Majority Vote, Winner Takes All, Borda Count, Bayesian Combination, Dynamically Averaged Networks, Stacked Generalization, and Stacked Generalization Plus are among the commonly investigated type of combination methods [73], which implement regarding the nature of the application for which the neural network ensemble is created. Moreover, three main algorithms, namely bootstrap re-sampling [74–77], Ada Boost [78, 79], and randomization [80] are applied to re-sample or partition the training data. Previous studies show that randomization method performs efficiently with a low level of noise; however, bagging performs successfully with a high level of it [80]. Different approaches such as GA [66], feasible sequential quadratic programming (SQP) [71], and systematized trial-and-error method with the same goal [81] are applied to find the optimal weights of stacking ANN. According to their results, the GA method proposes smaller networks with high generalization capacity [71] compared with other techniques such as Bagging [82] or Boosting [83].

1.2.4 Auto Encoder

AE is an unsupervised learning algorithm, which consists of two parts, including encoder and decoder (see Fig. 6). The encoder captures features from the input data without label information and maps input data to a hidden layer to compress input data [12]. In the case of highly nonlinear input data. It should be added more hidden layers to create deep AE. Stochastic gradient descent (SGD) is applied to obtain parameters and build auto-encoder by minimizing the objective loss function regarding the least square loss or cross-entropy loss. Several variants of AE have been proposed as follows:

1. DAE: DAE is applied isotropic Gaussian noise to input data and forced hidden layers to capture more features in order to reconstruct the stochastically corrupted input data [22].
2. SAE: SAE imposes sparsity constraints on the hidden units, even a large number of hidden layers exist [19, 20].
3. Contractive Auto Encoder (CAE): CAE learns more robust representations of the input data to overcome small perturbations of model [84].

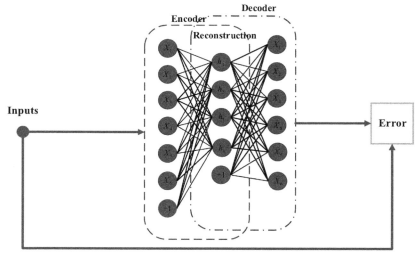

FIG. 6 The architecture of Autoencoder network.

1.2.5 Hybrid neural networks

Hybrid models integrate the phenomenology of systems with the empirical parameter of ANNs. That is to say, a hybrid model implements a simple mechanistic model to simulate a simple part of the process and ANNs to correct the errors of the mechanistic model or to model difficult part of the process. Curteanu and Leon [85] provide several methods such as a simplified phenomenological model and several feedforward ANNs to obtain hybrid models. Besides the ability to apply different topologies individually or in hybrid combination [25, 52, 86, 87], ANN can be integrated with other soft computing approaches, such as fuzzy systems and evolutionary algorithms. By doing so, the advantages of different methods are exploited, intending to obtain efficient, flexible, and robust techniques. Evolutionary algorithms are combined with ANNs to find the optimal topology of ANN, to estimate the optimal set of input parameters, or to optimize processes modeled with ANNs [88–90]. Hu [91] and Li [92] investigated the improvement of the prediction accuracy applying GA to obtain the optimal value of the RBFNN parameters. According to experimental results, the GA-RBF exhibits significant applicability and potential to obtain the nonlinear system modeling. Moreover, particle swarm optimization (PSO) [93] and harmony algorithm [94] are employed to set the weights of the BPNN model. Furthermore, hybrid neurofuzzy systems combine the explicit knowledge of fuzzy systems, which is easy to be understood, with implicit knowledge of ANNs [95, 96]. Adaptive network-based fuzzy inference system (ANFIS) model applies a number of adaptive nodes to capture and model the features of input data [97]. ANFIS minimizes the errors of the predictive model by employing the features of learning rules. In other words, the ANFIS model proposes a hybrid intelligent system where the advantages of both models are applied to a unified predictive model [98].

1.3 Application of neural network in chemical processes

CE is one of the main drivers of process and technology development in industries such as petrochemicals, oil and gas industries, biotechnology, manufacturing, food processing, health care, pulp and paper, polymers, nanotechnology, and environmental protection [99–104]. Traditional approaches to solving CE problems suffered from limitations on account of the highly complex and nonlinear process. Therefore, different approaches, such as ANN, have been proposed to overcome these problems [105–107]. The significant advantages of ANN model are that it does not need the knowledge of physical and chemical laws. Therefore, the ANN model has received much attention in the CE process since the last decade [108]. ANN model, however, requires a collection of examples to obtain the desired output [109]. Adequate trained ANN model can be successfully implemented to make predictions on input data, not used in developing the model [86]. Haykin [55] pointed out that the most indispensable properties of ANN are nonlinearity, input–output mapping, adaptability to provide a satisfactory input–output mapping, evidential response (remove ambiguous patterns and improve classification), contextual information, fault tolerance, and uniform analysis and design. ANN can provide a solution to address problems in sensor data assessment, fault detection, process identification, and control. Hoskins and Himmelblau [110] for the first time, implemented ANNs to detect an error of the chemical process. During the past decades, many reviews have been published on the application of ANNs in CE. Hussain [111] provide a review of applications and implementations of ANNs in control and simulation of a chemical process. Kalogirou and Bojic [112] argue that ANNs can be used in energy prediction and modeling of a chemical process. Zhang and Friedrich [113] investigated various ANN approaches for predicting specific properties of polymer composite materials. Venkatasubramanian et al. [114] summarized the application of ANN models for detecting and diagnosing a fault in process engineering. Du and Sun [115] reviewed studies on the application of ANNs for food quality evaluation. Roupas [116] provided a review of ANNs for the control of the dairy production process. Ludwig et al. [117] discussed the applications of ANNs for the oil production forecast. Qin et al. [118] summarized recent studies on developments, advancements, challenges, and barriers of ANNs approaches to simulate and optimize the control of petroleum waste management and site remediation. The use of different ANN modes in modeling and control of food analytical process is carried out by Marini [119]. Ahmad and Zhang [120] summarized novel techniques to improve ANN robustness for modeling and control of the nonlinear process. Noor et al. [121] provide an overview of ANNs application in modeling and control of the polymerization processes. A brief review of practical and straightforward procedures for selecting and training ANNs applied in the polymerization process are presented by Fernandes and Lona [25]. Curteanu and Cartwright [86] carried out a review on implementation methods of ANNs in CE applications. Pirdashti et al. [108] provided a comprehensive review of the application of ANNs in CE and showed a considerable potential of ANN to implement in simulation, optimization, and control of the chemical process. According to these studies, many parameters such as type of process, type and number of input and output variables, type of application (modeling, monitoring, optimization, or control), unique features of the system play a pivotal role to find the optimal type and architecture of ANN model. Although a surprising number of studies applied ANN in the CE system is continuously rising [122], further studies in this field should be carried out.

1.3.1 Application of artificial neural network in simulation, modeling, and optimization

1.3.1.1 Polymerization

It is difficult to provide a complete mechanistic model of polymerization reactors due to complex and numerous reactions and chemical species coinciding inside the reactor, complex determination of a large number of kinetic parameters, and the poor understanding of phenomena for mixtures involving polymers. Thus, ANNs are suitable alternatives to replace the phenomenological models in polymerization application. For instance, ANNs successfully simulate the free-radical polymerization of methyl methacrylate (MMA), associated with gel and glass effects [29]. Curteanu and Petrila [30] developed ANN model of semibatch and nonisothermal, which are two necessary polymerization conditions. The feedforward neural networks with two hidden layers are used for this polymerization modeling to tackle monomer conversion, initiator concentration, number, and weight average polymerization degrees. Besides, feedforward ANNs with two hidden layers were proposed to predict batch and semibatch, isothermal, and nonisothermal polymerization processes under different operation conditions. In this study, ANN methodology provides an outstanding representation of the suspension styrene polymerization under reaction conditions [30]. The stacked neural network was being applied to estimate a polymer quality [123], to assess the reactive impurities as well as the runaway of a discontinuous polymerization reactor [123], to obtain optimal control of polymerization reactor [65], and to model discontinuous polymerization process [68]. In another study, the stacked neural network is employed to model the gel effect, which is the most demanding process within the free radical polymerization process [71]. Curteanu and Leon [85] implemented the phenomenological model within MLP networks to evaluate the modeling of conversion and polymerization degrees of the polymerization process. Their hybrid model applied a simplified phenomenological model to describe the reaction kinetics without taking into account the gel and glass effects. MLP network employed to model this problematic part of the process. To put it another way, MLP simulates the residuals of conversion and polymerization degrees to modify the outputs of the simplified kinetic model [85]. Moreover, MLP models that part of the process in which gel and glass effect appears. The simulation using ANN indicates how semibatch operation or nonisothermal polymerization prevents the dead-end polymerization and also control molecular weight and polydispersity of the polystyrene as well as reaction time. The MLP model with one or two hidden layers can predict the thermostability of a series of new liquid crystalline ferrocene derivatives and similar phenyl compounds [124].

The Sugeno fuzzy inference technique and ANFIS are also implemented to evaluate polymer quality and to simulate the dynamic behavior of the polymer process by Chitanov et al. [125]. They also employed the ANFIS control for a batch polymerization of MMA using Sugeno fuzzy inference engine. They argued that the application of nonlinear Gaussian function for the estimation of the input space in nonlinear modeling could lead to convincing and reliable results. Karatas et al. [126] derived a new formula considering various injection parameters (cylinder temperature, injection pressure, injection flow rate, and mold temperature) applying ANNs to determine the yield length in the plastic molding of the commercial plastics, polyethylene, high-density polyethylene, polystyrene, and polypropylene. In this study, the ANN model uses a scaled conjugate gradient and LM learning algorithms and the logistic sigmoid transfer function. Multilayer feedforward, RBF, and GRNN models are employed by Koc et al. [127] to diagnosis the swelling behaviors of Ca^{2+}-alginate hydrogels under different conditions of pH and temperature. According to their results, the performance of GRNN is significantly better compared with the other ANN models in terms of the mean absolute error, the determination coefficient, and the standard error of predictions. Moreover, ANN models have superior performance than second-order swelling kinetics, quadratic, and cubic models of response surface methodology (RSM). In another research, the steric hindrance of polymers is modeled and simulated using a network with three inputs corresponding to three chemical descriptors, namely average polarizability, entropy, and dipole momentum [128]. The optimization of several properties of latex and rubber of styrene-butadiene rubber (SBR) emulsion polymerization such as solid content, Mooney viscosity, and polydispersity are conducted to find a specific grade of SBR using ANN with two hidden layers. This model can provide acceptable predictions under different operation conditions (batch and semibatch, isothermal and nonisothermal) for conversion and polymerization degrees. The ANN model performs efficiently to replicate the hardening behavior for the polyethylene terephthalate (PET) control samples and the mechanical behavior changes for the PET fibers with vapor grown carbon nanofibers (PET–VGCNF) samples. Gonzaga et al. [129] proposed an industrial ANN soft sensor to model the PET production process. Online prediction and redundant measurement of PET viscosity are provided by soft sensor being successfully integrated into a control system.

1.3.1.2 Fuel cell

The GFFNs model is developed to simulate the performance of a polybenzimidazole-based polymer electrolyte membrane fuel cell, taking into consideration the temperature in the conditioning period and temperature when the fuel cell was operated [32]. GFFN is also implemented to obtain the value of the tortuosity and the cell voltage of fuel cells, at a given current density regarding several properties such as Teflon content, air permeability, porosity, mean pore size, and hydrophobia level [32].

Recently, Guo et al. [130] proposed an ENN to identify parameters of proton exchange membrane fuel cell (PEMFC), which is optimized by a hybrid algorithm that combined the teaching–learning-based optimization (TLBO) and differential evolution (DE) algorithm, namely the TLBO-DE method. Moreover, a TLBO-DE-based ENN is applied to solve the nonlinear parameter identification problem of PEM fuel cell. Their result demonstrated that this network could predict the stack voltage reliably in various operational situations.

1.3.1.3 Combustion and fuel

Today, prediction of energy demands as well as resources, and its pollution play a pivotal role to obtain sustainability goals because of depleting fossil fuel resources and environmental issues of greenhouse gas emissions. ANNs can provide a powerful tool to model and optimize the process for many applications in this field. Recently, Kshirsagar and Anand [131] evaluated the forecasting capability of ANN regarding several, four-stroke engine performances and its emission (HC, CO_2, CO, NO, dry soot, and engine out O_2) using the mixture of Calophyllum inophyllum methyl ester (CIME) with neat diesel on a stationary with different blending percentages varied from 10% to 25%. The input variables of the ANN models consist of fuel blend, engine loading, fuel injection timing (FIT), and fuel injection pressure (FIP). Emission and engine performance are modeled applying two ANN models, which have a single hidden layer with 16 neurones, and two hidden layers with 14 neurones at each layer, respectively. A number of evaluation criteria such as root mean squared error (RMSE), correlation coefficient (R^2), mean absolute percentage error (MAPE), mean squared relative error (MSRE), NashSutcliffe Coefficient of Efficiency (NSE) as well as the THEIL U2 uncertainty evaluation, are implemented to assess the prediction performance of ANNs. Tang et al. [132] proposed a deep ANN model for the optimization of combustion system performance. The DBN method is applied to provide a model of the combustion efficiency and the NOx emission. The JAYA algorithm is employed to find the optimal solutions of the developed model. The developed optimization framework of the combustion process is validated using industrial data. According to their results, there is a significant potential to improve both of the combustion efficiency and NOx emission by settings control parameters of the combustion system.

Chakraborty et al. [133] evaluated the effect of engine speed and throttle position towards mass airflow, manifold pressure, fuel actual pulse width, exhaust gas temperature (EGT), engine temperature as well as NOx emission using a single hidden layer ANN model with ten neurons. In this study, statistical parameters such as the average RMSE, the average error, and the standard deviation are implemented to assess the performance of the ANN model. Their results indicate that the average RMSE is less than 3%. Yusaf et al. [134] proposed an ANN model of four-stroke Honda GX160 KI gasoline engine to obtain engine performance and gas exhaust emissions. The engine speed and the fuel type are considered as the input of the backpropagation ANN model, whilst the brake specific fuel consumption (BSFC), efficiency, exhaust temperature, engine body temperature, NOx, CO_2, CO, O_2, and UHC are selected as the output of the model. In this study, training is carried out with seventy percentage of experimental data, while the remaining 30% was used for model validation. Statistical criteria, namely MRE, the RMSE in addition to the correlation coefficient, R, are applied to evaluate the ANN model. The RMSE values are less than 4% for all parameters but for exhaust temperature and UHC. Moreover, the obtained MRE values are near 5% for all predicted output parameters but for BSFC, efficiency, NOx. Rao et al. [135] proposed a single hidden ANN to assess the effect of rice bran methyl ester with isopropanol additive on the performance and emissions of a cylinder four-stroke indirect diesel injection engine. The load, as well as the fuel, are considered as an input of neurones, while parameters such as EGT, BSFC, brake thermal efficiency (BTE), HC, O_2, CO_2, CO, NOx, and smoke are selected as an output of neurons, resulting in proposing nine network models. The trainlm learning algorithm is implemented to train the model using mean squared error (MSE) as assessment criteria.

In a recent study, Saree et al. [136] developed a single hidden layer ANN to investigate the performance and emission of a diesel engine using cerium oxide (CeO_2) nanoparticles diesel fuel mixture as a fuel. The nanoparticles were added to the base fuel, BSFC, and the engine speed are considered as the input parameter of the ANN model. The output parameters of ANN are the engine power, NOx, HC and CO. Totally, 70%, 20%, and 10% of input dataset are implemented in training, testing, and validation step. Training algorithms such as trainlm, trainscg, trainrp, and traingdx are applied in the training process. Besides, the learning rate and the momentum coefficients are adjusted to 0.4 and 0.8, respectively. Their results indicated that the trainlm training algorithm with 12 neurones results in the best prediction model. Seyed et al. [137] evaluated the H_2 dual fueled diesel engine applying an ANN model. The input variable of the ANN model is engine load and H_2 flow rate, whilst the output of it is performance and emission characteristics. In this study, seven training algorithms, namely with eight transfer function combinations, are employed. According to results, the best training algorithm considering regression coefficient are h Broyden, Fletcher, Goldfarb, and Shanno (BFGS) quasi-Newton backpropagation (trainbfg). Uzan [138] investigated the performance of a turbocharged inter-cooled diesel engine implementing the ANN model. Bietresato et al. [139] predicted torque and BSFC of a diesel engine, using the ANN model with various transfer functions. Gaussian transfer functions indicate the best performance in terms of the maximal regression coefficient. They recommended that H_2 used as dual fuel in a diesel engine because of improving engine performance and emissions. Gurgen et al. [140] put forward an ANN model with a single hidden layer to evaluate the effects of cyclic variability of a diesel engine using diesel fuel and butanol-diesel fuel in terms of engine performance and exhaust emissions of a four-stroke air-cooled direct injection diesel engine. Scaled conjugate gradient (trainscg) training algorithm indicated the best performance to create a model. The engine speed and fuel blend ratio are selected as input data, while the output of the model is the coefficient of variation of indicated mean sufficient pressure. Their results demonstrated that coefficient of determination of train, validation, and test are 0.9677, 0.86, and 0.737, respectively. Furthermore, the ANN model predicts the coefficient of variation of indicated mean adequate pressure accurately compared with experimental data.

Wu and Liu [141] developed a BP neural network (BPNN) to predict car fuel consumption. Although many factors affect the fuel consumption of a car, factors such as engine style, the weight of a car, vehicle type, and transmission system are considered as input information for the neural network training and fuel consumption prediction. The accuracy of the results applying ANN model was satisfactory. Pena et al. [142] proposed a probabilistic ANN and ANFIS model to assess the effectiveness of soot-blowing. These models are validated with a real case-study boiler, a 350 MWe Spanish power station. Heat flux measurements in the furnace water walls are carried out to provide training and testing dataset. Zahedi et al. [143] put forward an RBFNN model to provide a dynamic grey-box model (GBM) of ethylene oxide (EO) fixed bed reactor. In this model time, C_2H_4, C_2H_4O, CO_2, H_2O, and O_2 mole fractions, multiplication of reaction rate, and catalyst deactivation are considered as input parameters. Compared to the EO mechanistic model, RBFNN provides more accurate results. Balabin and Smirnov [144] investigated the performance of 16 different feature selection methods, (namely stepwise multiple linear regression (MLR), interval partial least squares (iPLS) regression, backward iPLS, forward iPLS, moving window partial least squares (MWPLS) regression, (modified) changeable size MWPLS, searching combination MWPLS, successive projections algorithm (SPA), uninformative variable elimination (UVE-SPA), simulated annealing, BP-ANN, Kohonen ANN, and GAs (GA-iPLS)) to obtain properties of biodiesel fuel, including density, viscosity, methanol content, and water concentration. Two calibration techniques, namely MLR and partial least squares (PLS) regression, are employed to assess biofuel properties. Eslamloueyan and Khademi [145] predicted the conductivity of pure gases using a feedforward three-layer ANN at atmospheric pressure and a wide range of temperatures considering their critical temperature, critical pressure, and molecular weight. The experimental conductivities of various gases, which some of them are not applied in the training process evaluated the accuracy of the model. The results of the comparison between the proposed model with conventional models indicated that the ANN model performs precisely compared with other alternative methods, in terms of accuracy, and extrapolation capabilities. Unlike conventional conductivity correlations, the ANN method covers a wide range of temperatures and substances. Erzin et al. [146] predicted soil thermal resistivity implementing an MLP model with the LM optimization algorithm. Experimental thermal resistivity values for clay, silt, silty-sand, fine-sand, and coarse-sand are considered as input parameters of the ANN model. The thermal resistivity of these soils obtained from ANN models concur thoroughly with experimental data. In addition, the ANN model predicts thermal resistivity more accurate than empirical relationships reported in the literature. Wang and Wan [147] proposed an ANN model to investigate the effects of temperature, initial pH and glucose concentration on fermentative hydrogen production. The model predicts substrate degradation efficiency, hydrogen yield, and the average hydrogen production rate with a coefficient of determination of 0.98, 0.99, and 0.98, respectively. Whiteman and Gueguim Kana [148] compared the results of RSM and ANN models to evaluate the effects of molasses concentration, initial pH, temperature, and inoculum concentration on the cumulative volume of hydrogen. R^2 values of RSM and ANN models are 0.75 and 0.91, respectively. Nasr et al. [149] proposed an ANN model to simulate hydrogen production applying initial pH, the initial substrate as well as biomass concentration, temperature, and time as the input data. ANN model gave an R^2 value of 0.99. They recommended that the application of conventional ANN model to diagnosis the influence of leading parameters can lead to obtaining the optimal set points for bio hydrogen production and the reduction in the process development time.

1.3.1.4 Petrochemical

Recently, there has been a growing interest in the development and evaluation of ANN models in petrochemicals, oil, and gas industry. The commonly investigated type of ANN models in this area are MLP; however, other types such as SVM, functional networks, GRNN, radial basis function neural networks (RBFNN) are applied, in a smaller proportion. This enumerated model has employed a supervised training algorithm, BP, LM, gradient descent with learning rate, resilient BP algorithm, and a quick prop algorithm as a learning rule. ANN models are commonly integrated with GA, fuzzy logic, PCA, ant colony optimization (ACO) to model and optimize systems and process in this area. Artun et al. [150] developed an ANN-proxy model to simulate a wide range of reservoir characteristics and design scenarios applying GA algorithm to search for the optimum design scenario by evaluating the objective function via proxy models. In other words, GA maximized the discounted oil volume produced per injected volume for a specified period of operation. Sadeghzadeh et al. [151] proposed a set of supervised ANN models to evaluate the performance of the oxidative coupling of methane at elevated pressures gathering reaction data of microreactor device. The optimal values of the operating conditions are obtained applying hybrid GA, which improves the performance of oxidative coupling of methane process at desired operating pressure. Safamirzaei and Modarress [152] applied an ANN model with the LM optimization algorithm to predict hydrogen solubility in heavy n-alkanes. The inputs of the model are temperature and pressure, and hydrogen solubility is considered as the output. The results of traditional ANN have slight deviation; however, the application of each trained network is restricted to the system for which it has been trained. In the second stage, a 3-4-1 ANN has been developed for all studied systems. The average relative deviation of 3-4-1 network is 1.66%. Ashena and Moghadasi [153] predicted the bottom-hole pressure as the flow parameter in an underbalanced drilling program applying an MLP model with ACO and GA optimization algorithm. In this study, the ANN model contains seven neurons in the hidden layer to model the nonstraightforward problem of two-phase flow in the annulus (ANN-BP). The ANN model optimized with two other algorithms, namely, ACO-BP, and GA-BP. The ACO-BP and GA-BP outperform ANN-BP for predicting bottom-hole pressure. Azadeh et al. [154] predicted a long-term natural gas consumption using ANFIS-stochastic frontier analysis (SFA) approach. Their results indicated that ANFIS-SFA approach could deal with complexity, uncertainty, and randomness of the gas consumption prediction in a series of countries for the period 2008 to 2015 regarding the data gathered in the interval 1980 to 2007. Asadisaghandi and Tahmasebi [155] provide a comparison study related to the performance of various learning algorithms of ANNs implemented in the prediction of pressure-volume-temperature (PVT) oil properties. Alhajree et al. [156] developed a two multilayer feedforward ANNs model with BP training algorithm for optimization of an industrial hydrocracker unit. Parameters such as fresh feed and recycle hydrogen flow rate, the temperature of reactors, the mole percentage of H_2 and H_2S, feed flow rate and temperature of debutanizer, pressure of debutanizer receiver, top and bottom temperatures of fractionator column, and pressure of fractionator column are considered as inputs of the model. The first network is used to predict the specific gravity of gas oil, kerosene, light naphtha (LN), heavy naphtha (HN), gas oil and kerosene flashpoint, and gas oil pour point. The second one calculated the volume per cent of C4, LN, HN and kerosene, gas oil, and fractionator column residual. Nikravesh and Aminazadeh [157] put forward an ANFIS methodology for diagnosing the nonlinear relationship and mapping between well logs/rock properties and seismic information and extract rock properties, relevant reservoir information, and rules (knowledge) from these databases. Souza et al. [158] employed static neural networks (SNNs) to optimize the reaction condition of the catalytic cracking of natural gasoline over HZSM-5 zeolite. Anifowose and Abdulraheem [159] estimated some vital oil and gas reservoir properties, such as porosity and permeability of oil and gas reservoir characterization applying hybrid SVM and functional network with a fuzzy logic model. Wang et al. [96] modeled the secondary reaction of fluidized bed catalytically cracked gasoline employing fuzzy neural network (FNN) with GA. Olatunji et al. [160] predicted the PVT properties of crude oil systems using hybrid ANFIS with supervised training type and BP learning algorithm.

1.3.1.5 Environment

Zaqoot et al. [161] provided an RBFNN model for predicting the dissolved oxygen concentration in the Mediterranean Sea water along with Gaza within two weeks. Inputs of network consist of water temperature, wind speed, turbidity, pH, and conductivity. According to their results, RBFNN model can be applied to predict dissolved oxygen with arbitrary accuracy. Previous studies, however, indicated that MLP improved by LM optimization performs slightly superior to RBFNN [162, 163]. Li et al. [164] developed an ELM network to predict pond dissolved oxygen concentration of different voltage pumps and improve dissolved oxygen distribution based on artificial push flow. The mean square error and the correlation coefficient R^2 of this model were 0.0394, and 0.9823, respectively. These values were lower than those of the BPNN model. Ta [165] put forward the CNN model to overcome the dissolved oxygen prediction problem. According to experimental results, the backward understanding of the CNN model obtains better prediction than the BPNN model. Aghav et al. [166] evaluate the application of three-layer feedforward ANN model with BP algorithm to model the performances in competitive adsorption of phenol and resorcinol from aqueous solution by conventional carbonaceous adsorbent materials, namely activated carbon, wood charcoal, and rice husk ash. The input parameters of the model are the amount of adsorbent, initial concentrations of phenol and resorcinol, contact time, and pH. The outputs of the model include the removal efficiencies of phenol and resorcinol. The application of traditional BPNN in water quality management is investigated in recent studies [167–170]. In this regard, Zou and Wang [171] evaluated a BPNN model to predict the water quality of the downstream detection section of the river. The BPNN model showed an acceptable performance for long-distance prediction of short river sections. Chen et al. [172] predicted dissolved oxygen in aquaculture, applying PCA and LSTM models. The LSTM model shows the highest prediction accuracy compared with PCA model. Liu [173] provided an ENN model with 7 or 8 hidden layers to model the dissolved oxygen concentration in the Hyriopsis Cumingii pond. He et al. [174] addressed the complexity of predicting Water quality parameter using the MLP model with partial mutual information input selection algorithm for selecting a set of input parameters. The MLP method exhibited superior performance compared with ANNs using partial correlation to obtain input. Chen et al. [175] integrated the RBFNN model with subtractive clustering (SC) method and K-means clustering method to improve the accuracy of the model in water quality prediction problems.

Gadehar et al. [176] combined RSM and ANN approach for optimization of disperse dye removal by adsorption using water treatment residuals (WTR). The RSM–ANN approach has an accurate performance to optimize the color removal process. Lotfan et al. [177] investigated reduction in CO and NOx emissions of a direct injection dual-fuel engine applying MLP and nondominated sorting genetic algorithm II (NSGA-II). The input data of the MLP model include controllable parameters, namely engine speed, output power, intake temperature, mass flow rate of diesel fuel, and mass flow rate of the gaseous fuel. In this study, uncertainty assessment is also carried out to identify the uncertainties of experiments and ANN model. NSGA-II algorithm applied to achieve optimal values of intake temperature, mass flow rate of diesel and gaseous fuels, resulting in the Pareto-optimal set of designs under any combination of engine speed and output power. Akbaş et al. [178] predicted biogas production from wastewater treatment developing MLP and indicated the applicability of MLP to model the process.

Peng et al. [179] improved the accuracy to solve the problem of poor robustness of traditional approaches by proposing a hybrid PCA and GRNN model to predict dissolved oxygen for aquaculture. This study proposed the primary components of the PCA-based aquatic eco-environmental indicators. Their model is capable of predicting with an error of 5%, which indicates its feasibility. Wan et al. [180] developed an ANFIS model for evaluation of suspended solids and chemical oxygen demand (COD) removal of a full-scale wastewater treatment plant in a paper mill industry. The optimal model structure and fuzzy rules are identified applying fuzzy SC method; meanwhile, the reduction in input variable dimensionality is accomplished employing the PCA method. This hybrid model can predict and control the performance of the developed system accurately. Curteanu et al. [181] applied various type of ANN such as MLPs, GFFNs, modular neural networks, and Jordan-Elman networks to simulate an electrolysis process in wastewater treatment. Bhatti et al. [182] developed RSM and ANN model for diagnosis the performance of electrocoagulation system for the removal of copper from synthetic wastewater using aluminum electrode pair. Metal removal efficiency and energy consumption are considered as outputs of the model. Moreover, GA multiobjective optimization based on ANN developed model is conducted and resulted in the creation of nondominated optimal points, which obtain the optimal operating conditions of the process. Al-Abri et al. [183] proposed the BPNN model for the comparison between humic substance retention and membrane fouling based on previous experimental data. Sadrzadeh et al. [184] developed an MLP model LM optimization technique, and a mathematical model to assess the separation per cent of Pb^{2+} ions in terms of concentration, temperature, flow rate, and voltage. The correlation coefficients of the mathematical model and ANN to predict the performance of electrodialysis desalination are 0.97 and 0.99, respectively. According to their results, although the ANN model predicts the nonlinear behavior of the electrodialysis process more accurate than the mathematical model, However, it shows less efficient at higher feed flow rates and lower voltages, temperatures, and feed concentrations. Gulbag et al. [185] evaluated the comparison of seven different types of binary volatile organic gas mixtures implementing a hybrid probabilistic neural network (PNN) and MLP model. This hybrid model is employed to classify the binary gas mixtures and to recognize individual gas concentrations in their gas mixtures. Training data are obtained using the steady-state sensor responses of quartz crystal microbalance-type sensors. Sahoo et al. [186] investigated on pesticide concentrations predictions in groundwater monitoring wells employing a BPNN model. The parameters such as the categorical indices of depth to aquifer material, pesticide leaching class, aquifer sensitivity to pesticide contamination, time of sample collection, well depth, depth to water from the land surface, and additional travel distance in the saturated zone are considered as the input variables, whilst the output variables are the total pesticide concentration detected in the well. The developed BPNN model can identify parameters for improving maps of aquifer sensitivity to pesticide contamination. Chellam [187] simulated membrane fouling caused by polydispersed feed suspensions in crossflow microfiltration using ANN models. According to results, the ANN model can simulate time-variant specific fluxes efficiently for several feed suspensions. ANNs outperform a complicated mathematical model to identify the effects of colloidal transport and deposition mechanisms, as well as cake morphology changes and hydrodynamics resistance. In another study, flux declines and time during crossflow microfiltration of phosphate and fly ash mixture is evaluated by applying an ANN and Koltuniewicz's method (KM). Input parameters of models include fly ash dosages, PO_4 concentrations, transmembrane pressures, and two membrane types. ANN has a better performance than KM to simulate the fluxes in terms of correlation values [188]. Yangali-Quintanilla et al. [189] predicted and evaluated the rejection of neutral organic compounds by polyamide nanofiltration and reverse osmosis membranes according to the solute size and hydrophobicity descriptors employing ANN models. In this study, various ANN structures such as using a hyperbolic tangent sigmoid function (HTSF) in the hidden layer, using HTSF or linear function in the output layer, and using two different training methods with adaptation learning functions are investigated. Kuzniz et al. [190] developed ANN model for real-time estimation of pollutant concentration based on absorption and fluorescent measurements in the visible spectral range. The authors argued that their research is an initial step in the development of a rapid warning system. In the comprehensive study of Emad and Malay [191], ANNs are applied to simulate and predict the oxidation performance in wastewater treatment, which include COD removal from antibiotic aqueous solution using Fenton process, oxidation reduction in color removal wastewater, dissolved organic carbon removal from polyvinyl alcohol aqueous solution, decolorization of acid orange 52 dye solution applying the UV/H_2O_2 process, methyl *tert*-butyl ether degradation, nitrogen oxides removal by TiO_2 photocatalysis, and humic substance removal from aqueous solution by ozonation.

1.3.1.6 Biotechnology

Biotechnology is a ground-breaking topic in ANN applications. Smith et al. [192] conducted a deterministic global optimization of an ANN-embedded problem with one hidden layer and three neurons for simulation of a flooded bed algae bioreactor. Maran and Priya [193] carried out a comparison between the performance of the MLP network and the RSM model to assess fatty acid methyl esters (FAME) conversion process in biodiesel production. They indicated that the MLP model outperforms according to efficiency value. Drăgoi et al. [194] obtained the optimal configurations of individual and staked neural networks using the DE algorithm to evaluate mass oxygen transfer in stirred bioreactors. According to their results, the ANN model results perform more accurate than Friedman multivariate adaptive regression spline method. Chang et al. [195] evaluated biomass enzymatic digestibility over different steam pretreatment conditions using three methods, namely, BPNN, MLR, and PLS regression. Steam explosion (temperature, time, and particle size) are considered as the input parameters, while the enzymatic digestibility is the output. The results indicated that the BPNN model provides the most efficient performance. Moreover, it is identified that the steam explosion temperature is the most significant parameter among the enumerated ones, followed by the steam explosion time. Desai et al. [196] conducted an investigation on dry yeast cell mass and its glucan content applying ANN model. The dry cell mass and yield of glucan are crucial parameters in process optimization and control; however, there are not well-established mathematical relationship between them.

Becker et al. [197] developed a dynamic ANN model for online optimization of industrial fermentation. This model can provide a future prediction of the batch fermentation once 12 h of process data are collected. ANN model is used to optimize the temperature trajectory in terms of technical and technological conditions of the brewery, and therefore, to reduce the processing time. Another study quantified the influence of n-dodecane addition, used as oxygen-vector, on oxygen transfer in stirred bioreactors, employing ANN [90]. In this study, DE—a global optimization algorithm is applied to obtain the optimal architecture and the best internal parameters of the ANN model. ANN model results indicated that its performance in modeling, classification, and optimization of the mentioned system are acceptable, and the DE algorithm results in the near-optimal ANN structure. Elnekave et al. [198] developed an ANN model, namely BPNN, RBF, and GRNN, to evaluate biogas production. Compared with other models, the BPNN has the best prediction performance with an average deviation in the range of 6.4%–15.6% from the experimental data. Mahanty et al. [199] applied the BPNN model for evaluation of industrial waste co-digestion from different sources, including paper, chemical, petrochemical, automobile and food. Their result showed that industrial chemical waste and automobile waste have the highest and lowest effect on specific methane yield. Moreover, the ANN model performs more accurately than the regression model regarding prediction ability and the significance analysis. Wu and Shi [200] evaluated the impact of glucose concentration on biomass concentration with applying a hybrid ANN model and a deterministic kinetic model. The optimal value of biomass concentrations and maximum productivity obtained by hybrid ANN model are 10% and 40%, respectively, which provide better results than the deterministic kinetic model. Prakasham et al. [201] optimized a biohydrogen production with an increase of 16% using ANN combined with GA. Their results indicated that the pH of the medium, carbon source, inoculum age, and concentration have a considerable impact on the metabolic activity of the hydrogen-producing bacteria and the overall hydrogen yield. They also recommend ANN–GA model for bioprocess optimization. Shi et al. [202] presented a SDAE deep learning network model to assess the performance of a two-stage biofilm system. The input variables of model contain concentrations of COD, ammonia (NH_4^+-N) and total nitrogen (TN) of biofilm system influent, concentrations of COD, NH_4^+-N and TN of anoxic biofilm reactor effluent, influent flow, and reflux ratio of biofilm system. The concentrations of COD, NH_4^+-N and TN of biofilm system effluent are considered as output variables for COD, NH_4^+-N, and TN prediction model, respectively. SDAE deep learning network outperforms in the prediction of the biofilm process compared with backpropagation neural network, support vector regression, extreme learning machine, and gradient boosting decision tree.

1.3.1.7 Nanotechnology

Given the novelty of this field, the application of ANN has been receiving much attention in recent years. The research in this area is often investigated on the prediction of processing parameters and morphologic characteristics of nanoparticles samples in the experimental environments, applying an MLP model with BP learning algorithm. Salehi et al. [203] developed a GA-ANN model to assess the heat transfer of a silver/water nanofluid in a two-phase closed thermosyphon improved thermally by a magnetic field. Input variables of model contain inlet power, the magnetic field strength, and volume fraction of nanofluid in water, whilst the thermal efficiency and thermal resistance are considered as the outputs. The GA-ANN method indicated a robust performance to predict thermosyphon behavior. Khanmohammadi et al. [204] developed a BPANN model to estimate particle size employed the relationship between particle size and diffuse reflectance (DR) spectra in the near-infrared region. In another study, Ma et al. [205] evaluated correlations between processing parameters and the morphologic characteristics of nanocomposite WC-18 at.% MgO powders using the BPNN model. The parameters such as ball-to-powder weight ratio, milling speed, and milling ball diameter are selected as input variables, and the output variables of three individual BP network models include crystallite size, specific surface area, and median particle size, respectively. They noted that developed models could be implemented to predict properties of composite WC-MgO powders at various milling parameters, and to optimize processing and ball milling parameters. Haciismailoglu et al. [206] applied an ANN model with a delta-bar-delta learning algorithm to predict dynamic hysteresis of toroidal nanocrystalline cores. The parameters, namely the geometrical dimensions of cores, peak magnetic induction, and magnetizing frequency, are the inputs of the model. The results indicated that the ANN model performs accurately in this process. Santra et al. [207] developed an ANN model with resilient-propagation (RPROP) training algorithm to predict and assess the heat transfer of laminar natural convection of copper water non-Newtonian nanofluid in a differentially heated square cavity. They also carried out the numerical simulation, where the finite volume approach with the SIMPLER algorithm is employed to solve numerically transport equations to obtain the required training data of the ANN model. Comparing simulation and RPROP-based ANN results, it has been indicated that the ANN outperforms more efficient to predict the heat transfer.

1.3.1.8 Mineral

Few types of research have addressed the application of ANN model in minerals field. Capdevila et al. [208] developed an ANN to derive a formula to calculate the chemical austenitizing temperature. In order to predict changes of tendency in the slopes of the austenitization temperature, consistent with the existence of a eutectoid composition, a wide range of compositions are tested. Zeng et al. [209] developed a hybrid model by integrating experiments and ANNs with the extended delta-bar-delta training algorithm to simulate and optimize flow injection system for the spectrophotometric determination of Ru(III), with m acetylchlorophosphonazo (CPA-mA). Five output parameters of this model are NaOH concentration, CPA-mA concentration, KIO_4 concentration, microwave irradiation time, and the time interval between the injection and the starting of microwave irradiation.

1.3.1.9 Other application

Schewidtmann et al. [210] developed a hybrid ANN-deterministic global process optimization model to learn fundamental thermodynamic properties. The results indicated that a hybrid ANN model is capable of proposing a prediction of thermodynamic properties favorably. Further studies indicated that the ANN size effects significantly on the optimization computational time, in other words, a smaller number of neurons make a relaxation tighter and result in reduced computational times while providing a high prediction accuracies [211]. Schewidtmann et al. [212] develop an optimization approach to optimize the ANN model of an illustrative function, a fermentation process, a compressor plant and a chemical process using in-house deterministic global solver. The results indicated adequate computational time compared to a state-of-the-art global general-purpose optimization solver. The developed optimization method provides optimal single shallow ANNs with up to 700 neurons in under 30 s. Li et al. [213] developed a deep learning ANN model to simulate a post-combustion CO_2 capture process. In the unsupervised pre-training phase of deep learning, a DBN with many layers of RBM is applied to achieve initial weights of the subsequent supervised phase. According to their results, the DBN model predicts CO_2 production rate and CO_2 capture level using inlet flue gas flow rate, CO_2 concentration in inlet flue gas, the pressure of flue gas, the temperature of flue gas, lean solvent flow rate, monoethamine (MEA) concentration, and temperature of lean solvent as an input parameter. A greedy layer-wise unsupervised learning algorithm is employed to optimize the DBN model, resulting in better generalization compared with a single hidden layer neural network. The developed ANN model can be applied in the optimization of the CO_2 capture process. Chang et al. [214] proposed the hybrid neural network rate-function model (HNNRF) to obtain chemical reactor processes. In this study, the HNNRF modeling can perform accurately over a broad operating region, and in the face of noisy measurements. An HNNRF model with four FANNs was obtained reliability using 8 and 15 batches information. Regarding noise-corrupted measurements, the developed HNNRF model provides acceptable performance. The developed model also could be applied to identify the optimal feed loadings and operating conditions of complex processes.

Elsayed and Lacor [215] proposed a two RBFNNs model to predict the complex nonlinear relationships between the performance parameters of the gas cyclone separator (pressure drop and cut-off diameter) and the geometrical dimensions accurately, and therefore optimize gas cyclone. The training process of ANNs is conducted with experimental data available in the literature for the pressure drop and Iozia and Leith model for the cut-off diameter. Moreover, the RSM has been implemented to evaluate the interaction between geometrical parameters, fitting a second-order polynomial to the RBFNN. In another study, the drying rate of seedy grapes is identified using a feedforward neural network (FNN) [216]. Comparing the FNN model results with those of nonlinear and linear regression models, the first variant of modeling of the FNN model performs more consistently. The results indicated that the developed FNN model could be implemented as a powerful tool for analyzing nonlinear behavior of drying and for applications to various engineering studies. Kamali and Mousavi [217] developed three modeling approaches, including the dense gas model with Peng Robinson (PR) equation of state as an analytical model, a three-layer feedforward neural network, and a hybrid analytical-neural network to address the thermodynamic modeling of the supercritical extraction of α-pinene using supercritical CO_2. Their results demonstrated that the neural network provides a suitable prediction in the training region; however, its predictions are not proper in the exploratory region. The hybrid model predicts efficiently in both training and testing regions. Jazayeri-Rad [218] conducted a study on nonlinear multi-input multi-output processes in a distillation column with time delays applying a combination of multiple neural networks to model.

Galván et al. [219] carried out an investigation on predictions of nonlinear neural models when acting as process simulators, considering parallel identification models. According to their results, the predictions of the developed model are not accurate because of employing multilayer feedforward networks to estimate the parameters of parallel identification models. Then the particular RNN is implemented to estimate the parameters of parallel models, which can identify the parameter sets of the parallel model to obtain process simulators. The results indicated that parallel models using RNN could be applied as a perfect alternative to phenomenological models for simulating the dynamic behavior of the heating/cooling circuits.

1.3.2   Application of artificial neural network in control

Neural networks can be employed as a predictive tool in control schemes such as the model predictive control (MPC) to enhance the performance of the control schemes [220–223]. The development of chemical process plants results in onerous to control, and—the increase of global competition in controlling and monitoring enumerated processes. Therefore, new modeling approaches were developed to design and evaluate nonlinear industrial processes [224]. By so doing, process safety, product quality, product efficiency, consistency, and even the productiveness of all running equipment is obtained, which are the goal of the novel developed model. A recent trend in control scheme has led to a proliferation of studies that the development of control and monitoring models for linear and nonlinear processes considering various identification approaches [225, 226]. The development of a mathematical model for controlling real industrial processes is faced with various challenges [227]. ANN models, which historically have a close connection to control concept [228], have demonstrated their ability to control linear and nonlinear process, and their accurate performance compared to other methods, namely proportional integral derivative (PID) controller [229–231], the sliding-mode controller [232], or fuzzy logic controller [233]. Moreover, the remarkable development of ANN because of technological advances leads to achieving particular international notability applying deep learning and reinforcement learning approaches [234–238] and proposing a powerful tool in control design [239, 240]. In recent years, dynamic programming methods using deep ANN model indicated precise learning and faster convergence [241]. Although, the dynamic programming controller provides outstanding versatility, flexibility, and operational robustness. It requires an essential computational time, and a stable state of the system to be initialized. Therefore, the dual dynamic programming class is proposed to overcome the problem of having a stable state when initializing the controller. Recently, several review studies on adaptive neural control have been published, one of which is conducted by Jiang et al. [242] summarizing recent progress on the applications of ANN on the control scheme.

1.3.2.1 Polymerization

In 1993, Mauricio and De Souza [220] developed integration of ANN and the MPC for a continuous polymerization reactor. The conjugate gradient training method is employed to train the ANN model with one hidden layer. Linear and sigmoidal functions are used in output and other layers of ANN, respectively. The number of neurons in the hidden layer is obtained by cross-validation method. In this research, the performance of the integrated model is evaluated and compared with those of the conventional PID controller. According to their results, developed control model outperforms PID regarding control of the reactor's temperature and stability of the process. Zhang and Yue [243] proposed a B-spline ANN model to model probability density function (PDF) with the least square algorithm, and therefore, to control molecular weight distribution in the styrene polymerization process. In this study, B-spine ANN through a recursive least square algorithm is applied to model PDF in single input single output (SISO) system. However, the least-square algorithm is employed in multiple inputs and single output (MISO) to model PDF. According to their results, both modeling and control of SISO and MISO are performed efficiently. Cao et al. [244] applied a B-spline ANN combined with a linear RNN as an indicator to model temperature distribution in a tubular polymerization reaction, and therefore, to monitor polymer molecular weight distribution. That is to say, dynamic and static ANNs—using a high-dimensional dataset—are applied to identify the time relationship of the distribution function as well as obtained variables, and to identify the algebraic relationship of temperature distribution and position in the tubular reactor, respectively.

Gonzagaa et al. [129] implemented the ANN-based soft-sensor (ANN-SS) for PET polymerization. The ANN-SS obtains online predictions of the PET viscosity, which is used to control process. The development of online SS-ANN to monitor viscosity proposes an alternative method to overcome the operational failure occurring with the viscometers. The proposed ANN-SS provides satisfactory results to estimate the polymer viscosity. The Levenberg–Marquardt optimization algorithm with the cross-validation-based early stopping mechanism to avoid over-fitting are applied as a training method. Su and McAvoy [221] developed an FNN-MPC and RNN-MPC models with long-range training method to control polymerization reactor and distillation column. In this research, the modified nonlinear auto-regressive with exogenous inputs (NARX) perceptron is developed to achieve an excellent long-range predictive model. The process includes a PI temperature control loop, which controls—the desired polymer property. A pseudo-random multi-state signal (PRMS) is employed for data generation. The prediction horizon provides large enough to be able to consider the uncertainty of measurement delay. The results indicated that the RNN model outperforms the FFN model according to control performance and straightforwardly obtaining an appropriate set of parameters. Eventually, the FNN provides better performance than the proportional integral

(PI) controller. Ibrahim et al. [245] conducted an investigation on the integration of ANN-MPC into a mathematical model to control the temperature of the polyethylene system in the polymerization reactor. This model takes into account the presence of particles participating in the reaction with emulsion and catalyst phases which depend on the superficial velocity and catalyst feed. The results of ANN-MPC are compared with PID controller through studies under setpoint tracking. According to this comparison, ANN-MPC provides an acceptable performance because of process response oscillation and minimal offset compared to the PID controller. In another study, staked neural network model are applied to achieve optimal control of a batch polymerization reactor by Zhang et al. [65]. The stacked neural network is employed as a result of the intricacy of standard mathematical models for batch processes, and the enhancement of generalization capability. The staked neural network may not be proposed the optimal control policy when employed to the real process owing to model mismatches and the presence of unknown disturbances. In this research, the stacked neural network model compared with a single neural network. Moreover, the effects of unknown disturbances are investigated on the stacked neural network model by proposing reactive impurities into the system. Compared with a single neural network, the staked neural network provides an effective batch-to-batch control performance, especially close to the desired results. Regarding the optimal configuration of the stacked neural network, it is demonstrated that the stability of stacked neural network model is obtained after stacking 10–20 networks.

Tian et al. [71] developed a simplified mathematical and stacked RNN to model and control a batch polymerization reactor. The gel effect, which is one of the most challenging parts of polymerization modeling, is considered only in stacked RNN. Comparative with the single neural network, stacked networks outperform, and improve robustness. Ng and Hussain [246] applied an inverse neural network combined with a first principle model to control a nonlinear semi-batch polymerization process directly. The setpoint of the temperature of the polymerization reactor under the nominal condition and with various disturbances are determined using a hybrid developed model. The performance of the hybrid model is accurate to model these highly nonlinear models compared with the Kalman filter (EKF) method.

1.3.2.2 Fuel cell

Recently, Xu et al. [247] proposed a constrained self-tuning PID using an RBF to control a solid oxide fuel cell. The RBF is applied to predict the Jacobian information, and PID parameters are obtained applying a gradient descent (GD) method. Comparing with PID controller, RBF indicates, a fewer overshoot under change of current load but brings small oscillations in transitional periods. The integration of the control method with an active fault-tolerant control (AFTC) strategy, namely neural model-based fault diagnosis and a reconfiguration mechanism is developed in another research [248]. Li et al. [249] developed a stoichiometry recurrent fuzzy ANN for the compressor voltage to prevent oxygen starvation. GD method is employed to achieve fuzzy learning rates and neural weights. Their results indicated that fuzzy ANN model provides robustness to unanticipated external current load disturbances, and more efficient performances compared to the PI controller. Four steps of current disturbances $0.1–0.3 \, A/cm^2$ are evaluated with transient times of up to $10 \, s$. The method can maintain anode/cathode difference pressure; however, its response time for stoichiometry control is remarkable. Sedighizadeh and Rezazadeh [250] carried out a study on wavelet NN with a local infinite impulse response (IIR) block to control the output power of a $1 \, kW$ PEMFC by changing the hydrogen humidity. The wavelet NN is able to identify the complex nonlinear system due to implementing wavelet transform embedded in the hidden units. The IIR block provides double local network architecture, providing fast learning and convergence of the model. For nonlinear systems with high uncertainties, Shafiq [251] developed a direct inverse controller applying two different MLP models. A first one is employed to predict the system and present an estimation of the output. BP algorithm is applied to adjust the weights of the MLP controller. Then, the output of this first model is considered as a residual error for the second model. By doing so, robustness properties against disturbances and model mismatches are obtained.

Rakhtala et al. [252] integrated an inverse MLP model with GD learning algorithm and a PID controller to control voltage of $6 \, kW$ PEMFC using the inlet air pressure. The MLP model faces fewer fluctuations in the transient response of the voltage. Rezazadeh et al. [253] presented an RBF model and PID controller to regulate methane flow for a $5 \, kW$ PEMFC. The Lyapunov theory is applied to evaluate stability, and RBF weights are determined with an adaptation law. According to their results, it shows acceptable robustness with a variance of 1% in the presence of an additive load current noise.

1.3.2.3 Fuel and combustion

Nikhil et al. [254] proposed an ANN model to manage the operation of a pilot-scale hydrogen production system. The parameters such as acids and alcohol concentration, alkalinity, biomass concentration, hydraulic retention time, recycle ratio, sucrose concentration and degradation, pH, oxidation-reduction potential (ORP) are considered as an input of ANN. The results of ANN model are in line with the experimental value, and able to achieve the nonlinear interactions between the process parameters and the hydrogen production rate. Kavchak and Budman [255] evaluated an adaptive controller with RBF model for a continuous stirred tank reactor (CSTR). The RBF model, applying dilation adaptation with an overall prediction error, provide more accurate results compared to the original controller with a fixed dilation. The considerable improvement in accuracy results from applying the original Cannon's method.

Köni et al. [256] proposed two ANFIS models to develop a smart control system of baker's yeast during fluidized bed drying by adjusting the temperature and dry matter content. Their results are in line with the simulation results and industrial-scale databases, which proves its applicability.

1.3.2.4   Petroleum

Jordanou et al. [257] proposed Echo State Network (ESN) model with a recursive least square (LS) learning approach to control a bottom-hole pressure for an oil well. The developed model presents a few overshoots. The pressure setpoint tracking and disturbance rejection are considered as the goal of the controller. Their results indicated that ESN model is able to deal with more substantial changes in pressure and impressive disturbance rejection, with no oscillations or larger overshoots. They argued that the inverse neural structure could mainly control a nonlinear system with uncertainties. However, the command signals and response times of this model are somewhat large. With respect to developing advanced ANN model, several studies are devoted to reduce overshoots and improve the robustness against disturbances. Bahar and Özgen [258] developed an Elman network integrated with the feedback inferential control algorithm to estimate the product composition values of the distillation column from temperature measured. The estimated compositions and the reflux ratio are considered as inputs to the controller. Their results indicated that the ANN model is able to control the compositions in this dynamically complex system. Three applications of ANN in petroleum industry can be perfectly demonstrated in [259–261], where the ANN performance markedly improved by a metaheuristic optimization algorithm for the sake of prediction of challenging factors, including free-flowing porosity, permeability, minimum horizontal stress and capillarity pressure.

1.3.2.5   Environment

Mingzhi et al. [262] investigated on the coagulation process of wastewater treatment in a paper mill. An ANFIS predictive control scheme for studying applying adaptive FNN. This ANFIS model can provide a model for nonlinear relationships between the removal rate of pollutants and the chemical dosages, resulting in the adaptation of the system to a variety of operating conditions. The developed controller indicates acceptable predictions and control performances. De Veaux et al. [263] integrated a first principle differential equations model with an ANN to control biological treatment for water contamination. Compared to the ANN model, this hybrid model required fewer data in the trained process and extrapolated more accurately. In another study, a self-organizing RBFNN model predictive control (SORBF-MPC) method applied to control the dissolved oxygen (DO) concentration of wastewater treatment plants [264]. The remarkable advantages of SORBF are a dynamic change of its structure to retain prediction accuracy. In this regard, this RBF model can provide accurate results without adjusting the weight by adding or removing the hidden nodes in the RBF based on node activity and mutual information (MI). Three ESN control models, namely Critic ESN, Actor ESN, and Predict ESN, are developed for dissolved oxygen concentration in the wastewater treatment process by Bo and Zhang [265]. LS training algorithm is employed to adjust ESN weights online. Compared with a PID controller, the ESN models revealed a smoother control, smaller deviation, smaller overshoots, and better adaptability. Moreover, adjusting time is also reduced from 0.65 days for a PID controller to 0.45 days for the developed controller.

1.3.2.6   Biotechnology

Xiong et al. [266] developed the RNN model to achieve a batch-to-batch iterative optimal control strategy for batch processes. The model prediction errors from previous batch runs are applied to improve the performance of the ANN model for the current batch and to overcome model-plant mismatches as well as unmeasured disturbances. A closed-loop control system based on the ANN model is developed by Andra Sik et al. [267] to investigate nonlinear control of a continuous flow stirred biochemical reactor. A gradient steepest descent method is applied to train ANN model, and also to accelerate the speed of convergence. In this research, the control system is composed of two ANNs to provide a predictive hybrid model of the controlled plant and to be used as a neural PID-like controller. The second ANN has been pre-trained offline as an inverse black-box model of the controlled process. Due to issues such as process uncertainties and time-varying parameters, additional online tuning of the neural controller is applied to address mentioned issues. Dong et al. [268] proposed PLS-NN models to obtain the desired conversion and molecular weight by solving a SQP problem. In this study, a batch-to-batch adaptation method is employed to address model errors. Tay and Zhang [269] proposed an ANFIS model for the complex process of anaerobic biological treatment of wastewater. Their results showed the capability of the ANFIS model to control accurately two case studies, namely, up flow anaerobic sludge blanket and anaerobic fluidized bed reactor. They argued that the dependence of the model on the quality of the training data is one of the main drawbacks of this model.

Chen and Chang [270] developed an ANFIS method for modeling the complex process of aeration in a submerged biofilm wastewater treatment process. Holubar et al. [271] used ANN model with BP algorithm to control methane production in anaerobic digesters. The model was trained applying obtained data from four anaerobic CSTRs operating at steady state. In a study conducted by Nagy et al. [272], ANN-MPC model with BP learning algorithm is evaluated and implemented to provide a dynamic modeling and temperature control of a continuous yeast fermentation bioreactor. The analytical model of this nonlinear process is applied to achieve the training data. A structure of the ANN model is optimized to avoid over-fitting of the data and obtain the best prediction with the simplest structure. Patnaik [273] developed a hybrid model by combining four neural networks with a kinetic model and a bioreactor model to provide accurate results with a reduction in the required amount of training process. Sabharwal et al. [274] developed an ANN-based controller for a xylene separation column. The nominal production data and plant data from a response to a step function are obtained to train the ANN model. ANN trained using plant data provide more acceptable and accurate control performances.

1.3.2.7 Other application

Kimaev et al. [275] proposed a data-driven model based on ANN to develop a shrinking horizon nonlinear MPC of a computationally intensive stochastic multiscale system, which is a simulation of thin-film formation by chemical vapor deposition. They noted that the capability of ANNs as powerful tools for online optimization and control of computationally demanding multiscale process systems because of their computational efficiency, accuracy, and the ability to remove disturbances. Wu et al. [276] developed a staked RNN prediction model to develop a control Lyapunov-barrier function (CLBF)-based economic model predictive control (EMPC) system to optimize process economics, stability, and operational safety simultaneously. In this research, the performance of RNN-based CLBF-EMPC method is evaluated through a chemical process with a bounded and an unbounded unsafe region.

Tian et al. [277] developed a hybrid RNN model for online reactive impurity estimation and reoptimization control policy of batch MMA reactor. The current off-line optimal control strategy will be no longer optimal due to an unknown amount of reactive impurities. The developed ANN-based inverse model is applied to predict online the number of reactive impurities during the early stage of a batch. Then online reoptimization is carried out to identify the optimal reactor temperature profile for the remaining time period of the batch reactor operation according to the estimated amount of reactive impurities. To overcome some of the difficulties associated with multivariable process control, Adebiyi and Corripio [278] proposed a strategy for open-loop identification of multivariable chemical processes implementing the dynamic neural networks PLS (DNNPLS). The DNNPLS identification strategy is applied for a fluid catalytic cracking unit and an isothermal reactor. According to their results, the DNNPLS method can provide an accurate model for the dynamics of the chemical processes in both cases. Cabrera et al. [279] developed a control strategy based on ANNs to obtain the optimal operation for a small scale Sea and Brackish Water Reverse Osmosis (SWRO) desalination plant. Roehl et al. [280] applied an ANN model to simulate a membrane fouling mechanism for a large scale reverse osmosis (RO) desalination plant. He and Jagannathan [281] developed an adaptive-critic-base discrete-time NN controller with the reinforcement learning to obtain desired tracking performance for nonlinear systems in the presence of actuator constraints. Vasickaninova et al. [282]

developed a neural network predictive control (NNPC) structure to control thermal processes, resulting in energy savings. Compared to results of PID control, the NNPC outperforms due to smaller consumption of heating medium.

1.3.3 Application of network in fault detection and diagnosis

In the CE process, intelligence is required not only in the production process but also in safety operation. That is to say, fault detection and diagnosis (FDD) should be considered in chemical industries. Therefore, the past few decades have witnessed a huge growth in the development of FDD. For the past decade, there has been a rapid rise in the implementation and evaluation of data-driven FDD methods [283]. Data-driven FDD methods include statistical methods and ANN methods. Statistical methods consists of PCA [284], PLS [285], Fisher discriminant analysis (FDA) [286], independent component analysis (ICA) [287], SVM [288], and qualitative trend analysis (QTA) [289, 290]. These methods perform accurately with high dimensional, noisy, and highly correlated data from chemical processes. Chemical process data, however, have other intricate features such as high dimensionality, multimode behaviors, nonlinearity, non-Gaussian, and time-varying [291]. Thus, several researchers proposed kernel-based methods for nonlinearity, namely kernel PCA [292] and kernel PLS [293], a neural net PLS for nonlinearity [294], and dynamic-based methods for temporal dependencies, namely dynamic PCA [295], dynamic PLS [296], dynamic inner PCA [297], dynamic inner PLS [297].

There are a considerable number of studies has been published on the automated diagnosis over the last 2 decades. However, it is not widely accepted in the industrial purpose from an economic point of view. Fenton et al. [298] conducted a review of rule-based, model-based, and case-based approaches as well as their applications. The need for automated diagnosis stems from Increasing costs, shorter product lifecycles, and rapid changes in technology. Although several studies have been carried out over the last 2 decades, further research needs to be done to improve the developed methods and their application and to cope with current and future technologies.

Chen et al. [299] studied on FDD of multivariate processes by proposing a new deep neural network (DNN) model with 1D convolutional auto-encoder (1D-CAE). Efficient feature extraction is obtain using auto-encoder integrated with convolutional kernels, and pooling units, which play a vital role in FDD in multivariate processes. Compared with other typical DNN models, the developed model outperforms for FDD on the Tennessee Eastman Process (TEP) and Fed-batch fermentation penicillin process. They noted that this method could be applied for deep-learning-based process FDD of multivariate processes. Laribi et al. [300] developed an MLP model with BP training algorithms to evaluate and diagnose PEM fuel cell failure modes, namely flooding and drying, among the physical parameters of the electrochemical impedance model. Furthermore, the ANN model implemented to the impedance model applied to predict the physical parameters of the electrochemical impedance model of the PEMFC; flooding and drying of the fuel cell heart. Yuan et al. [301] developed multiscale convolutional DNN proposes with the discrete wavelet transform to monitor intelligently industrial process and diagnosis a fault. The extracted multiscale features with a CNN are integrated by the LSTM network to reduce useless data and maintain useful data. They noted that the advantages of developed method are the ability to learn nonlinear, high-dimensional fault features due to the hierarchical learning structure; and to achieve a piece of complementary diagnosis information at different scales in view of the multiscale feature learning. The Tennessee Eastman benchmark process and the p-xylene oxidation reaction process are considered to evaluate their advanced method.

Zhao et al. [302] proposed an LSTM fault diagnosis method to identify the dynamic information in raw data and to classify the raw data directly without feature extraction and classifier design. The Tennessee Eastman (TE) benchmark process is selected to evaluate the performance of the LSTM fault diagnosis model. The experimental results indicated that LSTM shows superior fault detection performance compared to previous models. Hoskins and Himmelblau [110] and Hoskins et al. [303] implemented a multilayer feedforward analogue perception model for FDD in a large complex chemical plant. They argued that ANN is capable of learning nonlinear mappings with noisy inputs. Moreover, ANNs could indicate the rule-following behavior of knowledge-based expert systems without any explicit representations of the rules. Ruiz et al. [304] proposed an ANN-based model with a knowledge-based expert system to detect the fault in batch chemical plants. The historical database of the batch plant performances, a hazard and operability assessment, and the batch plant model are applied in the fault detection system. The faults signals are considered as the output of ANN. The developed system provides an accurate fault detection for the operation of a batch reactor. Yuan et al. [305] applied the GRU network to develop a three-stage fault diagnosis method. Batch normalization (BN) algorithm are employed to exploit the dynamic feature from the sequence units. Dynamic features are applied to classify faults using the softmax regression method. Tennessee Eastman (TE) benchmark process, as well as the para-xylene (PX) oxidation process, are selected to validate the developed method. The diagnosis results of TE application indicated that this method outperforms conventional methods. However, the results of the PX oxidation application demonstrated unusual behavior of method when historical data is not sufficient. Shao et al. [306] developed a novel fault diagnosis method applying multichannel LSTM-CNN (MCLSTM-CNN). LSTM is applied to achieve the output of the hidden layer using fault data as an input variable. Convolutional kernels with different sizes are employed to exploit the output features of a hidden layer simultaneously. Compared with LSTM-CNN, LSTM, CNN, random forest (RF) and KPCA + SVM models, MCLSTM-CNN model provides more accurate fault detection and classification in the Tennessee Eastman (TE) chemical process. Wu and Zhao developed a CNN model for fault diagnosis and detection of chemical processes, which can exploit fault features and classify them accurately [307].

Kim et al. [308] put forward a Hamming neural network to diagnosis the health state of 20 PEMFCs cells, using the PEMFC measured electrical performance (voltage, current). Hamming neural network can adjust the optimal parameters of an equivalent circuit model, from which the PEMFC failure can be identified and classified. Silva et al. [309] applied adaptive ANFIS system to develop a new method to estimate the reduction in the output voltage results from the PEMFC nominal degradation during long term operation. Shao et al. [310] have been implemented efficient technology combining an ANN ensemble method (BP-ANN) with the PEMFC dynamic model for fault detection in PEMFC systems. The results of this study showed that the accuracy of PEMFC fault detection is varied between 75.24% and 85.62%. Yousfi et al. [311] evaluated the PEMFC health state applying a diagnostic method based on ANN to identify the drying and flooding problems of the PEMFCs. The pressure drop and stack voltage are considered as an objective function. Tang et al. [312] implemented a DBN model to exploit and detect the fault features of the chemical industry. The quadratic programming method is employed to predict the sparse coefficients simultaneously class by class. You et al. [313] detected a battery state using an RNN in electric vehicle systems. The developed model is able to identify the battery replacement time or to evaluate the driving mileage.

Tayyebi et al. [314] developed a novel ANN diagnosis framework using augmented input containing steady-state characteristic data and newly defined dynamic characteristic data. The Tennessee Eastman (TE) chemical process is selected to evaluate the proposed model, which consists of large numbers of measurements and control parameters and overlapping faults. Comparing with the conventional ANN diagnostic system, the proposed system provides a more accurate and efficient performance than the conventional diagnoser to detect multiple concurrent faults and various combinations of concurrent faults of the TE process. Liu et al. [315] developed a combination of t-distributed stochastic neighbor with bidirectional LSTM to obtain data dimensionality reduction and exploit feature vectors for the fault diagnosis of PEMFCs. To remove the effects of dimensional differences of different parameters, the developed method employed the normalization strategy. Applying a t-SNE method result in the reduction in the dimensionality of normalized data to predict intrinsic dimensionality, and to exploit key characteristic variables. According to their results, The BiLSTM-tSNE method has the capability of the sequence fault diagnosis of the PEMFC water management subsystem with an accuracy of 96.88% and an operation time of 24 s. Their results demonstrated that the hybrid model provides an accurate diagnostic performance than conventionally applied SVM algorithm. Wu et al. [316] proposed a CNN model with transfer learning to develop an FDD method for multimode chemical processes. Their results indicated that transfer learning had marked improvement for the fault detection of the benchmark TE process. According to experimental results, the CNN model outperforms LSTM regarding transferability and generality from one mode to another mode.

Zhang et al. [317] developed a hierarchical BiRNN-based FDD system using advanced deep learning techniques for complex chemical processes. The sophisticated RNN cells are employed in the BiRNN structure to enhance the FDD performance. The resultant BiLSTM and BiGRU provide more impressive performances compared with other advanced FDD models based on ANN, which has significant potential for practical complex chemical processes. Disadvantages of these models, however, are highly dependent on historical fault data to train models and to be labelled by experts, which is a heavy workload. They recommended that a semi-supervised training method can be applied by integrating little labeled historical fault data with much unlabeled historical data in future research. Golizadeh et al. [318] integrated the extended Kalman filter (EKF) and ANFIS model to propose a new fault detection and identification (FDI) algorithm for nonlinear systems. The fuzzy rules are considered to evaluate the validity of the models. They showed that their integrated developed method is independent of pre-designing a bank of observers in the model-based methods. Moreover, their developed model did not require feature extraction of the signals without any physical insight and computational complexity in the data-driven approaches. The developed FDI scheme is applied to a chemical plant, namely, CSTR process to evaluate its fault detection performance. The results indicated an accurate performance of the developed model to identify different faults modes.

Yang et al. [319] developed a PNN to diagnosis faults. Support vector machine recursive feature elimination (SVMRFE) is applied to reduce the effect of useless features. Moreover, the modified bat algorithm (MBA) is employed to optimize the PNN parameter, especially its hidden layer element smoothing factor, which has a considerable effect on the diagnostic performance. The performance of the MBA showed better optimization results and more acceptable global convergence compared to the conventional optimization algorithm. The validation of the developed model to identify faults are carried out in terms of accuracy and F1-score for the Tennessee Eastman (TE) process dataset. They provided the charts to present the fault diagnostic results and classification for the different models. Their results indicated that their proposed method outperforms other diagnostic methods to identify and classify the TE process fault diagnosis accurately.

Zhao et al. [320] put forward the neighborhood preserving neural network (NPNN), which is a novel statistical feature extraction method. That is to say, this novel method is a nonlinear data-driven fault detection approach retaining the local geometrical structure of normal process data. Orthogonal constraints in the objective function are employed in NPNN to exploit uncorrelated and faithful features. Moreover, the BP and Eigen decomposition (ED) technique are applied in NPNN to extract low-dimensional features from original high-dimensional process data. The results of theoretical analysis and experimental assessment of the TE benchmark process indicated that NPNN could perform efficiently according to missed detection rate (MDR) and false alarm rate (FAR). Li et al. [321] integrated the remarkable ability of CNN in feature extraction and those of DAE in classification to develop a hybrid fault-diagnosis model for the distillation process of depolarization. According to their results, the hybrid model is able to provide an admirable performance compared to other models, regarding the accuracy of fault detection. Zhang et al. [322] developed an extensible DBN to detect faults. DBN applied mutual information technology to exploit fault features in both spatial and temporal domains. In this study, fault classification is conducted using a global two-layer BP network. The results showed that the developed model performs accurately to identify faults of the benchmarked TEP. Hu et al. [323] developed a novel incremental imbalance modified DNN (incremental-IMDNN) to cope with chemical imbalanced data streams. The performance of the developed model is validated in TE dataset. The results showed that the incremental IM-DNN provide more robust and accurate performance compared to the current method in chemical plant fault diagnosis. Mod Noor et al. [324] integrated multiscale Kernel Fisher discriminant analysis dimensionality reduction method with ANFIS classification model to propose a novel multiscale FDD framework for chemical process system. They noted that the developed model could be applied in a broad range of chemical process systems such as the TEP with accurate results.

Ren et al. [325] applied deep residual convolutional neural network (DRCNN) to put forward a novel fault detection and classification method. The DRCNN are able to exploit the in-depth process features resulting in the deep fault information and learn the latent fault patterns. The results demonstrated that the proposed model is able to provide better performance compared with advanced methods using the data set of the TEP, a chemical industrial process benchmark. Jiang et al. [326] developed a deep structure DNN with active learning and multiple SDAE for chemical fault diagnosis applying chemical sensor data. The results indicated that the developed model outperforms the advanced methods of two well-known datasets. They argued that the proposed method shows remarkable potential for fault diagnosis in complex chemical systems with multidimensional heterogeneous sensory signal and nonlinear relationships between the real senor data and model results.

2 Fuzzy logic

2.1 Introduction of fuzzy

Zadeh introduced Fuzzy set theory in 1965 [327]. Initially, the main aim of fuzzy logic was to deal with uncertain and qualitative problems. A short while later, Mamedani used the fuzzy set theory in control applications [328]. The typical configuration of a fuzzy controller is depicted in Fig. 7.

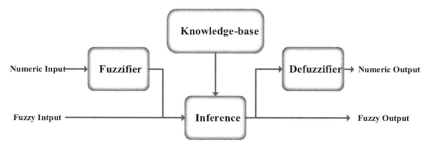

FIG. 7 The typical structure of the fuzzy controller.

The structure is divided into four main parts.

Fuzzifier: In this part, Inputs of the system are broken down into one or more fuzzy sets. The membership functions of the fuzzy set could have various forms such as triangle and trapezoidal [329]. As an illustration, Fig. 8 presents inputs/outputs membership functions of a fuzzy controller.

Knowledge-based: This part includes one or more IF-Then rules which have two main types, fuzzy control rules for state evaluation, and fuzzy control rules for object evaluation. Rules of a MISO system, for instance, are given in the following statements:

R1: if x is A_1, and y is B_1 Then z is C_1
R2: if x is A_2, and y is B_2 Then z is C_2
$\vdots$
Rn: if x is A_n, and y is B_n Then z is C_n

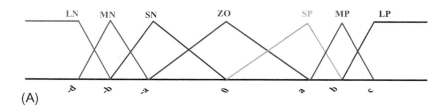

(A)

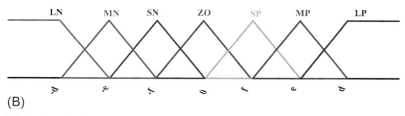

(B)

FIG. 8 Membership functions of a fuzzy controller.

Inference: Outputs of the fuzzy system are generated from inputs and rules in this section.

Defuzzifier: Defuzzifier is an inverse transformation which converts fuzzy outputs into crisp outputs. Several popular forms of defuzzification are Center of Area, Center of Maximum, and Mean-of-Maximum.

2.2 Application of fuzzy controller in process control

Fuzzy controllers have received significant consideration in process control over the past few years. For instance, Pathmanathan et al. [330] implemented a fuzzy logic controller for flow control. Furthermore, they analyzed the controller performance between the fuzzy controller and a conventional PID controller. As a result of their research, the fuzzy controller can be an attractive alternative to conventional controllers in process control. As another example, the importance of the fuzzy logic controller for temperature process control of a furnace system has been evaluated in [331]. The performance of the fuzzy controller has been compared with PID controllers which have been tuned by various tuning methods. The results illustrated that the fuzzy controller improves the robustness and dynamic performance of the system. It is worth noting that a body of research has been carried out to improve PID control efficiency by means of optimization, while the necessity of model is a major downside compared to a fuzzy logic controller [332].

Although the fuzzy controller has a satisfactory response to the nonlinearities of the industrial chemical process, it can have some difficulties to minimize the negative effect of system uncertainties [333]. To overcome this problem, the fuzzy type-2 controller has been suggested, recently. As an illustration, both types of controllers mentioned above have been utilized for the control of a binary distillation column in [334]. The simulation results confirmed that the robustness improved considerably by the fuzzy type-2 controller. A comprehensive experimental investigation of fuzzy control in process control has been proposed in [335]. In that work, PID-fuzzy methods were developed and compared to PID controllers for a couple of processes: temperature of a polymerization reactor, temperature of a refrigeration system, and overhead composition of a batch distillation column. A body of research has been used fuzzy logic in fault detection, diagnosis, and tolerant control [318, 336, 337].

2.3 Simulation: A sample of process control by fuzzy controller

Temperature process control of an industrial furnace is considered as a plant for applying a fuzzy controller. Fig. 9 shows a schematic of this process, where T_1C and T_2C represent primary and secondary controllers, respectively, and T_1T and T_2T describe temperature transmitters. Furthermore, it is evident from the figure that the set point of the secondary controller is given from the output of the primary controller. The block diagram of this process is shown in Fig. 10. It is

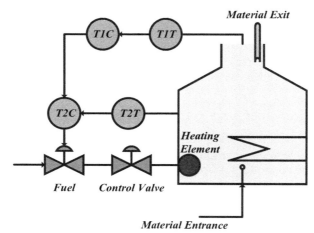

FIG. 9 schematic of the temperature process control of an industrial furnace [331].

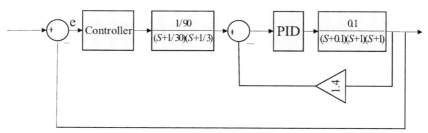

FIG. 10 The block diagram of this process.

clear that the cascade control strategy is recommended for controlling the process, as mentioned earlier.

The internal controller is a fixed PID controller, and only the external controller can be changed in this research. At first, PID controllers which are tuned by Ziegler-Nichols and try-and-error tuning methods are considered. The PID controllers' parameters are given in Table 1. Moreover, the step results of them are depicted in Fig. 11.

In the next step, a fuzzy controller is designed for this process. The membership function and rules of the proposed fuzzy controller are given in Fig. 12 and Table 2, respectively. The step response of the fuzzy controller is shown in Fig. 13. Furthermore, for a detailed comparison between controllers, Table 3 is indicated. It is evident that the overshoot of step response sees a dramatic decrease by applying the proposed fuzzy controller. Also, raise time has been declined from 27 to 18 s, and settling time has decreased from 6 to 5 s. Improvement in all of the metrics, as mentioned above, shows that the fuzzy controller has better performance than conventional controllers.

TABLE 1 PID controllers' parameters.

	K_P	K_I	K_D
Ziegler-Nichols	24.53	4.26	31.68
Try-and-error	18.12	0.88	28.54

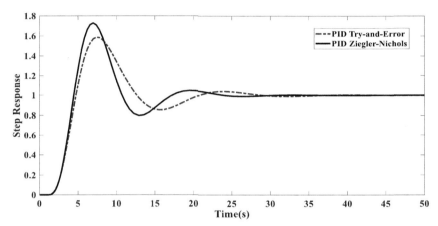

FIG. 11   Step response of PID controllers.

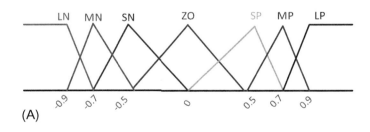

(A)

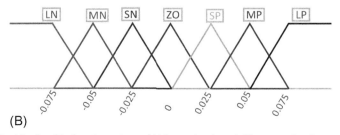

(B)

FIG. 12   Membership function patterns (A) input signals and (B) output signal.

TABLE 2 Fuzzy rules for the proposed fuzzy logic-based controller.

		$\Delta e/\Delta t$						
		LN	MN	SN	ZO	SP	MP	LP
Δe	LN	LP	LP	LP	MP	MP	SP	ZO
	MN	LP	MP	MP	MP	SP	ZO	SN
	SN	LP	MP	SP	SP	ZO	SN	MN
	ZO	MP	MP	SP	ZO	SN	MN	MN
	SP	MP	SP	ZO	SN	SN	MN	LN
	MP	SP	ZO	SN	MN	MN	MN	LN
	LP	ZO	SN	MN	MN	LN	LN	LN

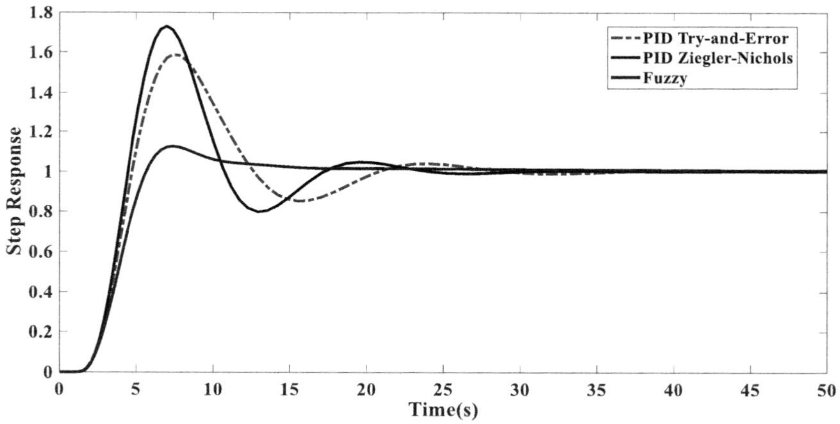

FIG. 13 Comparison between controllers.

TABLE 3 A detailed comparison between controllers.

	Overshot	Settling time (s)	Rise time (s)
Ziegler-Nichols	76%	27	6
Try-and-error	58%	27	7
Fuzzy	11%	18	5

3 Support vector machine

3.1 Introduction

The SVM has been presented by Vapnik, as a solution in machine learning and pattern recognition. Application of SVM in fault diagnosis and fault-tolerant control has received further consideration over the last decades [338–341]. The forecasting in this method is done by a linear composition of Kernel functions that are acting on a set of training data. Learning by SVM always finds the global minimum. The characteristics of SVM depend on the choice of Kernel function. In this section, a brief review of the basic concept of SVM is presented.

The machine learning is done in two cases: supervised and unsupervised methods. In most of the supervised methods, a set of input vectors $X = \{x_n\}$ and output vectors $T = \{t_n\}$ are given, and the objective is that the machine to be learned such that for a new training input x, the new output t to be forecasted. For this approach, two cases of problems are considered: regression and classification. The machine learning problem mathematically can be regarded as a mapping problem such that $x_i \rightarrow y_i$. In other words, the machine is defined as the set of possible mapping $x \rightarrow f(x, \alpha)$, in which the function $f(x, \alpha)$ is adjusted by α. It is assumed that for an input x there is α such that the output $f(x, \alpha)$ can be obtained. The opting of α is exactly what a machine does. For example, a neural network that α is corresponded with its weights and biases, is a machine learning [342].

3.2 SVM classifiers

Vapnik proposed SVM in 1979. In the purest form, namely linear from, as shown in Fig. 14, SVM is a line that separates a set of positive and negative samples with the most possible margin [343].

Generally, this problem can be considered in n-dimensional space that the data samples are in two sets that can be well separated. In this case, instead of

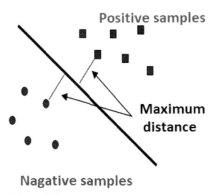

FIG. 14 An example of linear SVM.

lines, the hyperplanes are used. The classification problem by SVM is presented in the following circumstances [344]:

a. Linear SVM
   1. The data samples are two entirely separable sets.
   2. The data samples are two sets that are nonseparable.
b. Nonlinear SVM
c. SVM for multi-classifications

3.2.1 Separable case

The simplest case that machine is learned on the separable data. It should be noted that the equation of this case is expanded to other cases. As shown in Fig. 15, the data can be separated by several lines (or planes in the n-dimensional space). To choose the best hyperplane, the concept of maximum margin should be applied. In the linear case, the margin is defined as the distance of hyperplane to the nearest positive or negative sample. Each line in 2D space is written as [343]:

$$w_1 x_1 + w_2 x_2 + b = 0 \tag{1}$$

In Fig. 16, the x_i with labels $-1, +1$ are shown.

As shown in Fig. 17, In 3-dimension space, one has:

$$w_1 x_1 + w_2 x_2 + w_3 x_3 + b = 0 \tag{2}$$

where it is the equation of a plane. Generally, one has:

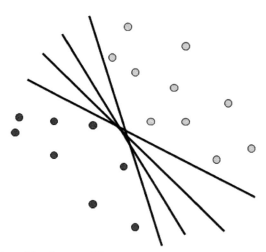

FIG. 15 Separating of data by several lines.

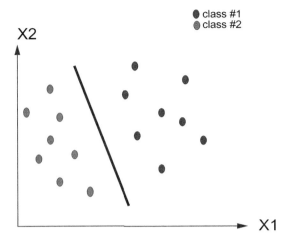

FIG. 16 Separable data in 2-dimension space.

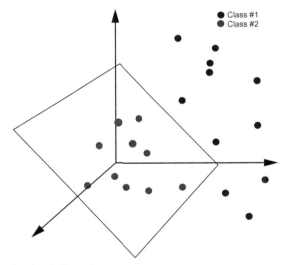

FIG. 17 Separating in a 3-dimension space.

$$\sum_i w_i x_i + b = 0 \tag{3}$$

In the vector form, it can be written as:

$$u = \vec{w} \cdot \vec{x} + b \tag{4}$$

where w is the perpendicular vector of weights to the hyperplane and b is bias. In this representation, $u = 0$ is the separating plane and the nearest points are on the plane $u = \pm 1$. Indeed, by the assumption that the data are separable in two classes, the marginal vectors are the following planes:

$$\vec{w} \cdot \vec{x} + b = \pm 1 \tag{5}$$

The space between two planes is called as Margin. Then space is divided into two class as:

$$\vec{w} \cdot \vec{x} + b \geq +1 \quad \text{for} \quad y_i = 1 \tag{6}$$

$$\vec{w} \cdot \vec{x} + b \leq -1 \quad \text{for} \quad y_i = -1 \tag{7}$$

It can be rewritten as:

$$y_i \left(\vec{w} \cdot \vec{x} + b \right) - 1 \geq 0 \; \forall i \tag{8}$$

In this case, the distance up to origin as perpendicular (least distance) for the points that are on the plane $x \cdot w + b = +1$ is:

$$\frac{|-1 - b|}{\|w\|} \tag{9}$$

Similarly, the distance up to origin as perpendicular for the points that are on the plane $x \cdot w + b = -1$ is:

$$\frac{|-1 - b|}{\|w\|} \tag{10}$$

Moreover, the distance between origin and plane is:

$$\frac{|b|}{\|w\|} \tag{11}$$

Then the least distance between the plane and the aforementioned planes is:

$$d_+ = d_- = \frac{1}{\|w\|} \tag{12}$$

Then the margin is:

$$d_+ + d_- = \frac{2}{\|w\|} \tag{13}$$

Then the maximize of margin cane written as the following constrained optimization problem:

$$\min \frac{1}{2} \|w\|^2 \text{ subject to } y_i \left(\vec{w} \cdot \vec{x} + b \right) - 1 \geq 0 \; \forall i \tag{14}$$

The problem can be rewritten as:

$$L_p = \min \frac{1}{2} \|w\|^2 - y_i \left(\vec{w} \cdot \vec{x} + b \right) - 1 \geq 0 \tag{15}$$

By applying the Karush-Kuhn-Tucker (KKT) conditions, one has:

$$w = \sum_i \alpha_i x_i y_i \tag{16}$$

$$0 = \sum_i \alpha_i y_i \tag{17}$$

Where the KKT conditions are:

$$\frac{\partial}{\partial b} L_p = w_v - \sum_i \alpha_i y_i x_{iv}, v = 1, \ldots, d \tag{18}$$

$$\frac{\partial}{\partial b} L_p = -\sum_i \alpha_i y_i = 0 \tag{19}$$

$$y_i \left(\vec{w} . \vec{x} + b \right) - 1 \geq 0 \ i = 1, \ldots, l \ \alpha_i \geq 0 \ \forall i \tag{20}$$

$$\alpha_i \left(y_i \left(\vec{w} . \vec{x} + b \right) - 1 \right) \geq 0 \ \forall i \tag{21}$$

By substituting into (15), one has:

$$L_D = \sum_i \alpha_i - \sum_i \sum_j \alpha_i \alpha_j y_i y_j x_i x_j \tag{22}$$

After addressing this problem, the support vectors are obtained, as shown in Fig. 18.

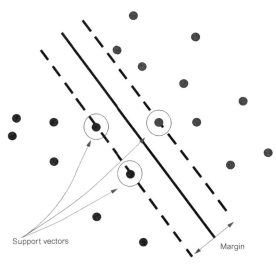

FIG. 18 Support vectors.

3.3 SVM for the nonlinear and nonseparable cases

In these cases, it is assumed that the data cannot be separated into two class, simply. This assumption is near to the working condition. Because the data are always noisy and SVM should overcome this condition. This situation, by definition, as soft margin can be solved. To develop the basic idea in the previous section, a new variable is added. Then the problem is reformulated as:

$$x_i.w + b \geq +1 - \xi_i \quad \text{for} \quad y_i = 1 \tag{23}$$

$$x_i.w + b \leq -1 + \xi_i \quad \text{for} \quad y_i = -1 \tag{24}$$

$$\xi_i \geq 0 \forall i \tag{25}$$

Then for any error, ξ_i should be appropriately increased. Then $\sum_i \xi_i$ is an upper bound for the number of training errors. If so, the optimization problem is:

$$\text{maximize} \, L_D = \sum_i \alpha_i - \frac{1}{2} \sum_i \sum_j \alpha_i \alpha_j y_i y_j \vec{x}_i \vec{x}_j \tag{26}$$

$$\text{subject to:}$$

$$0 \leq \alpha_i \leq C,$$
$$\sum_i \alpha_i y_i = 0 \tag{27}$$

Then the solution is:

$$w = \sum_{i=1}^{N_s} \alpha_i y_i x_i = 0 \tag{28}$$

Then the main difference is that the optimal hyperplane has upper and lower bounds, as we can see in Fig. 19.

The primal Lagrange, in this context, is written as:

$$L_p = \frac{\|w\|^2}{2} + C \left(\sum_i \xi_i \right)^k - \sum_i \alpha_i [y_i(x_i.w + b) - 1 + \xi_i] - \sum_i \mu_i \xi_i \tag{29}$$

where ξ_i is the Lagrange coefficient. By applying the KKT condition on the primal problem, one has:

$$\frac{\partial}{\partial w_v} L_p = w_v - \sum_i \alpha_i y_i x_{iv}, \quad v = 1, ..., d \tag{30}$$

$$\frac{\partial}{\partial w_v} L_p = - \sum_i \alpha_i y_i = 0 \tag{31}$$

$$\frac{\partial}{\partial \xi_i} L_p = C - \alpha_i - \mu_i = 0 \tag{32}$$

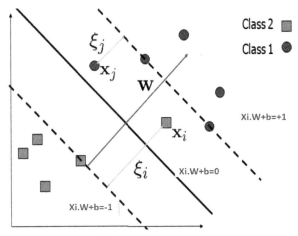

FIG. 19 The data in the nonseparable case.

3.4 Nonlinear SVMs

In the case that data cannot be separated simply, a linear SVM cannot be valid. As shown in Fig. 20, in this case, the data should be mapped into space with one more dimensional.

As it is seen from Fig. 20, after mapping the classification is possible. In the previous cases, the internal multiplication of training input data and support vectors is used to construct the separating as a hyperplane. On this matter, data, in the same way, is mapped into Euclidean space with higher dimension before the internal multiplication is applied as:

$$\Phi : R^d \rightarrow H \tag{33}$$

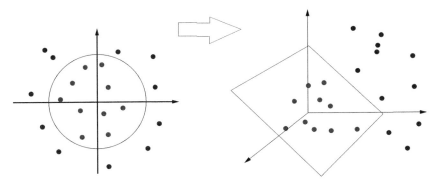

FIG. 20 Mapping data into space with one more dimension.

$$K\left(x_i, x_j\right) = \Phi(x_i)\Phi\left(x_j\right) \tag{34}$$

The popular Kernel functions are:

$$K\left(x_i, x_j\right) = \left(x_i . x_j + 1\right)^p$$

$$K\left(x_i, x_j\right) = e^{-\frac{\left\|x_i - x_j\right\|^2}{2\sigma^2}}$$

$$K\left(x_i, x_j\right) = \tanh\left(\beta x_i x_j + \delta\right) \tag{35}$$

The primal problem, by the same token, is written as:

$$\text{maximize}\, L_D = \sum_i \alpha_i - \frac{1}{2}\sum_i\sum_j \alpha_i\alpha_j y_i y_j K\left(x_i, x_j\right) \tag{36}$$

subject to:

$$0 \leq \alpha_i \leq C, \\ \sum_i \alpha_i y_i = 0 \tag{37}$$

References

[1] A.E. Rodrigues, M. Minceva, Modelling and simulation in chemical engineering: tools for process innovation, Comput. Chem. Eng. 29 (6) (2005) 1167–1183.

[2] J.D. Hoffman, S. Frankel, Numerical Methods for Engineers and Scientists, CRC Press, 2018.

[3] P. Fritzson, Introduction to Modeling and Simulation of Technical and Physical Systems With Modelica, John Wiley & Sons, 2011.

[4] W.S. McCulloch, W. Pitts, A logical calculus of the ideas immanent in nervous activity, Bull. Math. Biophys. 5 (4) (1943) 115–133.

[5] A.L. Samuel, Some studies in machine learning using the game of checkers, IBM J. Res. Dev. 44 (12) (2000) 206–226.

[6] F. Rosenblatt, Perceptron simulation experiments, Proc. IRE 48 (3) (1960) 301–309.

[7] B. Widrow, M.E. Hoff, Adaptive Switching Circuits, Stanford Univ CA Stanford Electronics Labs, 1960.

[8] M. Minsky, S.A. Papert, Perceptrons: An Introduction to Computational Geometry, MIT Press, 2017.

[9] D.W. Tank, J.J. Hopfield, Neural computation by concentrating information in time, Proc. Natl. Acad. Sci. 84 (7) (1987) 1896–1900.

[10] P.J. Werbos, Backpropagation through time: what it does and how to do it, Proc. IEEE 78 (10) (1990) 1550–1560.

[11] P. Smolensky, Information Processing in Dynamical Systems: Foundations of Harmony Theory, Colorado Univ at Boulder Dept of Computer Science, 1986.

[12] D.E. Rumelhart, G.E. Hinton, R.J. Williams, Learning representations by back-propagating errors, Nature 323 (6088) (1986) 533–536.

[13] S. El Hihi, Y. Bengio, Hierarchical recurrent neural networks for long-term dependencies, Adv. Neural Inf. Process. Syst. (1996) 493–499.

[14] S. Hochreiter, J. Schmidhuber, Long short-term memory, Neural Comput. 9 (8) (1997) 1735–1780.

[15] Y. LeCun, L. Bottou, Y. Bengio, P. Haffner, Gradient-based learning applied to document recognition, Proc. IEEE 86 (11) (1998) 2278–2324.

[16] G.E. Hinton, R.R. Salakhutdinov, Reducing the dimensionality of data with neural networks, Science (80-.) 313 (5786) (2006) 504–507.

[17] G.E. Hinton, S. Osindero, Y.-W. Teh, A fast learning algorithm for deep belief nets, Neural Comput. 18 (7) (2006) 1527–1554.

[18] L. Deng, M.L. Seltzer, D. Yu, A. Acero, A. Mohamed, G. Hinton, Binary coding of speech spectrograms using a deep auto-encoder, in: Eleventh Annual Conference of the International Speech Communication Association, 2010.

[19] M. Ranzato, C. Poultney, S. Chopra, Y.L. Cun, Efficient learning of sparse representations with an energy-based model, Adv. Neural Inf. Process. Syst. (2007) 1137–1144.

[20] M. Ranzato, Y.-L. Boureau, Y.L. Cun, Sparse feature learning for deep belief networks, Adv. Neural Inf. Process. Syst. (2008) 1185–1192.

[21] R. Salakhutdinov, G. Hinton, Deep Boltzmann machines, Artificial intelligence and statistics (2009) 448–455.

[22] P. Vincent, H. Larochelle, I. Lajoie, Y. Bengio, P.-A. Manzagol, L. Bottou, Stacked denoising autoencoders: Learning useful representations in a deep network with a local denoising criterion, J. Mach. Learn. Res. 11 (12) (2010).

[23] A. Krizhevsky, I. Sutskever, G.E. Hinton, Imagenet classification with deep convolutional neural networks, Adv. Neural Inf. Process. Syst. (2012) 1097–1105.

[24] I. Goodfellow, et al., Generative adversarial nets, Adv. Neural Inf. Process. Syst. (2014) 2672–2680.

[25] F.A.N. Fernandes, L.M.F. Lona, Neural network applications in polymerization processes, Brazilian J. Chem. Eng. 22 (3) (2005) 401–418.

[26] R. Ranawana, V. Palade, A neural network based multi-classifier system for gene identification in DNA sequences, Neural Comput. Applic. 14 (2) (2005) 122–131.

[27] T. Levstek, M. Lakota, The use of artificial neural networks for compounds prediction in biogas from anaerobic digestion—a review, Agri 7 (2010) 15–22.

[28] C.M. Bishop, Neural Networks for Pattern Recognition, Oxford University Press, 1995.

[29] S. Curteanu, Direct and inverse neural network modeling in free radical polymerization, Open Chem. 2 (1) (2004) 113–140.

[30] S. Curteanu, C. Petrila, Neural network-based modeling for semi-batch and nonisothermal free radical polymerization, Int. J. Quantum Chem. 106 (6) (2006) 1445–1456.

[31] S. Curteanu, A. Dumitrescu, C. Mihăilescu, B. Simionescu, Neural network modeling applied to polyacrylamide based hydrogels synthetized by single step process, Polym.-Plast. Technol. Eng. 47 (10) (2008) 1061–1071.

[32] J. Lobato, P. Cañizares, M.A. Rodrigo, J.J. Linares, C.-G. Piuleac, S. Curteanu, The neural networks based modeling of a polybenzimidazole-based polymer electrolyte membrane fuel cell: effect of temperature, J. Power Sources 192 (1) (2009) 190–194.

[33] C.G. Piuleac, M.A. Rodrigo, P. Cañizares, S. Curteanu, C. Sáez, Ten steps modeling of electrolysis processes by using neural networks, Environ. Model. Software 25 (1) (2010) 74–81.

[34] S. Curteanu, C. Racles, V. Cozan, Prediction of the liquid crystalline property for polyazomethines using modular neural networks, J. Optoelectron. Adv. Mater. 10 (12) (2008) 3382–3391.

[35] F.A. Caliman, S. Curteanu, C. Betianu, M. Gavrilescu, I. Poulios, Neural networks and genetic algorithms optimization of the photocatalytic degradation of Alcian Blue 8GX, J. Adv. Oxid. Technol. 11 (2) (2008) 316–326.

[36] S. Curteanu, A. Dumitrescu, C. Mihailescu, B.C. Simionescu, The synthesis of polyacrylamide-based multi-component hydrogels. A neural network modeling, J. Macromol. Sci. Pt A Pure Appl. Chem. 46 (4) (2009) 368–380.

[37] C.G. Piuleac, I. Poulios, F. Leon, S. Curteanu, A. Kouras, Modeling methodology based on stacked neural networks applied to the photocatalytic degradation of triclopyr, Sep. Sci. Technol. 45 (11) (2010) 1644–1650.

[38] P.G. Benardos, G.C. Vosniakos, Prediction of surface roughness in CNC face milling using neural networks and Taguchi's design of experiments, Robot. Comput. Integr. Manuf. 18 (5–6) (2002) 343–354.

[39] W. Sukthomya, J. Tannock, The optimisation of neural network parameters using Taguchi's design of experiments approach: an application in manufacturing process modelling, Neural Comput. Applic. 14 (4) (2005) 337–344.

[40] P.A. Castillo, J.J. Merelo, M.G. Arenas, G. Romero, Comparing evolutionary hybrid systems for design and optimization of multilayer perceptron structure along training parameters, Inf. Sci. (NY) 177 (14) (2007) 2884–2905.

[41] L. Ma, K. Khorasani, New training strategies for constructive neural networks with application to regression problems, Neural Netw. 17 (4) (2004) 589–609.

[42] R. Delogu, A. Fanni, A. Montisci, Geometrical synthesis of MLP neural networks, Neurocomputing 71 (4–6) (2008) 919–930.

[43] H.-J. Xing, B.-G. Hu, Two-phase construction of multilayer perceptrons using information theory, IEEE Trans. Neural Netw. 20 (4) (2009) 715–721.

[44] X. Yao, Evolving artificial neural networks, Proc. IEEE 87 (9) (1999) 1423–1447.

[45] X. Yao, Y. Liu, A new evolutionary system for evolving artificial neural networks, IEEE Trans. Neural Netw. 8 (3) (1997) 694–713.

[46] X. Yao, Y. Liu, Towards designing artificial neural networks by evolution, Appl. Math Comput. 91 (1) (1998) 83–90.

[47] H.A. Abbass, A memetic pareto evolutionary approach to artificial neural networks, in: Australian Joint Conference on Artificial Intelligence, 2001, pp. 1–12.

[48] M. Suchorzewski, Evolving scalable and modular adaptive networks with developmental symbolic encoding, Evol. Intell. 4 (3) (2011) 145–163.

[49] S. Mizuta, T. Sato, D. Lao, M. Ikeda, T. Shimizu, Structure design of neural networks using genetic algorithms, Complex Syst. 13 (2) (2001) 161–176.

[50] P. Kordík, J. Koutník, J. Drchal, O. Kovářík, M. Čepek, M. Šnorek, Meta-learning approach to neural network optimization, Neural Netw. 23 (4) (2010) 568–582.

[51] B.M. Wilamowski, Neural network architectures and learning algorithms, IEEE Ind. Electron. Mag. 3 (4) (2009) 56–63.

[52] J.C. Principe, N.R. Euliano, W.C. Lefebvre, Neural and Adaptive Systems: Fundamentals Through Simulations, vol. 672, Wiley, New York, 2000.

[53] T. Ince, S. Kiranyaz, L. Eren, M. Askar, M. Gabbouj, Real-time motor fault detection by 1-D convolutional neural networks, IEEE Trans. Ind. Electron. 63 (11) (2016) 7067–7075.

[54] J. Hertz, A. Krogh, R.G. Palmer, H. Horner, Introduction to the theory of neural computation, PhT 44 (12) (1991) 70.

[55] S. Haykin, Neural Networks, a Comprehensive Foundation, 7458, Prentice-Hall Inc, Upper Saddle River, NJ, 1999, pp. 161–175.

[56] J.L. Elman, Finding structure in time, Cogn. Sci. 14 (2) (1990) 179–211.

[57] A.C. Tsoi, A.D. Back, Locally recurrent globally feedforward networks: a critical review of architectures, IEEE Trans. Neural Netw. 5 (2) (1994) 229–239.

[58] M.I. Jordan, Supervised Learning and Systems With Excess Degrees of Freedom, University of Massachusetts at Amherst, Department of Computer Science, 1988.

[59] O. Nelles, Nonlinear System Identification: From Classical Approaches to Neural Networks and Fuzzy Models, Springer Science & Business Media, 2013.

[60] M.I. Jordan, Attractor dynamics and parallelism in a connectionist sequential machine, in: *Artificial Neural Networks: Concept Learning*, 1990, pp. 112–127.

[61] W.S. Stornetta, T. Hogg, B.A. Huberman, A dynamical approach to temporal pattern processing, Neural Inf. Process. Syst. (1988) 750–759.

[62] L. Xing, D.T. Pham, Neural Networks for Identification, Prediction, and Control, Springer-Verlag, 1995.

[63] J.M. Zamarreño, P. Vega, State space neural network. Properties and application, Neural Netw. 11 (6) (Aug. 1998) 1099–1112.

[64] K. Cho, et al., Learning phrase representations using RNN encoder-decoder for statistical machine translation, arXiv (2014). Prepr. arXiv1406.1078.

[65] J. Zhang, Batch-to-batch optimal control of a batch polymerisation process based on stacked neural network models, Chem. Eng. Sci. 63 (5) (2008) 1273–1281.

[66] Z.-H. Zhou, J. Wu, W. Tang, Ensembling neural networks: many could be better than all, Artif. Intell. 137 (1–2) (May 2002) 239–263.

[67] M.H. Nguyen, H.A. Abbass, R.I. McKay, Stopping criteria for ensemble of evolutionary artificial neural networks, Appl. Soft Comput. 6 (1) (Nov. 2005) 100–107.

[68] A. Mukherjee, J. Zhang, A reliable multi-objective control strategy for batch processes based on bootstrap aggregated neural network models, J. Process Control 18 (7–8) (Aug. 2008) 720–734.

[69] D.V. Sridhar, E.B. Bartlett, R.C. Seagrave, An information theoretic approach for combining neural network process models, Neural Netw. 12 (6) (Jul. 1999) 915–926.

[70] D.H. Wolpert, Stacked generalization, Neural Netw. 5 (2) (1992) 241–259.

[71] Y. Tian, J. Zhang, J. Morris, Modeling and optimal control of a batch polymerization reactor using a hybrid stacked recurrent neural network model, Ind. Eng. Chem. Res. 40 (21) (2001) 4525–4535.

[72] F. Herrera, J. Zhang, Optimal control of batch processes using particle swam optimisation with stacked neural network models, Comput. Chem. Eng. 33 (10) (Oct. 2009) 1593–1601.

[73] J. Torres-Sospedra, C. Hernández-Espinosa, M. Fernández-Redondo, Combining MF networks: a comparison among statistical methods and stacked generalization, in: *IAPR Workshop on Artificial Neural Networks in Pattern Recognition*, 2006, pp. 210–220.

[74] L. Breiman, Bagging predictors, Mach. Learn. 24 (2) (1996) 123–140.

[75] J. Zhang, Developing robust non-linear models through bootstrap aggregated neural networks, Neurocomputing 25 (1–3) (1999) 93–113.

[76] P. Cunningham, J. Carney, S. Jacob, Stability problems with artificial neural networks and the ensemble solution, Artif. Intell. Med. 20 (3) (2000) 217–225.

[77] R. Wehrens, H. Putter, L.M.C. Buydens, The bootstrap: a tutorial, Chemom. Intel. Lab. Syst. 54 (1) (2000) 35–52.

[78] Y. Freund, R.E. Schapire, Experiments with a new boosting algorithm, icml 96 (1996) 148–156.

[79] H. Schwenk, Y. Bengio, Boosting neural networks, Neural Comput. 12 (8) (2000) 1869–1887.

[80] T.G. Dietterich, An experimental comparison of three methods for constructing ensembles of decision trees: bagging, boosting, and randomization, Mach. Learn. 40 (2) (2000) 139–157.

[81] F. Leon, C.G. Piuleac, S. Curteanu, Stacked neural network modeling applied to the synthesis of polyacrylamide-based multicomponent hydrogels, Macromol. React. Eng. 4 (9–10) (Jul. 2010) 591–598.

[82] M.F. Amin, M.M. Islam, K. Murase, Ensemble of single-layered complex-valued neural networks for classification tasks, Neurocomputing 72 (10–12) (2009) 2227–2234.

[83] H. Drucker, Boosting using neural networks, in: Combining Artificial Neural Nets, 1999.

[84] A. Hassanzadeh, A. Kaarna, T. Kauranne, Unsupervised multi-manifold classification of hyperspectral remote sensing images with contractive Autoencoder, in: Scandinavian Conference on Image Analysis, 2017, pp. 169–180.

[85] S. Curteanu, F. Leon, Hybrid neural network models applied to a free radical polymerization process, Polym.-Plast. Technol. Eng. 45 (9) (Sep. 2006) 1013–1023.

[86] S. Curteanu, H. Cartwright, Neural networks applied in chemistry. I. Determination of the optimal topology of multilayer perceptron neural networks, J. Chemometr. 25 (10) (Oct. 2011) 527–549.

[87] P.H. Shah, B. Bhalja, Laboratory prototype to understand miscoordination of relays in radial network in the presence of distributed generation, Int. J. Artif. Intell 9 (2012) 26–36.

[88] M. Dam, D.N. Saraf, Design of neural networks using genetic algorithm for on-line property estimation of crude fractionator products, Comput. Chem. Eng. 30 (4) (Feb. 2006) 722–729.

[89] P.G. Benardos, G.-C. Vosniakos, Optimizing feedforward artificial neural network architecture, Eng. Appl. Artif. Intel. 20 (3) (Apr. 2007) 365–382.

[90] E.-N. Dragoi, S. Curteanu, A.-I. Galaction, D. Cascaval, Optimization methodology based on neural networks and self-adaptive differential evolution algorithm applied to an aerobic fermentation process, Appl. Soft Comput. 13 (1) (Jan. 2013) 222–238.

[91] X. Hu, Y. Hu, X. Yu, The soft measure model of dissolved oxygen based on RBF network in ponds, in: 2011 Fourth International Conference on Information and Computing, 2011, pp. 38–41.

[92] B. Li, L. Cong, W. Zhang, Research on optimized RBF neural network based on GA for sewage treatment, in: 2013 5th International Conference on Intelligent Human-Machine Systems and Cybernetics, 2013, pp. 520–523.

[93] M. Li, W. Wu, B. Chen, L. Guan, Y. Wu, Water quality evaluation using back propagation artificial neural network based on self-adaptive particle swarm optimization algorithm and chaos theory, Comput. Water, Energy, Environ. Eng. 06 (03) (2017) 229–242.

[94] W. Yingbo, W. Lin, D. Deng, Harmony search algorithm optimized BP network and application in water's quality evaluation, Comput. Meas. Control 20 (2012) 1931–1933.

[95] E. Li, L. Jia, J. Yu, A genetic neural fuzzy system and its application in quality prediction in the injection process, Chem. Eng. Commun. 191 (3) (Mar. 2004) 335–355.

[96] Z. Wang, B. Yang, C. Chen, J. Yuan, L. Wang, Modeling and optimization for the secondary reaction of FCC gasoline based on the fuzzy neural network and genetic algorithm, Chem. Eng. Process. Process Intensif. 46 (3) (Mar. 2007) 175–180.

[97] Z.M. Yaseen, et al., Rainfall pattern forecasting using novel hybrid intelligent model based ANFIS-FFA, Water Resour. Manag. 32 (1) (2018) 105–122.

[98] S. Faizollahzadeh Ardabili, B. Najafi, H. Ghaebi, S. Shamshirband, A. Mostafaeipour, A novel enhanced exergy method in analyzing HVAC system using soft computing approaches: a case study on mushroom growing hall, J. Build. Eng. 13 (Sep. 2017) 309–318.

[99] J. Villermaux, Future challenges for basic research in chemical engineering, Chem. Eng. Sci. 48 (14) (1993) 2525–2535.

[100] E. Favre, L. Marchal-Heusler, M. Kind, Chemical Product Engineering: Research and Educational Challenges, Chem. Eng. Res. Des. 80 (1) (Jan. 2002) 65–74.

[101] E. Favre, V. Falk, C. Roizard, E. Schaer, Trends in chemical engineering education: process, product and sustainable chemical engineering challenges, Educ. Chem. Eng. 3 (1) (Jun. 2008) e22–e27.

[102] J.D. Perkins, Chemical engineering—the first 100 years, in: *Chemical engineering: Visions of the world*, Elsevier, 2003, pp. 11–40.

[103] F. CECCHI, Chemical engineering for the environment. Mediterranean congress, Ind. Eng. Chem. Res. 46 (21) (2007) 6646–6830.

[104] S.W. Churchill, Role of universalities in chemical engineering, Ind. Eng. Chem. Res. 46 (24) (2007) 7851–7869.

[105] M. Shacham, N. Brauner, Preventing oscillatory behavior in error control for ODEs, Comput. Chem. Eng. 32 (3) (Mar. 2008) 409–419.

[106] R.-E. Precup, S. Preitl, E.M. Petriu, J.K. Tar, M.L. Tomescu, C. Pozna, Generic two-degree-of-freedom linear and fuzzy controllers for integral processes, J. Franklin Inst. 346 (10) (Dec. 2009) 980–1003.

[107] W.J. Cole, K.M. Powell, T.F. Edgar, Optimization and advanced control of thermal energy storage systems, Rev. Chem. Eng. 28 (2–3) (2012) 81–99.

[108] M. Pirdashti, S. Curteanu, M.H. Kamangar, M.H. Hassim, M.A. Khatami, Artificial neural networks: applications in chemical engineering, Rev. Chem. Eng. 29 (4) (2013) 205–239.

[109] J. Heaton, Introduction to Neural Networks With Java, Heaton Research Inc, Chesterfield, MO, 2008.

[110] J.C. Hoskins, D.M. Himmelblau, Artificial neural network models of knowledge representation in chemical engineering, Comput. Chem. Eng. 12 (9–10) (Sep. 1988) 881–890.

[111] M. Azlan Hussain, Review of the applications of neural networks in chemical process control—simulation and online implementation, Artif. Intell. Eng. 13 (1) (Jan. 1999) 55–68.

[112] S. Kalogirou, Artificial neural networks for the prediction of the energy consumption of a passive solar building, Energy 25 (5) (May 2000) 479–491.

[113] Z. Zhang, K. Friedrich, Artificial neural networks applied to polymer composites: a review, Compos. Sci. Technol. 63 (14) (Nov. 2003) 2029–2044.

[114] V. Venkatasubramanian, R. Rengaswamy, K. Yin, S.N. Kavuri, A review of process fault detection and diagnosis, Comput. Chem. Eng. 27 (3) (Mar. 2003) 293–311.

[115] C.-J. Du, D.-W. Sun, Learning techniques used in computer vision for food quality evaluation: a review, J. Food Eng. 72 (1) (2006) 39–55.

[116] P. Roupas, Predictive modelling of dairy manufacturing processes, Int. Dairy J. 18 (7) (Jul. 2008) 741–753.

[117] O. Ludwig, U. Nunes, R. Araújo, L. Schnitman, H.A. Lepikson, Applications of information theory, genetic algorithms, and neural models to predict oil flow, Commun. Nonlinear Sci. Numer. Simul. 14 (7) (Jul. 2009) 2870–2885.

[118] X.S. Qin, G.H. Huang, L. He, Simulation and optimization technologies for petroleum waste management and remediation process control, J. Environ. Manage. 90 (1) (2009) 54–76.

[119] F. Marini, Artificial neural networks in foodstuff analyses: trends and perspectives. A review, Anal. Chim. Acta 635 (2) (Mar. 2009) 121–131.

[120] Z. Ahmad, J. Zhang, Selective combination of multiple neural networks for improving model prediction in nonlinear systems modelling through forward selection and backward elimination, Neurocomputing 72 (4–6) (Jan. 2009) 1198–1204.

[121] R.A.M. Noor, Z. Ahmad, M.M. Don, M.H. Uzir, Modelling and control of different types of polymerization processes using neural networks technique: a review, Can. J. Chem. Eng. 88 (6) (Dec. 2010) 1065–1084.

[122] J. Zupan, Basics of Artificial Neural Networks, 2003, pp. 199–229.

[123] J. Zhang, Inferential estimation of polymer quality using bootstrap aggregated neural networks, Neural Netw. 12 (6) (Jul. 1999) 927–938.

[124] G. Lisa, D.A. Wilson, S. Curteanu, C. Lisa, C.-G. Piuleac, V. Bulacovschi, Ferrocene derivatives thermostability prediction using neural networks and genetic algorithms, Thermochim. Acta 521 (1–2) (Jul. 2011) 26–36.

[125] V. Chitanov, C. Kiparissides, M. Petrov, Neural-fuzzy modelling of polymer quality in batch polymerization reactors, in: 2004 2nd International IEEE Conference on "Intelligent Systems". Proceedings (IEEE Cat. No.04EX791), 2004, pp. 67–72.

[126] Ç. Karataş, A. Sözen, E. Arcaklioğlu, S. Ergüney, Modelling of yield length in the mould of commercial plastics using artificial neural networks, Mater. Des. 28 (1) (2007) 278–286.

[127] M.L. Koç, Ü. Özdemir, D. İmren, Prediction of the pH and the temperature-dependent swelling behavior of Ca2+-alginate hydrogels by artificial neural networks, Chem. Eng. Sci. 63 (11) (2008) 2913–2919.

[128] X. Yu, W. Yu, B. Yi, X. Wang, Artificial neural network prediction of steric hindrance parameter of polymers, Chem. Pap. 63 (4) (Jan. 2009).

[129] J.C.B. Gonzaga, L.A. Meleiro, C. Kiang, R. Maciel Filho, ANN-based soft-sensor for real-time process monitoring and control of an industrial polymerization process, Comput. Chem. Eng. 33 (1) (Jan. 2009) 43–49.

[130] C. Guo, J. Lu, Z. Tian, W. Guo, A. Darvishan, Optimization of critical parameters of PEM fuel cell using TLBO-DE based on Elman neural network, Energ. Conver. Manage. 183 (Mar. 2019) 149–158.

[131] C.M. Kshirsagar, R. Anand, Artificial neural network applied forecast on a parametric study of Calophyllum inophyllum methyl ester-diesel engine out responses, Appl. Energy 189 (Mar. 2017) 555–567.

[132] Z. Tang, Z. Zhang, The multi-objective optimization of combustion system operations based on deep data-driven models, Energy 182 (Sep. 2019) 37–47.

[133] A. Chakraborty, S. Roy, R. Banerjee, An experimental based ANN approach in mapping performance-emission characteristics of a diesel engine operating in dual-fuel mode with LPG, J. Nat. Gas Sci. Eng. 28 (Jan. 2016) 15–30.

[134] T. Yusaf, K.H. Saleh, M.A. Said, Engine performance and emission analysis of LPG-SI engine with the aid of artificial neural network, Proc. Inst. Mech. Eng. Pt A J. Power Energy 225 (5) (2011) 591–600.

[135] K.P. Rao, T.V. Babu, G. Anuradha, B.V.A. Rao, IDI diesel engine performance and exhaust emission analysis using biodiesel with an artificial neural network (ANN), Egypt. J. Pet. 26 (3) (2017) 593–600.

[136] H.S. Saraee, H. Taghavifar, S. Jafarmadar, Experimental and numerical consideration of the effect of CeO2 nanoparticles on diesel engine performance and exhaust emission with the aid of artificial neural network, Appl. Therm. Eng. 113 (2017) 663–672.

[137] J. Syed, R.U. Baig, S. Algarni, Y.V.V.S. Murthy, M. Masood, M. Inamurrahman, Artificial neural network modeling of a hydrogen dual fueled diesel engine characteristics: an experiment approach, Int. J. Hydrogen Energy 42 (21) (May 2017) 14750–14774.

[138] A. Uzun, A parametric study for specific fuel consumption of an intercooled diesel engine using a neural network, Fuel 93 (Mar. 2012) 189–199.

[139] M. Bietresato, A. Calcante, F. Mazzetto, A neural network approach for indirectly estimating farm tractors engine performances, Fuel 143 (Mar. 2015) 144–154.

[140] S. Gürgen, B. Ünver, İ. Altın, Prediction of cyclic variability in a diesel engine fueled with n-butanol and diesel fuel blends using artificial neural network, Renew. Energy 117 (Mar. 2018) 538–544.

[141] J.-D. Wu, J.-C. Liu, Development of a predictive system for car fuel consumption using an artificial neural network, Expert Syst. Appl. 38 (5) (May 2011) 4967–4971.

[142] B. Peña, E. Teruel, L.I. Díez, Soft-computing models for soot-blowing optimization in coal-fired utility boilers, Appl. Soft Comput. 11 (2) (Mar. 2011) 1657–1668.

[143] G. Zahedi, A. Lohi, K.A. Mahdi, Hybrid modeling of ethylene to ethylene oxide heterogeneous reactor, Fuel Process. Technol. 92 (9) (Sep. 2011) 1725–1732.

[144] R.M. Balabin, S.V. Smirnov, Variable selection in near-infrared spectroscopy: benchmarking of feature selection methods on biodiesel data, Anal. Chim. Acta 692 (1–2) (2011) 63–72.

[145] R. Eslamloueyan, M.H. Khademi, Estimation of thermal conductivity of pure gases by using artificial neural networks, Int. J. Therm. Sci. 48 (6) (2009) 1094–1101.

[146] Y. Erzin, B.H. Rao, D.N. Singh, Artificial neural network models for predicting soil thermal resistivity, Int. J. Therm. Sci. 47 (10) (Oct. 2008) 1347–1358.

[147] J. Wang, W. Wan, Application of desirability function based on neural network for optimizing biohydrogen production process, Int. J. Hydrogen Energy 34 (3) (Feb. 2009) 1253–1259.

[148] J.K. Whiteman, E.B. Gueguim Kana, Comparative assessment of the artificial neural network and response surface modelling efficiencies for biohydrogen production on sugar cane molasses, Bioenergy Res. 7 (1) (Mar. 2014) 295–305.

[149] N. Nasr, H. Hafez, M.H. El Naggar, G. Nakhla, Application of artificial neural networks for modeling of biohydrogen production, Int. J. Hydrogen Energy 38 (8) (Mar. 2013) 3189–3195.

[150] E. Artun, T. Ertekin, R.W. Watson, M.A. Al-Wadhahi, Development of universal proxy models for screening and optimization of cyclic pressure pulsing in naturally fractured reservoirs, in: International Petroleum Technology Conference, 2009.

[151] J.S. Ahari, M.T. Sadeghi, S.Z. Pashne, Optimization of OCM reaction conditions over Na–W–Mn/SiO2 catalyst at elevated pressure, J. Taiwan Inst. Chem. Eng. 42 (5) (Sep. 2011) 751–759.

[152] M. Safamirzaei, H. Modarress, Hydrogen solubility in heavy n-alkanes; modeling and prediction by artificial neural network, Fluid Phase Equilib. 310 (1–2) (Nov. 2011) 150–155.

[153] R. Ashena, J. Moghadasi, Bottom hole pressure estimation using evolved neural networks by real coded ant colony optimization and genetic algorithm, J. Petrol. Sci. Eng. 77 (3–4) (Jun. 2011) 375–385.

[154] A. Azadeh, S.M. Asadzadeh, M. Saberi, V. Nadimi, A. Tajvidi, M. Sheikalishahi, A neuro-fuzzy-stochastic frontier analysis approach for long-term natural gas consumption forecasting and behavior analysis: the cases of Bahrain, Saudi Arabia, Syria, and UAE, Appl. Energy 88 (11) (Nov. 2011) 3850–3859.

[155] J. Asadisaghandi, P. Tahmasebi, Comparative evaluation of back-propagation neural network learning algorithms and empirical correlations for prediction of oil PVT properties in Iran oilfields, J. Petrol. Sci. Eng. 78 (2) (Aug. 2011) 464–475.

[156] I. Alhajree, G. Zahedi, Z.A. Manan, S.M. Zadeh, Modeling and optimization of an industrial hydrocracker plant, J. Petrol. Sci. Eng. 78 (3–4) (Sep. 2011) 627–636.

[157] M. Nikravesh, F. Aminzadeh, Mining and fusion of petroleum data with fuzzy logic and neural network agents, J. Petrol. Sci. Eng. 29 (3–4) (May 2001) 221–238.

[158] M.J.B. Souza, F.A.N. Fernandes, A.M.G. Pedrosa, A.S. Araujo, Selective cracking of natural gasoline over HZSM-5 zeolite, Fuel Process. Technol. 89 (9) (Sep. 2008) 819–827.

[159] F. Anifowose, A. Abdulraheem, Fuzzy logic-driven and SVM-driven hybrid computational intelligence models applied to oil and gas reservoir characterization, J. Nat. Gas Sci. Eng. 3 (3) (Jul. 2011) 505–517.

[160] S.O. Olatunji, A. Selamat, A.A.A. Raheem, Predicting correlations properties of crude oil systems using type-2 fuzzy logic systems, Expert Syst. Appl. 38 (9) (Sep. 2011) 10911–10922.

[161] H.A. Zaqoot, Water Quality Assessment Model of the Mediterranean Sea Along Gaza-Palestine, 2011.

[162] H.R. Maier, A. Jain, G.C. Dandy, K.P. Sudheer, Methods used for the development of neural networks for the prediction of water resource variables in river systems: current status and future directions, Environ. Model. Software 25 (8) (Aug. 2010) 891–909.

[163] S. Palani, S.-Y. Liong, P. Tkalich, An ANN application for water quality forecasting, Mar. Pollut. Bull. 56 (9) (Sep. 2008) 1586–1597.

[164] X. Li, J. Ai, C. Lin, H. Guan, Prediction model of dissolved oxygen in ponds based on ELM neural network, IOP Conf. Ser. Earth Environ. Sci. 121 (Feb. 2018), 022003.

[165] X. Ta, Y. Wei, Research on a dissolved oxygen prediction method for recirculating aquaculture systems based on a convolution neural network, Comput. Electron. Agric. 145 (2018) 302–310.

[166] R.M. Aghav, S. Kumar, S.N. Mukherjee, Artificial neural network modeling in competitive adsorption of phenol and resorcinol from water environment using some carbonaceous adsorbents, J. Hazard. Mater. 188 (1–3) (2011) 67–77.

[167] Q. Guo, Z. He, L. Li, H. Li, Application of BP neural network model for prediction of water pollutants concentration in Taihu Lake, J. South. Agric. 42 (10) (2011) 1303–1306.

[168] G. Wu, J. Qin, Q. Wan, Water quality evaluation of Xiangjiang estuary based on BP neural network model, J. Nat. Sci. Hunan Norm. Univ. 36 (5) (2013) 92–95.

[169] Y. Huang, D. Ji, Experimental study on seawater-pipeline internal corrosion monitoring system, Sens. Actuators B 135 (1) (2008) 375–380.

[170] Z. Xiao, L. Peng, Y. Chen, H. Liu, J. Wang, Y. Nie, The dissolved oxygen prediction method based on neural network, Complexity 2017 (2017) 1–6.

[171] Z.H. Zou, X.L. Wang, The errors analysis for river water quality prediction based-on BP-modeling, Acta Sci. Circumstantiae 27 (6) (2007) 1038–1042.

[172] Y. Chen, Q. Cheng, X. Fang, H. Yu, D. Li, Principal component analysis and long short-term memory neural network for predicting dissolved oxygen in water for aquaculture, Trans. Chin. Soc. Agric. Eng. 34 (2018) 183–191.

[173] S. Liu, M. Yan, H. Tai, L. Xu, D. Li, Prediction of Dissolved Oxygen Content in Aquaculture of Hyriopsis Cumingii Using Elman Neural Network, 2012, pp. 508–518.

[174] J. He, C. Valeo, A. Chu, N.F. Neumann, Prediction of event-based stormwater runoff quantity and quality by ANNs developed using PMI-based input selection, J. Hydrol. 400 (1–2) (Mar. 2011) 10–23.

[175] Y. Chen, H. Yu, Y. Cheng, Q. Cheng, D. Li, A hybrid intelligent method for three-dimensional short-term prediction of dissolved oxygen content in aquaculture, PLoS One 13 (2) (Feb. 2018), e0192456.

[176] M.R. Gadekar, M.M. Ahammed, Modelling dye removal by adsorption onto water treatment residuals using combined response surface methodology-artificial neural network approach, J. Environ. Manage. 231 (2019) 241–248.

[177] S. Lotfan, R.A. Ghiasi, M. Fallah, M.H. Sadeghi, ANN-based modeling and reducing dual-fuel engine's challenging emissions by multi-objective evolutionary algorithm NSGA-II, Appl. Energy 175 (Aug. 2016) 91–99.

[178] H. Akbaş, B. Bilgen, A.M. Turhan, An integrated prediction and optimization model of biogas production system at a wastewater treatment facility, Bioresour. Technol. 196 (Nov. 2015) 566–576.

[179] X. Peng, L. Chen, Y. Yu, D. Wang, PCA-GRNN-GA based pH value prediction model applied in penaeus orientalis culture, in: 2016 6th International Conference on Digital Home (ICDH), 2016, pp. 227–232.

[180] J. Wan, et al., Prediction of effluent quality of a paper mill wastewater treatment using an adaptive network-based fuzzy inference system, Appl. Soft Comput. 11 (3) (2011) 3238–3246.

[181] S. Curteanu, C.G. Piuleac, K. Godini, G. Azaryan, Modeling of electrolysis process in wastewater treatment using different types of neural networks, Chem. Eng. J. 172 (1) (2011) 267–276.

[182] M.S. Bhatti, D. Kapoor, R.K. Kalia, A.S. Reddy, A.K. Thukral, RSM and ANN modeling for electrocoagulation of copper from simulated wastewater: multi objective optimization using genetic algorithm approach, Desalination 274 (1–3) (Jul. 2011) 74–80.

[183] M. Al-Abri, K. Al Anezi, A. Dakheel, N. Hilal, Humic substance coagulation: artificial neural network simulation, Desalination 253 (1–3) (Apr. 2010) 153–157.

[184] M. Sadrzadeh, T. Mohammadi, J. Ivakpour, N. Kasiri, Separation of lead ions from wastewater using electrodialysis: comparing mathematical and neural network modeling, Chem. Eng. J. 144 (3) (Nov. 2008) 431–441.

[185] A. Gulbag, F. Temurtas, I. Yusubov, Quantitative discrimination of the binary gas mixtures using a combinational structure of the probabilistic and multilayer neural networks, Sens. Actuators B 131 (1) (Apr. 2008) 196–204.

[186] G.B. Sahoo, C. Ray, E. Mehnert, D.A. Keefer, Application of artificial neural networks to assess pesticide contamination in shallow groundwater, Sci. Total Environ. 367 (1) (Aug. 2006) 234–251.

[187] S. Chellam, Artificial neural network model for transient crossflow microfiltration of polydispersed suspensions, J. Membr. Sci. 258 (1–2) (Aug. 2005) 35–42.

[188] C. Aydiner, I. Demir, E. Yildiz, Modeling of flux decline in crossflow microfiltration using neural networks: the case of phosphate removal, J. Membr. Sci. 248 (1–2) (2005) 53–62.

[189] V. Yangali-Quintanilla, A. Verliefde, T.-U. Kim, A. Sadmani, M. Kennedy, G. Amy, Artificial neural network models based on QSAR for predicting rejection of neutral organic compounds by polyamide nanofiltration and reverse osmosis membranes, J. Membr. Sci. 342 (1–2) (Oct. 2009) 251–262.

[190] T. KUZNIZ, et al., Instrumentation for the monitoring of toxic pollutants in water resources by means of neural network analysis of absorption and fluorescence spectra, Sens. Actuators B 121 (1) (Jan. 2007) 231–237.

[191] E.S. Elmolla, M. Chaudhuri, The use of artificial neural network (ANN) for modelling, simulation and prediction of advanced oxidation process performance in recalcitrant wastewater treatment, in: Artificial Neural Networks—Application, InTech, 2011.

[192] J.D. Smith, A.A. Neto, S. Cremaschi, D.W. Crunkleton, CFD-based optimization of a flooded bed algae bioreactor, Ind. Eng. Chem. Res. 52 (22) (Jun. 2013) 7181–7188.

[193] J.P. Maran, B. Priya, Comparison of response surface methodology and artificial neural network approach towards efficient ultrasound-assisted biodiesel production from muskmelon oil, Ultrason. Sonochem. 23 (Mar. 2015) 192–200.

[194] E.-N. Dragoi, S. Curteanu, F. Leon, A.-I. Galaction, D. Cascaval, Modeling of oxygen mass transfer in the presence of oxygen-vectors using neural networks developed by differential evolution algorithm, Eng. Appl. Artif. Intel. 24 (7) (2011) 1214–1226.

[195] C.-W. Chang, W.-C. Yu, W.-J. Chen, R.-F. Chang, W.-S. Kao, A study on the enzymatic hydrolysis of steam exploded napiergrass with alkaline treatment using artificial neural networks and regression analysis, J. Taiwan Inst. Chem. Eng. 42 (6) (Nov. 2011) 889–894.

[196] K.M. Desai, B.K. Vaidya, R.S. Singhal, S.S. Bhagwat, Use of an artificial neural network in modeling yeast biomass and yield of β-glucan, Process Biochem. 40 (5) (Apr. 2005) 1617–1626.

[197] B. T., E. T., and D. A., Dynamic neural networks as a tool for the online optimization of industrial fermentation, Bioprocess Biosyst. Eng. 24 (6) (Mar. 2002) 347–354.

[198] M. Elnekave, S.O. Celik, M. Tatlier, N. Tufekci, Artificial neural network predictions of up-flow anaerobic sludge blanket (UASB) reactor performance in the treatment of citrus juice wastewater, Polish J. Environ. Stud. 21 (1) (2012).

[199] B. Mahanty, M. Zafar, H.-S. Park, Characterization of co-digestion of industrial sludges for biogas production by artificial neural network and statistical regression models, Environ. Technol. 34 (13–14) (Jul. 2013) 2145–2153.

[200] Z. Wu, X. Shi, Optimization for high-density cultivation of heterotrophic Chlorella based on a hybrid neural network model, Lett. Appl. Microbiol. 44 (1) (Jan. 2007) 13–18.

[201] R.S. Prakasham, T. Sathish, P. Brahmaiah, Imperative role of neural networks coupled genetic algorithm on optimization of biohydrogen yield, Int. J. Hydrogen Energy 36 (7) (Apr. 2011) 4332–4339.

[202] S. Shi, G. Xu, Novel performance prediction model of a biofilm system treating domestic wastewater based on stacked denoising auto-encoders deep learning network, Chem. Eng. J. 347 (2018) 280–290.

[203] H. Salehi, S. Zeinali Heris, M. Koolivand Salooki, S.H. Noei, Designing a neural network for closed thermosyphon with nanofluid using a genetic algorithm, Brazilian J. Chem. Eng. 28 (1) (Mar. 2011) 157–168.

[204] M. Khanmohammadi, A.B. Garmarudi, N. Khoddami, K. Shabani, M. Khanlari, A novel technique based on diffuse reflectance near-infrared spectrometry and back-propagation artificial neural network for estimation of particle size in TiO2 nano particle samples, Microchem. J. 95 (2) (Jul. 2010) 337–340.

[205] J. Ma, S.G. Zhu, C.X. Wu, M.L. Zhang, Application of back-propagation neural network technique to high-energy planetary ball milling process for synthesizing nanocomposite WC–MgO powders, Mater. Des. 30 (8) (Sep. 2009) 2867–2874.

[206] M.C. Haciismailoglu, I. Kucuk, N. Derebasi, Prediction of dynamic hysteresis loops of nano-crystalline cores, Expert Syst. Appl. 36 (2) (Mar. 2009) 2225–2227.

[207] A.K. Santra, N. Chakraborty, S. Sen, Prediction of heat transfer due to presence of copper–water nanofluid using resilient-propagation neural network, Int. J. Therm. Sci. 48 (7) (Jul. 2009) 1311–1318.

[208] C. Capdevila, C. García-Mateo, F.G. Caballero, C.G. de Andrés, Proposal of an empirical formula for the austenitising temperature, Mater. Sci. Eng. A 386 (1–2) (Nov. 2004) 354–361.

[209] Y.-B. Zeng, et al., Application of artificial neural networks in multifactor optimization of an on-line microwave FIA system for catalytic kinetic determination of ruthenium (III), Talanta 54 (4) (2001) 603–609.

[210] A.M. Schweidtmann, W.R. Huster, J.T. Lüthje, A. Mitsos, Deterministic global process optimization: accurate (single-species) properties via artificial neural networks, Comput. Chem. Eng. 121 (2019) 67–74.

[211] D. Bongartz, A. Mitsos, Deterministic global optimization of process flowsheets in a reduced space using McCormick relaxations, J. Glob. Optim. 69 (4) (2017) 761–796.

[212] A.M. Schweidtmann, A. Mitsos, Deterministic global optimization with artificial neural networks embedded, J. Optim. Theory Appl. 180 (3) (Mar. 2019) 925–948.

[213] F. Li, J. Zhang, C. Shang, D. Huang, E. Oko, M. Wang, Modelling of a post-combustion CO_2 capture process using deep belief network, Appl. Therm. Eng. 130 (2018) 997–1003.

[214] J.-S. Chang, S.-C. Lu, Y.-L. Chiu, Dynamic modeling of batch polymerization reactors via the hybrid neural-network rate-function approach, Chem. Eng. J. 130 (1) (May 2007) 19–28.

[215] K. Elsayed, C. Lacor, Modeling and pareto optimization of gas cyclone separator performance using RBF type artificial neural networks and genetic algorithms, Powder Technol. 217 (Feb. 2012) 84–99.

[216] G. Çakmak, C. Yıldız, The prediction of seedy grape drying rate using a neural network method, Comput. Electron. Agric. 75 (1) (Jan. 2011) 132–138.

[217] M.J. Kamali, M. Mousavi, Analytic, neural network, and hybrid modeling of supercritical extraction of α-pinene, J. Supercrit. Fluids 47 (2) (Dec. 2008) 168–173.

[218] H. Jazayeri-Rad, The nonlinear model-predictive control of a chemical plant using multiple neural networks, Neural Comput. Applic. 13 (1) (2004) 2–15.

[219] I.M. Galván, P. Isasi, J.M. Zaldı́var, PNNARMA model: an alternative to phenomenological models in chemical reactors, Eng. Appl. Artif. Intel. 14 (2) (2001) 139–154.

[220] M.B. De Souza, J.C. Pinto, E.L. Lima, Neural net based model predictive control of a chaotic continuous solution polymerization reactor, in: Proceedings of 1993 International Conference on Neural Networks (IJCNN-93-Nagoya, Japan). vol. 2, 1993, pp. 1777–1780.

[221] J.J. Song, S. Park, Neural model predictive control for nonlinear chemical processes, J. Chem. Eng. Japan 26 (4) (1993) 347–354.

[222] M. Hadian, N. AliAkbari, M. Karami, Using artificial neural network predictive controller optimized with Cuckoo Algorithm for pressure tracking in gas distribution network, J. Nat. Gas Sci. Eng. 27 (2015) 1446–1454.

[223] M. Hadian, M. Mehrshadian, M. Karami, A. Biglary Makvand, Event-based neural network predictive controller application for a distillation column, Asian J. Control (2019).

[224] G. Wang, W. Tang, J. Xia, J. Chu, H. Noorman, W.M. van Gulik, Integration of microbial kinetics and fluid dynamics toward model-driven scale-up of industrial bioprocesses, Eng. Life Sci. 15 (1) (Jan. 2015) 20–29.

[225] H. Chen, P. Tiňo, X. Yao, Cognitive fault diagnosis in Tennessee Eastman Process using learning in the model space, Comput. Chem. Eng. 67 (2014) 33–42.

[226] H. Faris, A. Sheta, Identification of the tennessee eastman chemical process reactor using genetic programming, Int. J. Adv. Sci. Technol. 50 (2013) 121–140.

[227] A.F. Sheta, M. Braik, E. Öznergiz, A. Ayesh, M. Masud, Design and Automation for Manufacturing Processes: An Intelligent Business Modeling Using Adaptive Neuro-Fuzzy Inference Systems, 2013, pp. 191–208.

[228] F. Lamnabhi-Lagarrigue, et al., Systems & control for the future of humanity, research agenda: current and future roles, impact and grand challenges, Annu. Rev. Control 43 (2017) 1–64.

[229] E. Buzi, P. Marango, A comparison of conventional and nonconventional methods of DC motor speed control, IFAC Proc. Vol. 46 (8) (2013) 50–53.

[230] B. Dehghan, B.W. Surgenor, Comparison of fuzzy and neural network adaptive methods for the position control of a pneumatic system, in: 2013 26th IEEE Canadian Conference on Electrical and Computer Engineering (CCECE), 2013, pp. 1–4.

[231] L. Sun, Analysis and comparison of variable structure fuzzy neural network control and the PID algorithm, in: *2017 Chinese Automation Congress (CAC)*, 2017, pp. 3347–3350.

[232] F.-J. Lin, R.-F. Fung, R.-J. Wai, Comparison of sliding-mode and fuzzy neural network control for motor-toggle servomechanism, IEEE/ASME Trans. Mechatronics 3 (4) (1998) 302–318.

[233] R.Y. Adhitya, et al., Comparison methods of fuzzy logic control and feed forward neural network in automatic operating temperature and humidity control system (Oyster Mushroom Farm House) using microcontroller, in: *2016 International Symposium on Electronics and Smart Devices (ISESD)*, 2016, pp. 168–173.

[234] P. Bawane, S. Gadariye, S. Chaturvedi, A.A. Khurshid, Object and character recognition using spiking neural network, Mater. Today Proc. 5 (1) (2018) 360–366.

[235] L. Buşoniu, T. de Bruin, D. Tolić, J. Kober, I. Palunko, Reinforcement learning for control: performance, stability, and deep approximators, Annu. Rev. Control 46 (2018) 8–28.

[236] P.L. Galdámez, W. Raveane, A. González Arrieta, A brief review of the ear recognition process using deep neural networks, J. Appl. Log. 24 (Nov. 2017) 62–70.

[237] J.H. Lee, J. Shin, M.J. Realff, Machine learning: overview of the recent progresses and implications for the process systems engineering field, Comput. Chem. Eng. 114 (2018) 111–121.

[238] W. Yang, et al., Down image recognition based on deep convolutional neural network, Inf. Process. Agric. 5 (2) (Jun. 2018) 246–252.

[239] H. Kazmi, F. Mehmood, S. Lodeweyckx, J. Driesen, Gigawatt-hour scale savings on a budget of zero: deep reinforcement learning based optimal control of hot water systems, Energy 144 (2018) 159–168.

[240] W. Liu, G. Qin, Y. He, F. Jiang, Distributed cooperative reinforcement learning-based traffic signal control that integrates V2X networks' dynamic clustering, IEEE Trans. Veh. Technol. 66 (10) (2017) 8667–8681.

[241] N. Wu, H. Wang, Deep learning adaptive dynamic programming for real time energy management and control strategy of micro-grid, J. Clean. Prod. 204 (Dec. 2018) 1169–1177.

[242] Y. Jiang, C. Yang, J. Na, G. Li, Y. Li, J. Zhong, A brief review of neural networks based learning and control and their applications for robots, Complexity 2017 (2017) 1–14.

[243] J. Zhang, H. Yue, Steady-State Modeling and Control of Molecular Weight Distributions in a Styrene Polymerization Process Based on B-Spline Neural Networks, 2007, pp. 329–338.

[244] L. Cao, D. Li, C. Zhang, H. Wu, Control and modeling of temperature distribution in a tubular polymerization process, Comput. Chem. Eng. 31 (11) (Nov. 2007) 1516–1524.

[245] A.S. Ibrehem, M.A. Hussain, N.M. Ghasem, Mathematical model and advanced control for gas-phase olefin polymerization in fluidized-bed catalytic reactors, Chin. J. Chem. Eng. 16 (1) (2008) 84–89.

[246] C.W. Ng, M.A. Hussain, Hybrid neural network—prior knowledge model in temperature control of a semi-batch polymerization process, Chem. Eng. Process. Process Intensif. 43 (4) (Apr. 2004) 559–570.

[247] D. Xu, W. Yan, N. Ji, RBF neural network based adaptive constrained PID control of a solid oxide fuel cell, in: *2016 Chinese control and decision conference (CCDC)*, 2016, pp. 3986–3991.

[248] C. Lebreton, et al., Fault tolerant control strategy applied to PEMFC water management, Int. J. Hydrogen Energy 40 (33) (Sep. 2015) 10636–10646.

[249] C. Li, X. Zhu, S. Sui, W. Hu, M. Hu, Adaptive inverse control of air supply flow for proton exchange membrane fuel cell systems, J. Shanghai Univ. (English Ed.) 13 (6) (2009) 474.

[250] M. Sedighizadeh, A. Rezazadeh, Adaptive self-tuning wavelet neural network controller for a proton exchange membrane fuel cell, in: Applications of Neural Networks in High Assurance Systems, Springer, 2010, pp. 221–245.

[251] M.A. Shafiq, Direct adaptive inverse control of nonlinear plants using neural networks, in: *2016 Future Technologies Conference (FTC)*, 2016, pp. 827–830.

[252] S.M. Rakhtala, R. Ghaderi, A.R. Noei, Proton exchange membrane fuel cell voltage-tracking using artificial neural networks, J. Zhejiang Univ. Sci. C 12 (4) (2011) 338–344.

[253] A. Rezazadeh, A. Askarzadeh, M. Sedighizadeh, Adaptive inverse control of proton exchange membrane fuel cell using RBF neural network, Int. J. Electrochem. Sci. 6 (2011) 3105–3117.

[254] B.O. Nikhil, A. Visa, C.-Y. Lin, J.A. Puhakka, O. Yli-Harja, An artificial neural network based model for predicting H2 production rates in a sucrose-based bioreactor system, World Acad. Sci. Eng. Technol. 37 (2008) 20–25.

[255] M. Kavchak, H. Budman, Adaptive neural network structures for non-linear process estimation and control, Comput. Chem. Eng. 23 (9) (1999) 1209–1228.

[256] M. Köni, U. Yüzgeç, M. Türker, H. Dinçer, Adaptive neuro-fuzzy-based control of drying of baker's yeast in batch fluidized bed, Drying Technol. 28 (2) (Mar. 2010) 205–213.

[257] J.P. Jordanou, E.A. Antonelo, E. Camponogara, M.A.S. de Aguiar, Recurrent neural network based control of an oil well, in: *Brazilian Symposium on Intelligent Automation, Porto Alegre 1-4 October 2017*, 2017, pp. 924–931.

[258] A. Bahar, C. Özgen, State estimation for a reactive batch distillation column, IFAC Proc. Vol. 41 (2) (2008) 3304–3309.

[259] M. Jamshidian, M. Hadian, M.M. Zadeh, Z. Kazempoor, P. Bazargan, H. Salehi, Prediction of free flowing porosity and permeability based on conventional well logging data using artificial neural networks optimized by imperialist competitive algorithm—a case study in the South Pars Gas field, J. Nat. Gas Sci. Eng. 24 (2015) 89–98.

[260] M. Jamshidian, M. Mansouri Zadeh, M. Hadian, S. Nekoeian, M. Mansouri Zadeh, Estimation of minimum horizontal stress, geomechanical modeling and hybrid neural network based on conventional well logging data—a case study, Geosystem Eng. 20 (2) (2017) 88–103.

[261] M. Jamshidian, M.M. Zadeh, M. Hadian, R. Moghadasi, O. Mohammadzadeh, A novel estimation method for capillary pressure curves based on routine core analysis data using artificial neural networks optimized by Cuckoo algorithm—a case study, Fuel 220 (2018) 363–378.

[262] H. Mingzhi, Y. Ma, W. Jinquan, W. Yan, Simulation of a paper mill wastewater treatment using a fuzzy neural network, Expert Syst. Appl. 36 (3) (Apr. 2009) 5064–5070.

[263] R.D. De Veaux, R. Bain, L.H. Ungar, Hybrid neural network models for environmental process control (The 1998 Hunter Lecture), Environmetrics Off. J. Int. Environmetrics Soc. 10 (3) (1999) 225–236.

[264] H.-G. Han, J.-F. Qiao, Q.-L. Chen, Model predictive control of dissolved oxygen concentration based on a self-organizing RBF neural network, Control Eng. Pract. 20 (4) (2012) 465–476.

[265] Y.-C. Bo, X. Zhang, Online adaptive dynamic programming based on echo state networks for dissolved oxygen control, Appl. Soft Comput. 62 (2018) 830–839.

[266] Z. Xiong, J. Zhang, A batch-to-batch iterative optimal control strategy based on recurrent neural network models, J. Process Control 15 (1) (2005) 11–21.

[267] A. Andrášik, A. Mészáros, S.F. de Azevedo, On-line tuning of a neural PID controller based on plant hybrid modeling, Comput. Chem. Eng. 28 (8) (Jul. 2004) 1499–1509.

[268] D. Dong, T.J. McAvoy, E. Zafiriou, Batch-to-batch optimization using neural network models, Ind. Eng. Chem. Res. 35 (7) (1996) 2269–2276.

[269] J. Tay, A fast predicting neural fuzzy model for high-rate anaerobic wastewater treatment systems, Water Res. 34 (11) (Aug. 2000) 2849–2860.

[270] J.-C. Chen, N.-B. Chang, Mining the fuzzy control rules of aeration in a submerged biofilm wastewater treatment process, Eng. Appl. Artif. Intel. 20 (7) (Oct. 2007) 959–969.

[271] P. Holubar, Advanced controlling of anaerobic digestion by means of hierarchical neural networks, Water Res. 36 (10) (May 2002) 2582–2588.

[272] Z.K. Nagy, Model based control of a yeast fermentation bioreactor using optimally designed artificial neural networks, Chem. Eng. J. 127 (1–3) (2007) 95–109.

[273] P.R. Patnaik, Neural and hybrid neural modeling and control of fed-batch fermentation for streptokinase: comparative evaluation under nonideal conditions, Can. J. Chem. Eng. 82 (3) (2004) 599–606.

[274] A. Sabharwal, W.Y. Svrcek, D.E. Seborg, Hybrid neural net, physical modeling applied to a xylene splitter, IFAC Proc. Vol. 32 (2) (Jul. 1999) 6799–6805.

[275] G. Kimaev, L.A. Ricardez-Sandoval, Nonlinear model predictive control of a multiscale thin film deposition process using artificial neural networks, Chem. Eng. Sci. 207 (2019) 1230–1245.

[276] Z. Wu, P.D. Christofides, Optimizing process economics and operational safety via economic MPC using barrier functions and recurrent neural network models, Chem. Eng. Res. Des. 152 (Dec. 2019) 455–465.

[277] Y. Tian, J. Zhang, J. Morris, Dynamic on-line reoptimization control of a batch MMA polymerization reactor using hybrid neural network models, Chem. Eng. Technol. Ind. Chem. Equip. Process Eng. 27 (9) (2004) 1030–1038.

[278] O.A. Adebiyi, A.B. Corripio, Dynamic neural networks partial least squares (DNNPLS) identification of multivariable processes, Comput. Chem. Eng. 27 (2) (Feb. 2003) 143–155.

[279] P. Cabrera, J.A. Carta, J. González, G. Melián, Artificial neural networks applied to manage the variable operation of a simple seawater reverse osmosis plant, Desalination 416 (Aug. 2017) 140–156.

[280] E.A. Roehl, et al., Modeling fouling in a large RO system with artificial neural networks, J. Membr. Sci. 552 (Apr. 2018) 95–106.

[281] P. He, S. Jagannathan, Reinforcement learning neural-network-based controller for nonlinear discrete-time systems with input constraints, IEEE Trans. Syst. Man, Cybern. Pt B 37 (2) (2007) 425–436.

[282] A. Vasičkaninová, M. Bakošová, A. Mészáros, J.J. Klemeš, Neural network predictive control of a heat exchanger, Appl. Therm. Eng. 31 (13) (Sep. 2011) 2094–2100.

[283] S. Yin, S.X. Ding, A. Haghani, H. Hao, P. Zhang, A comparison study of basic data-driven fault diagnosis and process monitoring methods on the benchmark Tennessee Eastman process, J. Process Control 22 (9) (Oct. 2012) 1567–1581.

[284] B.M. Wise, N.L. Ricker, D.F. Veltkamp, B.R. Kowalski, A theoretical basis for the use of principal component models for monitoring multivariate processes, Process Control Qual. 1 (1) (1990) 41–51.

[285] J.F. MacGregor, C. Jaeckle, C. Kiparissides, M. Koutoudi, Process monitoring and diagnosis by multiblock PLS methods, AIChE J. 40 (5) (1994) 826–838.

[286] L.H. Chiang, E.L. Russell, R.D. Braatz, Fault diagnosis in chemical processes using Fisher discriminant analysis, discriminant partial least squares, and principal component analysis, Chemom. Intel. Lab. Syst. 50 (2) (Mar. 2000) 243–252.

[287] M. Kano, S. Tanaka, S. Hasebe, I. Hashimoto, H. Ohno, Monitoring independent components for fault detection, AIChE J. 49 (4) (2003) 969–976.

[288] L.H. Chiang, M.E. Kotanchek, A.K. Kordon, Fault diagnosis based on Fisher discriminant analysis and support vector machines, Comput. Chem. Eng. 28 (8) (Jul. 2004) 1389–1401.

[289] M.R. Maurya, R. Rengaswamy, V. Venkatasubramanian, Fault diagnosis by qualitative trend analysis of the principal components, Chem. Eng. Res. Des. 83 (9) (Sep. 2005) 1122–1132.

[290] M.R. Maurya, R. Rengaswamy, V. Venkatasubramanian, A signed directed graph and qualitative trend analysis-based framework for incipient fault diagnosis, Chem. Eng. Res. Des. 85 (10) (Jan. 2007) 1407–1422.

[291] Z. Ge, Z. Song, F. Gao, Review of recent research on data-based process monitoring, Ind. Eng. Chem. Res. 52 (10) (Mar. 2013) 3543–3562.

[292] J.-H. Cho, J.-M. Lee, S.W. Choi, D. Lee, I.-B. Lee, Fault identification for process monitoring using kernel principal component analysis, Chem. Eng. Sci. 60 (1) (2005) 279–288.

[293] R. Rosipal, L.J. Trejo, Kernel partial least squares regression in reproducing kernel hilbert space, J. Mach. Learn. Res. 2 (Dec) (2001) 97–123.

[294] S.J. Qin, T.J. McAvoy, Nonlinear PLS modeling using neural networks, Comput. Chem. Eng. 16 (4) (1992) 379–391.

[295] W. Ku, R.H. Storer, C. Georgakis, Disturbance detection and isolation by dynamic principal component analysis, Chemom. Intel. Lab. Syst. 30 (1) (Nov. 1995) 179–196.

[296] M.H. Kaspar, W.H. Ray, Dynamic PLS modelling for process control, Chem. Eng. Sci. 48 (20) (1993) 3447–3461.

[297] Y. Dong, S.J. Qin, Dynamic-inner partial least squares for dynamic data modeling, IFAC-PapersOnLine 48 (8) (2015) 117–122.

[298] W.G. Fenton, T.M. McGinnity, L.P. Maguire, Fault diagnosis of electronic systems using intelligent techniques: a review, IEEE Trans. Syst. Man, Cybern. Pt C (Appl. Rev.) 31 (3) (2001) 269–281.

[299] S. Chen, J. Yu, S. Wang, One-dimensional convolutional auto-encoder-based feature learning for fault diagnosis of multivariate processes, J. Process Control 87 (2020) 54–67.

[300] S. Laribi, K. Mammar, Y. Sahli, K. Koussa, Analysis and diagnosis of PEM fuel cell failure modes (flooding & drying) across the physical parameters of electrochemical impedance model: using neural networks method, Sustain. Energy Technol. Assessments 34 (2019) 35–42.

[301] J. Yuan, Y. Tian, A multiscale feature learning scheme based on deep learning for industrial process monitoring and fault diagnosis, IEEE Access 7 (2019) 151189–151202.

[302] H. Zhao, S. Sun, B. Jin, Sequential fault diagnosis based on LSTM neural network, IEEE Access 6 (2018) 12929–12939.

[303] J.C. Hoskins, K.M. Kaliyur, D.M. Himmelblau, Fault diagnosis in complex chemical plants using artificial neural networks, AIChE J. 37 (1) (1991) 137–141.

[304] D. Ruiz, J. Nougués, Z. Calderón, A. Espuña, L. Puigjaner, Neural network based framework for fault diagnosis in batch chemical plants, Comput. Chem. Eng. 24 (2–7) (Jul. 2000) 777–784.

[305] J. Yuan, Y. Tian, An intelligent fault diagnosis method using GRU neural network towards sequential data in dynamic processes, Processes 7 (3) (2019) 152.

[306] B. Shao, X. Hu, G. Bian, Y. Zhao, A multichannel LSTM-CNN method for fault diagnosis of chemical process, Math. Probl. Eng. 2019 (Dec. 2019) 1–14.

[307] H. Wu, J. Zhao, Deep convolutional neural network model based chemical process fault diagnosis, Comput. Chem. Eng. 115 (2018) 185–197.

[308] J. Kim, I. Lee, Y. Tak, B.H. Cho, State-of-health diagnosis based on hamming neural network using output voltage pattern recognition for a PEM fuel cell, Int. J. Hydrogen Energy 37 (5) (2012) 4280–4289.

[309] R.E. Silva, et al., Proton exchange membrane fuel cell degradation prediction based on adaptive neuro-fuzzy inference systems, Int. J. Hydrogen Energy 39 (21) (2014) 11128–11144.

[310] M. Shao, X.-J. Zhu, H.-F. Cao, H.-F. Shen, An artificial neural network ensemble method for fault diagnosis of proton exchange membrane fuel cell system, Energy 67 (2014) 268–275.

[311] N.Y. Steiner, D. Hissel, P. Moçotéguy, D. Candusso, Diagnosis of polymer electrolyte fuel cells failure modes (flooding & drying out) by neural networks modeling, Int. J. Hydrogen Energy 36 (4) (2011) 3067–3075.

[312] Q. Tang, Y. Chai, J. Qu, H. Ren, Fisher discriminative sparse representation based on DBN for fault diagnosis of complex system, Appl. Sci. 8 (5) (2018) 795.

[313] G.-W. You, S. Park, D. Oh, Diagnosis of electric vehicle batteries using recurrent neural networks, IEEE Trans. Ind. Electron. 64 (6) (Jun. 2017) 4885–4893.

[314] S. Tayyebi, R. Bozorgmehry Boozarjomehry, M. Shahrokhi, Neuromorphic multiple-fault diagnosing system based on plant dynamic characteristics, Ind. Eng. Chem. Res. 52 (36) (Sep. 2013) 12927–12936.

[315] J. Liu, Q. Li, H. Yang, Y. Han, S. Jiang, W. Chen, Sequence fault diagnosis for PEMFC water management subsystem using deep learning with t-SNE, IEEE Access 7 (2019) 92009–92019.

[316] H. Wu, J. Zhao, Fault detection and diagnosis based on transfer learning for multimode chemical processes, Comput. Chem. Eng. 135 (Apr. 2020) 106731.

[317] S. Zhang, K. Bi, T. Qiu, Bidirectional recurrent neural network-based chemical process fault diagnosis, Ind. Eng. Chem. Res. 59 (2) (Jan. 2020) 824–834.

[318] M. Gholizadeh, A. Yazdizadeh, H. Mohammad-Bagherpour, Fault detection and identification using combination of ekf and neuro-fuzzy network applied to a chemical process (cstr), Pattern Anal. Appl. 22 (2) (2019) 359–373.

[319] X. Yang, J. Zhou, Z. Xie, G. Ke, Chemical process fault diagnosis based on enchanted machine-learning approach, Can. J. Chem. Eng. 97 (12) (2019) 3074–3086.

[320] H. Zhao, Z. Lai, Y. Chen, Global-and-local-structure-based neural network for fault detection, Neural Netw. 118 (Oct. 2019) 43–53.

[321] C. Li, D. Zhao, S. Mu, W. Zhang, N. Shi, L. Li, Fault diagnosis for distillation process based on CNN–DAE, Chin. J. Chem. Eng. 27 (3) (Mar. 2019) 598–604.

[322] Z. Zhang, J. Zhao, A deep belief network based fault diagnosis model for complex chemical processes, Comput. Chem. Eng. 107 (Dec. 2017) 395–407.

[323] Z. Hu, P. Jiang, An imbalance modified deep neural network with dynamical incremental learning for chemical fault diagnosis, IEEE Trans. Ind. Electron. 66 (1) (Jan. 2019) 540–550.

[324] N. Md Nor, M.A. Hussain, C.R. Che Hassan, Multi-scale kernel Fisher discriminant analysis with adaptive neuro-fuzzy inference system (ANFIS) in fault detection and diagnosis framework for chemical process systems, Neural Comput. Applic. 32 (13) (Jul. 2020) 9283–9297.

[325] X. Ren, Y. Zou, Z. Zhang, Fault detection and classification with feature representation based on deep residual convolutional neural network, J. Chemometr. 33 (9) (Sep. 2019).

[326] P. Jiang, Z. Hu, J. Liu, S. Yu, F. Wu, Fault diagnosis based on chemical sensor data with an active deep neural network, Sensors 16 (10) (Oct. 2016) 1695.

[327] L.A. Zadeh, Fuzzy sets, Infect. Control 8 (3) (1965) 338–353.

[328] E.H. Mamdani, Application of fuzzy algorithms for control of simple dynamic plant, Proc. Inst. Electr. Eng. 121 (12) (1974) 1585–1588.

[329] L.A. Zadeh, Fuzzy logic, Computer (Long. Beach. Calif.) 21 (4) (1988) 83–93.

[330] E. Pathmanathan, R. Ibrahim, Development and implementation of fuzzy logic controller for flow control application, in: 2010 International Conference on Intelligent and Advanced Systems, 2010, pp. 1–6.

[331] B.V. Murthy, Y.V.P. Kumar, U.V.R. Kumari, Fuzzy logic intelligent controlling concepts in industrial furnace temperature process control, in: *2012 IEEE International Conference on Advanced Communication Control and Computing Technologies (ICACCCT)*, 2012, pp. 353–358.

[332] A. Aarabi, M. Shahbazian, M. Hadian, Improved closed loop performance and control signal using evolutionary algorithms based PID controller, in: Proceedings of the 2015 16th International Carpathian Control Conference (ICCC), 2015, pp. 1–6.

[333] A.N. Fard, M. Shahbazian, M. Hadian, Adaptive fuzzy controller based on cuckoo optimization algorithm for a distillation column, in: *2016 International Conference on Computational Intelligence and Applications (ICCIA)*, 2016, pp. 93–97.

[334] M. Miccio, B. Cosenza, Control of a distillation column by type-2 and type-1 fuzzy logic PID controllers, J. Process Control 24 (5) (2014) 475–484.

[335] A.M.F. Fileti, A.J.B. Antunes, F.V. Silva, V. Silveira, J.A.F.R. Pereira, Experimental investigations on fuzzy logic for process control, Control Eng. Pract. 15 (9) (Sep. 2007) 1149–1160.

[336] S. Dash, R. Rengaswamy, V. Venkatasubramanian, Fuzzy-logic based trend classification for fault diagnosis of chemical processes, Comput. Chem. Eng. 27 (3) (2003) 347–362.

[337] H. Ma, Q. Zhou, L. Bai, H. Liang, Observer-based adaptive fuzzy fault-tolerant control for stochastic nonstrict-feedback nonlinear systems with input quantization, IEEE Trans. Syst. Man, Cybern. Syst. 49 (2) (2018) 287–298.

[338] S.S. Rodil, M.J. Fuente, Fault tolerance in the framework of support vector machines based model predictive control, Eng. Appl. Artif. Intel. 23 (7) (2010) 1127–1139.

[339] Z. Yin, J. Hou, Recent advances on SVM based fault diagnosis and process monitoring in complicated industrial processes, Neurocomputing 174 (2016) 643–650.

[340] J. MacGregor, A. Cinar, Monitoring, fault diagnosis, fault-tolerant control and optimization: data driven methods, Comput. Chem. Eng. 47 (2012) 111–120.

[341] N. Pooyan, M. Shahbazian, M. Hadian, Simultaneous fault diagnosis using multi class support vector machine in a dew point process, J. Nat. Gas Sci. Eng. 23 (2015) 373–379.

[342] Y.-J. Lee, O.L. Mangasarian, RSVM: reduced support vector machines, in: *Proceedings of the 2001 SIAM International Conference on Data Mining*, 2001, pp. 1–17.

[343] C.J.C. Burges, A tutorial on support vector machines for pattern recognition, Data Min. Knowl. Discov. 2 (2) (1998) 121–167.

[344] H. Byun, S.-W. Lee, Applications of support vector machines for pattern recognition: a survey, in: *International Workshop on Support Vector Machines*, 2002, pp. 213–236.

Chapter 12

Integrated machine learning framework for computer-aided chemical product design

Qilei Liu, Haitao Mao, Lei Zhang, Linlin Liu, and Jian Du
Institute of Chemical Process System Engineering, School of Chemical Engineering, Dalian University of Technology, Dalian, China

1 Introduction

Modern society requires many chemical-based products for its survival, such as fuels, plastics, detergents, medicines, and food [1]. As the increasing requirements for sustainable and human-centered developments, chemical product design has been paid wide attentions from the scientific communities and chemical industries. Traditionally, chemical products are designed and developed through heuristic rule-based and/or trial-and-error experiment-based approaches. Although these kinds of approaches often lead to safe and reliable product designs, it is not practically feasible to evaluate all alternatives or to obtain the optimal solution. Recently, the use of model-based design methods has been gaining increased attention as they have the potential to generate and/or screen feasible product candidates in a much larger design space, and at the same time, reduce the time and costs for their development [2]. If required data and the models giving reliable estimations for product properties and functions are available, it is possible to develop and use model-based chemical product design approaches with the advances in computer-aided technologies. During last three decades, efforts have been made to develop databases, design methods and associated software tools for chemical products. Review articles for the design methods of chemical products can be found in Zhang et al. [1–3]. Although good progress has been made in model-based approaches for chemical product design, due to the multidisciplinary and multiscale nature of chemical products, challenges still exist which impede the development and application of chemical product design methodologies. The multidisciplinary and multiscale nature of chemical product design makes it difficult to understand basic scientific issues of chemical products, which hinders the

Applications of Artificial Intelligence in Process Systems Engineering
https://doi.org/10.1016/B978-0-12-821092-5.00004-8

establishment of reliable and quantitative physicochemical property models. Thus, these hurdles prevent the use of model-based approaches for the design of chemical products. Nowadays, with the development of the data-driven methods, machine learning (ML) has been regarded as an alternative solution to the above-mentioned challenges.

ML has experienced successful resurgence during the last decades. The explosive growth of data with sophisticated ML algorithms make it possible to establish the ML models, which show promising potentials in applications of chemistry and chemical industry [5, 6]. Nowadays, the fundamental paradigm of statistical analysis has changed from system identification to predictive modeling. ML can model complex chemical properties due to its abilities in autonomously learning data characteristics and trends [7]. Because of the adoptable generalization power, the established ML can be used in more general scenarios and thus has abilities in designing chemical products. Moreover, increasing efforts have been made to interpret ML models, for obtaining the insights of ML from basic scientific issues to chemical products. As a result, ML is possible to be one of the major research trends in chemical product design. Nowadays, ML methods have been widely applied in different aspects of chemical product design. One of the most important roles of ML methods is establishing the quantitative structure-property relationships (QSPRs). Growing number of new potentially useful ML methods in chemistry, such as the use of artificial neural networks (ANNs) [8], in quantitative structure-activity relationships (QSARs) [9] and in ligand-based virtual screening [10]. The ML ability to predict key properties of a product has been highlighted by Zonouz et al. [11] who developed an ANN coupled with a genetic algorithm for the modeling and optimization of toluene oxidation over perovskite-type nano-catalysts. In the area of crystallization, Velásco-Mejía et al. [12] employed ANN combined with a genetic algorithm in modeling and optimization of process design. Liu et al. [13] outlined the typical modes and basic procedures for applying ML in materials science. Pankajakshan et al. [14] provided conceptual scheme to obtain chemical insights into complex phenomena and development of predictive models for material design. Similarly, Vanhaelen et al. [15] emphasized ML-based workflow in drug design.

In this chapter, our recent works on ML-based chemical product design (focus on molecular product design) are presented. In Section 2, an overall ML-CAMD framework is presented. In Section 3, the ML-CAMD framework is discussed in detail for the establishment of ML models for property prediction as well as chemical product design. In Section 4, two case studies are presented for the applications of the proposed framework.

2 An integrated ML framework for computer-aided molecular design

The structure-property relationships are essential in property prediction and chemical product design as these relationships serve as the "bridge" between the molecular structure/product constitution and the desired property. In this section, an overall ML-CAMD framework is presented to assist in the establishment of the ML models for the missing structure-property relationships for property prediction and chemical product design. The diagrammatic sketch of the ML-CAMD framework is shown in Fig. 1, which consists of four steps: data collection, data preprocessing and feature engineering, model establishment, and chemical product design.

In the data collection step, a set of chemical products (molecules, mixtures, blends, etc.) as well as their product descriptors (groups, compositions, etc.) and known properties from experiments, literatures, etc. are organized as a database for the establishment of the ML models. Then, the established database is sent to the data preprocessing and feature engineering step to eliminate the abnormal values, complement the missing data and avoid high computation load and overfitting issues in model establishment. Finally, the ML model is established for property prediction through model selection, training and validation, as well as chemical product design through efficient mathematical optimization algorithms.

3 Establishment of ML model for computer-aided molecular design

In this section, the steps in the proposed ML-CAMD framework are applied to establish the ML models for computer-aided molecular design (CAMD). These steps are discussed in detail in the following content.

3.1 Data collection

Create database. It is essential to establish a database before establishing ML models. Many molecular databases have already been available online, such as NIST Chemistry webbook (www.webbook.nist.gov/chemistry/), PubChem (www.pubchem.ncbi.nlm.nih.gov), ZINC (www.zinc.docking.org), DRUG-BANK (www.drugbank.ca), ChEMBL (https://www.ebi.ac.uk/chembl/), and ProCAPD database (www.pseforspeed.com/procapd), etc. For the design of different types of products, separate databases with their separate ontologies are required. For examples, separate databases are needed for solvents, aroma compounds, active ingredients for different types of functional products, refrigerants, membranes, adsorbents, and many more. Therefore, the elements (molecules, ingredients, etc.) need to be carefully selected in the database to establish a balanced tailor-made ML model between generalization and accuracy for a fixed type of product.

Select product descriptors. The data in the database contain descriptors (e.g., groups for molecules, ingredient mole fractions for mixtures, bonding-based graphical representation for crystals). Descriptor is one of the key factors to determine the performance of ML models. The selection of descriptors varies

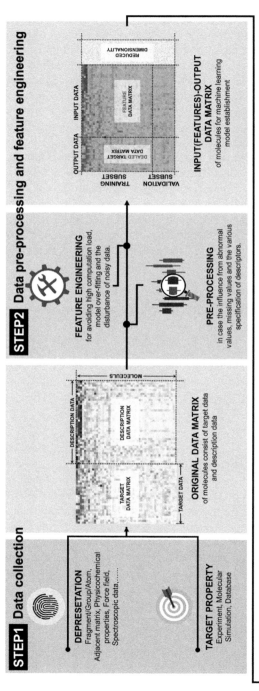
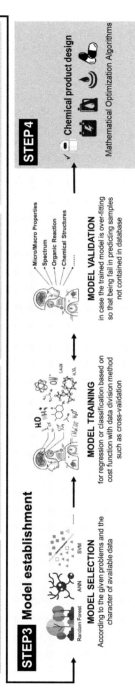

FIG. 1 The diagrammatic sketch of the ML-CAMD framework.

TABLE 1 Commonly used descriptors in chemical product design problems.

Categories	Descriptors	Application examples
Practical values	• Experimental physicochemical descriptors • ...	• Design of catalysts [16]
Groups	• Functional groups (CH_3, CH_2, OH, ...) • Adjacency matrix • ...	• Prediction of pH [17] • Estimation of cetane and octane numbers [18] • Odor prediction [4]
Chemoinformatic	• Constitutional descriptors • Topological descriptors • Physicochemical descriptors • Structural descriptors • ...	• Drug design [19] • Material design [20]
Molecular fingerprints	• Hashed fingerprint [21] • Extended-connectivity fingerprints [22] • ...	• Organic chemistry reaction prediction [23]
Molecular spectrum	• IR, UR • NMR, Raman • Material image • ...	• Material property prediction [24]
Computational chemistry	• Initial nuclear velocities • Chemical environment • Molecular orbitals • ...	• Catalyst prediction [25] • Ab initio molecular dynamics simulations [26] • Predict chemical reactions [27]
Graph convolution	• 2D molecular graph • ...	• Drug efficacy and photovoltaic prediction [28] • Properties of different biological classes [29]
Text description	• SMILES [30] • SMARTS [21] • ...	• Drug design [31]

with the product types and design problems. Table 1 shows some commonly used descriptors in chemical product design problems.

The advantages and disadvantages of different descriptors are discussed as follows:

➤ *Practical values*: The advantage is that practical values are accurate to measure molecular physicochemical properties, and thereby leading to reliable ML-based property prediction models. The disadvantage is that practical values are not easily accessible.

➢ *Groups*: The advantage is that groups are structural descriptors, which is easily obtained with a minor computational cost. However, the definitions of reasonable groups are highly dependent on expert knowledge.

➢ *Chemoinformatic*: The advantage is that chemoinformatic has a large number of structural and property descriptors and is easily available through theoretical calculations. The disadvantage is that chemoinformatic descriptors have the issue of inconsistent dimensions, which may result in poor training results for ML models.

➢ *Molecular fingerprints*: The advantage is that fingerprints contain molecular topological information and is easily available through theoretical calculations. The disadvantage is that fingerprints are 2-dimensional descriptors, and they cannot correlate 3-dimensional properties, for example, molecular docking calculations between drugs and target proteins.

➢ *Molecular spectrum*: The advantage is that molecular spectrum is able to represent molecular features and integrate with graph neural networks. The disadvantage is that molecular spectra are not easily accessible.

➢ *Computational chemistry*: The advantage is that computational chemistry descriptors are pseudo experimental data and is obtained by theoretical calculations. However, the high computational cost for computational chemistry descriptors hinders the high-throughput design and/or selection of chemical products.

➢ *Graph convolution*: The advantage is that graph convolution is able to summarize the local chemical environments of atoms and intramolecular connectivity, which is suitable for establishing graph neural networks. The disadvantage is that they have difficulties in representing stereoscopic characteristics of molecules.

➢ *Text description*: The advantage is that text descriptors have a strong expansibility and wide applicability for ML modeling. The disadvantage is that text descriptors (e.g., SMILES) cannot represent conformational isomers.

Collect product properties. The data in the database include product properties (e.g., normal boiling point, solubility parameter, odor, color). The property data can be collected from experiments, literatures, and molecular databases. If it is hard to perform experiments or the data are unavailable in the literatures/databases, computation/simulation tools are alternatives to complementing the missing data.

3.2 Data preprocessing and feature engineering

With the accumulation of historical data and the development of computation/simulation tools, the data available for ML is often sufficient. However, these data are not always valid when directly applied in the establishment of ML models. On the one hand, systematic errors (noises) are inevitable during experiments. On the other hand, issues such as inconsistent dimensions, inconsistent order of magnitude, high dimensionality, and irrelevant or redundant data often makes poor training results of the ML models. Therefore, it is necessary to employ data preprocessing and feature engineering methods to process the original data. Note that ML also includes deep learning without feature engineering, which is not further discussed in this chapter. More details about deep learning can be found in LeCun's work [32].

Data preprocessing is essential for the experimental data (errors are often caused by human or equipment, as well as disturbances from the environment) and the issues of inconsistent dimensions and inconsistent order of magnitude. For chemical product design problems, one of the issues is the required descriptors and/or the product properties are not always available in the database. The abnormal and missing data of the descriptors and/or properties can be replaced by the mean, median, or the most frequent values based on different preprocessing methods. These data preprocessing methods are summarized in Table 2.

Feature engineering is an indispensable technique to identify the irrelevant, redundant, and noisy data from the raw data (high dimensional data) and convert them to new features (low dimensional data), which retain the prominent characteristics of a system and contribute to the efficient learning process of ML models. For chemical product design problems, if the collected molecular descriptors (raw data) are high dimensional, it is necessary to perform feature engineering before ML modeling. Feature engineering generally consists of two methods, namely feature extraction and feature selection.

Feature extraction, for example, principal components analysis (PCA), is able to extract critical information from the original feature space (raw data) and thereby reduce the data dimension. This technique is commonly used in image processing [33] and speaker recognition [34]. For chemical product design problems, feature extraction is popular with spectra recognitions, for example, proteomic mass spectra [35] and nuclear magnetic resonance [36], as spectra are suitable descriptors to represent unique molecular structures. After feature extraction, the significant features in high dimensional spectra are identified and further associated with product properties through ML modeling.

Different from feature extraction, feature selection is a process to remove irrelevant, redundant, and noisy descriptors from the original feature space and adopt the rest as essential features for ML modeling. This technique is more promising for the estimation of product properties as the selected descriptors retain physical significance, which leads to an interpretability of ML models. Feature selection generally has three categories: (1) Filter: select descriptors as features without any ML involved; (2) Wrapper: employ a ML model to evaluate the selected features; (3) Embedding: combine the feature selection and the training process of ML model. Detailed advantages and disadvantages of the above three techniques and their corresponding specific methods are summarized in Table 3.

TABLE 2 Summary of data preprocessing methods.

	Function	Description	Advantage	Disadvantage	Formula
Standardization scaler	Nondimensionalization	Scale using the mean and standard values of one descriptor.	Remove the dimension and retain the original distribution characteristics of the data.	N/A	$x^* = \frac{x - \bar{x}}{s}$ $\bar{x} = \frac{1}{n}\sum_{i=1}^{n} x_i$ $s^2 = \frac{1}{n}\sum_{i=1}^{n}(x_i - \bar{x})^2$
Interval scaler	Nondimensionalization	Scale based on the maximum and minimum values of one descriptor.	Scale flexibly according to the data.	Affected if the distribution is not uniform severely.	$x^* = \frac{x - \min}{\max - \min}$ $\max = \mathrm{Max}\{x_1, x_2, \dots, x_n\}$ $\min = \mathrm{Min}\{x_1, x_2, \dots, x_n\}$
Normalization	Compute the similarity among samples	Calculate the p-norm for each sample and scale the data of a sample by dividing the corresponding norm.	A uniform standard is obtained by processing data of samples within the same row into unit vector.	The difference of dimensions in units of descriptors remains.	$x^* = \frac{x}{\left(\sum_i x_i^p\right)^{\frac{1}{p}}}$
Binarization	Process the continuous descriptors into the categorical values for classification.	Transform the values of descriptors into binary variety.	N/A	N/A	$x^* = \begin{cases} 1, & x > threshold \\ 0, & x \le threshold \end{cases}$

Encoding categorical features	Process the descriptors not being given as continuous values but categorical.	Transform categorical features with n possible values into n binary features.	N/A	The dimension might be enormous if the values of descriptors are widely distributed.	$X = [x_1 \ ... \ x_n]$ $X^* = I_{n \times n} = f([x_1 \ ... \ x_n])$
Generating polynomial features	Add complexity to the model by considering nonlinear features of the input data.	Calculate the descriptors' high-order and interaction terms as a supplement.	More feasible for modeling complexity relationship linking descriptors and properties.	Higher computational complexity during modelling the machine learning.	$X = [x_1, x_2]$ $X^* = [1, x_1, x_2, x_1 x_2, x_1^2, x_2^2]$
Inferring them from the known part of the data	Imputation of missing values	Replace missing values, either using the mean, the median or the most frequent value of the descriptor the missing values are located.	Enable the model to be formulated without the hindrance from missing data.	Affected by the distribution of data severely.	N/A

TABLE 3 Advantages and disadvantages of filter, wrapper, and embedding techniques and their specific methods.

Technique	Filter		Wrapper		Embedding	
Common advantage	√ Easily implement with a certain evaluation measure. √ Flexibly control the result using the criteria. √ Independent of the classifier.		√ Interacts with the learning machine. √ Model feature dependencies.		√ Better computationally complexity than wrapper. √ Interacts with the learning machine. √ Model feature dependencies.	
Common disadvantage	× Ignoring the interaction with the model. × Fail to distinguish the redundant descriptors.		× Computationally intractable. × Overfitting risk. × Model dependent selection.		× Model dependent selection.	
Specific method	• Univariate	• Multivariate	• Deterministic	• Randomized	• Nested subset methods	• Direct objective optimization
Detailed advantage	✓ Fast and Scalable	✓ Models feature dependencies	✓ Less computationally intensive ✓ Less overfitting risk	✓ Less prone to local optima	✓ Less computationally complexity	✓ Less computationally intensive
Detailed disadvantage	× Fail to distinguish the redundant descriptors	× Slower and less scalable than univariate.	× Prone to local optima	× Higher overfitting risk × Higher computationally intensive	× Higher computationally intensive	× Higher computationally complexity

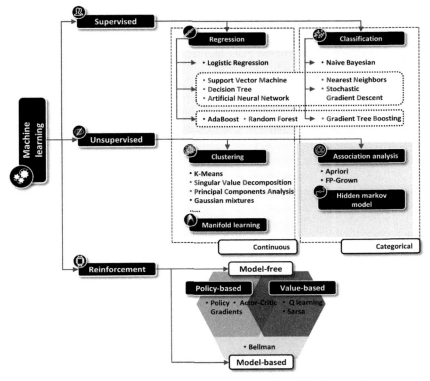

FIG. 2 The classification architecture of ML models and their application scenarios.

3.3   Model establishment

After data collection, preprocessing, and feature engineering, the features are generated and prepared for the establishment of ML model, which includes model selection, model training, and validation process.

Model selection. The model selection depends on the application scenarios, which are generally categorized into supervised learning, unsupervised learning and reinforcement learning (RL). Fig. 2 shows the commonly used ML models and their application scenarios.

Supervised learning is employed to solve two problems of regression and classification, while unsupervised learning is used to solve the problems such as clustering and association analysis. Note that the purpose of unsupervised learning is similar to dimensionality reduction, which aims to discover the potential relationships among variables. RL is the process of training the model through sequences of state-action pairs, observing the rewards that result, and adapting the model predictions to those rewards using policy iteration or value iteration until it accurately predicts the best results. For chemical product design problems, supervised learning is a preferred method for the estimations of product properties as both continuous and discrete properties are predicted by the descriptors through regression and classification methods, respectively. Thus, supervised learning is further discussed in the following.

Currently, three major ML algorithms of supervised learning, support vector machine (SVM), decision tree (DT)/random forest (RF), and ANN, are popular with scientific research and industry practice. These algorithms have been proved possessing excellent abilities in prediction and classification. Table 4 lists several improvements and wide applications of the above ML algorithms of supervised learning.

Model training and validation process. After the type of supervised learning algorithms is confirmed, it could be built/trained by following pseudo Eqs. (1), (2).

$$\min f_{loss}(\mathbf{p}^{pre}, \mathbf{p}^{tar}) \tag{1}$$

$$\mathbf{p}^{pre} = F(\mathbf{D}, \mathbf{P}) \tag{2}$$

where f_{loss} is the loss function to quantify the difference between the prediction outputs $\mathbf{p}^{pre}$ and the target outputs $\mathbf{p}^{tar}$ (e.g., mean squared error (MSE) for prediction problems or cross entropy (CE) for classification problems), $\mathbf{D}$ is the input dataset (i.e., generated features after data preprocessing and feature engineering), $\mathbf{P}$ are the set of hyperparameters (e.g., number of hidden layers in ANN) and parameters (e.g., weights and biases in ANN) for the ML model, and F is the optimization algorithm (e.g., adaptive moment estimation (Adam) algorithm [61]) which enables the model to "learn" the relationships between inputs and outputs. The training process is actually to employ F to minimize f_{loss}. Here, the used dataset is called the training dataset.

A diagrammatic sketch of establishing a ML model is shown in Fig. 3. A well-trained ML model is always unacceptable if large prediction errors are identified among new samples, which is called generalization error or overfitting. To avoid this problem, the original dataset is divided into three subsets, namely training, validation, and testing datasets. Cross-validation methods (e.g., K-folds cross-validation) [67] are usually employed to select the training and validation datasets.

The training dataset is used to train the model and tune the ML parameters (e.g., weights and biases in ANN). If the trained ML model has poor predictions on the validation dataset, hyperparameters need to be adjusted based on knowledge or systematical methods (e.g., the grid method or the random method [7]) to prevent the overfitting problem. During the training process, introducing dropout [68] and regularization [69] layers also contributes to

TABLE 4 Abstract of typical supervised learning algorithms: remarkable improvement, advantages, disadvantages and applications.

Algorithm	Remarkable improvement	Advantage	Disadvantage	Application
Support vector machine	1. Soft margin [37]. 2. Support vector regression [38]. 3. Kernel function [39]. 4. Multiclass classification [40].	√ Separating linear indivisible problem. √ The global optimal can be found. √ Good at the classification of small sample.	× Difficult to solve problems with large. × Hard to determine a suitable kernel function. × SVM performances poor in multiclass problem.	• Prediction of viscosity [41]. • Classification of fragrance properties [42]
Decision tree/random forest	1. Information entropy [43]. 2. Gini index [44]. 3. Pruning [45]. 4. Multivariate decision tree [46]. 5. Random tree [47].	**Decision tree** √ Comprehensible. √ Easy to construct. **Random forest** √ The error rate is significantly reduced. √ Less risk to overfitting.	**Decision tree** × Liable to be overfitting otherwise less accuracy. × Possible combination of features explosively increases. **Random forest** × Complex and time-consuming to construct. × Less intuitive.	**Decision tree** • Rate constant prediction [48]. **Random forest** • Microkinetics models [49]. • Prediction of toxicity [50]

TABLE 4 Abstract of typical supervised learning algorithms: remarkable improvement, advantages, disadvantages and applications—cont'd

Algorithm	Remarkable improvement	Advantage	Disadvantage	Application
Artificial neural network	**Categories of neural network** 1. Radial basis function network [51]. 2. Recurrent neural network [52]. 3. Extremely learning machine [53]. 4. Deep belief network [54]. 5. Transfer learning [55]. 6. Convolution neural network [56].	√ It is able to approximate any function, regardless of its linearity. √ Great for complex/abstract problems. √ It is able to significantly out-perform other models when the conditions are right (lots of high quality labeled data). √ Being robust to an unstable environment due to its adaptability. √ Capable of dealing big data and extracting features, based on which a high-generalization model is formulated.	**For both** ✕ It is limited by the application. ✕ Unsuitable in cases where simpler solutions like linear regression would be best. ✕ Requires a spate of training data. ✕ Increasing accuracy by a few of percent need to bump up the scale by several magnitudes. ✕ Computationally intensive and expensive. ✕ No uniform cognition about how to configure the model or tune the parameters. ✕ Hard to interpret the model though the data of each neuron layer could be obtained.	**Prediction to mixture** • Ternary mixture [62] • Binary mixtures [63] **Design and optimization of chemical product** • Catalysts [16] • Crystallization process [12] **Identification the structure of chemicals** • Crystal [64] • Material [14] • Catalyst [65] **Prediction to the properties involving interaction between molecules**

7. Generative adversarial networks [57]. **Techniques for neural network** 1. Regularization 2. Back-propagation algorithm [52]. 3. Bagging [58]. 4. Dropout [59]. 5. RMSProp [60]. 6. Adam [61].	• Catalyst [66] • Partition coefficient **Predict bio-chemical properties** • Fragrance [3, 4]

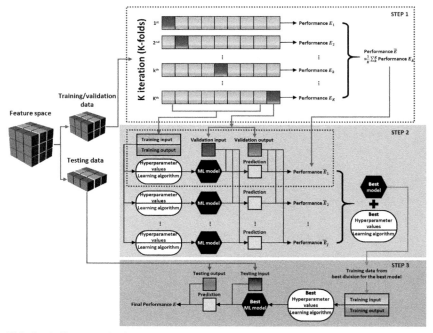

FIG. 3   A diagrammatic sketch of establishing a ML model.

avoiding the overfitting problems. Afterwards, the testing dataset is used to make a final test on the prediction accuracy and generalization ability of the ML model.

3.4   Chemical product design

In this chapter, the chemical product design problems focus on the CAMD problems. With the obtained ML models, the CAMD problem is formulated as a mixed-integer nonlinear programming (MINLP) model as follows:

$$\max/\min_{n_i(i\in\{G\})} F_{obj}(n_i) \text{ (according to specific CAMD problems)}$$

Subject to

- Molecular structure constraints: octet rule, valence bond rule and complexity constraints (linear part).

$$f(n_i) = 0, \quad n_i \in \mathbb{N}^+ \tag{3}$$

- Chemical product property constraints: GC-based properties T_m, T_b, δ, $-\log(LC_{50})FM$, μ, etc. that are calculated by the GC methods (linear part).

$$P_k^L \leq p_k(n_i) \leq P_k^U \tag{4}$$

- Structural descriptor conversion constraints: convert the functional group sets to other molecular representations (e.g., SMILES, fingerprints, etc.) (nonlinear part).

$$f_{conv}\left(n_i, mol_{rep}\right) = 0 \qquad (5)$$

- Chemical product property constraints: ML-based properties that are predicted by the ML models (nonlinear part).

$$p_{k,ML} = f_{ML}\left(mol_{rep}\right) \qquad (6)$$

The ML models generally consist of nonlinear equations, which make it difficult to search for the optimal solution with other constraints simultaneously. So, a decomposition-based solution strategy [70] is used to decompose the MINLP model into an ordered set of four subproblems:

Subproblem (1): Molecular structure constraints and GC-based property constraints are first restricted to design a certain number (N_1) of feasible molecular candidates (group sets) from an ergodic combination of all functional groups by mathematical programming method.

Subproblem (2): Structural descriptor conversion constraints are then used to generate N_2 molecular representations (e.g., SMILES naming molecules) based on group sets.

Subproblem (3): Based on the molecular representations, ML models are employed to fast predict ML-based chemical product properties.

Subproblem (4): Rank the designed products based on the objective function, and a portion of top products are further verified by database, rigorous models, and/or experiments.

4 Case studies

4.1 A ML-based atom contribution methodology for the prediction of charge density profiles and crystallization solvent design

This case study refers to our previous work [83]. In this work, an optimization-based ML-CAMD framework is established for crystallization solvent design, where a novel ML-based atom contribution (MLAC) methodology is developed to correlate the weighted atom-centered symmetry functions (wACSFs) with the atomic surface charge density profiles (atomic σ-profiles, $p_{atom}(\sigma)$) using a high-dimensional neural network (HDNN) model (a kind of ML model), successfully leading to a high prediction accuracy in molecular σ-profiles ($p(\sigma)$) and an ability of isomer identification. Then, the MLAC methodology is integrated with the CAMD problem for crystallization solvent design by formulating and solving a MINLP model, where model complexities are managed with a decomposition-based solution strategy.

- **Data collection**

Create database. A $p_{atom}(\sigma)$ database is prepared for the construction of the HDNN model, where 1120 solvents containing H, C, N, O elements are collected from the Virginia Tech database [71]. Note that the samples are atoms in each solvent.

Select product descriptors. The 3-dimensional atomic descriptors, wACSFs [72], are employed to establish the HDNN model for $p_{atom}(\sigma)$ predictions. The wACSFs represent the local atomic environment of a centered atom i via the functions of radial (G_i^{rad}) and angular (G_i^{ang}) distributions of the surrounding atoms inside a cutoff sphere, which is able to describe the complex intramolecular interactions. Besides, the wACSFs are numerically calculated from the stereoscopic cartesian coordinates, and therefore are able to identify isomers (specifically, all constitutional and cis-trans isomers). More detailed information about the wACSFs can be found in Gastegger et al.'s work [72].

Collect product properties. The $p_{atom}(\sigma)$ of all solvents (product properties) are prepared with the Gaussian 09W software (http://www.gaussian.com/) and the conductor like screening model—segment activity coefficient (COSMO-SAC) model [73]. The reason for selecting $p_{atom}(\sigma)$ as the HDNN output rather than molecular $p(\sigma)$ is that the number of output samples needs to be consistent with the number of input samples (i.e., atoms) when establishing ML models. Finally, a database of $p_{atom}(\sigma)$ is established as the outputs of HDNN model, where the number of samples for H, C, N, O atoms is 15,535, 9108, 305, and 1215, respectively. In this work, the solid-liquid equilibrium (SLE) behavior in cooling crystallization process is predicted with the solvent property activity coefficients γ, which is further predicted by the COSMO-SAC model using the solvent $p(\sigma)$ (a linear addition of $p_{atom}(\sigma)$ in each solvent).

- **Data preprocessing and feature engineering**

Inconsistent order of magnitude exists among the functions of radial (G_i^{rad}) and angular (G_i^{ang}) distributions in the wACSFs, which makes poor training results of the ML models. Therefore, a standardization method is used for data preprocessing. Considering that the wACSFs are numerically calculated from the stereoscopic cartesian coordinates and are all significant in representing the local atomic environments, there is no need to employ feature engineering techniques to identify the irrelevant, redundant and noisy data from the wACSFs.

- **Model establishment**

With the obtained input (wACSFs) and output data ($p_{atom}(\sigma)$), a HDNN model (a kind of ML model) is established.

Model selection.

The HDNN model is made up of four separate element-based (H, C, N, O) ANNs. ANN is selected to correlate the wACSFs with $p_{atom}(\sigma)$ due to the following reasons:

√ ANN has a strong ability to fit complex nonlinear relationships among data, which provides an opportunity to correlate the wACSFs with $p_{atom}(\sigma)$ since their relationships are complex, and traditional linear/nonlinear fitting methods usually fail to account for such relationships.

√ ANN is also an efficient surrogate model with high-throughput calculation speed, which is suitable for an efficient ML-CAMD framework for solvent design.

Model training and validation process.

The molecular $p(\sigma)$ is a sum of $p_{atom}(\sigma)$ that are predicted by the HDNN model. A diagrammatic sketch of HDNN model is shown in Fig. 4. In each ANN model, the optimizer, loss function, metrics function, and activation function are Adam [61], MSE, coefficient of determination R^2 and ReLu [74], respectively. For each element (e.g., H element), atom samples are randomly divided into the training, validation, and test sample, the radios of which are 6:1:1. The size of the input vector (wACSFs) and output vector ($p_{atom}(\sigma)$) are 152 and 51, respectively. To ensure $p_{atom}(\sigma)$ nonnegative, a ReLu activation function ($f(x) = \max(0,x)$) is added to the output layer. Dropout layers are also added to the hidden layers to overcome the overfitting problem in the training process [68]. The hyperparameters (hidden layer number, neuron number in each layer, epochs, batch size, dropout ratio) are determined (as shown in Table 5) by the empirical knowledge to ensure each element-based ANN model possess good generalization (extrapolation) ability while keeping concise. The HDNN model is established on the Keras platform [75] using the Python language [76].

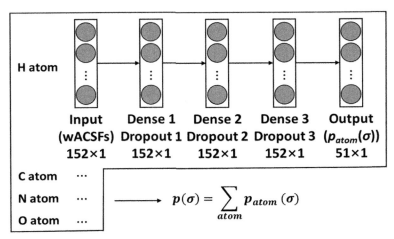

FIG. 4 A diagrammatic sketch of HDNN model.

TABLE 5 The hyperparameters of HDNN model.

Element	Sample size (training/ validation/ test)	Hidden layer number	Neuron number in each layer	Epochs	Batch size	Dropout ratio
H	11,653/ 1941/1941	3	152	1000	2000	0.05
C	6832/1138/ 1138	3	152	1000	1000	0.05
N	229/38/38	3	152	500	20	0.2
O	913/151/ 151	3	152	500	100	0.2

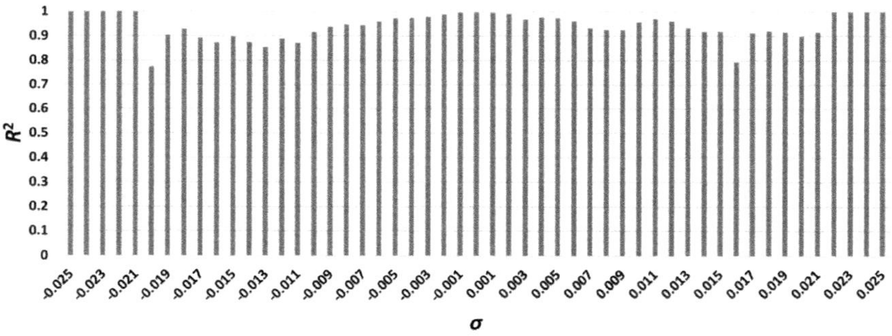

FIG. 5 The R^2 results of the predicted $p(\sigma)$ for each discrete σ interval in the MLAC method.

Finally, the metrics of training sample R^2_{train}, validation sample R^2_{val} and test sample R^2_{test} are 0.964, 0.918, 0.907 for H element, 0.975, 0.931, 0.931 for C element, 0.950, 0.889, 0.865 for N element, 0.935, 0.867, 0.902 for O element. All these results satisfy the fitting criterion $\frac{R^2_{train}-R^2_{test}}{R^2_{train}} < 0.1$ ($\frac{R^2_{train}-R^2_{test}}{R^2_{train}} \geq 0.1$ indicates overfitting) [77], indicating that the ANNs for H, C, N, O elements are reliable for $p_{atom}(\sigma)$ predictions.

To further demonstrate the feasibility and effectiveness of the HDNN model, the differences of predicted $p(\sigma)$ between the MLAC method and the density functional theory (DFT) method (benchmark) are evaluated with the criterion R^2. The R^2 results of the predicted $p(\sigma)$ for each discrete σ interval in the MLAC method are shown in Fig. 5.

Furthermore, the MLAC method is employed to provide molecular $p(\sigma)$ for the predictions of infinite dilution activity coefficients $\gamma^{\infty} = f(p(\sigma), V_C)$ based on the COSMO-SAC model, where the molecular cavity volume V_C is predicted by the group contribution (GC) method using the MG1 group sets [78] with the fitting result $R^2 = 0.9998$. The γ^{∞} predictions of the MLAC method are compared with those predicted by the DFT calculated $p(\sigma)$ and V_C. Sixteen solutes (denoted as "c1 ∼ c16") and 1120 solvents that composed of H, C, N, O elements from the Virginia Tech database (solvents are classified into 13 categories and denoted as "s1 ∼ s13") are selected for γ^{∞} calculations. The average absolute percent error (AAPE) criterion is used to evaluate the differences of predicted γ^{∞} between the DFT (benchmark) and MLAC method, as shown in Eq. (7).

$$AAPE = \frac{1}{T} \sum_{t=1}^{T} \frac{\left| \gamma_t^{\infty, est} - \gamma_t^{\infty, DFT} \right|}{\left| \gamma_t^{\infty, DFT} \right|} \times 100\% \tag{7}$$

where $\gamma_t^{\infty, \, est}$ is the estimated infinite dilution activity coefficients using the MLAC method (generate $p(\sigma)$) and the GC method (generate V_C), $\gamma_t^{\infty, \, DFT}$ is the DFT calculated infinite dilution activity coefficients, and T is the total number of data points. The smaller AAPE indicates the better prediction ability. The AAPEs among 16 types of solutes and 13 types of solvents using the MLAC method are shown in Fig. 6.

It is shown that most of the AAPEs in Fig. 6 are acceptable with minor prediction errors. Also, the overall AAPE for the total number of 17,920 data points (1120 solvents × 16 solutes) is calculated using the MLAC method and the result 6.6% confirms that the developed MLAC methodology is feasible and reliable to provide molecular $p(\sigma)$ (a sum of $p_{atom}(\sigma)$ that are predicted by the HDNN model) to the COSMO-SAC model for the predictions of γ^{∞}.

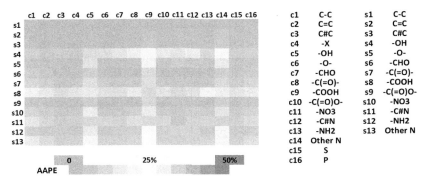

FIG. 6 The heap map of AAPEs among 16 types of solutes and 13 types of solvents using the MLAC method.

TABLE 6 List of structure and property constraints for crystallization solvent design in ibuprofen cooling crystallization process.

Property	Constraint
Number of groups	$2 \leq N_G \leq 8$
Number of same groups	$N_S \leq 8$
Number of functional groups	$1 \leq N_F \leq 8$
Hildebrand solubility parameter at 298 K	$17 \leq \delta \leq 19$ MPa$^{1/2}$
Hydrogen bonding solubility parameter	$\delta_H \geq 8$ MPa$^{1/2}$
Flash point	$T_f \geq 323$ K
Toxicity	$-\log(LC_{50})FM \leq 3.3 - \log(\text{mol/L})$
Normal melting point	$T_m \leq 270$ K
Normal boiling point	$T_b \geq 340$ K
Viscosity	$\mu \leq 1$ cP
Solid-liquid equilibrium (SLE)	$\ln x_i^{Sat} - \frac{\Delta H_{fus,i}}{RT_{m,i}}(1 - T_{m,i}/T) + \ln \gamma_i^{Sat} = 0$
	$\Delta H_{fus,\ Ibuprofen} = 27.94$ kJ/mol
	$T_{m,\ Ibuprofen} = 347.6$ K
Molar fraction normalization	$x_1 + x_2 = 1$
Crystallization temperature range	$260 \leq T \leq 320$ K
Objective function (crystallization case study)	$PR\% = \frac{100}{1 - X_L}\left(1 - \frac{X_L}{X_H}\right)$

More details about the above constraints can be found in Karunanithi's work [79].

- **Chemical product design**

With the obtained HDNN model (ML model), the CAMD problem of designing a cooling crystallization solvent for Ibuprofen is formulated as a MINLP model. The following groups are selected: CH_3, CH_2, CH, C, OH, CH_3CO, CH_2CO, CHO, CH_3COO, CH_2COO, HCOO, CH_3O, CH_2O, CH—O, COOH, COO. The lower and upper bonds for structure and property constraints are given in Table 6. The objective for this case study is to design a crystallization solvent with the highest potential recovery $PR\%$.

Through using the decomposition-based algorithm, $N_1 = 272$ feasible molecular candidates (group sets) are obtained at the first step. Then, $N_2 = 6723$ SMILES-based isomers are generated using the SMILES-based isomer generation algorithm (a kind of structural descriptor conversion algorithm). After that, among N_2 solvent candidates and Ibuprofen solute, $PR\%$ are individually calculated and arranged in descending order with the key property γ_i^{Sat} that

TABLE 7 The top two designed crystallization solvents for the Ibuprofen cooling crystallization process.

	1. CC(=O)OCC(=O) OCOCOC(C)C	2. COOCC(CC(=O)OC) C(=O)OC
SMILES		
Molecular structure		
PR% (MLAC method)	95.91%	95.91%
PR% (DFT method)	96.04%	95.84%
δ (MPa$^{1/2}$)	18.470	18.902
δ_H (MPa$^{1/2}$)	11.018	11.329
T_f (K)	400.172	387.476
$-\log(LC_{50})FM$ ($-\log$(mol/L))	2.939	2.770
T_m (K)	266.773	268.948
T_b (K)	523.919	514.742
μ (cP)	0.577	0.394

is estimated by the MLAC method (generate $p(\sigma)$, a sum of $p_{atom}(\sigma)$ that are calculated by the HDNN model), GC method (generate V_C) and COSMO-SAC model. Finally, the top two designed crystallization solvents are given in Table 7.

Although the DFT-based $PR\%$ of the best designed **solvent 1** in Table 7 (96.04%) has made a minor improvement (1.15% = (96.04%–94.95%)/ 94.95% × 100%) compared with Karunanithi's solvent (94.95%) [79], our best designed **solvent 1** is safer (T_f=400.172 K) and lower toxic ($-\log(LC_{50})$ $FM = 2.939 - \log$(mol/L)) than Karunanithi's one (T_f= 354.290 K and $-\log$ $(LC_{50})FM = 3.040 - \log$(mol/L)). Further experimental verifications will be performed in the future to confirm the rationality of the top two designed solvents.

4.2 A ML-based computer-aided molecular design/screening methodology for fragrance molecules

This case study refers to our previous work [4]. In this work, an optimization-based ML-CAMD framework is developed for the design of fragrance molecules, where the odor of the molecules are predicted using a data-driven ML approach, while a GC-based method is employed for prediction of important physical properties, such as, vapor pressure, solubility parameter and viscosity [3, 4]. A MINLP model is established for the design of fragrance molecules. Decomposition-based solution approach is used to obtain the optimal result.

- **Data collection**

Create database. In this case study, the database developed by Keller et al. [80] is used. This database has 480 molecules. The molecules have between 1 and 28 nonhydrogen atoms, and, include 29 amines and 45 carboxylic acids. Two molecules contain halogen atoms, 53 have sulfur atoms, 73 have nitrogen atoms, and 420 have oxygen atoms. The molecules are structurally and chemically diverse, and many of them have unfamiliar smells, some have never been used in prior psychophysical experiments.

Select product descriptors. Fragment (group)-based representation is commonly used in GC methods and group-based QSPR methods. It has been shown that the properties of a molecule can be determined with relatively high accuracy by summation of the contributions of the associated groups. In this case study, 50 groups are selected as descriptors for ML modeling.

Collect product properties. Here, the odor pleasantness and odor characters are selected as the required key properties for a fragrance product, which are defined as follows. The odor pleasantness is a scale from the rating of people for a certain molecule, from 0 to 100; the odor characters are classified in terms of the following 20 categories based on people's perception [81], namely "edible," "bakery," "sweet," "fruit," "fish," "garlic," "spices," "cold," "sour," "burnt," "acid," "warm," "musky," "sweaty," "ammonia/urinous," "decayed," "wood," "grass," "flower," and "chemical." These 20 categories of odor characters cover most of the odors for the design of fragrance products in industry. To simplify the problem, only the key odor character is reserved for each molecule.

- **Data preprocessing and feature engineering**

As the input data (groups) have significant physical meaning without any data issue, data preprocessing, and feature engineering are not performed in this case study.

- **Model establishment**

With the obtained input (groups) and output data (odor pleasantness and odor characters), two convolutional neural networks (CNNs) models are established.
Model selection
CNNs are selected to correlate groups with the odor pleasantness and odor characters due to the following reasons:

√ As is well-known, there is a unique parameter in CNN, called the kernel function. Since every input variable g_i is multiplied by a weight factor according to the convolutional calculation, the learning algorithm is capable of extracting features from the kernel. Thus, the output of CNN is also called a feature map. With this CNN character, essential groups, which map the odor of a molecule, can be selected by using the learning algorithm. Hence, the odor prediction models can be implemented without the disturbance of unimportant groups.

√ Besides, the convolution calculation involves some insights, which could enhance the model performance, such as sparse interactions and parameter sharing. Sparse interactions mean the output is only affected by a few variables (groups), which could enlarge their influence on the odorant via the learning algorithm. Parameter sharing ensures the odorant prediction results are affected by all the molecular structural parameters as a whole.

Therefore, the above traits of CNN make it more suitable for the odor prediction than other neural network models.

Model training and validation process

Here, Python Keras [75] is used for the development of the CNN odor prediction models. The input to the model is a 50×1 vector of groups, and the output is odor properties, including odor characters and odor pleasantness. The layer information is shown in Fig. 7. In Keras, the embedding layers and the flatten layers are used for reshaping the data; the dropout layer is used to prevent overfitting; the dense layer is a fully connected layer, so the neurons in the layer are connected to those in the next layer. As shown in Fig. 7, the established CNN model structure consists a 50×64 embedding layer, a 47×128 convolutional layer, a 44×128 convolutional layer, a 22×128 max-pooling layer, a 22×128 dropout layer, a 2816×1 flatten layer, a 128×1 dense layer, another 128×1 dropout layer and 20×1 dense layer. Finally, the properties of odor characters and odor pleasantness are predicted using this model structure, with different trained parameters.

The 480 molecules in the database are trained using the established CNN models. The predicted results of odor characters and odor pleasantness are compared with the experimental data as shown in Figs. 8 and 9. It is not necessary to obtain continuous values of odor pleasantness, due to the odor properties are diverse among different people. Therefore, the odor pleasantness is discretized into 5 levels from the original odor pleasantness values (e.g., level 1: 0–20; level 2: 20–40, and so forth) to make the model more representative and applicability. Therefore, the prediction results shown in Fig. 8 are also discretized into five levels.

From the results in Figs. 8 and 9, it is seen that both the prediction of odor characters and pleasantness are accurate using the developed CNN models, which are trained using the 480 molecules in the database. The average correctness of odor characters is 92.9%, while the average prediction error of odor

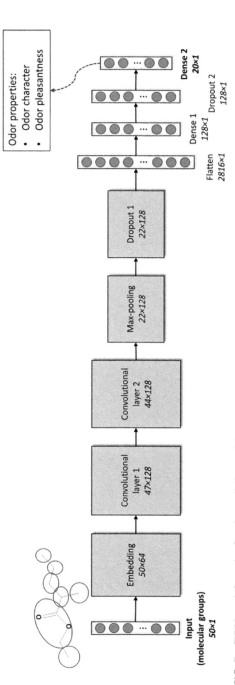

FIG. 7 CNN layer information for odor prediction model.

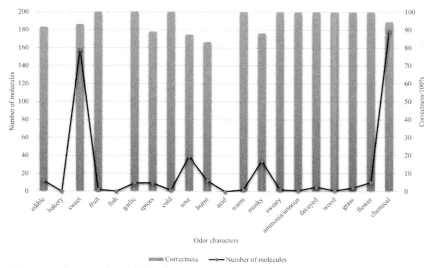

FIG. 8 Predicted results of 480 molecules in the database for odor characters.

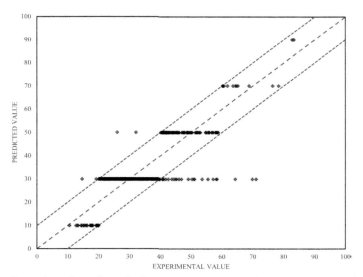

FIG. 9 Comparison of experimental values and predicted results of odor pleasantness (scale from 0 to 100) for the 480 molecules in the database.

pleasantness is 18.4%. In Fig. 8, the black line shows the number of molecules for a typical odor character in the database, while the bar shows the correctness of the model prediction. In the database, character "sweet" and "chemical" possess the largest amount of the molecules, while other ones possess smaller but sufficient numbers of molecules. In Fig. 9, the two dashed lines indicate the

TABLE 8 Predicted results of several commonly used fragrance molecules using the developed CNN models.

Molecule	Smells	Odor	Predicted odor character	Predicted odor pleasantness (scale from 0 to 100)	Correct?
Limonene	CC1= CCC(CC1) C(=C)C	A strong smell of oranges	Musky	20–40	N
Geraniol	OCC= C(O) CCC= C(O)C	Floral	Sweet	40–60	Y
Vanillin	O= Cc1ccc(O) c(OC)c1	Vanilla	Sweet	80–100	Y
Linalool	C=CC(O) (C)CCC= C(O)C	Sweet, floral, petitgrain-like	Sweet	40–60	Y

acceptable range for the predicted properties, that is, the predicted property (indicated by dots) must be inside region covered by the dashed lines. From Fig. 9, 27 molecules are out of the acceptable range. Since the odor pleasantness experimental values are obtained from the rating of people, the data may not be quite accurate. Therefore, although most of the characters and pleasantness have a satisfactory correctness, the prediction of odor characters for molecules outside the database has to be reevaluated. In Table 8, several fragrance molecules outside the database, which are commonly used in our daily life, are evaluated using the ML model. The evaluation results of Table 8 show roughly 75% correctness for molecules outside the database using the developed CNN models, which indicates that the trained CNN models are not overfitting.

- **Chemical product design**

The objective of this case study is to find suitable fragrance molecules as additives for shampoo, where the odor of the molecules is predicted using the developed CNN models while GC-based models are included to predict the rest of the needed physical properties, such as vapor pressure, solubility parameter and viscosity. The CAMD problem is formulated as a MINLP model for the design of fragrance molecules. The decomposition-based solution approach [70] is used

TABLE 9 Properties and constraints for fragrance molecule design.

Properties	Constraints
Total group number	$4 \leq n \leq 10$
Repeat group number	$n_i \leq 4$
Functional group number	$1 \leq n_F \leq 3$
Odor character	$OC=$ sweet, fruit or flower
Odor pleasantness	$OP \geq 40$
Diffusion coefficient (m²/h)	$D \geq 0.15$
Vapor pressure (Pa)	$P^{sat} \geq 100$
Normal Boiling point (K)	$T_b \geq 440$
Normal melting point (K)	$T_m \leq 293.15$
Solubility parameter (MPa$^{1/2}$)	$15 \leq S_p \leq 17$
Viscosity (cP)	$\eta \leq 2$
Density (g/cm³)	$0.8 \leq \rho \leq 1$
$-\log(LC_{50})FM$ ($-\log$(mol/L))	$-\log(LC_{50})$ FM ≤ 4.2

to obtain the optimal result. The following groups are selected: CH_3, CH_2, CH, C, $CH_2=CH$, $CH=CH$, $CH_2=C$, $CH=C$, OH, CH_3CO, CH_2CO, CH_3COO, CH_2COO, CH_3O, CH_2O. The lower and upper bonds for structure and property constraints are given in Table 9. More detailed information can be found in our previous work [4].

First, feasible candidates are generated by matching constraints T_b, T_m, S_p, η, ρ and $-\log(LC_{50})FM$ as the model equations for these properties are linear. 40 Feasible molecules are generated in this subproblem using the OptCAMD software [82]. Then, constraints D and P_{sat} are added to evaluate each generated candidate to check if they satisfy these additional constraints. 26 Molecules are selected in this subproblem. Then, the 26 molecules are tested using the CNN model for odor character prediction, to test if these molecules are "sweet," "fruit," or "flower" (as defined in Table 9), and 8 molecules are found to match these constraints. The odor pleasantness CNN model is then used for the screening of these 8 molecules, which finds 6 molecules matching this constraint. The final solution is the molecule which has the highest odor pleasantness within these 6 molecules. The 6 generated molecules satisfying all property constraints are listed in Table 10, together with their properties.

TABLE 10 The generated feasible candidates.

No.	1	2	3	4	5	6
Formula	$C_8H_{16}O$	$C_8H_{16}O_2$	$C_7H_{12}O_2$	$C_7H_{12}O_3$	$C_8H_{14}O_2$	$C_9H_{18}O_2$
Groups	2 CH$_3$ 4 CH$_2$ 1 CH$_2$CO	1 CH$_3$ 4 CH$_2$ 1 CH$_2$CO 1 CH$_3$O	1 CH$_3$ 1 CH 1 CH$_2$=C 1 CH$_3$CO 1 CH$_3$O	1 CH$_3$ 1 CH 1 CH$_2$=CH 1 CH$_2$COO 1 CH$_3$O	1 CH$_3$ 2 CH 1 CH$_2$=CH 1 CH$_3$CO 1 CH$_3$O	3 CH$_3$ 3 CH$_2$ 1 C 1 CH$_3$COO
T_m/K	244	265	253	217	253	240
T_b/K	443	469	443	442	459	458
S_p/Mpa$^{1/2}$	16.47	16.88	16.55	16.49	16.32	15.23
η/CP	1.08	0.91	0.21	0.2	0.15	0.89
ρ/g/cm^3	0.82	0.9	0.96	1	0.94	0.9
$-\log(LC_{50})$	3	2.58	2.83	3.53	3.58	3.13
P^{sat}/Pa	1003.8	138.2	838.4	1298.3	318.4	501.7
D/m^2/h	0.17	0.16	0.17	0.17	0.16	0.16
OC	Sweet	Sweet	Sweet	Sweet	Sweet	Sweet
OP	40	40	40	40	40	60
Available in database?	Y	Y	N	N	N	N
CAS number	111-13-7	106-73-0	–	–	–	–
Molecular structure			–	–	–	–
Odor in literature	Cheese-like, dairy nuances	Fruity	–	–	–	–

From the optimization result, molecule $C_9H_{18}O_2$ has the highest odor pleasantness. Therefore, it is selected as the best potential fragrance molecule in this case study. Database search has been performed for all the six feasible molecules. The optimal molecule, however, is not found in any database as fragrance and therefore, it needs to be evaluated through experiments to verify if the odor properties are the same as predicted. The molecules $C_8H_{16}O$ (CAS number: 111-13-7) and $C_8H_{16}O_2$ (CAS number: 106-73-0) are found in the database as commonly used fragrances for various purposes, which confirms the effectiveness of the fragrance molecule design model and its solution.

5 Conclusions

In this chapter, an optimization-based ML-CAMD framework is discussed in detail for the establishment of ML models for property prediction through model selection, training, and validation, as well as chemical product design through efficient mathematical optimization algorithms. Two case studies are presented for the applications of the proposed framework, where HDNN and CNN models are respectively established for the predictions of the atomic surface charge density profiles and the odor pleasantness/characters. Afterwards, the ML models are successfully incorporated into the mixed-integer nonlinear programming models for computer-aided molecular design design. The model complexity is managed by a decomposition-based algorithm. Both case studies obtained the optimal products (crystallization solvents for case study 4.1 and fragrance molecules for case study 4.2) with satisfied performances, which will be further verified by experiments in our future work.

Acknowledgments

The authors are grateful for the financial support of National Natural Science Foundation of China (22078041, 21808025) and "the Fundamental Research Funds for the Central Universities (DUT20JC41)."

References

[1] L. Zhang, D.K. Babi, R. Gani, New vistas in chemical product and process design, Annu. Rev. Chem. Biomol. Eng. 7 (2016) 557–582.

[2] L. Zhang, H. Mao, Q. Liu, R. Gani, Chemical product design—recent advances and perspectives, Curr. Opin. Chem. Eng. 2020 (27) (2020) 22–34.

[3] L. Zhang, K.Y. Fung, C. Wibowo, R. Gani, Advances in chemical product design, Rev. Chem. Eng. 34 (3) (2018) 319–340.

[4] L. Zhang, H. Mao, L. Liu, J. Du, R. Gani, A machine learning based computer-aided molecular design/screening methodology for fragrance molecules, Comput. Chem. Eng. 115 (2018) 295–308.

[5] H. Cartwright, Development and Uses of Artificial Intelligence in Chemistry, Reviews in Computational Chemistry, John Wiley & Sons, Inc., New York, 2007, pp. 349–390.

[6] M.N.O. Sadiku, S.M. Musa, O.M. Musa, Machine learning in chemistry industry, Int. J. Adv. Sci. Res. Eng. 3 (10) (2017) 12–15.

[7] I. Goodfellow, Y. Bengio, A. Courville, Deep Learning, MIT Press, 2016.

[8] V.N. Vapnik, The Nature of Statistical Learning Theory, Springer, 2000.

[9] O. Ivanciuc, Applications of support vector machines in chemistry, Rev. Comput. Chem. 23 (2007) 291.

[10] H. Geppert, M. Vogt, J. Bajorath, Current trends in ligand-based virtual screening: molecular representations, data mining methods, new application areas, and performance evaluation, J. Chem. Inf. Model. 50 (2010) 205–216.

[11] P.R. Zonouz, A. Niaei, A. Tarjomannejad, Modeling and optimization of toluene oxidation over perovskite-type nanocatalysts using a hybrid artificial neural network-genetic algorithm method, J. Taiwan Inst. Chem. Eng. 65 (2016) 276–285.

[12] A. Velásco-Mejía, V. Vallejo-Becerra, A.U. Chávez-Ramírez, J. Torres-González, Y. Reyes-Vidal, F. Castañeda-Zaldivar, Modeling and optimization of a pharmaceutical crystallization process by using neural networks and genetic algorithms, Powder Technol. 292 (2016) 122–128.

[13] Y. Liu, T. Zhao, W. Ju, S. Shi, Materials discovery and design using machine learning, J. Mater. 3 (3) (2017) 159–177.

[14] P. Pankajakshan, S. Sanyal, O.E. de Noord, I. Bhattacharya, A. Bhattacharyya, U. Waghmare, Machine learning and statistical analysis for materials science: stability and transferability of fingerprint descriptors and chemical insights, Chem. Mater. 29 (10) (2017) 4190–4201.

[15] Q. Vanhaelen, A.M. Aliper, A. Zhavoronkov, A comparative review of computational methods for pathway perturbation analysis: dynamical and topological perspectives, Mol. BioSyst. 13 (1) (2017) 1692–1704.

[16] N. Hadi, A. Niaei, S.R. Nabavi, R. Alizadeh, M.N. Shirazi, B. Izadkhah, An intelligent approach to design and optimization of M-Mn/H-ZSM-5 (M: Ce, Cr, Fe, Ni) catalysts in conversion of methanol to propylene, J. Taiwan Inst. Chem. Eng. 59 (2016) 173–185.

[17] T. Zhou, S. Jhamb, X. Liang, K. Sundmacher, R. Gani, Prediction of acid dissociation constants of organic compounds using group contribution methods, Chem. Eng. Sci. 183 (2018) 95–105.

[18] W.L. Kubic Jr., R.W. Jenkins, C.M. Moore, T.A. Semelsberger, A.D. Sutton, Artificial neural network based group contribution method for estimating cetane and octane numbers of hydrocarbons and oxygenated organic compounds, Ind. Eng. Chem. Res. 56 (2017) 12236–12245.

[19] Y. Lo, S.E. Rensi, W. Torng, R.B. Altman, Machine learning in chemoinformatics and drug discovery, Drug Discov. Today 23 (8) (2018) 1538–1546.

[20] T. Le, V.C. Epa, F.R. Burden, D.A. Winkler, Quantitative structure–property relationship modeling of diverse materials properties, Chem. Rev. 112 (5) (2012) 2889–2919.

[21] Daylight Theory Manual, Chemical Information Systems, Inc., 2019. https://www.daylight.com/dayhtml/doc/theory/theory.finger.html. (Accessed 10 August 2019).

[22] D. Rogers, M. Hahn, Extended-connectivity fingerprints, J. Chem. Inf. Model. 50 (5) (2010) 742–754.

[23] J.N. Wei, D. Duvenaud, A. Aspuru-Guzik, Neural networks for the prediction of organic chemistry reactions, ACS Cent. Sci. 2 (10) (2016) 725–732.

[24] H.S. Stein, D. Guevarra, P.F. Newhouse, E. Soedarmadji, J.M. Gregoire, Machine learning of optical properties of materials—predicting spectra from images and images from spectra, Chem. Sci. 10 (2019) 47–55.

[25] B. Meyer, B. Sawatlon, S. Heinen, O.A. von Lilienfeld, C. Corminboeuf, Machine learning meets volcano plots: computational discovery of cross-coupling catalysts, Chem. Sci. 9 (35) (2018) 7069–7077.

[26] F. Häse, I.F. Galván, A. Aspuru-Guzik, R. Lindhb, M. Vacher, How machine learning can assist the interpretation of ab initio molecular dynamics simulations and conceptual understanding of chemistry, Chem. Sci. 10 (2019) 2298–2307.

[27] M.A. Kayala, C. Azencott, J.H. Chen, P. Baldi, Learning to predict chemical reactions, J. Chem. Inf. Model. 51 (9) (2011) 2209–2222.

[28] D. Duvenaud, D. Maclaurin, J. Aguilera-Iparraguirre, R. Gómez-Bombarelli, T. Hirzel, A. Aspuru-Guzik, R.P. Adams, Convolutional Networks on Graphs for Learning Molecular Fingerprints, 2015. arXiv preprint. arXiv:1509.09292.

[29] S. Kearnes, K. Mccloskey, M. Berndl, V. Pande, P. Riley, Molecular graph convolutions: moving beyond fingerprints, J. Comput. Aided Mol. Des. 30 (8) (2016) 595–608.

[30] D. Weininger, SMILES, a chemical language and information system. 1. Introduction to methodology and encoding rules, J. Chem. Inf. Model. 28 (1988) 31–36.

[31] M.H. Segler, T. Kogej, C. Tyrchan, M.P. Waller, Generating focused molecule libraries for drug discovery with recurrent neural networks, ACS Cent. Sci. 4 (1) (2018) 120–131.

[32] Y. LeCun, Y. Bengio, G. Hinton, Deep learning, Nature 521 (7553) (2015) 436–444.

[33] M.S. Nixon, A.S. Aguado, Feature Extraction & Image Processing for Computer Vision, Academic Press, 2012.

[34] G. Chaudhary, S. Srivastava, S. Bhardwaj, Feature extraction methods for speaker recognition: a review, Int. J. Pattern Recognit. Artif. Intell. 31 (12) (2017) 1750041.

[35] I. Levner, V. Bulitko, G. Lin, Feature extraction for classification of proteomic mass spectra: a comparative study, in: Feature Extraction, Springer, Berlin, Heidelberg, 2006, pp. 607–624.

[36] A.R. Tate, D. Watson, S. Eglen, T.N. Arvanitis, E.L. Thomas, J.D. Bell, Automated feature extraction for the classification of human in vivo 13C NMR spectra using statistical pattern recognition and wavelets, Magn. Reson. Med. Off. J. Soc. Magn. Reson. Med. 35 (6) (2010) 834–840.

[37] C. Cortes, V. Vapnik, Support-vector networks, Mach. Learn. 20 (3) (1995) 273–297.

[38] H.D. Drucker, C.J.C. Burges, L. Kaufman, A. Smola, V. Vapnik, Support vector regression machines, in: M.C. Mozer, M.I. Jordan, T. Petsche (Eds.), Advances in Neural Information Processing Systems, 9, Morgan Kaufmann, San Mateo, 1997, pp. 155–161.

[39] B. Schölkopf, A.J. Smola, Learning With Kernels: Support Vector Machines, Regularization, Optimization, and Beyond, MIT Press, Cambridge, MA, 2002.

[40] C.W. Hsu, C.J. Lin, A comparison of methods for multiclass support vector machines, IEEE Trans. Neural Netw. 13 (2) (2002) 415–425.

[41] Y. Zhao, X. Zhang, L. Deng, S. Zhang, Prediction of viscosity of imidazolium-based ionic liquids using MLP and SVM algorithms, Comput. Chem. Eng. 92 (2016) 37–42.

[42] F. Luan, H.T. Liu, Y.Y. Wen, X.Y. Zhang, Classification of the fragrance properties of chemical compounds based on support vector machine and linear discriminant analysis, Flavour Fragr. J. 23 (4) (2008) 232–238.

[43] J.R. Quinlan, Induction of decision trees, Mach. Learn. 1 (1) (1986) 81–106.

[44] L. Breiman, J.H. Friedman, R.A. Olshen, C.J. Stone, Classification and Regression Trees, Chapman & Hall/CRC, Boca Raton, FL, 1984.

[45] R.J. Quinlan, C4.5: Programs for Machine Learning, Morgan Kaufmann, San Mateo, CA, 1993.

[46] S.K. Murthy, S. Kasif, S. Salzberg, A system for induction of oblique decision trees, J. Artif. Intell. Res. 2 (1994) 1–32.

[47] L. Breiman, Random forests, Mach. Learn. 45 (1) (2001) 5–32.

[48] S. Datta, V.A. Dev, M.R. Eden, Hybrid genetic algorithm-decision tree approach for rate constant prediction using structures of reactants and solvent for Diels-Alder reaction, Comput. Chem. Eng. 106 (2017) 690–698.

[49] B. Partopour, R.C. Paffenroth, A.G. Dixon, Random forests for mapping and analysis of microkinetics models, Comput. Chem. Eng. (2018).

[50] D.S. Cao, Y.N. Yang, J.C. Zhao, J. Yan, S. Liu, Q.N. Hu, Y.Z. Liang, Computer-aided prediction of toxicity with substructure pattern and random forest, J. Chemom. 26 (1–2) (2012) 7–15.

[51] D.S. Broomhead, D. Lowe, Multivariable functional interpolation and adaptive networks, Complex Syst. 2 (3) (1988) 321–355.

[52] D.E. Rumelhart, G.E. Hinton, R.J. Williams, Learning representations by backpropagating errors, Nature 323 (1986) 533–536.

[53] G.B. Huang, Q.Y. Zhu, C.K. Siew, Extreme learning machine: a new learning scheme of feedforward neural networks, in: 2004 IEEE International Joint Conference on Neural Networks, 2004. Proceedings, vol. 2, IEEE, 2004, pp. 985–990.

[54] G.E. Hinton, S. Osindero, Y.W. Teh, A fast learning algorithm for deep belief nets, Neural Comput. 18 (7) (2006) 1527–1554.

[55] S.J. Pan, Q. Yang, A survey on transfer learning, IEEE Trans. Knowl. Data Eng. 22 (10) (2010) 1345–1359.

[56] Y. LeCun, Y. Bengio, Convolutional networks for images, speech, and time series, in: M.A. Arbib (Ed.), The Handbook of Brain Theory and Neural Networks, MIT Press, Cambridge, MA, 1995.

[57] I. Goodfellow, J. Pouget-Abadie, M. Mirza, B. Xu, D. Warde-Farley, S. Ozair, A. Courville, Y. Bengio, Generative adversarial nets, in: Advances in Neural Information Processing Systems, Springer, Berlin, 2014, pp. 2672–2680.

[58] H. Schwenk, Y. Bengio, Training methods for adaptive boosting of neural networks, in: Advances in Neural Information Processing Systems, MIT Press, Cambridge, MA, 1998, pp. 647–653.

[59] N. Srivastava, G. Hinton, A. Krizhevsky, I. Sutskever, R. Salakhutdinov, Dropout: a simple way to prevent neural networks from overfitting, J. Mach. Learn. Res. 15 (1) (2014) 1929–1958.

[60] G.E. Hinton, Tutorial on Deep Learning, IPAM Graduate Summer School: Deep Learning, Feature Learning, 2012.

[61] D. Kingma, J.B.J.C. Science, Adam: A Method for Stochastic Optimization, 2014. arXiv e-prints. arXiv:1412.6980.

[62] A.Z. Hezave, M. Lashkarbolooki, S. Raeissi, Using artificial neural network to predict the ternary electrical conductivity of ionic liquid systems, Fluid Phase Equilib. 314 (2012) 128–133.

[63] P. Díaz-Rodríguez, J.C. Cancilla, G. Matute, J.S. Torrecilla, Viscosity estimation of binary mixtures of ionic liquids through a multi-layer perceptron model, J. Ind. Eng. Chem. 21 (2015) 1350–1353.

[64] M. Spellings, S.C. Glotzer, Machine learning for crystal identification and discovery, AICHE J. 64 (6) (2018) 2198–2206.

[65] T. Gao, J.R. Kitchin, Modeling palladium surfaces with density functional theory, neural networks and molecular dynamics, Catal. Today 312 (2018) 132–140.

[66] J.R. Boes, J.R. Kitchin, Neural network predictions of oxygen interactions on a dynamic Pd surface, Mol. Simul. 43 (5–6) (2017) 346–354.

[67] A. Ethem, Design and analysis of machine learning experiments, in: T. Dietterich (Ed.), Introduction of Machine Learning, second ed., The MIT Press, Cambridge, Massachusetts London, England, 2009, pp. 475–514.

[68] G.E. Hinton, N. Srivastava, A. Krizhevsky, I. Sutskever, R.R. Salakhutdinov, Improving Neural Networks by Preventing Co-Adaptation of Feature Detectors, 2012. arXiv e-prints. arXiv:1207.0580.

[69] A.Y. Ng, Feature selection, L1 vs. L2 regularization, and rotational invariance, in: Proceedings of the Twenty-First International Conference on Machine Learning, 2004, p. 78.

[70] A.T. Karunanithi, L.E. Achenie, R. Gani, A new decomposition-based computer-aided molecular/mixture design methodology for the design of optimal solvents and solvent mixtures, Ind. Eng. Chem. Res. 44 (13) (2005) 4785–4797.

[71] E. Mullins, R. Oldland, Y.A. Liu, et al., Sigma-profile database for using COSMO-based thermodynamic methods, Ind. Eng. Chem. Res. 45 (12) (2006) 4389–4415.

[72] M. Gastegger, L. Schwiedrzik, M. Bittermann, F. Berzsenyi, P. Marquetand, wACSF—weighted atom-centered symmetry functions as descriptors in machine learning potentials, J. Chem. Phys. 148 (24) (2018) 241709.

[73] C.-M. Hsieh, S.I. Sandler, S.-T. Lin, Improvements of COSMO-SAC for vapor–liquid and liquid–liquid equilibrium predictions, Fluid Phase Equilib. 297 (1) (2010) 90–97.

[74] G. Xavier, B. Antoine, B. Yoshua, Deep sparse rectifier neural networks, in: Proceedings of the Fourteenth International Conference on Artificial Intelligence and Statistics: PMLR, 2011, pp. 315–323.

[75] F. Chollet, et al., Keras: The Python Deep Learning Library, Astrophysics Source Code Library, 2018. ascl:1806.1022.

[76] T.E. Oliphant, Python for scientific computing, Comput. Sci. Eng. 9 (3) (2007) 10–20.

[77] Y. Zhao, J. Chen, Q. Liu, Y. Li, Profiling the structural determinants of aryl benzamide derivatives as negative allosteric modulators of mGluR5 by in Silico study, Molecules 25 (2) (2020).

[78] A.S. Hukkerikar, B. Sarup, A. Ten Kate, J. Abildskov, G. Sin, R. Gani, Group-contribution+ (GC+) based estimation of properties of pure components: improved property estimation and uncertainty analysis, Fluid Phase Equilib. 321 (2012) 25–43.

[79] A.T. Karunanithi, L.E. Achenie, R. Gani, A computer-aided molecular design framework for crystallization solvent design, Chem. Eng. Sci. 61 (4) (2006) 1247–1260.

[80] A. Keller, R.C. Gerkin, Y. Guan, A. Dhurandhar, G. Turu, B. Szalai, J.D. Mainland, Y. Ihara, C.W. Yu, R. Wolfinger, Predicting human olfactory perception from chemical features of odor molecules, Science 355 (2017) 820.

[81] A. Keller, L.B. Vosshall, Olfactory perception of chemically diverse molecules, BMC Neurosci. 17 (2016) 55.

[82] Q. Liu, L. Zhang, L. Liu, J. Du, A.K. Tula, M. Eden, R. Gani, OptCAMD: an optimization-based framework and tool for molecular and mixture product design, Comput. Chem. Eng. 124 (2019) 285–301.

[83] Q. Liu, L. Zhang, K. Tang, L. Liu, J. Du, Q. Meng, R. Gani, Machine learning-based atom contribution method for the prediction of surface charge density profiles and solvent design, AIChE J. 67 (2) (2021) e17110.

Chapter 13

Machine learning methods in drug delivery

Rania M. Hathout
Department of Pharmaceutics and Industrial Pharmacy, Faculty of Pharmacy, Ain Shams University, Cairo, Egypt

1 Introduction

With figures approaching an average of $2.6 billion and time-periods reaching over 10 years to develop new medicines, the pharmaceutical industry currently faces huge expenses and intense pressure to reduce resources and power costs and to reduce the number of new active and in-active pharmaceutical ingredients and excipients used in formulation and manufacturing. In this context, the pharmaceutical industry should adopt more efficient and systematic processes in both drug formulation and development [1].

Machine learning methods are a branch of artificial intelligence that utilizes certain software algorithms in order to make computers "learn" and to make hard and critical decision functions from representative data samples. They are specifically useful when certain software that is needed to perform a specific task or reach a targeted goal is unavailable or unfeasible.

With the advent of the internet and the enormous growth of data that are obtained from huge number of sources, the recent advances in various statistical and programming tools and the continuous innovations in machine learning algorithms have resulted in a rapid increase in new applications in the different areas of pharmaceutics and drug delivery disciplines [2]. Over the past 20 years, this new approach of data analysis has been extensively used in drug development and delivery.

Machine learning methods offer several advantages over the other conventional statistical methods [3]. First, the majority of these methods can model nonlinear relationships that are hard-to-model using the common quantitative structure-bioavailability or structure–property relationships methods. Second, they can model incomplete data or nontrending data and the users do not have to suggest any models or particular deigns before use. In other words, no restrictions are encountered while implementing machine learning algorithms where

Applications of Artificial Intelligence in Process Systems Engineering
https://doi.org/10.1016/B978-0-12-821092-5.00007-3

all types of data whether binary classification, multiple classes and continuous data can be analyzed and modeled. And finally, these methods can cluster the data into meaningful groups beside the powerful predicting power they offer [4,5]. Integrating the machine learning methods into drug delivery saves costs, resources, time, and effort.

In this chapter, different types of the machine learning methods that were exploited in drug delivery aspects and applications will be demonstrated and discussed. These methods are currently considered important elements of the new approach of computer-aided drug formulation design (CADFD) and computer-aided pharmaceutical technology. The cross-disciplinary integration of drug delivery and machine learning methods as a branch of artificial intelligence may shift the paradigm of pharmaceutical research from experience-dependent investigations to data-driven studies [6].

Importantly, the selection of the appropriate machine learning method is crucial. This usually depends on various factors, including the type of the data and the size of the dataset. Accordingly, the choice of the implemented machine learning method can be considered task specific. With a sufficient amount of carefully curated data, many applications using advanced machine learning algorithms may become a good practice that can solve several drug delivery challenges such as poor solubility, heterogeneous particle size, low drug loading levels, slow release, and poor stability [7]. To this end, this branch of computational Pharmaceutics has a great potential to cause huge shifts in drug delivery research in the future. Soon, the computer would be a very important tool for formulators just as the chemicals and the preparation tools.

2 Types of machine learning methods used in drug delivery

According to the learning algorithms, the machine learning methods that were employed in drug delivery included the unsupervised and the supervised at the same level [8]. The unsupervised methods are those that deal with the x-variables only and are used to classify and cluster the data in one hand such as the principal component analysis (PCA) and hierarchical clustering analysis (HCA) or reduce the number of variables and dimensionality of the data associated with feature(s) selection such as the PCA [9]. Nevertheless, the supervised counterparts are mainly involved in correlating inputs and outputs, that is, x and y variables and they are mainly associated with regression and predictions [10]. Amongst these methods are the artificial neural networks, Gaussian processes, Support vector machines, and partial least squares (see Fig. 1).

In another classification, the machine learning methods that were used in drug discovery and development were divided into three categories: regression analyses methods such as PLS, ANNs, and GPs, classifier methods such as SVM and finally the clustering methods such as PCA and HCA [11].

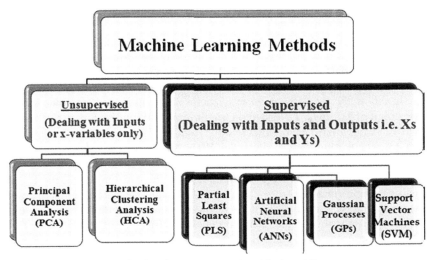

FIG. 1   The types of machine learning methods employed in drug delivery.

3   Applications of machine learning methods in drug delivery

The applications of machine learning methods as a type of artificial intelligence are many and diverse. They encompass all the stages of drug delivery and development from loading on carriers to the stability of formulations and passing by the drug release and the subsequent absorption. The machine learning tools can provide more risk management and a more hassle-free experience during drug development and scale-up.

3.1   Artificial neural networks (ANNs) in modeling the drug loading in lipophilic nanoparticles

The artificial neural networks (ANNs) are computational methods that were biologically inspired. They consist of a large number of interconnected nodes that perform summation and thresholding, in great analogy with the neurons of the brain. The ANNs weights that are given to the inputs (x-variables) simulate the biological neurons synapsis (synaptic strength) where the amount of released neurotransmitters is usually responsible for the magnitude of strength of the signals. The artificial neural networks do not explicitly model the underlying distribution but in certain cases the outputs are treated in a robust probabilistic way. The artificial neural networks have been successfully applied in many useful settings, including speech recognition, handwritten digit recognition, and car driving [12]. They were also successfully utilized as a useful tool for many areas of drug design [13] and have exhibited good performance in optimizing chemical libraries [14] and SAR analysis [15]. However, their use in the drug delivery and optimizing drug carriers and formulation aspects is still considered in its infancy.

In 2015, the concept of computer-assisted drug formulation design was introduced as a new approach in the drug delivery [16]. This new approach encompasses data mining, molecular dynamics and docking experiments and machine learning methods such as the PCA, HCA …etc. (see Fig. 2).

In this study, the entrapment efficiencies and total loaded mass per 100 mg for a total of 21 drugs; 11 loaded on tripalmitin solid lipid nanoparticles and 10 on PLGA polymeric nanoparticles were gathered from the scientific literature databases using PubMed, Scopus, and Web of science. The carriers were then virtually simulated using all-atom molecular dynamics using GROMACS v.6.5 open source software where the parameters for the tripalmitin and the PLGA matrices were obtained from CgenFF that is available online (https://cgenff. paramchem.org/). Afterwards, the drugs were molecularly docked on their corresponding virtual simulated nanoparticulate matrices using ArgusLab v.4.0.1 (Mark Thompson and Planaria Software LCC, Seattle, WA) and AutoDock vina (Molecular Graphics Laboratory, The Scripps Research Group, La Jolla, CA) and the resultant binding docking energies (ΔG) were recorded. Mathematical relationships were constructed between the loaded masses and the binding energies. Moreover, the drugs were projected to their main constitutional, topological, and electronic descriptors, namely, the molecular weight, xLogP, total polar surface area, and fragment complexity using the open-source software Bioclipse v.2.6.2 developed by the Bioclipse project [17]. Using the ANNs of the JMP 7.0 (SAS, Cary, NC) two networks were constructed for each of the tripalmitin and PLGA modeling. The networks consisted of inputs (x-factors) as the four main descriptors and three hidden nodes and an output

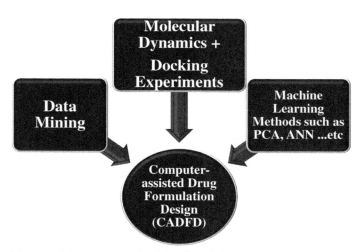

FIG. 2 Elements of the computer-assisted drug formulation design (CADFD) approach.

(the binding energy). The number of generated tours was 16 and the maximum used iterations were 50 with an overfit penalty of 0.001. k-Fold cross validation equivalent to 5 was utilized for the two constructed networks. Fig. 3 presents a schematic diagram of the constructed ANNs.

The constructed ANNs could successfully model the relation between the main drugs descriptors and the drugs binding energies. The 3D generated response surface plots demonstrated the effect of decreasing the molecular weight and the total polar surface area (TPSA) in decreasing the binding energies to more negative values (means favorable binding between the drug and the carrier). On the other hand, increasing the fragment complexity led to decreasing the binding energy values in case of tripalmitin while an optimum xLogP range (1.5–4) and an optimum value (around 1000) were responsible for the least obtained binding energies in the tripalmitin matrix (see Fig. 4).

Moreover, the ability of the ANNs to predict the binding energies of three drugs; resveratrol, lomefloxacin and testosterone on tripalmitin and prednisolone and resveratrol on PLGA using the software prediction profile tool was high scoring a percentage bias of only $8.09\% \pm 4.32$ (see Table 1).

It is worth mentioning, that the success in predicting the binding energies of drugs on their carriers would enable the good estimation of their loaded masses per 100 mg carrier through the generated mathematical model relating the binding energies with the loaded masses. The exploitation of this approach would lead to a great save in resources, time, and efforts exerted in experimenting new drugs on the different matrices.

3.2 Gaussian processes (GPs) in modeling the drug loading in lipophilic nanoparticles

Similar to the artificial neural networks, the Gaussian Processes are machine learning methods that include black-box algorithms and the inputs-outputs relationship equations remain unknown. In the process of prediction of a new output (y)

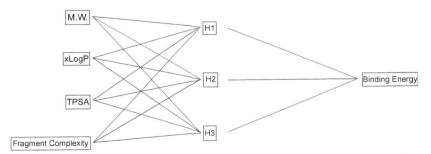

FIG. 3 A schematic diagram of the two constructed ANNs of tripalmitin and PLGA nanoparticles. *(Reprinted with permission from A.A. Metwally, R.M. Hathout, Computer-assisted drug formulation design: novel approach in drug delivery, Mol. Pharm. 12 (2015) 2800–2810, copyright 2020, ACS through RightsLink®.)*

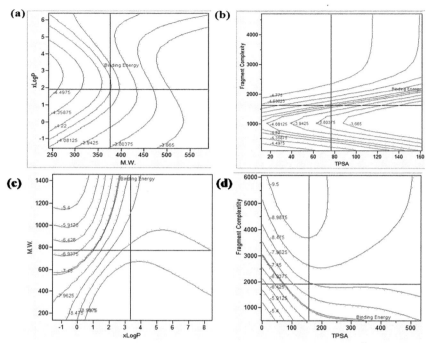

FIG. 4 3D response surface plots obtained from the constructed ANNs of drugs-tripalmitin systems (A and B) and drugs-PLGA system (C and D) drug-descriptor. *(Reprinted with permission from A.A. Metwally, R.M. Hathout, Computer-assisted drug formulation design: novel approach in drug delivery, Mol. Pharm. 12 (2015) 2800–2810, copyright 2020, ACS through RightsLink®.)*

TABLE 1 Percentages Bias between the actual and predicted docking binding energies using ANNs.

Carrier	Drug	Actual ΔG (kcal/mol)	Predicted ΔG (kcal/mol)	% Bias
Tripalmitin SLN	Resveratrol	−4.11	−3.95	3.66
	Lomefloxacin	−4.20	−4.42	5.25
	Testosterone	−4.72	−4.33	7.87
PLGA nanoparticles	Prednisolone	−8.00	−7.30	8.75
	Resveratrol	−7.80	−6.64	14.9

Retaken with permission from A.A. Metwally, R.M. Hathout, Computer-assisted drug formulation design: novel approach in drug delivery, Mol. Pharm. 12 (2015) 2800–2810, copyright 2020, ACS through RightsLink®.

that corresponds to a new input(s) (x_*), the distances of the questioned novel point to the training set of points are usually calculated. Hence, the training points are granted weights according to their distances such that the near points are largely weighted while the further points are given lower weights [18]. Accordingly, that the set of values of $y(x)$ evaluated at an arbitrary set of points $x1,.. xN_{trn}$ (N_{trn} is the number of training points) will acquire a Gaussian distribution as well. An important merit is that the variance of the predicted (y) is additionally calculated indicating the confidence of the calculated function.

The mean at (x_*) is given by the following equation:

$$E[y*] = k_*^T \left(K + \sigma_n^2 I \right)^{-1} y \tag{1}$$

where, k_* denotes the vector of covariances (distances) between the test point and N_{trn} (the training data), K represents the covariance matrix of the training data which expresses the expected correlation between the values of $f(x)$ at two points in two dimensions defining nearness or similarity between data points, σ_n^2 is the variance of an independent identically distributed noise or error multiplied by the identity matrix (I). k_*^T is the transpose of k_* while y denotes the vector of training targets (column vector representing the y values corresponding to the training x-values). Therefore, k_*^T is a row vector multiplied by the covariance matrix (which is a symmetric $(d \times d)$ matrix) multiplied by a column vector (y) leading to a scalar value corresponding to the probability (position) of finding the test point and its corresponding $(y*)$. The equation states that the position of any new output (y) in the space can be estimated from the distances of the test points to the training points multiplied by the calculated weights and multiplied by the corresponding y values of the training data, taking the noise (error) in consideration.

And the variance is given by:

$$var\,[y*] = k(x*, x*) - k_*^T \left(K + \sigma_n^2 I \right)^{-1} k* \tag{2}$$

where, $k(x*, x*)$ is a kernel denoting the variance of $y*$.

Frequently, the mean is used as the estimated predicted value while the variance refers to the error bars of the prediction.

the Lam et al., 2010 [19] compared different techniques in developing predictive models of percutaneous absorption where the Gaussian process machine learning methods produced more robust models than the quantitative structure property relationships (QSPR) or the single linear networks (SLN) (see Fig. 5). It was implied that ignoring the ionization state of the permeated drugs has limited the validity and accuracy of all the previously studied models.

Again, the loading of drugs on the tripalmitin solid lipid nanoparticles and PLGA polymeric nanoparticles were modeled in 2016. Correlating the binding energies (ΔG) as outputs resulting from the molecular docking of drug on the carriers matrices with the main drugs descriptors (inputs) was adopted using Gaussian Processes modeling in JMP® 7.0 (SAS, Cary, NC).

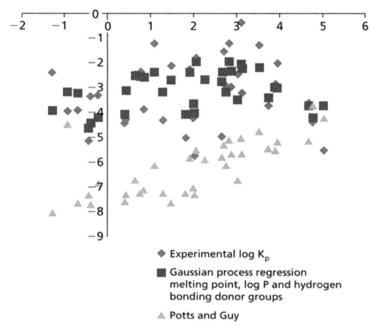

◆ Experimental log K_p

■ Gaussian process regression melting point, log P and hydrogen bonding donor groups

▲ Potts and Guy

FIG. 5 Comparison of the predictive ability of Gaussian process models with the quantitative structure–permeability relationship model proposed by Potts and Guy [20] across a wide range of lipophilicities. Data points shown were obtained from a subset of the overall dataset of Lam et al. [19]. *(Reprinted with permission from Wiley ® through RightsLink® license number: 4943241261948.)*

The percentage bias was calculated between the predicted and the actual docking values of the investigated drugs. The obtained percentage bias was 3.35% lower than that previously scored using Artificial Neural Networks method (see Table 2) [21].

It is worth noting, that the masses of loaded drugs could be predicted in turn through a mathematical relationship relating the binding energies with the masses of loaded drugs per 100 mg of the carriers. Fig. 6 depicts the usefulness of the Gaussian Processes and the sequence of the used steps in the prediction of drug loading.

3.3 Utilizing unsupervised machine learning methods in detecting the stability of soft nanoscale drug carriers

Of the most important unsupervised machine learning methods are the PCA and the HCA. In both techniques, a correlation matrix is usually generated between the different input variables (factors). In PCA, in particular, the number of the generated principal components is equivalent to the number of input variables.

TABLE 2 Calculation of percentage Bias between the actual docking binding energies of several drugs on molecular dynamics simulated solid lipid nanoparticles and those predicted using Gaussian Processes based on the main constitutional, electronic and topological descriptors of the investigated drugs (Inputs in the used Gaussian Processes).

Drug	SMILES	M.W.	xLogP	TPSA	Fragment complexity	Actual ΔG (kcal/mol)	Predicted ΔG (kcal/mol)	% Bias
Aceclofenac	C1=CC=C(C(=C1)CC(=O)OCC(=O)O)NC2=C(C=CC=C2Cl)Cl	353.02	1.92	75.63	863.07	−21.17	−21.49	1.51
Artmether	C[C@@H]1CC[C@H]2[C@H]([C@H](O[C@H]3[C@@]24[C@H]1CCC(O3)(OO4)C)OC)C	298.18	3.33	46.15	2080.05	−28.86	−28	2.98
Mepivacaine	CC1=C(C(=CC=C1)C)NC(=O)C2CCCCN2C	246.17	1.46	32.34	1375.03	−22.96	−22.82	0.61
Praziquental	C1CC(CC1)C(=O)N2C3C4=CC=CC=C4CCN3C(=O)C2	482.12	0.86	155.14	2531.1	−26.65	−27.52	3.26
Raloxifene	C1CCN(CC1)CCOC2=CC=C(C=C2)C(=O)C3=C(SC4=C3C=CC(=C4)O)C5=CC=C(C=C5)O	473.17	2.72	98.24	3103.06	−26.17	−26.01	0.61
Clozapine	CN1CCN(CC1)C2=C3C=CC=CC3=NC4=C(N2)C=C(C=C4)Cl	326.13	2.51	30.87	1519.05	−26.05	−26.2	0.58
Piroxicam	CN1C(=C(C2=CC=CC=C2S1(=O)=O)O)C(=O)NC3=CC=CC=N3	331.06	0.43	107.98	119.08	−21.35	−21.49	0.19

Continued

TABLE 2 Calculation of percentage Bias between the actual docking binding energies of several drugs on molecular dynamics simulated solid lipid nanoparticles and those predicted using Gaussian Processes based on the main constitutional, electronic and topological descriptors of the investigated drugs (Inputs in the used Gaussian Processes) —cont'd

Drug	SMILES	M.W.	xLogP	TPSA	Fragment complexity	Actual ΔG (kcal/mol)	Predicted ΔG (kcal/mol)	% Bias
Silibenin	COC1=C(C=CC(=C1)[C@@H]2[C@H](OC3=C(O2)C=C(C=C3)[C@@H]4[C@H](C(=O)C5=C(C=C(C=C5O4)O)O)CO)O)	482.12	0.86	155.14	2531.1	−28.46	−27.52	3.30
Tamoxifen	CC/C(=C(\C1=CC=CC=C1)/C2=CC=C(C=C2)OCCN(C)C)/C3=CC=CC=C3	371.2	6.34	12.47	144	−27.37	−27.41	0.51
Vinpocetine	CC[C@@]12CCCN3[C@@H]1C4=C(CC3)C5=CC=CC=C5N4C(=C2)C(=O)OCC	350.2	3.296	34.47	311.04	−27.30	−29.11	6.63
Resveratrol	C1=CC(=CC=C1/C=C/C2=CC(=CC(=C2)O)O)O	228.08	2.05	60.69	628.03	−18.57	−21.87	17.77
Lomefloxacin	CCN1C=C(C(=O)C2=CC(=C(C=C21)F)N3CCNC(C3)C)F)C(=O)O	351.14	2.27	72.88	129.08	−24.11	−23.03	4.48
Testosterone	C[C@]12CC[C@H]3[C@H]([C@@H]1CC[C@@H]2O)CCC4=CC(=O)CC[C@]34C	288.21	3.63	37.3	309.02	−26.7	−27.1	1.49

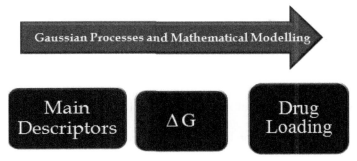

FIG. 6 Role of Gaussian processes as a machine learning method in the prediction of drug loading.

In an important preliminary step, the variation in the investigated data values and scale ranges is usually normalized to unit variance (1/SD) prior to analysis [22]. The correlation matrix is utilized to determine the data principal components which are actually the eigen vectors of the matrix. Consequently, the computed eigen vectors are ranked according to their eigen values (scalar values) and accordingly, the eigen vector (which is actually a direction) with the largest eigen value is selected as the main principal component followed by the vector possessing the second eigen value in magnitude and so on.

The correlation matrix is actually a co-variance matrix that represents all the co-variances (correlations) between the input variables (dimensions). For an n-dimensional data set, you can calculate $\frac{n!}{(n-2)!*2}$ different covariance values. For example, in case of three input variables (three-dimensional data; x, y, and z), there would be three covariance values for the three dimensions (variables) studied.

And the covariance square and fortunately symmetrical matrix will look like this:

$$A = \begin{pmatrix} cov(x,x) & cov(x,y) & cov(x,z) \\ cov(y,x) & cov(y,y) & cov(y,z) \\ cov(z,x) & cov(z,y) & cov(z,z) \end{pmatrix}$$

Where the covariance between any two dimensions is usually calculated as follows:

$$cov(x,y) = \frac{\sum_{i=1}^{n} (X_i - \overline{X})(Y_i - \overline{Y})}{(n-1)} \tag{3}$$

The eigen vector can be calculated for the covariance matrix (A) simply by solving the following characteristic equation:

$$\det (A - \lambda I) = 0 \tag{4}$$

Where I is the Identity matrix $\begin{pmatrix} 1 & 0 & 0 \\ 0 & 1 & 0 \\ 0 & 0 & 1. \end{pmatrix}$

For the obtained 3×3 covariance matrix, this will lead to a cubic polynomial equation the solution of which (the roots) yields three values for λ which are the eigen values (λ_1, λ_2, and λ_3). The solution can be obtained by mathematical packages such as Wlofram Alpha or statistical programming languages such as $R^\circledR$ or Python$^\circledR$.

Following, the equation to find the first eigen vector would be:

$$\begin{pmatrix} cov(x,x) & cov(x,y) & cov(x,z) \\ cov(y,x) & cov(y,y) & cov(y,z) \\ cov(z,x) & cov(z,y) & cov(z,z) \end{pmatrix} \begin{pmatrix} a \\ b \\ c \end{pmatrix} = \lambda_1 \begin{pmatrix} a \\ b \\ c \end{pmatrix} \tag{5}$$

These will yield three linear equations for a, b and c which can be easily solved to obtain a relation between them. Any vector in the form $\begin{pmatrix} a \\ b \\ c \end{pmatrix}$ satisfying the relation between a, b, and c is an eigen vector for the eigen value λ_1.

Similarly, for the second eigen vector,

$$\begin{pmatrix} cov(x,x) & cov(x,y) & cov(x,z) \\ cov(y,x) & cov(y,y) & cov(y,z) \\ cov(z,x) & cov(z,y) & cov(z,z) \end{pmatrix} \begin{pmatrix} a \\ b \\ c \end{pmatrix} = \lambda_2 \begin{pmatrix} a \\ b \\ c \end{pmatrix} \tag{6}$$

Eq. 5 will yield three linear equations giving another relationship between a, b, and c and consequently, another eigen vector. The same calculations go for λ_3.

A feature vector representing the main principal component (the one possessing the highest eigen value) is then selected [23].

The new derived data "final data" according to the obtained feature vector is derived as follows:

$$\text{Final data} = \text{Row feature vector} \times \text{Row Zero Mean data} \tag{7}$$

In other words,

$$\text{Final data} = \text{Feature vector}^T \times \text{Mean adjusted data}^T \tag{8}$$

Where the row feature vector is the feature vector transposed
And,

Row zero mean data is the mean-adjusted data transposed. The mean adjusted data is obtained by calculating the mean of each data set then subtracting it from all the data values.

The first attempts to utilize the clustering power of unsupervised machine learning methods such as the PCA and HCA in detecting the stable microemulsion formulations (nanoscale soft carriers possessing droplet size of 10–150 nm [24]) goes back to 2010 where lovastatin and glibenclamide loaded self microemulsifying and nanoemulsifying drug delivery systems (SMEDDS) and (SNEDDS), respectively, were investigated [25,26]. PCA and HCA were used to further characterize the formed microemulsion formulations in presence of water with respect to the similarity of particle size distribution obtained at 2 and 24 h.

Similarly, at 2015 two finasteride SMEDDS systems (one comprised Capryol 90®, Tween 80® and Transcutol® while the other was composed of Labrafac cc®, Tween 80® and Labrasol®) were prepared. Different formulations of the two systems were prepared and characterized for particle size and polydispersity index at 30 min and 24 h postpreparation [27]. PCA and HCA were utilized to detect the stable formulations as indicated by clustering the same formulation at 30 min and 24 h at the same cluster (see Fig. 7).

It was concluded accordingly that the microemulsion formulations 2, 4 and 5 of the first system Capryol 90®/Tween 80®/Transcutol were proven stable. Likewise, the microemulsion formulations of the Labrafac cc®/Tween 80®/ Labrasol® system 1, 2, 4 and 6 were considered stable [27].

More research focus on utilizing machine learning methods to aid in selecting promising formulations possessing good stability would be a great asset from the industrial point of view. The concept can be projected to any drug delivery system as well.

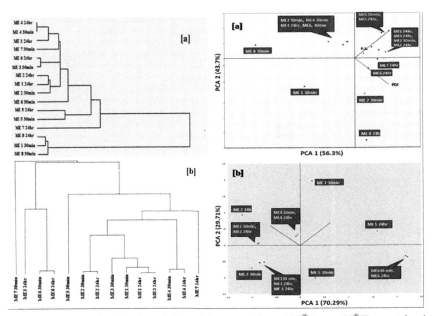

FIG. 7   Clustering of (A) the formulations of the system Capryol 90®/Tween 80®/Transcutol and the formulations of the system Labrafac cc®/Tween 80®/Labrasol® (B). Left panel represents HCA while the right panel represents PCA according to two principal components. *(Modified with permission from Future Medicine Ltd.® through Copyright Clearance Center® license ID: 1074183-1.)*

3.4 Capturing the highly loaded drugs on gelatin nanoparticles using different machine learning methods

Gelatin matrices (micro and nanoparticles) are gaining high interests due to the high biocompatibility, safety, availability and abundancy of the reactive functional groups that can be exploited to conjugate wide varieties of drugs and ligands [28,29] of gelatin as a protein carrier [30–33]. Accordingly, the machine learning methods; the partial least squares (PLS), PCA and HCA were successfully utilized to cluster and/or capture the highly loaded drugs on the gelatin matrix (see Fig. 8).

The partial least squares is a supervised machine learning method that employs principal component analysis at both the inputs and outputs matrices. It selects feature vectors out of both and correlates with them generating what is

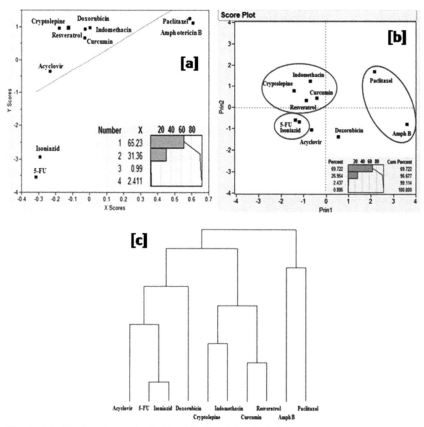

FIG. 8 Machine learning methods (A) PLS, (B) PCA and (C) HCA capturing the highest loaded drugs on the gelatin matrix; 5-FU (5-fluorouracil) and isoniazid. *(Reprinted with permission from Hathout et al., ACS OMEGA 2020, 5, 1549–1556, copyright 2020, ACS through RightsLink®.)*

called "x-scores" and "y-scores" for every point in the experimental space [22,34] and consequently a distinct plot for the PLS is obtained (see Fig. 8).

The results were further explored and interpreted using a bioinformatics approach.

The obtained results were attributed to the properties of gelatin as a protein carrier with a relatively balanced hydrophilic/hydrophobic nature where its structure comprises numerous hydrogen bond donor and acceptor groups with a repetitive units of amino acids; -Ala-Gly-Pro-Arg-Gly-Glu-4Hyp-Gly-Pro-that represent its backbone structure [31]. This structure can be translated to numerical values that could specify each amino acid; "z-scale descriptors" [35] that were derived from PCA analysis of several experimental and physico-chemical properties of the 20 natural amino acids; $z1$, $z2$, and $z3$ and which represent the amino acids hydrophobicity, steric properties, and polarity, respectively. Additionally, they were useful in the quantitative structure activity relationship analysis of peptides where they have proven effective in predicting different physiological activities [36–38]. An extended amino acids scale (including 67 more artificial and derivatized amino acids) [39] was utilized due to the presence of 4-hydroxyproline in the gelatin structure.

The use of the first descriptor ($z1$) was extended to predict the drug loading properties on gelatin nanospheres. The first scale ($z1$) was selected to represent the lipophilicity scale that encompasses several variables (amino acid descriptors) such as: the thin layer chromatography (TLC) variables, log P, nonpolar surface area (Snp) and polar surface area (Spol) in addition to the number of proton accepting electrons in the side chain (HACCR) [40]. In this scale, a large negative value of $z1$ corresponds to a lipophilic amino acid, while a large positive $z1$ value corresponds to a polar, hydrophilic amino acid. Therefore, the gelatin typical structure amino acids (-Ala-Gly-Pro-Arg-Gly-Glu-4Hyp-Gly-Pro-) can be represented by their $z1$ values as follows: (0.24), (2.05), (−1.66), (3.52), (2.05), (3.11), (−0.24), (2.05) and (−1.66). Accordingly, an overall topological description of the repetitive sequence was computed by encoding the $z1$ descriptors of each amino acid into one auto covariance variable [41] that was first introduced by Wold et al. [42]. The auto covariance value (AC) was calculated as follows:

$$AC_{z.lag} = \sum_{i=1}^{N-lag} \frac{V_{z.i} \times V_{z,i+lag}}{N - lag} \qquad (9)$$

where AC represents autocovariances of the same property (z-scale); $i = 1, 2, 3, \ldots$; N is the number of amino acids; $lag = 1, 2, 3, \ldots L$ (where L is the maximum lag which is the longest sequence used and V is the scale value) [37].

Therefore, the AC value for the gelatin typical structure sequence was determined utilizing a lag of 1 scoring a value approximating to zero (0.028) indicating a balanced hydrophobicity/hydrophilicity property of the gelatine sequence. In this context, the high loading of the drugs; 5FU and Isoniazid;

in particular was ascribed to their amphiphilic nature similar to gelatin with Log P values approaching 0, and further due to the presence of several hydrogen bond donors and acceptors groups in their structure that aids in the entrapment in a protein matrix like that of gelatin nanospheres.

3.5 The application of support vector machines as tools to identify and classify compounds with potential as transdermal enhancers

Usually support vector machines classify the data into only two classes. They are considered intelligent classifiers. And though they suffer from an important limitation of dependency on the quality of the input data; however, it is still very popular due to the excellent performance it demonstrates in binary classifications [43]. The SVMs work by constructing a hyperplane that separates two different classes with the maximum margin as possible [44].

Support vector machines (SVMs) are one of the supervised learning techniques that are employed in the analysis of data aiming for classifying data or for pattern recognition [45]. SVMs increasingly find use in regression analyses in addition (see Fig. 9). The general form of the decision function for SVM is represented by:

$$y(x) = \sum_{n=1}^{N} \alpha^n t^n k(x, x^n) + b \tag{10}$$

$$\text{with constraints}: \sum_{n=1}^{N} \alpha^n t^n = 0 \tag{11}$$

and $0 \leqslant \alpha^n \leqslant A$, where b is a threshold; α^n are Lagrange multipliers introduced in a constrained optimization problem; A is a constant to determine the trade-off

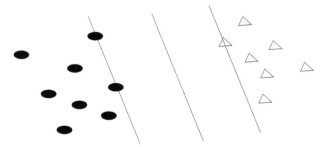

FIG. 9 A binary classification case of support vector machines: The goal of this SVMs is to separate the *dots* from the *triangles*. The *solid line* represents the optimal constructed hyperplane. The *dashed lines* parallel to the *solid* one show how much one can move the decision hyperplane without misclassification of the data. The patterns on the *dotted lines* are the support vectors. *(Reprinted with permission from Elsevier ® through RightsLink® license number: 4945510532522.)*

between minimizing the training error and maximizing the margin; $k(x, x^n)$ is a kernel function, which defines a similarity measure for x and x^n [46].

Accordingly, Moss et al. 2012 [46] used SVMs in order to classify 73 transdermal penetration enhancers into "good" and "bad" enhancers according to five physico-chemical descriptors and compared the results to several other machine learning methods [47,48]. SVMs were proven superior in classifying the data according to the Enhancement Ratio "ER."

In a similar study, support vector regression (SVR) which is an extension of the support vector machines as a pioneering classification algorithm was employed to model and predict the solvent effect on human skin permeability. SVR projects the descriptor matrix from the input space into a high-dimensional feature space via kernel functions and it was proven superior to random forest (RF) method as well [41].

Since the associated side effects, particularly irritation, that are usually encountered with the penetration enhancers or solvents usage [49], then the ability of the machine learning methods to accurately and robustly classify the chemicals that are highly efficient at enhancing the percutaneous absorption would obviously pose high potential for developing efficient formulations which can minimize the side effects [50]. Therefore, the main practical finding and benefit of this study is the provision of a proof-of-concept that the adopted machine learning methods were able to significantly reduce the false positive classification rate, which has significant implications for formulation design and efficiency. Reducing the waste associated with the false positive studies is a great asset from the pharmaceutical point of view.

4 Conclusion

The machine learning methods pose valuable tools in the different aspects of drug delivery. Their clustering power could group drugs into meaningful groups according to their main constitutional, electronic and physico-chemical properties and could interpret similarities in their loading on different carriers and matrices. Moreover, PCA and HCA, in particular, could capture the most stable formulations amongst others. The prediction utilities of methods such as partial least squares, artificial neural networks, and Gaussian processes could also help in the prediction and the determination of the loading behavior of various drugs in addition to their permeability coefficients through the different body barriers such as the skin. Moreover, support vector machines originated as successful tools in classifying penetration enhancers into "good" or "bad" and hence determine the safe ones from the pharmaceutical point of view. The different applications of the demonstrated machine learning methods can lead to great impact and achievements in the drug delivery field saving resources, efforts and time of wet lab experimentation and performing un-planned trials.

References

[1] S.S. Naguib, R.M. Hathout, S. Mansour, Optimizing novel penetration enhancing hybridized vesicles for augmenting the in-vivo effect of an anti-glaucoma drug, Drug Deliv. 24 (1) (2017) 99–108.

[2] S.A. Damiati, Digital Pharmaceutical Sciences, AAPS PharmSciTech 21 (2020) 206.

[3] A.P. Ferreira, M. Tobyn, Multivariate analysis in the pharmaceutical industry: enabling process understanding and improvement in the PAT and QbD era, Pharm. Dev. Technol. (2014) 1–15.

[4] H.A. Gad, S.H. El-Ahmady, M.I. Abou-Shoer, M.M. Al-Azizi, Application of chemometrics in authentication of herbal medicines: a review, Phytochem. Anal. 24 (2013) 1–24.

[5] R.M. Hathout, S.H. El-Ahmady, A.A. Metwally, Curcumin or bisdemethoxycurcumin for nose-to-brain treatment of Alzheimer disease? A bio/chemo-informatics case study, Nat. Prod. Res. 32 (2018) 2873–2881.

[6] Y. Yang, Z. Ye, Y. Su, Q. Zhao, X. Li, D. Ouyang, Deep learning for in vitro prediction of pharmaceutical formulations, Acta Pharm. Sin. B 9 (2019) 177–185.

[7] D.E. Jones, H. Ghandehari, J.C. Facelli, A review of the applications of data mining and machine learning for the prediction of biomedical properties of nanoparticles, Comput. Methods Prog. Biomed. 132 (2016) 93–103.

[8] S. Chan, V. Reddy, B. Myers, Q. Thibodeaux, N. Brownstone, W. Liao, Machine learning in dermatology: current applications, opportunities, and limitations, Dermatol. Ther. 10 (2020) 365–386.

[9] R.M. Hathout, Using principal component analysis in studying the transdermal delivery of a lipophilic drug from soft nano-colloidal carriers to develop a quantitative composition effect permeability relationship, Pharm. Dev. Technol. 19 (2014) 598–604.

[10] S. Chibani, F.X. Coudert, Machine learning approaches for the prediction of materials properties, APL Mater. 8 (2020), 080701.

[11] J. Vamathevan, D. Clark, P. Czodrowski, I. Dunham, E. Ferran, G. Lee, B. Li, A. Madabhushi, P. Shah, M. Spitzer, S. Zhao, Applications of machine learning in drug discovery and development, Nat. Rev. Drug Discov. 18 (2019) 463–477.

[12] R. Burbidge, M. Trotter, B. Buxton, S. Holden, Drug design by machine learning: support vector machines for pharmaceutical data analysis, Comput. Chem. 26 (2001) 5–14.

[13] G. Schneider, Neural networks are useful tools for drug design, Neural Netw. 13 (2000) 15–16.

[14] J. Sadowski, Optimization of chemical libraries by neural networks, Curr. Opin. Chem. Biol. 4 (2000) 280–282.

[15] I. Kövesdi, M. Dominguez, L. Orfi, G. Náray-Szabó, A. Varró, J.G. Papp, P. Mátyus, Application of neural networks in structure-activity relationships, Med. Res. Rev. 19 (1999) 249–269.

[16] A.A. Metwally, R.M. Hathout, Computer-assisted drug formulation design: novel approach in drug delivery, Mol. Pharm. 12 (2015) 2800–2810.

[17] O. Spjuth, T. Helmus, E.L. Willighagen, S. Kuhn, M. Eklund, J. Wagener, P. Murray-Rust, C. Steinbeck, J.E. Wikberg, Bioclipse: an open source workbench for chemo- and bioinformatics, BMC. Bioinf. 8 (2007) 59.

[18] P. Ashrafi, G.P. Moss, S.C. Wilkinson, N. Davey, Y. Sun, The application of machine learning to the modelling of percutaneous absorption: an overview and guide, SAR QSAR Environ. Res. 26 (2015) 181–204.

[19] L.T. Lam, Y. Sun, N. Davey, R. Adams, M. Prapopoulou, M.B. Brown, G.P. Moss, The application of feature selection to the development of Gaussian process models for percutaneous absorption, J. Pharm. Pharmacol. 62 (2010) 738–749.

[20] R.O. Potts, R.H. Guy, Predicting skin permeability, Pharm. Res. 9 (1992) 663–669.
[21] R.M. Hathout, A.A. Metwally, Towards better modelling of drug-loading in solid lipid nano-particles: molecular dynamics, docking experiments and Gaussian processes machine learn-ing, Eur. J. Pharm. Biopharm. 108 (2016) 262–268.
[22] S. Martins, I. Tho, E. Souto, D. Ferreira, M. Brandl, Multivariate design for the evaluation of lipid and surfactant composition effect for optimisation of lipid nanoparticles, Eur. J. Pharm. Sci. 45 (2012) 613–623.
[23] M. Ringnér, What is principal component analysis? Nat. Biotechnol. 26 (2008) 303–304.
[24] R.M. Hathout, T.J. Woodman, Applications of NMR in the characterization of pharmaceutical microemulsions, J. Control. Release 161 (2012) 62–72.
[25] S.K. Singh, P.R. Verma, B. Razdan, Development and characterization of a lovastatin-loaded self-microemulsifying drug delivery system, Pharm. Dev. Technol. 15 (2010) 469–483.
[26] S.K. Singh, P.R. Verma, B. Razdan, Glibenclamide-loaded self-nanoemulsifying drug delivery system: development and characterization, Drug Dev. Ind. Pharm. 36 (2010) 933–945.
[27] W. Fagir, R.M. Hathout, O.A. Sammour, A.H. ElShafeey, Self-microemulsifying systems of finasteride with enhanced oral bioavailability: multivariate statistical evaluation, characteriza-tion, spray-drying and in vivo studies in human volunteers, Nanomedicine (Lond) 10 (2015) 3373–3389.
[28] H.A. Gad, R.M. Hathout, Can the docking experiments select the optimum natural bio-macromolecule for doxorubicin delivery? J. Clust. Sci. (2020).
[29] S. Karthikeyan, N. Rajendra Prasad, A. Ganamani, E. Balamurugan, Anticancer activity of resveratrol-loaded gelatin nanoparticles on NCI-H460 non-small cell lung cancer cells, Biomed. Prev. Nutr. 3 (2013) 64–73.
[30] R.M. Hathout, S.G. Abdelhamid, A.A. Metwally, Chloroquine and hydroxychloroquine for combating COVID-19: investigating efficacy and hypothesizing new formulations using bio/chemoinformatics tools, Inform. Med. Unlocked 21 (2020) 100446.
[31] R.M. Hathout, A.A. Metwally, Gelatin nanoparticles, Methods Mol. Biol. 2000 (2019) 71–78.
[32] M. Shokry, R.M. Hathout, S. Mansour, Exploring gelatin nanoparticles as novel nanocarriers for Timolol maleate: augmented in-vivo efficacy and safe histological profile, Int. J. Pharm. 545 (2018) 229–239.
[33] H. Abdelrady, R.M. Hathout, R. Osman, I. Saleem, N.D. Mortada, Exploiting gelatin nanocar-riers in the pulmonary delivery of methotrexate for lung cancer therapy, Eur. J. Pharm. Sci. 133 (2019) 115–126.
[34] A. Malzert-Freon, D. Hennequin, S. Rault, Partial least squares analysis and mixture Design for the Study of the influence of composition variables on lipidic nanoparticle characteristics, J. Pharm. Sci. 99 (2010) 4603–4615.
[35] S. Hellberg, M. Sjoestroem, B. Skagerberg, S. Wold, Peptide quantitative structure-activity relationships, a multivariate approach, J. Med. Chem. 30 (1987) 1126–1135.
[36] M. Junaid, M. Lapins, M. Eklund, O. Spjuth, J.E. Wikberg, Proteochemometric modeling of the susceptibility of mutated variants of the HIV-1 virus to reverse transcriptase inhibitors, PLoS One 5 (2010) e14353.
[37] M. Lapins, J.E. Wikberg, Kinome-wide interaction modelling using alignment-based and alignment-independent approaches for kinase description and linear and non-linear data anal-ysis techniques, BMC Bioinf. 11 (2010) 339.
[38] H. Strombergsson, M. Lapins, G.J. Kleywegt, J.E. Wikberg, Towards proteome-wide interac-tion models using the proteochemometrics approach, Mol. Inform. 29 (2010) 499–508.
[39] M. Sandberg, L. Eriksson, J. Jonsson, M. Sjostrom, S. Wold, New chemical descriptors rele-vant for the design of biologically active peptides. A multivariate characterization of 87 amino acids, J. Med. Chem. 41 (1998) 2481–2491.

[40] G. Maccari, L.M. Di, R. Nifosi, F. Cardarelli, G. Signore, C. Boccardi, A. Bifone, Antimicrobial peptides design by evolutionary multiobjective optimization, PLoS Comput. Biol. 9 (2013) e1003212.

[41] H. Baba, J.i. Takahara, F. Yamashita, M. Hashida, Modeling and prediction of solvent effect on human skin permeability using support vector regression and random forest, Pharm. Res. 32 (2015) 3604–3617.

[42] S. Wold, J. Jonsson, M. Sjörström, M. Sandberg, S. Rännar, DNA and peptide sequences and chemical processes multivariately modelled by principal component analysis and partial least-squares projections to latent structures, Anal. Chim. Acta 277 (1993) 239–253.

[43] H.H. Lin, D. Ouyang, Y. Hu, Intelligent classifier: a tool to impel drug technology transfer from academia to industry, J. Pharm. Innov. 14 (2019) 28–34.

[44] A. Ose, K. Toshimoto, K. Ikeda, K. Maeda, S. Yoshida, F. Yamashita, M. Hashida, T. Ishida, Y. Akiyama, Y. Sugiyama, Development of a support vector machine-based system to predict whether a compound is a substrate of a given drug transporter using its chemical structure, J. Pharm. Sci. 105 (2016) 2222–2230.

[45] Q. Li, W. Zhou, D. Wang, S. Wang, Q. Li, Prediction of anticancer peptides using a low-dimensional feature model, Front. Bioeng. Biotechnol. 8 (2020) 892.

[46] G.P. Moss, A.J. Shah, R.G. Adams, N. Davey, S.C. Wilkinson, W.J. Pugh, Y. Sun, The application of discriminant analysis and machine learning methods as tools to identify and classify compounds with potential as transdermal enhancers, Eur. J. Pharm. Sci. 45 (2012) 116–127.

[47] W.J. Pugh, R. Wong, F. Falson, B.B. Michniak, G.P. Moss, Discriminant analysis as a tool to identify compounds with potential as transdermal enhancers, J. Pharm. Pharmacol. 57 (2005) 1389–1396.

[48] H. Baba, J.i. Takahara, H. Mamitsuka, In silico predictions of human skin permeability using nonlinear quantitative structure-property relationship models, Pharm. Res. 32 (2015) 2360–2371.

[49] A.C. Williams, B.W. Barry, Penetration enhancers, Adv. Drug Deliv. Rev. 56 (2004) 603–618.

[50] I. Furxhi, F. Murphy, M. Mullins, A. Arvanitis, C.A. Poland, Practices and trends of machine learning application in nanotoxicology, Nanomaterials 10 (1) (2020) 116.

Chapter 14

On the robust and stable flowshop scheduling under stochastic and dynamic disruptions

Zhaolong Yang[a], Fuqiang Li[b], and Feng Liu[b]
[a]*Hospital of Clinical Medicine Engineering Department, Deyang People's Hospital, Deyang, PR China, [b]School of Management Science and Engineering, Dongbei University of Finance and Economics, Dalian, PR China*

1 Introduction

As one of the most imperative processes in the manufacturing and service systems, scheduling centers on allocating limited resources (e.g., machines or tools) to a set of tasks (e.g., jobs or operations) during a visible planning horizon, such that one or more objectives with respect to tasks' completion times are optimized [1]. The output of scheduling is a baseline schedule that specifies the time/machine/operation assignments. Among the existing scheduling problems in various machine environments, the one in the flowshop environment has attracted an enormous attention from both academia and practice due to its challenging theoretical complexities and wide applications in assembling/manufacturing industries as well as medical operations [2–6].

The classic flowshop scheduling problems in existing literature mainly focus on generating a baseline schedule in a static and deterministic environment [7]. However, in practice, there also exist flowshop scheduling problems that involve stochastic and dynamic disruptions [8–10]. The scheduling process of the surgical operating theaters is a good example of such flowshop scheduling problem with stochastic and dynamic disruptions [2–5, 8]. When hospitals have to undergo the economic pressure or resource shortages, the efficiency of the surgical suite utilization is of great importance due to a high turnover rate of the medical equipment and materials and the patients' unwillingness of waiting a long time for the surgical operations. In such scheduling process, the medical devices are prone to break down stochastically, while the arrivals of emergency

Applications of Artificial Intelligence in Process Systems Engineering
https://doi.org/10.1016/B978-0-12-821092-5.00013-9

patients can be considered as dynamic events. Hospital managers must consider these stochastic and dynamic scenarios when they generate baseline schedules for the operating theaters and/or reschedule the subsequent operations according to the new or emergent patients' arrivals [4]. The overall goals of proactive-reactive scheduling and rescheduling include, but are not limited to, ensuring the patient safety and the optimal patient outcome, increasing the utilization of staff and equipment, reducing delays, and enhancing the overall staff, patients, and surgeon/physicians' satisfaction [4, 5]. Another example of flow-shop scheduling with stochastic and dynamic disruptions is the scheduling of *seru*—a cellular assembly system in the electronics industry [6]. Dynamic and high costs market conditions forced many electronic manufacturers to adopt the *seru* production system, which prioritizes responsiveness to the dynamic disruptions over cost reduction in setting a firm's operations strategy [6]. Since the baseline schedules are often infeasible due to these disruptions, the methods used in the classical flowshop scheduling problems must be adjusted in order to solve the abovementioned practical problems. Other similar examples can be found in steelmaking-continuous casting [11], TV fabrication, air conditioner manufacturers [12], petrochemical industry [13], etc. However, the results in literature considering stochastic and dynamic disruptions in the flowshop scheduling are very limited [2–6, 11, 12, 14–17].

To address this deficiency, we consider a flowshop scheduling problem with the stochastic and dynamic disruptions in both the planning and the execution stages. We adopt a *proactive-reactive* approach, that is, we first generate a baseline schedule under uncertainties, and then update it accordingly after a disruption occurs. In the problem setting, *robustness* and *stability* are considered simultaneously. *Robustness* refers to the insensitiveness of the schedule performance to the variation of the system parameters, while *stability* is used to capture the dissatisfactions of the system participants including the production manager, the shopfloor operator, and the customers. We consider human behaviors because they constitute a core part of the functioning and performance of a manufacturing or service system [10, 18, 19]. The contribution of this chapter can be summarized as follows. We propose a proactive-reactive approach to optimize the robustness and stability in a Pareto optimization manner, against stochastic machine breakdown and unexpected job-related events (dynamic new job arrival and job availability delay). We fill the gap in the flowshop scheduling field by incorporating the system participants' dissatisfactions. To the best of our knowledge, the research work that directly models human dissatisfactions and behaviors in the scheduling field is very limited (De Snoo et al. [19] is the only example we have found).

The remainder of this paper is organized as follows. Section 2 reviews the production scheduling literature that explicitly considers robustness and/or stability. In Section 3, the research problem is formally defined and formulated. In Section 4, we design the Evolutionary Multiobjective Optimization (EMO)-based algorithms to tackle the conflicts between the objectives in the proactive and reactive stages. Extensive computational experiments are conducted to demonstrate the effectiveness of our approach in Section 5. Finally, we conclude the paper with future research directions in Section 6.

2 Literature review

The scheduling models that explicitly optimize the robustness and/or stability for the baseline schedule against stochastic and dynamic disruptions are reviewed in this section. The relevant works are organized using the framework proposed in Goren and Sabuncuoglu [9], based on the manufacturing environment, the schedule generation, and the schedule evaluation, which optimizes the robustness and stability measures for the single-machine scheduling problems. Moreover, *"Multi-obj"* and *"BOM"* (Behavioral Operations Management) terms are added to fit our problem as in Table 1. *"Multi-obj"* indicates that the robustness and stability are optimized in a Pareto-optimization manner instead of using a linearly weighted combination of the duo. *"BOM"* indicates that the behaviors of the system participators are considered.

In the job shop environment, Leon et al. [20] were among the first to consider job shop scheduling with stochastic machine breakdowns. A convex combination of the expected C_{max} of the realized schedule and the expected performance degradation was used to measure robustness, which was approximated by a slack-based surrogate measure. Jensen [23] proposed another representative surrogate measure of robustness based on the current schedule's neighborhood, which was defined as the set of schedules generated from the incumbent by exchanging adjacent operations. Then the robustness for the schedule was measured by the average performance over its neighborhood. For the uncertain processing time, Zuo et al. [26] established a set of workflow models and accordingly proposed a multiobjective immune algorithm to obtain the Pareto-optimal robust schedules. Goren et al. [29] designed an exact branch-and-bound algorithm for the problem with a surrogate stability measure and proposed a Tabu Search method to deal with the stochastic machine breakdown. The above papers all assume a fixed job-machine assignment. Please refer to Xiong et al. [30] for more complicated flexible job shop scheduling problems that optimize both robustness and stability. Hamid et al. [33] proposed a new job shop scheduling problem which the objective is to minimize a weighted sum of make span and total outsourcing cost and examined two solution approaches of combinatorial optimization problems, that is, mathematical programming and constraint programming. Dai et al. [34] established a multiobjective optimization model to minimize energy consumption and make span for a flexible job shop scheduling problem with transportation constraints and then designed an enhanced genetic algorithm to solve the problem. Zhang et al. [35] thought we should transfer traditional scheduling into smart distributed scheduling (SDS) and explored a future research direction in SDS and constructed a framework on solving the JSP problem under Industry 4.0.

TABLE 1 Bibliography of scheduling literatures explicitly addressing robustness and/or stability for uncertain disruptions.

| | Environment | Static/ | Deterministic/ | | Schedule Generation And evaluation | | | | When to | How to |
	Initial	dynamic	stochastic	Method	robustness	stability	BOM	Multi-obj		
[20]	$J\|\|C_{max}$	Static	stoch brkdwn	GA	$\mathbb{E}(deviation)$	N/A	N/A	N/A	periodic	RS
[21]	$J\|\|L_{max}$	Static	stoch brkdwn	OSMH/LPH	N/A	$\|C_i^R - C_j^p\|$ based	N/A	N/A	periodic	RS
[22]	$1\|r_j\|\sum T_j$	Static	stoch brkdwn/p_i	OSMH ATC derivatives	N/A	$\|C_i^R - C_j^p\|$ based	N/A	N/A	continuous	RS/ATC derivatives
[23]	$1\|r_j\|C_{max}$	Static	brkdwn	GA	Neighborhood	N/A	N/A	N/A	continuous	RS/hillclimbing/reschedule
[24]	$1\|r_j\|\sum w_jU_j$	Static	stoch r_i	GA	Sample	Distance	N/A	Weighted sum	periodic	RS
[25]	$1\|r_j\|\sum w_jT_j$	Static	stoch brkdwn	GA	$\mathbb{E}(\sum w_jT_j)$	$\|C_i^R - C_j^p\|$ based	N/A	Weighted sum	periodic	RS
[9]	$1\|r_j\|C_{max}/\sum C_j/\sum T_j$	Static	stoch brkdwn	TS	$\mathbb{E}(C_{max}/\sum C_j/\sum T_j)$	$\|C_i^R - C_j^p\|$ based	N/A	Weighted sum	periodic/continuous	RS/reschedule
[26]	$J\|\|C_{max}$	Static	stoch p_i	VNIA	Standard deviation	N/A	N/A	N/A	periodic	RS
[27]	$1\|r_j\|\sum C_j/\sum T_j$	Static	stoch brkdwn/p_i	B&B/BS	$\mathbb{E}(\sum C_j/\sum T_j)$	$\|C_i^R - C_j^p\|$ based	N/A	N/A	periodic	RS

[28]	Hybrid $F\|\|C_{max}$	Static	scenarios of p_i	GA	Sample	N/A	N/A	N/A	periodic	RS
[29]	$J\|\|C_{max}$	Static	stoch brkdwn/p_i	B&B/TS	$\mathbb{E}(C_{max})$	$\|C_j^R - C_j^P\|$ based	N/A	Constraint	periodic	RS
[30]	Flexible $J\|\|C_{max}$	Static	stoch brkdwn	GA	Simulation/slack	N/A	N/A	N/A	periodic	RS
[31]	$F_2\|\|C_{max}$	Dynamic	stoch p_i	MIP	Regret	N/A In planning;$\|C_j^R - C_j^P\|$ based in reaction	N/A	Weighted sum	continuous	reschedule

stoch, stochastic; brkdwn, breakdown; B&B, branch-and-bound; BS, beam search; TS, tabu search; E($\bullet$), the expected value of a stochastic variable; p_j, w_j, r_j, job j's processing time, weight and release time; C_{max}, the make span; L_{max}, the maximum lateness; $\sum w_j U_j$, the total weighted number of tardy jobs; $\sum T_j$, the total tardiness; C_j^R and C_j^P, job j's completion time in the realized and planned baseline schedules. The absolute difference $|C_j^R - C_j^P|$ is commonly used to measure stability.

RS, Right-Shifting. "periodic" means to make decisions periodically. "continuous" means to reschedule all the unfinished jobs whenever a disruption occurs, also known as the event-driven rescheduling policy [32].

In the single machine environment, the model constraints often involve the job release time, and the objective is the due-date-related performance measure such as $\sum w_j U_j$ or $\sum w_j T_j$. O'Donovan et al. [22] proposed a proactive-reactive approach to deal with stochastic machine breakdowns and dynamic job processing time. Seavax et al. [24] proposed a hybrid genetic algorithm embedded with a simulation model to optimize the weighted sum of robustness and stability. Similarly, Liu et al. [25] calculated the weighted sum of robustness and stability using a surrogate measure of aggregated machine breakdowns. Goren and Sabuncuoglu [9] applied the surrogate measure proposed in Liu et al. [25] and compared different robustness and stability measures as well as rescheduling methods. Assuming job's processing time to be stochastic, Goren and Sabuncuoglu [27] analyzed the optimal properties of the problem's special cases, and thereafter designed an exact branch-and-bound algorithm for the small-sized problems and an effective Beam Search heuristic for the large-sized problems. After that, Liu et al. [36] considered the single-machine group scheduling problem with deterioration effect and ready times, then developed and exhaustively tested an algorithm based on enumeration, a heuristic algorithm and a branch-and-bound algorithm to determine the sequence of groups and the sequence of jobs to minimize the make span.

In the flowshop environment, Chaari et al. [28] used scenario modeling to consider jobs' stochastic processing time. Robustness was measured by the schedule's makespan deviation between all disrupted scenarios and the initial scenario without disruption. Rahmani and Heydari [31] built up a Mixed-Integer-Programming (MIP) model and proposed a predictive-reactive approach for dynamic arrivals of new jobs and uncertain processing times. In the predictive stage, a robust optimization method was used to obtain the baseline schedule, which was updated by rescheduling after unexpected jobs arrived in the reactive stage. Liu et al. [37] developed a branch-and-bound (B&B) algorithm to solve a two-machine flowshop problem with release dates in which the job processing times are variable according to a learning effect. The bicriterion is to minimize the weighted sum of make span and total completion time subject to release dates. Jose et al. [38] investigated how real-time information on the completion times of the jobs in a flowshop with variable processing times can be used to reschedule the jobs, then they showed that rescheduling policies pay off as long as the variability of the processing times is not very high, and only if the initially generated schedule is of good quality. Liao et al. [39] studied the permutation flow-shop scheduling problem with interval production time and established a min-max regret criterion-based robust scheduling model to decrease the infinite number of scenarios to a limited level. Mohanad et al. [40] used a multiobjective optimization model to minimize the effect of different real-time events in a scheduling problem of a permutation flowshop, after that, they demonstrated that the proposed multiobjective model is better than the other models in reducing the relative percentage deviation. Peng et al. [41] considered a hybrid flowshop rescheduling which contained simultaneously three types of dynamic events and established a mathematical model to minimize make span and system instability.

To summarize, through the literature review we can identify a few unsolved research questions: (1) most works only consider one type of disruptions, that is, machine availability. However, job-related events (unexpected arrival or delay) are also common in the flowshop scheduling problems. Models that consider all of these disruptions are needed; (2) only a few papers (Goren and Sabuncuoglu [9]; Seavax et al. [24]; Liu et al. [25]; Goren et al. [29]) have simultaneously considered robustness and stability in the proactive stage. However, as advocated by Goren et al. [29], more effective approaches are needed to see a trade-off between the two measures; (3) none of the reviewed articles consider the dissatisfactions of the system participants in reaction to disruptions, which however needs to be investigated and measured. Our research attempts to provide answers to these questions.

3 Problem formulation

The notations for parameters/variables used in defining the problem are summarized as follows. Notations in bold font denote vectors.

Notations	
i:	the machine index $\quad j$: $\quad\quad$ the job/customer index
k:	the system participator index. $k=1$ for the manager, $k=2$ for the shopfloor operator and $k=3$ for the customers
Parameters	
m:	the number of machines
n_0:	the number of jobs/customers for the initial problem before disruptions occur
p_{ij}:	the processing time of job j's operation on machine i, $1\leq i\leq m$, $1\leq j\leq n_0$
x:	the disruption for the system participator, that is, the degraded schedule performance for the manager, the variation of job sequence for the shopfloor operator, and the variation of jobs' completion time for the customers
$V(x)$:	the value variation of the system participator in a gaining or losing situation in Prospect Theory
α, β:	the degrees of risk-aversion and risk-preference in $V(x)$, when the behavioral subject faces gaining and losing situations, and $0<\alpha, \beta<1$
λ:	the risk-aversion coefficient in $V(x)$, and $\lambda>1$, meaning the subject is more sensitive to loss (see Kahneman and Tversky [20] for more details about the range of these parameters)
O_k:	the reference point for the system participators and $O_k\geq 0$, $k=1, 2, 3$
$\mu_k(x)$:	the dissatisfaction function of measuring stability for the system participators, $k=1, 2, 3$
Variables	
$\boldsymbol{\pi}=\{\pi_1,\pi_2,\ldots,\pi_{n_0}\}$:	a feasible schedule, where $\pi_j=[j]$, $[j]$ is the index of job in the jth position, $1\leq j\leq n_0$
$C_{i[j]}$:	the completion time of job $[j]$ on machine i in $\boldsymbol{\pi}$, $1\leq i\leq m$, $1\leq j\leq n_0$
C_{mj}^P:	the completion time of job j in the planned baseline schedule, $1\leq j\leq n_0$
C_{mj}^R:	the completion time of job j in the realized schedule, $1\leq j\leq n_0$

3.1 Deterministic flowshop problem

We begin our discussion with the deterministic flowshop scheduling problem to minimize the total flow time n_0 jobs, each of which corresponds to a customer, are to be processed on m machines. Each job must visit all machines exactly once in the same sequence. The buffer/storage capacities in between any two successive machines are supposed to be infinite, that is if a job finishes processing on a certain machine, it may leave immediately to start or wait for processing on the next machine. For each machine, only one operation of a job can be processed at one time. All machines and jobs are available at the beginning of the planning horizon.

For a given baseline schedule π, we have $C_{i[j]} = \max(C_{i,[j-1]}, C_{i-1,[j]}) + p_{i[j]}$, $2 \leq i \leq m$, $2 \leq j \leq n_0; C_{1[j]} = C_{1[j-1]} + p_{1[j]}$, $2 \leq j \leq n_0; C_{i[1]} = C_{i-1,[1]} + p_{i[1]}$, $2 \leq i \leq m$; and $C_{1[1]} = p_{1[1]}$. The completion time of each job in π is C_{mj}, $1 \leq j \leq n_0$, and the objective is to minimize the total flowtime $\sum_{j=1}^{n_0} C_{mj}$. This objective guarantees that jobs are rapidly turned around, and that the Work-in-Process inventory is minimized. Using the classic three-field representation scheme, we denote the deterministic problem as $F_m \| \sum_{j=1}^{n_0} C_{mj}$, which has been proved to be NP-hard in Garey et al. [42]. Many meta-heuristics have been designed to solve this problem (Yenisey and Yagmahan [7]), including Particle Swarm Optimization (PSO), Iterated Greedy algorithm (IGRIS), Discrete Differential Evolutionary algorithm (DDERLS), and Genetic Local Search algorithm (GLS). The more recent Quantum-Inspired Iterated Greedy (QIG) algorithm [43], which uses a hybrid representation of a job and a Q-bit for permutation and a new rotation gate to adaptively update the population, is very effective as it outperforms all the abovementioned algorithms in solving the Taillard benchmark instances. Therefore, we adopt QIG to solve our initial problem, and the obtained solutions will serve as benchmarks.

3.2 The proactive-reactive approach

According to their effects on the baseline schedule, we classify the disruptions into two categories: stochastic machine breakdown requiring repair and unexpected job-related disruptions (a number of jobs dynamically arrive or certain job in the baseline schedule is delayed with respect to its original starting time).

We adopt a proactive-reactive approach for this scenario. For a stochastic machine breakdown, we consider the general breakdown/repair planning horizon to be composed of several uptime and downtime (repair) periods. One uptime period is immediately followed by a downtime period and vice versa. Assume that the uptime and downtime periods are both subject to identical and independent distributions with probability density functions $g(t)$ and $h(t)$ obtained from historical data, respectively (Goren and Sabuncuoglu [9]). In the proactive stage, the objective is to minimize the robustness and stability of the baseline schedule. Preemption is not allowed unless the processing is interrupted by a certain machine breakdown. The *"Preempt-resume"* policy is adopted, which means partial processing is allowed when a job is interrupted, and only the remaining part can be processed after the machine is repaired. In contrast to a stochastic machine breakdown, we assume that the newly arriving jobs and the delay in job availability cannot be anticipated and can only be handled after they occur. Thus, we need to take appropriate rescheduling actions in the reactive stage to minimize the realized schedule performance and the deviation from the baseline schedule.

The formal robustness and stability measures of a given baseline schedule are defined. First, the stochastic machine breakdown and repair make the robustness objective measure also stochastic. In practice, the decision maker is often more concerned about C_{mj}^R than C_{mj}^P. Let $f_1 = \sum_{j=1}^{n_0} C_{mj}^R$ be the realized schedule performance. Following [9, 27], we use the expected realized performance $\mathbb{E}(f_1) = \mathbb{E}\left(\sum_{j=1}^{n_0} C_{mj}^R\right)$ as the robustness objective.

Stability measures the deviation between the realized and the baseline schedules when disruptions occur. We apply the Prospect Theory to quantify the dissatisfactions of the system participants to disruptions into the stability measure (Kahneman and Tversky [18]). As is shown in Fig. 1A, we define the value function $V(x)$ for the system participators

$$V(x) = \begin{cases} x^\alpha & x \geq 0 \\ -\lambda(-x)^\beta & x < 0 \end{cases} \tag{1}$$

where $0 < \alpha, \beta < 1, \lambda > 1.0$ is the reference point for judging whether a subject is losing or gaining. From Eq. (1), the behavioral subjects' reactions to gaining or losing are asymmetric, which makes our stability measure more practical and meaningful. The subject at the losing region is dissatisfied when the deviation is larger than what is expected. Thus, given the behavioral subject' deviation x from the baseline schedule and the reference pointO_k, we modify Eq. (1) by assuming that the subject is at the losing region for $x \geq O_k$.

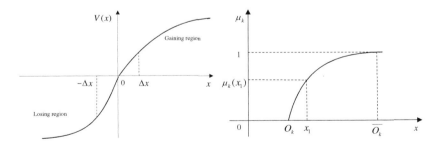

(a) Value function based on Prospect Theory   (b) Dissatisfaction for losing region

FIG. 1 Stability measure based on Prospect Theory from BOM perspective. (A) Value function based on Prospect Theory (B) Dissatisfaction for losing region.

The decision maker aims to mitigate the risks in the planning stage and reduce the deviations caused by disruptions in the execution stage. Because the deviation measures and reference points for the three behavioral subjects are different, Eq. (1) cannot be directly applied to measure the dissatisfactions. Therefore, we convert the value loss part of Eq. (1) $(x<0)$ to a function whose output is a real value number in $[0, 1]$, to represent the magnitude of dissatisfaction for a subject.

$$\mu_k(x) = \begin{cases} 1 & x \geq \overline{O_k} \\ \lambda_k(x/O_k - 1)^{\beta_k} & O_k \leq x < \overline{O_k} \\ 0 & 0 \leq x < O_k \end{cases} \tag{2}$$

where $\overline{O_k} = O_k\left(1 + \lambda_k^{-\beta_k^{-1}}\right)$ can be obtained by solving x from $\lambda_k(x/O_k - 1)^{\beta_k} = 1$ and the result defines the deviation value for a subject when he/she is totally dissatisfied, $k = 1, 2, 3$. 0 means full satisfaction and 1 means full dissatisfaction. The formula $\mu_k(x)$ is also illustrated in Fig. 1B. Refer to Kahneman and Tversky [18] for the details. With the increase of the deviation x, the subject's dissatisfaction $\mu_k(x)$ becomes stronger. When $x \geq \overline{O_k}$, the subject becomes totally dissatisfied. If the dissatisfaction accumulates over a long period without unleashing, we will lose the customers as well as the talented workers.

(1) *Manager's dissatisfaction*: the primary concern of the manager is the manufacturing cost, which is measured by $\sum_{j=1}^{n_0} C_{mj}^R$. The dissatisfaction of the manager is measured by

$$\mu_1\left(f_1^R\right) = \begin{cases} 1 & f_1^R \geq \overline{O_1} \\ \lambda_1\left(f_1^R/f_1^* - 1\right)^{\beta_1} & f_1^* \leq f_1^R < \overline{O_1} \\ 0 & 0 \leq f_1^R < f_1^* \end{cases} \tag{3}$$

where $\overline{O_1} = f_1^*\left(1 + \lambda_1^{-\beta_1^{-1}}\right)$, and $O_1 = f_1^*$ denotes the optimal manufacturing cost for the baseline schedule, and f_1^R denotes the manufacturing cost for the original n_0 jobs in the realized schedule obtained through QIG. For simplicity, Eq. (3) is abbreviated as μ_1.

(2) *Shopfloor operator's dissatisfaction*: the shopfloor operator mainly focuses on the resource reallocations and the shift rearrangements, which are reflected in the sequence variance with respect to the baseline plan (Hall and Potts [44]). Denote job j's positions in the planned and the realized schedules as s_j^P and s_j^R, respectively, $1 \leq j \leq n_0$. Taking job j's original position as the reference point, we can calculate the relative position variance for job j as $g_j = |s_j^P - s_j^R|/n_0$. It can be easily obtained that $0 \leq g_j < 1$. As the reference point for job j is $O_2 = 0$, the shopfloor operator's dissatisfaction for job j is measured by

$$\mu_2\left(g_j\right) = \begin{cases} 1 & g_j \geq \overline{O_2} \\ \lambda_2 g_j^{\beta_2} & 0 \leq g_j < \overline{O_2} \end{cases} \tag{4}$$

where $\overline{O_2} = \lambda_2^{-\beta_2^{-1}}$, and $1 \leq j \leq n_0$. For n_0 jobs in the proactive stage, the dissatisfaction of the shopfloor operator is measured by

$$\mu_2 = \frac{1}{n_0} \sum_{j=1}^{n_0} \mu_2\left(g_j\right) \tag{5}$$

(3) *Customer's dissatisfaction*: the customer's dissatisfaction is mainly caused by the deviation from the original delivery date (Hall and Potts [32]), which is C_{mj}^P in the baseline schedule, $1 \leq j \leq n_0$. The reference point for customer j is set as $O_3 = C_{mj}^P$. Following Kahneman and Tversky [18], we assume that customers are homogeneous, that is they use the same coefficients for risk-preference and risk-aversion. The dissatisfaction of customer j is measured by

$$\mu_3\left(C_{mj}^R\right) = \begin{cases} 1 & C_{mj}^R \geq \overline{O_3} \\ \lambda_3\left(C_{mj}^R / C_{mj}^P - 1\right)^{\beta_3} & C_{mj}^P \leq C_{mj}^R < \overline{O_3} \\ 0 & 0 \leq C_{mj}^R < C_{mj}^P \end{cases} \tag{6}$$

where $\overline{O_3} = C_{mj}^P\left(1 + \lambda_3^{-\beta_3^{-1}}\right)$, $1 \leq j \leq n_0$. The average dissatisfaction for all the customers is

$$\mu_3 = \frac{1}{n_0} \sum_{j=1}^{n_0} \mu_3\left(C_{mj}^R\right) \tag{7}$$

Since the stability is measured by the dissatisfaction degrees of the three participators, we use the average of the three functions Eqs. (5)–(7), $f_2 = (\mu_1 + \mu_2 + \mu_3)/3$, to represent the average dissatisfaction. Different weights can also be used to reflect different relative importance, but the results are largely the same. Incorporating the stochastic machine breakdown, we define the stability measure function based on Eqs. (3)–(7) as the expectation of f_2

$$\mathbb{E}(f_2) = \mathbb{E}((\mu_1 + \mu_2 + \mu_3)/3) \tag{8}$$

To summarize, our proactive-reactive approach works in the following way. First, a robust and stable schedule against the stochastic machine breakdown, is obtained by minimizing $\mathbb{E}(f_1)$ and $\mathbb{E}(f_2)$ in the proactive stage. Then the schedule is implemented. After new jobs arrive or the original jobs are delayed, we react to the dynamic events by re-optimizing the processing sequence of the original unfinished and new jobs, with two objectives of the schedule performance $\sum C_{mj}^R$ and the system participators' dissatisfaction $(\mu_1 + \mu_2 + \mu_3)/3$, using a hybridization-strategy-based EMO algorithm.

4 Hybridized evolutionary multiobjective optimization (EMO) methods

Our proposed problem is NP-hard because even a reduced form, that is the deterministic $F_m \| \sum_{j=1}^{n_o} C_{mj}$ is known to be NP-hard (Garey et al. [42]). Thus, meta-heuristics have attracted immense attention from the academia and industry [7, 45]. Please refer to Li and Yin [46] for an effective and efficient cuckoo search (CS)-based memetic algorithm, embedded with a largest-ranked-value (LRV)-rule-based random key and a NEH-based initialization. And also refer to Li and Yin [47] for a differential evolution (DE)-based memetic algorithm, which solves the benchmark problems well and finds 161 new upper bounds for the well-known difficult DMU problems. In order to examine the tradeoff between robustness and stability as suggested by Goren and Sabuncuoglu [9, 27] and Goren et al. [29], we design a hybridized Evolutionary Multiobjective Optimization (EMO) method to search for the Pareto front of robustness and stability. The most important advantage of this approach is that the decision maker can choose arbitrary solutions from a set of alternatives according to his/her preferences.

4.1 Evaluation of baseline schedule in the proactive stage

In the EMO approach, after a solution (a phenotype) is properly encoded into a genotype, the key issue is the evaluation of the genotype that can guild the searching process toward the desired regions with high-quality solutions. Genotype evaluations in the proactive and reactive stages are different. In the proactive stage, one cannot know exactly which machine would break down at what time and last for how long, whereas in the reactive stage all uncertain information has been revealed. The problem in the reactive stage is easier to solve because the realized schedules can be used to calculate the performance and the stability measure. Thus, the difficulty lies in the evaluation of baseline schedule in the proactive stage. To address this issue, we propose surrogate measures to estimate the impact of disruptions in the proactive stage. To this end, we make extensions to the existing measures and propose three pairs of surrogate measures, (R_i, S_i), $i = 1, 2, 3$, to approximate $(\mathbb{E}(f_1), \mathbb{E}(f_2))$.

4.1.1 Slack-based and neighborhood-based surrogates (R_1, S_1)

We extend the robustness measure based on the slack proposed by Leon et al. [20], and modify the neighborhood-based approach to measure the stability proposed by Jensen [23]. Given a baseline schedule π, let $s_{i[j]}^*$ and $C_{i[j]}^*, * \in \{e, l\}$, denote the starting time and completion time of the i-th operation of job $[j]$, where "e" in the superscript means earliest and "l" means latest. Since the original objective $\sum_{j=1}^{n_0} C_{mj}$ is regular, the current starting and completion times for all operations are $s_{i[j]}^e$ and $C_{i[j]}^e$. "Latest" indicates that any delay in an operation will not affect the completion times of the operations on the last machine, i.e., $\sum_{j=1}^{n_0} C_{mj}$ will not be affected. For each operation, we have $s_{i[j]}^l = C_{i[j]}^l - p_{i[j]}$, and the latest completion time of the operation is determined

$$C_{i[j]}^l = \begin{cases} \min\left\{ s_{i+1,[j]}^l, s_{i,[j+1]}^l \right\} & i = 1, \ldots, m-1, j = 2, \ldots, n_0 - 1 \\ s_{i+1,[j]}^l & i = 1, \ldots, m-1, j = n_0 \\ C_{i[j]}^e & i = 1, \ldots, m-1, j = 1; i = m, j = 1, \ldots, n_0 - 1 \end{cases}$$

(9)

In Xiong et al. [30], the slack of job $[j]$'s i-th operation $C_{i[j]}^l - C_{i[j]}^e$ is used to define the surrogate robustness measure $\sum_{i=1}^{m} \sum_{j=1}^{n_0} (C_{i[j]}^l - C_{i[j]}^e)$. We modify this measure by considering the probability of machine breakdown during $[C_{i[j]}^e, C_{i[j]}^l]$ to define R_1.

$$R_1 = \sum_{i=1}^{m} \sum_{j=1}^{n_0} \omega_{i[j]} \cdot \left(C_{i[j]}^l - C_{i[j]}^e \right)$$

(10)

where $\omega_{i[j]} = \int_{C_{i[j]}^e}^{C_{i[j]}^l} h(t) dt$, and $h(t)$ is the probability density function for the downtime.

Next, for a baseline schedule π to be evaluated, we define its neighborhood $N_1(\pi)$ as the set of schedules generated by exchanging any two adjacent jobs (this can be seen as a noise) in π, and $|N_1(\pi)| = n_0 - 1$. By modifying the mean–variance model for $N_1(\pi)$, we define the surrogate stability S_1 of π as

$$S_1 = \frac{1}{3} \left(\min_{\pi' \in N_1(\pi)} (f_2(\pi')) + \max_{\pi' \in N_1(\pi)} (f_2(\pi')) \right.$$
$$\left. + \text{mean}_{\pi' \in N_1(\pi)} (f_2(\pi')) \right) \cdot \text{var}_{\pi' \in N_1(\pi)} (f_2(\pi'))$$

(11)

where mean$(\cdot)$ and var$(\cdot)$ are the average value and the variance of the input variable.

This stability measure is used on $N_1(\pi)$ because there is no positive or apparent correlation between the mean and the variance. The first component of S_1 is used to depict the spread of performance, and to avoid over-estimation or under-estimation of the stability. The second component of S_1 penalizes the phenotype with a larger variance, so that the stable solutions are preferred. From Eqs. (10) and (11), both R_1 and S_1 can be regarded as the surrogate measures for evaluating a baseline schedule before uncertainty unfolds in the proactive stage.

4.1.2 Breakdown and repair estimation-based surrogates (R_2, S_2)

We apply the single machine estimation method used in Goren and Sabuncuoglu [9] to the flowshop case. This method enables us to quickly generate a realized schedule and calculate its performance for the baseline schedule. We construct an approximate schedule by inserting an estimated repair time into the baseline schedule after the machine has been busy for a given period. The underlying assumption is that the machine may possibly break down after a busy period and require repairing. The length of the period is given by $\omega L + (1 - \omega)U$, and the expected repair time is given by $\mathbb{E}(h(t))$. ω is a given weight in $[0, 1]$ (e.g., $\omega = 0.6$ in Goren and Sabuncuoglu [9]). L and U are the $(1000 \cdot \alpha/2)$-th and the $[1000 \cdot (1 - \alpha/2)]$-th tiles of the 1000-tile breakdown time distribution. The probability of machine breakdown during $[L, U]$ is $1 - \alpha$. In this chapter, we choose $\alpha = 0.05$. Given a baseline schedule π, we obtain a realized schedule π' after the repair insertion. Then the robustness and stability measures of π are estimated as $R_2 = f_1(\pi')$, $S_2 = f_2(\pi')$.

The repair time insertion method is different from the idle time insertion method used in Mehta and Uzsoy [21], where the authors did not use the machine breakdown information. We estimate the breakdown time using distribution tiles information.

4.1.3 Simulation-based surrogates (R_3, S_3)

We build a simple event simulation model to tackle uncertainty in the phenotype evaluation, as shown in Fig. 2. The *EMO meta-heuristic* passes each phenotype of the population to the *Simulation model* for evaluation. Uncertainty is simulated by variables that follow the machine breakdown and repair distributions $g(t)$ and $h(t)$. At the end of each run, a realized schedule can be obtained and the corresponding robustness and stability can be calculated. This process is repeated for N times in order to get a sample of robustness and stability, and the *Results analysis* returns the fitness of the phenotype.

For a given phenotype π, a sample of size N can be obtained through simulations, and R_3 and S_3 of π are then calculated through the mean-variance model as $R_3 = \frac{1}{3}(\min(f_1) + \max(f_1) + \text{mean}(f_1)) \cdot \text{var}(f_1)$, $S_3 = \frac{1}{3}(\min(f_2) + \max(f_2) + \text{mean}(f_2)) \cdot \text{var}(f_1)$.

Clearly different baseline schedule can be generated under a different measure (R_i, S_i), $i = 1, 2, 3$, and they need further performance evaluation, which will be elaborated in Section 5. From the computational perspective, (R_3, S_3) would

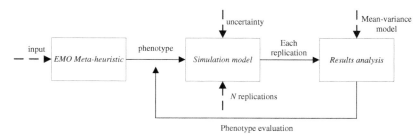

FIG. 2   Simulation-based stability and robustness approximation.

incur the highest complexity because each replication in the simulation can be seen as an implementation of (R_2, S_2). The complexity of (R_1, S_1) is also higher than (R_2, S_2) since the construction of the neighborhood involves $n_0 - 1$ times evaluations of the robustness.

4.2   A new hybridization strategy

According to Yenisey and Yagmahan [7], the meta-heuristics for solving permutation flowshop problems can be categorized into two classes: the population-based methods (Genetic Algorithm, Particle Swarming Algorithm, and Quantum-inspired Genetic Algorithm) and the trajectory methods (Iterated Greedy, Simulated Annealing, and Tabu Search). In the searching process, the population-based methods maintain and update a population with high parallelism, self-construction, self-learning, and self-adaptation. However, the local search ability needs to be strengthened. The trajectory methods start from a randomly generated solution, and then improve the initial solution iteratively according to various mechanisms until a near-optimal and satisfactory solution is found. Considering the strengths and weaknesses of the two types of methods, we propose a new hybridization strategy through initial solution improvement (Liu et al. [25]; Liu et al. [48]) and exploitation intensification (Pan et al. [49]). The quality of the initial solutions positively affect that of the final solutions. And exploitation intensification enhances the local search ability and helps escaping from the local optimum.

The hybridization strategy combines an inner loop for exploitation intensification with an outer loop for exploration. First, the initial solutions with high quality are generated and passed to the inner loop from the outer loop. Second, after receiving the initial solutions, the inner loop would exploit the solution space to balance exploration with exploitation using different searching mechanisms. Finally, the qualified solutions obtained from the two loops are selected and combined as the starting solutions for evolution in the next iteration. The hybrid strategy can lower the impact of the quality of the initial solutions on the algorithm performance and can also maintain those elite individuals by combining the results of the two loops. As is shown in Fig. 3, NSGA-II (Deb et al. [50]) is selected as the inner loop. The algorithm provides a unique Pareto front classification and crowding distance calculation which can maintain the proximity and diversity of the solutions during evolutions. When the outer loop method is population-based, the initial population in NSGA-II in each iteration is selected from the high-quality output in the outer loop. When the outer loop method is trajectory, the initial population is constituted by one of *popsize* − 1 randomly generated solutions, where *popsize* is NSGA-II's population size. In Fig. 3, we finally obtain four hybridized algorithms by implementing different searching mechanisms in the shaded rectangle.

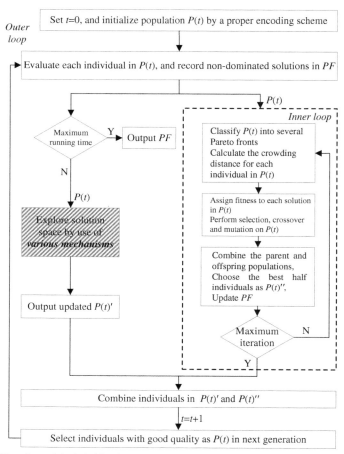

FIG. 3 Flowchart of the hybridization strategy.

4.3 Four hybridized EMO algorithms

For the population-based outer loop, the Quantum-inspired Genetic Algorithm (QGA) and the Particle Swarm Optimization (PSO) algorithm are selected. The unique Qubit encoding scheme in QGA is superior in maintaining the population diversity [45, 51]. Meanwhile, PSO algorithm searches the entire solution space randomly and iteratively, which is easy-to-apply with a low computational complexity. For the trajectory outer loop, the Simulate Annealing (SA) algorithm and the Quantum Iterated Greedy (QIG) algorithm are selected. SA is generally applicable for combinatorial optimization problems and is capable of escaping the local optimum through probabilistic jumping. Using a quantum-based encoding scheme, QIG, originally designed for solving $F_m||\sum_{j=1}^{n_0} C_{mj}$, searches the solution space iteratively in a greedy manner. It is anticipated that QIG can return promising results for our problem. For QGA, PSO, and SA in the outer loop that are not specially designed, the weighted linear combination is used to deal with the bi-objectives. The weights for the first and second objectives are $rand_1/(rand_1+rand_2)$ and $rand_2/(rand_1+rand_2)$, where $rand_1$ and $rand_2$ are uniformly generated from [0, 1]. So, the objective to be optimized in the outer loop is the weighted sum of the two given objectives.

4.3.1 Hybrid quantum genetic algorithm

In hybrid quantum genetic algorithm (HQGA), the outer loop is based on the QGA. First, the encoding scheme: $\begin{bmatrix} \alpha_1 & \alpha_2 & \cdots & \alpha_q \\ \beta_1 & \beta_2 & \cdots & \beta_q \end{bmatrix}$ which can be considered as a string of Qubit $(\alpha_l \ \beta_l)^T$ as the smallest information unit in the Quantum theory, is used to represent a job. α_l^2 and β_l^2 specify the probabilities that a Qubit is found at $|0\rangle$ state and $|1\rangle$ state, respectively, with $|\alpha_l|^2+|\beta_l|^2=1, l=1, 2, \ldots,$ q. Therefore, n_0 jobs will require a Qubit string of length n_0q, and $q=\lceil log_2n_0\rceil$, where the operator $\lceil\cdot\rceil$ returns the smallest integer that is no less than the input value.

As shown in Fig. 3, the input for the shaded rectangular in the outer loop of the HQGA is either an initialized population $P(t)$ through the Qubit encoding when $t=0$, or an updated population $P(t)$ when $t\geq0$. The population is obtained by classic genetic operations. First, all individuals in the population are ranked according to their fitness values. Then, the individual with the highest fitness is selected as one parent, while the other parent is determined by another randomly selected individual. A two-point crossover operator is implemented to produce an off-spring, and a swap or insert operation is used for mutation (Li and Wang [45]).

After genetic operations are completed, each Qubit is converted to 0 or 1 through the quantum collapse. For each job, a 0–1 binary string is obtained and then further converted to a decimal number. Following this process, a permutation of n_0 jobs can be achieved by applying a random-key representation to the n_0-length real value vector. During the evolution, each genotype is compared with the up-to-now best genotype. Then a rotation angle θ_l is obtained from the rotation angle table for performing the update operation $\begin{bmatrix} \alpha_l^{t+1} \\ \beta_l^{t+1} \end{bmatrix} = \begin{bmatrix} \cos(\theta_l) & -\sin(\theta_l) \\ \sin(\theta_l) & \cos(\theta_l) \end{bmatrix} \cdot \begin{bmatrix} \alpha_l^t \\ \beta_l^t \end{bmatrix}, l=1, 2, \ldots, q$, and t is the generation number (see Han and Kim [16] for more details).

4.3.2 Hybrid particle swarm optimization

In hybrid particle swarm optimization (HPSO), each permutation solution is encoded as a particle at a certain position which flies at a certain speed in the search space. For generation t, particle l's position vector is defined as $\mathbf{X}_l^t=\{x_{l1}^t,x_{l2}^t,\ldots,x_{ln_0}^t\}$, and its speed vector is given as $\mathbf{V}_l^t=\{v_{l1}^t,v_{l2}^t,\ldots,v_{ln_0}^t\}$,

$v_{lj} \leq V_{max}$, $1 \leq j \leq n_0$. $\mathbf{X}_l^t$ is real-value coded and can be converted to a jobs' sequence permutation through the random-key representation method. V_{max} is used to limit the region of particles. The j-th elements in $\mathbf{X}_l^t$ and $\mathbf{V}_l^t$ are defined for job [j], $1 \leq j \leq n_0$. Up to the t-th generation, the best position vector for particle l is recorded as $\mathbf{P}_l^t = \{p_{l1}^t, p_{l2}^t, ..., p_{ln_0}^t\}$. Let $\mathbf{g}_{best}$ be the best position vector of the entire particle population up to the t-th generation.

The speed and position of each particle must be updated to achieve information sharing, so that the entire population becomes regulated. The update operation is $\mathbf{V}_l^{t+1} = \omega \cdot \mathbf{V}_l^t + c_1 \cdot \gamma_1 \cdot (\mathbf{P}_l^t - \mathbf{X}_l^t) + c_2 \cdot \gamma_2 \cdot (\mathbf{g}_{best} - \mathbf{X}_l^t)$ and $\mathbf{X}_l^{t+1} = \mathbf{X}_l^t + \mathbf{V}_l^{t+1}$. ω is the coefficient for a particle to keep its current speed inertial, c_1 and c_2 are positive learning factors often assuming value 2, and γ_1 and γ_2 are random numbers sampled from a uniform distribution [0,1].

4.3.3 Hybrid quantum-inspired iteration greedy

The encoding scheme of hybrid quantum-inspired iteration greedy (HQIG) combines permutation and qubit as $\begin{bmatrix} 1 & 2 & ... & n_0 \\ \alpha_1 & \alpha_2 & ... & \alpha_{n_0} \\ \beta_1 & \beta_2 & ... & \beta_{n_0} \end{bmatrix}$ with $|\alpha_l|^2 + |\beta_l|^2 = 1$, $l = 1, 2, ..., n_0$. Unlike the encoding scheme in the HQGA, each Qubit in the HQIG stands for one job. All jobs are divided into two partial schedules through quantum collapse on the initial schedule $\boldsymbol{\varphi}_c$. Those jobs with state value 1 constitute a partial schedule $\boldsymbol{\varphi}_p$, whereas those jobs with 0 constitute another partial schedule Ω. Jobs in $\boldsymbol{\varphi}_p$ and Ω maintain their original sequences in $\boldsymbol{\varphi}_c$. Then the NEH procedure is applied: jobs in Ω are inserted sequentially into $\boldsymbol{\varphi}_p$ to form a complete schedule, which is then locally improved through the Variable Neighborhood Searching (VNS). When the NEH and VNS procedures are completed, a new schedule, $\boldsymbol{\varphi}_n$, is formed.

Let $\boldsymbol{\varphi}*$ be the best solution searched so far. $\boldsymbol{\varphi}_n$, $\boldsymbol{\varphi}_c$, and $\boldsymbol{\varphi}*$ are used to compute the rotation angle for the l-th Qubit: $\theta_l = sign(\alpha_l\beta_l) \cdot \{c_1\gamma_1[1 - f_1(\boldsymbol{\varphi}_n)/f_1(\boldsymbol{\varphi}_c)] + c_2\gamma_2[1 - f_1(\boldsymbol{\varphi}_n)/f_1(\boldsymbol{\varphi}*)]\}\pi$ with $c_1 = c_2 = 1$, γ_1, $\gamma_1 \in [0,1]$ and $sign(\cdot)$ returns the input's sign. We can see that the rotation angle is dynamically determined based on the relationship between solutions in HQIG. After the rotation angle is determined, the l-th Qubit is updated using the same rotation gate formula in HQGA.

4.3.4 Hybrid simulated annealing

In hybrid simulated annealing (HSA), the encoding scheme of the outer loop is permutation-based. The energy of the current permutation is E, which is defined as the objective value obtained by the weighted combination. A new permutation is obtained by perturbation, which is achieved through randomly designating two positions and swapping the jobs arranged on these two positions. The energy of the new permutation is denoted as E'. If $E - E' \geq 0$, the current permutation is replaced by the new one. Otherwise, whether or not to accept the new permutation depends on a probability $e^{(E-E')/T_t}$, and $T_t \geq 0$ is the annealing temperature for the iteration controller t.

A random number r is generated from $[0,1]$. If $e^{(E-E')/T_t} > r$, the new permutation is accepted, and T_{t+1} is updated as $T_{t+1} = \varepsilon \cdot T_t$ for next iteration ($\varepsilon \in [0,1]$ is the cooling down factor); otherwise, the new permutation is discarded. Based on the acceptance probability formula $e^{(E-E')/T_t}$, we can see that the higher the temperature is, the higher the probability of accepting an inferior solution is, and the more probable the SA jumps out of the local optimum. This iterative procedure terminates until the iteration controller t is larger than the given length of the Mapkob chain.

5 Computational study

The purpose of the computational study is to verify the hybridization strategy, to select the most effective one as the EMO method in the proactive-reactive approach, to evaluate the value of incorporating uncertainty into the baseline schedule, and to examine the impact of three pairs of the surrogate measures in the reactive stage when the stochastic breakdown and the unexpected events concurrently occur.

5.1 Testing instances

Since there exists no similar research that can be found in the literature, we are not able to directly compare our approach with any benchmark cases. Therefor we modify 90 flowshop instances in the TA class from the current Operations Research library for testing (Taillard [52]). We consider two types of disruption seriousness in two scenarios, only stochastic machine breakdown in scenario 1, and unexpected new job arrival, delay in job availability and machine breakdown in scenario 2.

In scenario 1, we use the Gamma distribution to describe the machine breakdown and the repair duration (Goren et al. [29]). Gamma distribution is determined by two parameters, shape, and scale. In this case, the shape parameter was set to be 0.7, and the scale was determined as $\delta \cdot \left(\frac{1}{mn_0}\sum_{i=1}^{m}\sum_{j=1}^{n_0}p_{ij}\right)/0.7$, where $\delta > 0$. A large value of δ indicates a long busy time and a low breakdown frequency. Similarly for the machine repair time, the shape parameter was set as 1.4, and the scale parameter was determined in a similar manner: $\varepsilon \cdot \left(\frac{1}{mn_0}\sum_{i=1}^{m}\sum_{j=1}^{n_0}p_{ij}\right)/1.4$, $\varepsilon > 0$, and a large value of ε indicates a long repair time. In scenario 2, for new job arrivals, let n_N be the number of new jobs. The problem becomes more difficult to solve with a larger n_N. The processing time of new jobs on each machine is randomly generated from the uniform distribution $[1,99]$. For delays in job availability, we randomly select one job in the baseline schedule and delay its planned starting time by a certain period. The length of that delay period is related to the average processing time that is adjusted by a positive coefficient $\tau > 0$: $\tau \cdot \left(\frac{1}{mn_0} \sum_{i=1}^{m} \sum_{j=1}^{n_0} p_{ij} \right)$.

The parameters in both scenarios are summarized as follows: in scenario 1, $\delta = 10$, $\varepsilon = 0.5$ for Type 1, and $\delta = 3$, $\varepsilon \doteq 1.5$ for Type 2. In scenario 2, $\delta = 10$, $\varepsilon = 0.5$, $n_N = 20$, $\tau = 0.5$ for Type 1, and $\delta = 3$, $\varepsilon = 1.5$, $n_N = 30$, $\tau = 1.5$ for Type 2. We can see that for each scenario, Type 2 implies a more difficult problem than Type 1.

5.2 Parameters setting

The parameters of NSGA-II are used both separately and in the inner loop shared by the four hybridized methods. In order to obtain high quality solutions during evolution, the probabilities of crossover mutation in the inner loop are both set to be 1. After genetic operations, the new and original populations are combined to one set, from which the good individuals are selected to form the population of the next generation. This process keeps good quality candidates as well as reduces the negative impact from the high crossover and mutation probabilities. The termination of the inner loop is controlled by a maximum iteration number set to be 10.

For HQGA, a string of Qubits with length q is used to represent a job. Since the maximum number of jobs in our paper is 130 (100 jobs in TA81-90 instances and maximum 30 new jobs), we have $2^7 < 130 < 2^8$. Thus, we set $q = 8$. The value of the rotation angle θ is determined by the rotation table in Han and Kim [51].

For HPSO, the speed inertial coefficient ω is 0.5. γ_1 and γ_1 are randomly sampled from [0,1]. The learning coefficients are set to be $c_1 = c_2 = 5$. The value of each particle position is in $[-100, 100]$, and its maximum speed is $V_{max} = 20$.

For HSA, the termination temperature is set to be 0.01, the cooling down factor is set to be 0.9, and the length of Mapkob chain is set to be 5. The tuning of the initial temperature T_0 is very important. Higher T_0 value incurs higher jumping probability, so high quality solutions might be missed. In the same time, lower T_0 values are prone to get the searching process trapped in the local optimum. After a pilot testing, we find that when $T_0 = 10$, the outer loop runs 7 iterations, the inner loop runs 10 iterations, and the temperature falls from 10 to 4 in the given running time. Thus, 10 is a reasonable value for T_0.

For HQIG, the initial solution is generated by applying the $LR(N/M)$ heuristic, and the parameters are set following Zhang and Li [43]: the number of local searches in the VNS is 1, and the probability of accepting an inferior solution is 0.01. The initialization of the Qubit encoding is $\alpha_l = \sqrt{1 - 9/n_0}$ and $\beta_l = \sqrt{9/n_0}$, $l = 1, \ldots, n_0$.

For all EMO methods, the running time is used to control the termination of the entire procedure. We follow the procedure in Nearchou [53] and set the population size of the NSGA-II, HQGA, and HPSO as $popsize = n_0$, and the maximum running time of each algorithm is $3n_0$ CPU seconds. All the algorithms are coded in C++ language, and are tested on a PC with a CPU of Intel (R) Core (TM) 2 Duo 3.00GHz and a 4G RAM.

Finally, we specify the parameters in the stability measure. For all three participators, the dissatisfaction parameters are set to be $\beta = 0.88$ and $\lambda = 2.25$ [18].

5.3 Performance metrics for the Pareto front

Since the EMO method is implemented in both proactive and reactive stages, the Pareto front of all nondominated solutions are calculated in two stages under various conditions. We outline several metrics in literature for measuring the Pareto front's performance (Li and Wang [45]). These metrics gauge the proximity to the optimal Pareto front and the diversity of the nondominated solutions.

First, the metrics for measuring the Pareto Front's proximity contain CM, D_{ave}, D_{max} and $Rate$. CM (C metric) is defined to quantify the dominance relationship between two Pareto fronts PF_1 and PF_2. $CM(PF_1, PF_2)$ maps an ordered pair (PF_1, PF_2) into a value in [0,1], which indicates the proportion of the solutions in PF_2 that are dominated by the solutions in PF_1. $CM(PF_1, PF_2) > CM(PF_2, PF_1)$ indicates PF_1 is better than PF_2. The distance metrics D_{ave} and D_{max} are defined as the average and the maximum distances between a given Pareto front PF and the optimal PF. The smaller the distance metrics are, the closer PF is to the optimal one. As the optimal PF is often unknown before hand. The common practice is to combine all the PFs for comparison into one set, and to construct the reference set $\mathcal{R}$ of the optimal PF from all the nondominated solutions in that set. D_{ave} and D_{max} are then calculated with respect to $\mathcal{R}$. $Rate$ is defined as the proportion of the number of solutions in PF that appear in $\mathcal{R}$. The larger the $Rate$ value, the better the proximity to the optimal PF.

Second, the metrics for measuring a PF's diversity contain the Overall Nondominated Vector Generation ($ONVG$) and the Tan's Spacing (TS). The $ONVG$ is defined as $|PF|$ for a given PF, meaning the number of nondominated solutions. The larger the $|PF|$ value is, the better diversity it has. The TS metric is calculated as the solution's Euclid distance to its nearest neighbor in the solution space for a given PF. The smaller the TS value is, the more uniformly distributed the solutions in the PF are, and the better diversity the PF has.

Third, the metrics for measuring the PF's average quality contain the Average Quality (AQ) metric, considering proximity, and diversity simultaneously. It is defined as the average scalarized function value over a representative weight sample. The smaller the AQ value is, the better average quality the given PF has.

5.4 Computational results

5.4.1 Analysis of the hybridization strategy

The benchmark problem instances TA081-090 are modified and tested with 30 new jobs arriving at the beginning of the baseline schedule. For each problem instance of TA081-090, independent cases are generated from the uniform processing time distribution [1, 99] of new jobs and solved by the EMO methods. In this paper we have developed four hybridized EMOs based on NSGA-II. Besides NSGA-II, we also make comparison with two state-of-the-art EMO algorithms, the multiobjective evolutionary algorithm based on decomposition (MOEA/D, Zhang and Li [54]) and the multiobjective memetic algorithms (MOMA, Ishibuchi et al. [55]). MOEA/D has successfully solved many difficult numerical testing problems, and outperforms NSGA-II and SPEA2 (the improved strength Pareto evolutionary algorithm) for the two objectives and three objectives benchmark flowshop scheduling instances (Chang et al. [56]). Note that the metrics in Section 5.3 are often used to compare two alternative Pareto fronts. Thus, we modify these metrics to ensure that seven Pareto fronts can be compared simultaneously. For $ONVG$, the metric for each method is $\frac{(ONVG_{max} - ONVG_*)}{ONVG_{max}} \cdot 100\%$, and for AQ, the metric is $\frac{(AQ_* - AQ_{min})}{AQ_{min}} \cdot 100\%$, where $* \in \{$HQIG, HSA, HPSO, HQGA, NSGA-II, MOEA/D, MOMA$\}$. TS is kept as originally designed because when the PF has only two elements, its value is zero. The rest metrics, CM, D_{ave}, D_{max}, and $Rate$, are calculated with respect to the reference set $\mathcal{R}$, which is formed by combining seven Pareto fronts and selecting the nondominated solutions in the set. $\mathcal{R}$ is used as the first input in CM's formula $CM(\mathcal{R}, \cdot)$. To summarize, only the $Rate$ metric, which is the proportion of solutions of each Pareto front in $\mathcal{R}$, is maximal and the other metrics are converted to be minimal. The comparison of PFs returned by seven algorithms for TA085 is shown in Fig. 4, and the rest results are similar. The detailed average values of all metrics are shown in Tables 2 and 3. The best value of seven algorithms for each metric is highlighted in bold font.

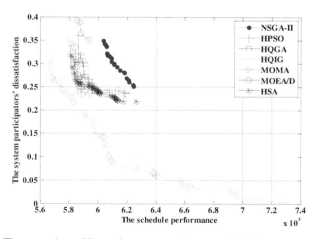

FIG. 4 The comparison of Pareto fronts returned by seven algorithms.

TABLE 2 Effectiveness of the hybridization strategy in ONVG, TS, and AQ.

Instances	ONVG							TS						
	N	I	S	P	G	D	E	N	I	S	P	G	D	E
TA081	0.14	**0.00**	0.30	0.56	0.48	0.95	0.59	**20.79**	36.06	21.52	32.80	27.43	84.15	19.66
TA082	0.57	**0.00**	0.54	0.44	0.50	0.96	0.92	17.63	42.17	23.97	26.35	20.45	**2.44**	108.22
TA083	0.35	**0.00**	0.35	0.38	0.52	0.94	0.37	20.76	64.74	28.72	19.56	25.66	28.44	**12.21**
TA084	0.50	**0.00**	0.27	0.77	0.50	0.94	0.64	25.81	62.69	19.07	11.52	28.21	**8.49**	16.26
TA085	0.71	**0.00**	0.35	0.69	0.70	0.95	0.70	25.50	39.46	36.02	24.59	**17.97**	39.17	47.30
TA086	0.65	**0.00**	0.61	0.68	0.66	0.95	0.86	27.55	42.63	**23.30**	115.92	33.21	38.88	84.57
TA087	0.56	**0.00**	0.56	0.59	0.46	0.93	0.77	20.22	104.5	**17.79**	34.78	17.94	48.42	62.72
TA088	0.69	**0.00**	0.55	0.49	0.43	0.95	0.91	51.97	26.43	20.46	41.51	**18.71**	39.00	101.50
TA089	0.29	**0.00**	0.27	0.58	0.45	0.95	0.95	**12.17**	54.60	52.10	31.86	25.68	40.85	171.24
TA090	0.67	**0.00**	0.44	0.43	0.51	0.97	0.71	33.80	24.72	29.05	10.36	25.61	**0.00**	106.12

Instances	AQ						
	N	I	S	P	G	D	E
TA081	0.03	**0.00**	0.03	0.04	0.04	0.06	0.01
TA082	0.04	**0.00**	0.03	0.05	0.03	0.04	0.01
TA083	0.03	**0.00**	0.02	0.02	0.02	0.05	0.00

Continued

TABLE 2 Effectiveness of the hybridization strategy in ONVG, TS, and AQ—cont'd

Instances	AQ						
	N	I	S	P	G	D	E
TA084	0.02	**0.00**	0.01	0.03	0.02	0.05	0.01
TA085	0.06	**0.00**	0.02	0.03	0.03	0.03	0.02
TA086	0.06	**0.00**	0.01	0.01	0.02	0.07	0.01
TA087	0.04	**0.00**	0.00	0.02	0.01	0.04	0.02
TA088	0.06	**0.00**	0.01	0.01	0.02	0.04	0.02
TA089	0.03	0.01	0.03	0.04	0.03	0.06	**0.00**
TA090	0.04	**0.00**	0.02	0.02	0.01	0.03	0.01

"N, I, S, P, G, D, E" stand for NSGA-II, HQIG, HSA, HPSO, HQGA, MOEA/D, and MOMA, respectively.

TABLE 3 Effectiveness of the hybridization strategy in CM, D_{ave}, D_{max} and Rate.

CM

Instances	N	I	S	P	G	D	E
TA081	0.58	**0.00**	0.72	0.53	0.50	0.70	0.13
TA082	0.64	**0.00**	0.45	0.51	0.48	0.26	0.29
TA083	0.53	**0.00**	0.56	0.64	0.44	0.53	0.55
TA084	0.58	**0.00**	0.36	0.33	0.39	0.36	0.19
TA085	0.51	**0.00**	0.52	0.49	0.39	0.29	0.29
TA086	0.82	**0.16**	0.40	0.36	0.77	0.73	0.37
TA087	0.51	**0.05**	0.27	0.35	0.33	0.38	0.54
TA088	0.79	**0.17**	0.37	0.68	0.59	0.65	0.49
TA089	0.57	0.39	**0.06**	0.56	0.56	0.62	0.36
TA090	0.57	**0.00**	0.43	0.41	0.30	0.38	0.43

D_{ave}

Instances	N	I	S	P	G	D	E
TA081	0.38	**0.00**	0.34	0.39	0.37	0.62	0.38
TA082	0.35	**0.00**	0.34	0.30	0.31	0.52	0.29
TA083	0.27	**0.00**	0.27	0.23	0.25	0.51	0.25
TA084	0.26	**0.00**	0.29	0.37	0.29	0.57	0.36
TA085	0.42	**0.00**	0.35	0.36	0.37	0.59	0.35
TA086	0.36	**0.00**	0.28	0.25	0.29	0.51	0.36
TA087	0.30	**0.00**	0.27	0.33	0.27	0.55	0.34
TA088	0.34	**0.00**	0.21	0.21	0.24	0.43	0.25
TA089	0.26	**0.01**	0.22	0.27	0.23	0.45	0.35
TA090	0.33	**0.00**	0.25	0.28	0.28	0.53	0.33

D_{max}

Instances	N	I	S	P	G	D	E
TA081	0.71	**0.00**	0.65	0.72	0.66	0.94	0.75
TA082	0.72	**0.00**	0.71	0.65	0.67	0.91	0.63
TA083	0.66	**0.00**	0.67	0.60	0.64	0.95	0.67
TA084	0.60	**0.00**	0.70	0.77	0.68	0.94	0.77

Rate

Instances	N	I	S	P	G	D	E
TA081	0.00	**1.00**	0.00	0.00	0.00	0.00	0.00
TA082	0.00	**1.00**	0.00	0.00	0.00	0.00	0.00
TA083	0.00	**0.95**	0.00	0.00	0.00	0.00	0.05
TA084	0.00	**1.00**	0.00	0.00	0.00	0.00	0.00

Continued

TABLE 3 Effectiveness of the hybridization strategy in CM, D_{ave}, D_{max} and Rate—cont'd

Instances	D_{max}							Rate						
	N	I	S	P	G	D	E	N	I	S	P	G	D	E
TA085	0.72	**0.00**	0.68	0.66	0.70	0.89	0.67	0.00	**1.00**	0.00	0.00	0.00	0.00	0.00
TA086	0.72	**0.02**	0.70	0.67	0.68	0.88	0.81	0.00	**0.84**	0.00	0.16	0.00	0.00	0.00
TA087	0.63	**0.00**	0.62	0.68	0.62	0.90	0.69	0.00	**0.94**	0.06	0.00	0.00	0.00	0.00
TA088	0.77	**0.02**	0.69	0.69	0.71	0.95	0.75	0.00	**0.79**	0.18	0.03	0.00	0.00	0.00
TA089	0.66	**0.07**	0.68	0.72	0.67	0.89	0.74	0.00	**0.60**	0.36	0.00	0.00	0.00	0.04
TA090	0.70	**0.00**	0.64	0.65	0.68	1.00	0.70	0.00	**0.88**	0.05	0.00	0.07	0.00	0.00

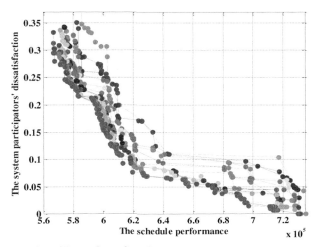

FIG. 5 The evolution of Pareto fronts for HQIG.

We explain the results of metrics in Tables 2 and 3. First, with respect to almost all metrics, HQIG shows the performance dominance, implying the Pareto front of HQIG has the best proximity and diversity. The evolution of Pareto fronts for HQIG in TA085 is shown in Fig. 5. This could be explained by that the original QIG is designed for $F_m || \sum_{j=1}^{n_0} C_{mj}$, and the rotation angle in HQIG is determined dynamically so that the perturbation strength during the evolution is very adaptive. This leads to better performances compared with the constant rotation angle method. The best *TS* metrics of 10 instances are evenly distributed among all methods. Second, some metric values of HSA, HQGA, and HPSO are comparable to those of NSGA-II, MOEA/D, and MOMA. See *ONVG*, *TS*, and *AQ*. It is implied that their average qualities are close to each other. However, the D_{ave} and D_{max} metrics of HSA, HQGA, and HPSO are better than NSGA-II, MOEA/D, and MOMA almost all instances, indicating closer distances to the reference set $\mathcal{R}$. To summarize, the effectiveness of the hybridization strategy is well demonstrated.

Because we used the algorithm running time as the termination controller, and the population-based meta-heuristics often take a long time to converge. It is highly likely that when a certain algorithm is forced to terminate at some point, the searching process is not completed. This case is shown in an instance of TA090 with 30 new jobs. If we allow the algorithm to run longer than the given running time, HQGA's and HPSO's performances would have been significantly improved. Thus, precisely speaking, only under the given running time, can HQIG and HSA provide better solutions than the other algorithms. Finally, comprehensively considering diversity, proximity, and efficiency, HQIG is selected as the EMO method for our proactive-reactive approach.

5.4.2 Analysis of robustness and stability measures against determined schedule

In this subsection, we evaluate the value of incorporating uncertainty into the baseline schedule. In particular, we compare the performance of the proactive schedule generated in (R_i, S_i), $i = 1, 2, 3$, with that of the determined schedule produced through QIG. As Scenario 1 only considers the stochastic machine breakdown, for which the surrogate measures of robustness and stability are designed, we perform experiments under Scenario 1 to evaluate the effectiveness of (R_i, S_i), $i = 1, 2, 3$, in obtaining the baseline schedule.

First, we apply HQIG to solve the 90 benchmark instances subject to the stochastic machine breakdown of types 1 and 2. (R_i, S_i), $i = 1, 2, 3$, can produce three Pareto fronts. Then, another simulation-based approach is applied. We randomly generate a sample of machine breakdowns using the gamma breakdown/repair distribution. This sample serves as a proactive schedule in each Pareto front. After that, we right-shift those affected operations until the machine is available again. The entire process is replicated 100 times. We then obtain two average values for the realized f_1 and f_2, which are used as approximations for $\mathbb{E}(f_1)$ and $\mathbb{E}(f_2)$ for evaluation purposes. For each approach, we generate the baseline schedules under (R_i, S_i), $i = 1, 2, 3$, and QIG. Thus, we have run a total of $90*2*100 = 18000$ simulation replications.

Note that the output of the proactive scheduling in the (R_i, S_i), $i = 1, 2, 3$, criteria contains a set of nondominated solutions. For the purpose of further analysis in the reactive stage, we apply weighted linear combination to determine the optimal solution from the set. An upper bound of f_1 first needs to be calculated for scalarization.

Denote the maximum processing time among all operations as p^{UB}- $= \max_{1 \le i \le m, \ 1 \le j \le n_0} \{p_{ij}\}$. The upper bound for the completion time of the first job on machine m can be calculated as mp^{UB}. The upper bound for the completion time of the second job is $(m+1)p^{UB}$, and so on so forth. Finally, the upper bound for f_1 is be $(n_0 m + 0.5 n_0 (n_0 - 1))p^{UB}$. We use the following linear combination to evaluate a baseline schedule: $\frac{1}{100} \sum_{num=1}^{100}$ $\left(0.2 f_1 ((n_0 m + 0.5 n_0 (n_0 - 1))p^{UB})^{-1} + 0.8 f_2 \right)_{num}$, where num denotes the num-th replication in the simulation, and we assume the stability is more of a concern than $\sum C_{mj}$.

The means and standard deviations of the four methods under various instance sizes and disruption types are shown in Fig. 6. The numbers in the horizontal axis, "1–9", represent TA20*5, TA20*10, TA20*20, TA50*5, TA50*10, TA50*20, TA100*5, TA100*10, and TA100*20 of type 1. Each group contains 10 instances. And numbers "10–18" represent the same 9 groups of the instances of type 2. From Fig. 6, we can see that for the problems of the same size, when disruptions from type 1 to type 2 become more severe, the average objective value increases for each baseline schedule. However, for each

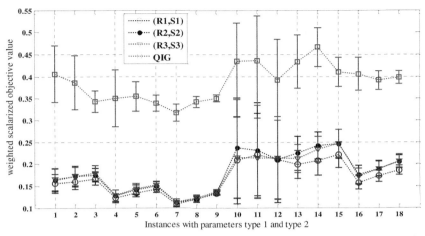

FIG. 6 The means and standard deviations of four methods on weighted scalarized objective under different instance size and disruption type.

objective function, as the problem size increases, no obvious trend can be observed because the upper bounds for different problems are different. In addition, the average objective values based on the (R_i, S_i), $i = 1, 2, 3$, criteria, have significant improvement in comparison with that of the baseline. The objective value can be reduced by more than 50% for all testing instances with either disruption type, by incorporating the uncertainty into the generation of the baseline schedules. What's more, the standard deviations derived by the four approaches are generally smaller than those derived by QIG, except for cases 10, 11, and 12. Thus, it can be concluded that the proactive approaches are generally more stable than the deterministic scheduling.

The variation trends of the weighted scalarized objective value are very close among (R_i, S_i), $i = 1, 2, 3$. However, these trends are apparently different from that of QIG. In order to explain this statistically, we ran the Friedman test on the 18,000-size sample. The results of the Friedman test are shown in Table 4. The P value equals 0.000 which is less than 0.1% indicates that the test results are acceptable at a 0.1% significance level. All three important parameters (Mean, Median, and Standard deviation) of (R_i, S_i), $i = 1, 2, 3$, are significantly better than those of QIG. Based on the Rank values, that is, 1.63 of (R_1, S_1), 2.15 of (R_2, S_2), 2.22 of (R_3, S_3), and 4.00 of QIG, we can conclude that the proactive scheduling approaches perform significantly better than the determined scheduling. The difference between (R_1, S_1) and (R_i, S_i), $i = 1, 2, 3$ is not significant, whereas no difference is observed between (R_2, S_2), and (R_3, S_3). If the running time is a primary concern for decision makers, (R_1, S_1) will be

TABLE 4 Statistical results of Friedman test for 3-related samples on weighted scalarized objective.

Methods	Mean	Stdev	Median	Rank	Sample size	χ^2	Degree of freedom	P value
(R_1, S_1)	0.17	0.06	0.16	1.63	18,000	34,526.03	3	0.000
(R_2, S_2)	0.18	0.06	0.17	2.15				
(R_3, S_3)	0.18	0.06	0.17	2.22				
QIG	0.39	0.07	0.38	4.00				

Stdev, standard deviation.

recommended as the robustness and stability measures; otherwise, (R_2, S_2) will be recommended.

5.4.3 Analysis of the reactive methods

In this subsection, we focus on a class of more complicated problems where various disruptions are handled simultaneously. The effectiveness of different reactions will also be evaluated accordingly. Recall in Scenario 2 of the testing environment, machine breakdown, new arriving jobs, and delay in job availability for $\tau \cdot \left(\frac{1}{mn_0} \sum_{i=1}^{m} \sum_{j=1}^{n_0} p_{ij} \right)$ are considered and simulated at the beginning of the planning horizon. For each combination of the scenario and parameter type, we apply the "RS" method by right-shifting the unfinished operations of jobs affected by the machine breakdown, adding the new arriving jobs at the end of the current queue, and inserting the delayed job into the available position.

For the purpose of comparisons, we reschedule all original and new jobs through HQIG, and repair the schedule if the delayed job is arranged before it is available using insertion in the RS. This rescheduling from the scratch approach is defined as "RES." The number of jobs in HQIG is updated from n_0 to $n_0 + n_N$. The population size is also updated accordingly, while the other parameters remain the same. For both RS and RES methods, we teste a total of $90*2*2 = 360$ problem instances. For each instance, 10 replications are simulated. One representative result of TA005 is shown in Fig. 7, where "SC" means scenario, RS and RES are the two reactions. The other results are similar, and therefore are not reported here.

From Fig. 7, we can observe that the RS often returns a solution with good stability, while the RES can effectively balance schedule performance with stability. However, Fig. 7C shows that the stability obtained from the RS is not necessarily optimal. Higher-level disruption parameters (type 2) always lead to larger dissatisfaction of the system participators than lower-level (type 1) parameters do. More alternative solutions are obtained with high quality. We can conclude that the proactive-reactive approach subject to both stochastic and dynamic disruptions is more appropriate for the manufacturing/service systems than the proactive scheduling approach.

6 Conclusions and future research directions

In this research, we study a permutation flowshop scheduling problem with the objective of minimizing robustness and stability. We develop a proactive-reactive approach to cope with stochastic and dynamic disruptions. Uncertain machine breakdown is tackled in the proactive stage by generating a robust and stable baseline schedule. The robustness measure is determined by the schedule performance, and the stability measure is determined by the behaviors of various participants using the Prospect Theory. Stochastic and dynamic disruptions are simultaneously considered in the reactive stage. In order to deal

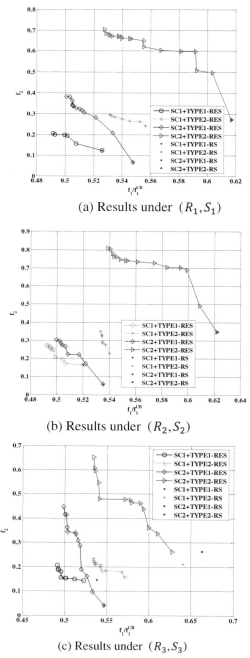

(a) Results under (R_1, S_1)

(b) Results under (R_2, S_2)

(c) Results under (R_3, S_3)

FIG. 7 Comparison of reactions to Scenarios 1 and 2 of disruptions types 1 and 2.

with the conflict between robustness and stability, we develop an algorithm hybridization strategy that is able to effectively search the Pareto front in each stage. The effectiveness of the proactive-reactive approach and algorithm hybridization strategy are demonstrated by extensive computational experiments on the Taillard flowshop benchmark instances.

Our research provides a contribution to the robustness and stability scheduling arena. Future research can be examined in the following two aspects. First, the proactive-reactive approach and the algorithm hybridization strategy can be adapted and tested in more practical flowshop problem settings, such as the distributed environment, the assemble line, and the no-idle permutation. Second, we apply the Prospect Theory in this paper to model the system participators' behaviors, but it might be more interesting to design behavioral lab experiments to study human reactions toward disruptions in the manufacturing/service systems. Behavioral operations management attracts significant intensions from researchers and practitioners in recent years, and we believe the parameters derived through this way would be more accurate.

Acknowledgments

This work was supported by the NSFC [grant number 71872033], the 2020 LiaoNing Revitalization Talents Program, and the Dalian High Level Talents Innovation Support Plan (2019RQ107).

References

[1] X. Li, M. Li, Multiobjective local search algorithm-based decomposition for multiobjective permutation flow shop scheduling problem, IEEE Trans. Eng. Manage. 62 (4) (2015) 544–557.

[2] B. Addis, G. Carello, A. Grosso, E. Tanfani, Operating room scheduling and rescheduling: a rolling horizon approach, Flex. Serv. Manuf. J. 28 (1–2) (2016) 206–232.

[3] K. Stuart, E. Kozan, Reactive scheduling model for the operating theatre, Flex. Serv. Manuf. J. 24 (4) (2012) 400–421.

[4] E. Erdem, X. Qu, J. Shi, Rescheduling of elective patients upon the arrival of emergency patients, Decis. Support. Syst. 54 (1) (2012) 551–563.

[5] M. Heydari, A. Soudi, Predictive / reactive planning and scheduling of a surgical suite with emergency patient arrival, J. Med. Syst. 40 (1) (2016) 1–9.

[6] Y. Yin, K.E. Stecke, M. Swink, I. Kaku, Lessons from seru production on manufacturing competitively in a high cost environment, J. Oper. Manag. (2017), https://doi.org/10.1016/j.jom.2017.01.003. Forthcoming.

[7] M.M. Yenisey, B. Yagmahan, Multi-objective permutation flow shop scheduling problem: literature review, classification and current trends, Omega 45 (2014) 119–135.

[8] K. Lee, L. Lei, M. Pinedo, S. Wang, Operations scheduling with multiple resources and transportation considerations, Int. J. Prod. Res. 51 (23–24) (2013) 7071–7090.

[9] S. Goren, I. Sabuncuoglu, Robustness and stability measures for scheduling: single-machine environment, IIE Trans. 40 (1) (2008) 66–83.

[10] F. Gino, G. Pisano, Toward a theory of behavioral operations, Manuf. Serv. Op. 10 (4) (2008) 676–691.

[11] J. Li, Q. Pan, K. Mao, A hybrid fruit fly optimization algorithm for the realistic hybrid flow-shop rescheduling problem in steelmaking systems, IEEE. T. Autom. Sci. Eng. 13 (2) (2016).

[12] H. Seidgar, M. Zandieh, H. Fazlollahtabar, I. Mahdavi, Simulated imperialist competitive algorithm in two-stage assembly flow shop with machine breakdowns and preventive maintenance, Proc. Inst. Mech. Eng. B. J. Eng. 230 (5) (2016) 934–953.

[13] D. Rahmani, R. Ramezanian, A stable reactive approach in dynamic flexible flow shop scheduling with unexpected disruptions: a case study, Comput. Ind. Eng. 98 (2016) 360–372.

[14] Y.H. Dong, J. Jang, Production rescheduling for machine breakdown at a job shop, Int. J. Prod. Res. 50 (10) (2012) 2681–2691.

[15] D. Tang, M. Dai, M. Salido, A. Giret, Energy-efficient dynamic scheduling for a flexible flow shop using an improved particle swarm optimization, Comput. Ind. 81 (2016) 82–95.

[16] E. Ahmadi, M. Zandieh, M. Farrokh, S.M. Emami, A multi-objective optimization approach for flexible job shop scheduling problem under random machine breakdown by evolutionary algorithms, Comput. Oper. Res. 73 (2016) 56–66.

[17] J. Framinan, P. Perez-Gonzalez, On heuristic solutions for the stochastic flowshop scheduling problem, Eur. J. Oper. Res. 246 (2) (2015) 413–420.

[18] D. Kahneman, A. Tversky, Choices, Values, and Frames, Cambridge University Press, 2000.

[19] C. De Snoo, W. Van Wezel, J.C. Wortmann, G.J. Gaalman, Coordination activities of human planners during rescheduling: case analysis and event handling procedure, Int. J. Prod. Res. 49 (7) (2011) 2101–2122.

[20] V.J. Leon, S.D. Wu, R.H. Storer, Robustness measures and robust scheduling for job shops, IIE Trans. 26 (5) (1994) 32–43.

[21] S.V. Mehta, R.M. Uzsoy, Predictable scheduling of a job shop subject to breakdowns, IEEE. Trans. Robotic. Autom. 14 (3) (1998) 365–378.

[22] R. O'Donovan, R. Uzsoy, K.N. McKay, Predictable scheduling of a single machine with breakdowns and sensitive jobs, Int. J. Prod. Res. 37 (18) (1999) 4217–4233.

[23] M.T. Jensen, Generating robust and flexible job shop schedules using genetic algorithms, IEEE. Trans. Evolut. Comput. 7 (3) (2003) 275–288.

[24] M. Sevaux, K.S. Rensen, A genetic algorithm for robust schedules in a one-machine environment with ready times and due dates, 4OR 2 (2) (2004).

[25] L. Liu, H. Gu, Y. Xi, Robust and stable scheduling of a single machine with random machine breakdowns, Int. J. Adv. Manuf. Tech. 31 (7–8) (2007) 645–654.

[26] X.Q. Zuo, H. Mo, J. Wu, A robust scheduling method based on a multi-objective immune algorithm, Inform. Sci. 179 (19) (2009) 3359–3369.

[27] S. Goren, I. Sabuncuoglu, Optimization of schedule robustness and stability under random machine breakdowns and processing time variability, IIE Trans. 42 (3) (2010) 203–220.

[28] T. Chaari, S. Chaabane, T. Loukil, D. Trentesaux, A genetic algorithm for robust hybrid flow shop scheduling, Int. J. Comput. Integ. Manuf. 24 (9) (2011) 821–833.

[29] S. Goren, I. Sabuncuoglu, U. Koc, Optimization of schedule stability and efficiency under processing time variability and random machine breakdowns in a job shop environment, Nav. Res. Log. 59 (1) (2012) 26–38.

[30] J. Xiong, L. Xing, Y. Chen, Robust scheduling for multi-objective flexible job-shop problems with random machine breakdowns, Int. J. Prod. Econ. 141 (1) (2013) 112–126.

[31] D. Rahmani, M. Heydari, Robust and stable flow shop scheduling with unexpected arrivals of new jobs and uncertain processing times, J. Manuf. Syst. 33 (1) (2014) 84–92.

[32] N.G. Hall, C.N. Potts, Rescheduling for job unavailability, Oper. Res. 58 (3) (2010) 746–755.

[33] H. Safarzadeh, F. Kianfar, Job shop scheduling with the option of jobs outsourcing, Int. J. Prod. Res. (2019), https://doi.org/10.1080/00207543.2019.1579934.

[34] D. Min, T. Dunbing, G. Adriana, A. Salido Miguel, Multi-objective optimization for energy-efficient flexible job shop scheduling problem with transportation constraints, Robot. Comput. Integr. Manuf. 59 (2019) 143–157.

[35] J. Zhang, G. Ding, Y. Zou, S. Qin, F. Jianlin, Review of job shop scheduling research and its new perspectives under industry 4.0, J. Intell. Manuf. 30 (2019) 1809–1830.

[36] F. Liu, J. Yang, Y.-Y. Lu, Solution algorithms for single-machine group scheduling with ready times and deteriorating jobs, Eng. Optim. (2019), https://doi.org/10.1080/0305215X.2018.1500562.

[37] J.-B. Wang, J. Xu, J. Yang, Bicriterion Optimization for Flow Shop with a Learning Effect Subject to Release Dates, 2018, p. 12. Article ID 9149510.

[38] J.M. Framinan, V. Fernandez-Viagas, P. Perez-Gonzalez, Using real-time information to reschedule jobs in a flowshop with variable processing times, Comput. Ind. Eng. 129 (2019) 113–125.

[39] W. Liao, F. Yanxiang, Min–max regret criterion-based robust model for the permutation flow-shop scheduling problem, Eng. Optim. (2020), https://doi.org/10.1080/0305215X.2019.1607848.

[40] M. Al-Behadili, D. Ouelhadj, D. Jones, Multi-objective biased randomised iterated greedy for robust permutation flow shop scheduling problem under disturbances, J. Oper. Res. Soc. (2019), https://doi.org/10.1080/01605682.2019.1630330.

[41] K. Peng, Q.-K. Pan, G. Liang, X. Li, S. Das, B. Zhang, A multi-start variable neighbourhood descent algorithm for hybrid flowshop rescheduling, Swarm Evol. Comput. 45 (2019) 92–112.

[42] M.R. Garey, D.S. Johnson, R. Sethi, The complexity of flowshop and jobshop scheduling, Math. Oper. Res. 1 (2) (1976) 117–129.

[43] Y. Zhang, X.P. Li, A quantum-inspired iterated greedy algorithm for permutation flowshops in a collaborative manufacturing environment, Int. J. Comput. Integ. Manuf. 25 (10) (2012) 924–933.

[44] N.G. Hall, C.N. Potts, Rescheduling for new orders, Oper. Res. 52 (3) (2004) 440–453.

[45] B. Li, L. Wang, A hybrid quantum-inspired genetic algorithm for multiobjective flow shop scheduling, IEEE. T. Syst. Man. Cyber. 37 (3) (2007) 576–591.

[46] X. Li, M. Yin, A hybrid cuckoo search via Lévy flights for the permutation flow shop scheduling problem, Int. J. Prod. Res. 51 (16) (2013) 4732–4754.

[47] X. Li, M. Yin, An opposition-based differential evolution algorithm for permutation flow shop scheduling based on diversity measure, Adv. Eng. Softw. 55 (2013) 10–31.

[48] F. Liu, J.J. Wang, D.L. Yang, Solving single machine scheduling under disruption with discounted costs by quantum-inspired hybrid heuristics, J. Manuf. Syst. 32 (4) (2013) 715–723.

[49] Q. Pan, R. Ruiz, An effective iterated greedy algorithm for the mixed no-idle permutation flowshop scheduling problem, Omega 44 (2014) 41–50.

[50] K. Deb, A. Pratap, S. Agarwal, T.A.M.T. Meyarivan, A fast and elitist multiobjective genetic algorithm: NSGA-II, IEEE. T. Evolut. Comput. 6 (2) (2002) 182–197.

[51] K. Han, J. Kim, Quantum-inspired evolutionary algorithm for a class of combinatorial optimization, IEEE. T. Evolut. Comput. 6 (6) (2002) 580–593.

[52] E. Taillard, Benchmarks for basic scheduling problems, Eur. J. Oper. Res. 64 (2) (1993) 278–285.

[53] A.C. Nearchou, Scheduling with controllable processing times and compression costs using population-based heuristics, Int. J. Prod. Res. 48 (23) (2010) 7043–7062.

[54] Q. Zhang, H. Li, MOEA/D: a multiobjective evolutionary algorithm based on de-composition, IEEE. Trans. Evolut. Comput. 11 (6) (2007) 712–731.

[55] H. Ishibuchi, T. Yoshida, T. Murata, Balance between genetic search and local search in memetic algorithms for multiobjective permutation flowshop scheduling, IEEE. Trans. Evolut. Comput. 7 (2) (2003) 204–223.

[56] P.C. Chang, S.H. Chen, Q. Zhang, J.L. Li, MOEA/D for flowshop scheduling problems, in: Proceedings of IEEE CEC, 2008, pp. 1433–1438.

Chapter 15

Bi-level model reductions for multiscale stochastic optimization of cooling water system

Qiping Zhu[a,b] and Chang He[a]
[a]*School of Materials Science and Engineering, Guangdong Engineering Centre for Petrochemical Energy Conservation, Sun Yat-sen University, Guangzhou, China,* [b]*College of Chemistry and Bioengineering, Guilin University of Technology, Guilin, China*

1 Introduction

In the thermo-electric and other energy-intensive sectors, the circulating water loss especially the evaporation and blowdown losses from open cooling towers contributes a major share of the consumption of freshwater during plant operation. Incorporating the newly developed water-saving cooling towers in the cooling water system has been recognized as a viable path to reduce water consumption [1–3]. A closed-wet cooling tower (CWCT) is a promising closed-type cooling device that is specifically designed for the cooling down of circulating water or process streams. Nevertheless, it remains a challenging task to gain a better understanding of the operation of CWCTs, as the flow behaviors inside CWCT s are characterized by the complex counter-current interactions between multiphase flow dynamics and heat transfer processes [4]. Besides, the CWCTs acting as the core of the closed-loop circulating water system are sensitive to the changes in outdoor weather conditions, which should be switchable to various operating modes in distinct seasons.

The acquisition of a better understanding of CWCT models can be realized through multiphase field numerical simulation of the original full-order model (FOM) with the advance of computational fluid dynamics (CFD). Nevertheless, the solution of the principled CFD models involves tracking the multiphase interactions and the spatial movement of each liquid droplet characterized by computationally expensive partial differential and algebraic equations. For instance, a single task of CFD simulation of 3D CWCT model normally takes

Applications of Artificial Intelligence in Process Systems Engineering
https://doi.org/10.1016/B978-0-12-821092-5.00016-4

417

more than 30 CPU hours using Reynolds average Navier-Stokes equations [2]. This computational burden becomes more critical as the FOMs are encountered in an equation-oriented optimization task because a large number of recourses to the numerical models have to be executed before converging to the optimum FOMs. In this regard, the benefits of FOMs must be weighed against the time necessary to obtain them, which is a major cost driver and impedes developing better processes.

In recent years, machine learning algorithms have been widely used in the process system engineering field for constructing the reduced-order models of complex FOMs. ROM is a model reduction technique that involves small degrees of freedom, which reproduces the behavior of an actual physical system or a reference FOM without loss of fidelity [5]. The ROM of a CFD-based model is a surrogate model for input-to-output mapping due to the computationally expensive nature of high-fidelity simulation which entails iterative calculations. It only approximates the input-output relation of a system is one of the practical solutions to construct a more tractable model as an alternative for the computationally expensive FOMs without loss of fidelity. For example, given input parameters (i.e., initial, boundary, and operational conditions), the quantities of interest for calculating flow at each mesh point, such as velocity, pressure, temperature, and moisture content can be obtained rapidly without conducting the principled CFD simulations.

Over the past decades, the benefits of ROMs have been evaluated for many chemical processes, for example, pressure swing adsorption [6], solid handling [7–9], fluidized bed adsorption [10], complex distillation [11], oil and gas [12–14], and fluids [15], etc. Generally, the model reduction approaches are roughly categorized into two classes: projection-based ROMs and data-fit ROMs. In projection-based ROMs, a reduced basis is extracted from the simulation data using an unsupervised learning technique, for example, principal component analysis (PCA), and the FOM operator is projected onto the subspace spanned by the reduced basis. PCA is based on the idea that a reduced number of orthonormal basis vectors in a subspace can be used to express a random vector of higher dimensionality without significant loss of information. The degrees of freedom of the system can be significantly reduced, and meanwhile, the underlying structure of the FOMs can be retained to a certain extent. Another way to enable rapid simulations is to build a data-fit model, where a response surface of the system is learned from the simulation data in a supervised manner. Recent work explores the application of ROMs in multi-scale integration by mapping process outputs to input variables using a meta-modeling approach such as polynomial response [16, 17], artificial neural network [18–20], and Kriging interpolation [11, 21, 22]. For example, it becomes essential to wrap reduced models (RMs) to fit multiscale models for material design and detailed unit operations (CFD, FOMs) into the overall process system. This multiscale integration further enables researchers to simultaneously consider the heterogeneous collection of micro- and macroscale model and the optimization of these integrated models, for example, supply chain system [23], coal gasification [20], and other process systems [24–26], using efficient nonlinear programming techniques [27–29]. General information about ROM-based multiscale process optimization can be found in the reviews by Floudas et al. [30] and Biegler [31]. To date, there are quite few studies focusing on the ROMs of cooling towers, such as those by Gao et al. [32], Wei et al. [33], and Qasim et al. [34].

The uncertain environmental parameters such as ambient temperature and relative humidity play a key role in determining the thermodynamic behaviors of a cooling tower under the same operating conditions. The existing studies typically assume that the weather conditions are perfectly known or even fixed in advance without seasonal variability, leading to an over-optimistic design or underestimation of the negative impacts on the performance of the cooling tower systems [35, 36]. In practice, as the core of the closed-loop cooling water system, the operation of a cooling tower is closely related to the fluctuating weather conditions, and multiple types of exogenous uncertainties (i.e., temperature and humidity) could strongly affect the characteristics of heat and mass transfer processes. The situation is worse especially in extreme weather conditions, such as over-heated or over-humid weather which usually happen during summer. In this case, there may be abnormal operations in which the cooling targets of the cooling tower system cannot be achieved, due to the relatively small evaporation on the surface of air-liquid contact units.

In this chapter, we develop an integrated framework based on model reductions for multiscale optimization of the CWCT-based cooling water system with consideration of the environmental variations. The model reduction methodology used for generating bi-level ROMs for CWCTs included four statistical steps: optimal DoE, multisample CFD simulation, reduced model construction, and model evaluation. A small-size circulating water system is illustrated to introduce the detailed constructions of the bi-level ROMs (Example I) and the multiscale stochastic optimization model (Example II).

This chapter is derived from the work by Zhu et al. [37].

2 Methodology

Given a geometry of an equipment unit, the construction of its bi-level ROMs involves three types of variable spaces: input space $\mathbf{X}$, output space $\mathbf{Y}$, and state-space $\mathbf{Z}$ in the fluid field. Fig. 1 shows the interconnection of these three variable spaces. As shown, the process geometry has n inputs and m outputs while s state variables are monitored in a total of p discretized elements. The relationships of the state and output variables corresponding to its input variables can be expressed by the following state-space model [20]:

$$\begin{bmatrix} y \\ z \end{bmatrix} = \begin{bmatrix} B \\ D \end{bmatrix} \times x \tag{1}$$

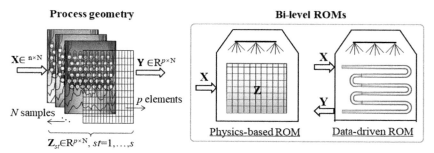

FIG. 1   Interconnection of the three variable spaces used in building bi-level ROMs.

where B and D are the coefficient matrices that can be derived using the snapshot PCA approach; x, y, and z are the input, output, and state vectors, respectively, which contain the relevant variables of interest. The state vector is spatially distributed and bounded by the geometry of the equipment.

In this study, there are two kinds of ROMs: data-driven ROM and physics-based ROM. The former is the mapping from the input-output stream data of the geometry of an equipment unit through field calculation. It can be used as a "surrogate module" to be wrapped to fit the modular framework of process flowsheet integration, or to be applied in iterative model-based optimization with advanced equation-oriented algorithms [7]. The latter displays and check the field profiles of the state variable, which assists researchers in observing process behavior inside the equipment. To develop the bi-level ROMs, much effort is required to collect and reduce the mapping data from a batch of rigorous CFD simulations of the equipment unit. Fig. 2 shows the proposed model reduction methodology for processing input data and obtaining bi-level ROMs of the CWCTs. It consists of four statistical steps.

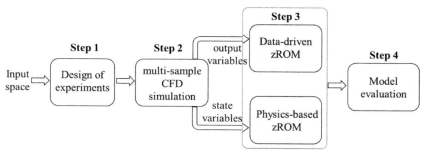

FIG. 2   The proposed methodology for bi-level model reduction.

Step 1: Implement a DoE in the specific range to obtain a finite set of samples over the input domain.

Step 2: Solve the CFD cases one by one with the obtained samples under the defined mesh system. The results of the state and output variables of interest are retrieved from the information stored in the CFD solutions.

Step 3.1: Implement the mapping $\mathbf{X} \to \mathbf{Y}$ and then formulate the data-driven ROM.

Step 3.2: PCA is performed for the dimension reduction of obtained $\mathbf{Z}$ to obtain the ranked principal components and PCA score Φ. Then, implement the mapping $\mathbf{X} \to \Phi$ and then formulate the physics-based ROM.

Step 4: Evaluate both the training and generation performances.

More detailed information about these four steps are as follows.

3 Optimal design of experiments

DoE is a well-suited approach to reduce resources and time, as it can maximize the amount of information obtained from a finite size of calculations by properly selecting experimental design points. As seen in Fig. 3, it starts with a parameter characterization for input space to obtain the probability

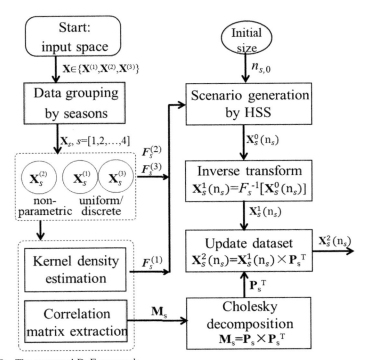

FIG. 3 The proposed DoE approach.

density functions (PDFs) of multivariate. Herein, the operating variables $\mathbf{X}^{(1)} = [v^a, m^{sw}, T^{sw}, T^{cw}, m^{cw}]$ can be directly characterized by uniform distributions with known bounds according to the operating constraints, while the design variables $\mathbf{X}^{(3)} = [d^{tube}, N^{tube}]$ are discrete points provided by the equipment manufacturer. Besides, note that the uncertain variables $\mathbf{X}^{(2)} = [T^{db}, \varphi^{rh}]$, depending on the actual weather conditions normally fall into distinct temperature-intervals in different seasons and thus are hard to be described by any standard parametric probability distribution models like Normal distribution. Thus, the original data in the input space requires pretreatment and segregation into four distinct groups by seasons $\mathbf{X}_s^{(2)}$, $s \in (1, ..., 4)$. These grouped datasets are then processed by the Gaussian Kernel density estimation stated in Eq. (2) that provides a nonparametric alternative to approximate the cumulative distribution function (CDF, F_s).

$$F_s^{(1)}(x) = \frac{1}{n_s \sigma_s} \left(\frac{4}{3n_s} \right)^{-1/5} \sum_{i=1}^{n_s} G\left[x - \mathbf{X}_s^{(2)}(i) \right] \ s \in \{1, 2, ..., 4\} \qquad (2)$$

$$G(x) = \int_{-\infty}^{x} \frac{1}{\sqrt{2\pi}} \exp\left(-\frac{1}{2}t^2 \right) dt \qquad (3)$$

where σ_s and n_s are the variance and size of $\mathbf{X}_s^{(2)}$.

Once all input variables are characterized, the developed probability distribution models are discretized into a finite set of distinct samples for performing stochastic CFD simulation. In this study, an efficient sampling technique, Hammersley sequence sampling (HSS) based on Hammersley points [38, 39] for evenly placing the sample points on a k-dimensional hypercube is employed to ensure that the sample set is more representative of the original data. Note that the uncertain variables of interest have a positive symmetric correlation, for example, the profile of saturated humidity generally has an evident decrease as the temperature is reduced from warm to cold seasons. This would badly distort the actual multivariate distribution as well as the uniform property of the sampled points over the input space. Thus, the sample set $\mathbf{X}_s^0$ should be rearranged via the implementation of rank correlations to guarantee the independence of the dataset and to obtain almost the same probability distribution for each variable with the historical data. This work utilizes the Pearson correlation coefficient (PCC, ρ) to define the correlation structure $\mathbf{M}_s$ among all uncertain variables of interest. At this point, $\mathbf{X}_s^0$ is mapped to the inverse CDF, resulting in an inverse transformed dataset $\mathbf{X}_s^{\,1}$. Multiplying dataset $\mathbf{X}_s^{\,1}$ with the lower triangular matrix $\mathbf{P}_s^T$ (by Cholesky decomposition [40] of $\mathbf{M}_s$, $\mathbf{M}_s = \mathbf{P}_s \times \mathbf{P}_s^T$), we can obtain $\mathbf{X}_s^2$ as the solution of DoE.

$$\rho_s(i, j) = \frac{\text{cov}\left[\mathbf{X}_s^{(2)}(i), \mathbf{X}_s^{(2)}(j) \right]}{\sigma\left[\mathbf{X}_s^{(2)}(i) \right] \sigma\left[\mathbf{X}_s^{(2)}(j) \right]} \ \forall i \neq j \in \{T^{db}, T^{dw}, \varphi^{rh}\}; \ s \in \{1, 2, ..., 4\} \qquad (4)$$

$$\mathbf{M}_s = \begin{bmatrix} \rho_s(1,1) & \rho_s(1,2) & \rho_s(1,3) \\ \rho_s(2,1) & \rho_s(2,2) & \rho_s(2,3) \\ \rho_s(3,1) & \rho_s(3,2) & \rho_s(3,3) \end{bmatrix} \qquad (5)$$

4 Multisample CFD simulation

Fig. 4A shows the experimental set-up of the staggered tube bundle CWCT under investigation, which mainly consists of the tube bundle, eliminator, fan, pump, and tank [2, 4]. The spray water is pumped out of a tank and split equally to the nozzles on top of the tower, then flows through the staggered tubes and returns to the tank. Air flows in the opposite direction from the bottom-up and successively traverses the blind window, outside tubes and mist eliminator, and finally out of the tower with suction from the draft fan. The hot circulating water flows into the CWCT from the left side, and then passes through the staggered tubes and goes back to the boiler.

In CWCTs, the heat and mass transfer processes of water film are mainly sustained by the joint effect of counter-current spray water and ambient air. A small proportion of spray water diffuses together with the upward airflow due to evaporation, and the remaining portion leaving the water film is collected in a water tank and finally returns to the top nozzles through the spray pump. Thus, the circulating water relies on the film evaporation to achieve the required cooling target, such as temperature drop and heat dissipation. This study focuses on the heat and mass transfer processes of the water film outside the tube bundle of CWCTs, which are simulated by a 2-D CFD model of flow, temperature, and pressure fields. The cross-section of the tube bundle is in a triangular arrangement and the arbitrary adjacent tubes in each row have the same centerline spacing, as shown in Fig. 4B.

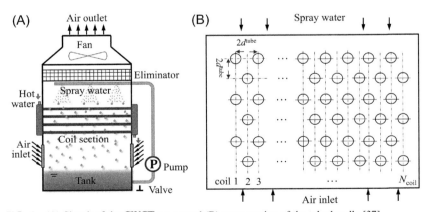

FIG. 4   (A) Sketch of the CWCT set-up and (B) cross section of the tube bundle [37].

To simplify the calculation, several assumptions are made as follows: (1) the outer wall of the coil tubes have no slips; (2) the circulating water and inlet air are both incompressible fluids; (3) the surface of the coil tubes is evenly covered by water film, while the airflow and spray water are uniformly distributed throughout the tower; (4) the temperature of the water film is equal to the average temperature of circulating water; and (5) the thermal radiation between the tower and the surroundings is neglected. Besides, note that a commercial scale CWCT has a relatively small cooling capacity. It is approximated by taking advantage of the fact that several parts like coil tubes are geometrically standardized using certain a finite set of standard diameters and thicknesses. In 2-D modeling, the cooling capacity can be expanded approximately in proportion to the number of tubes to meet the actual requirements.

To obtain the datasets for the development of ROMs, the discretized PDAEs for CWCTs are solved by using FLUENT software for each of the samples. The process geometry of the tube bundle section shown in Fig. 4B, firstly drawn using the Design Modeler package, is based on the standard diameter and length listed in Table 1. The generated preliminary sketch is meshed to adequately capture the change of fluids using the Meshing package. Note that the meshes near to the tube walls are refined to improve the model accuracy. Besides, the element size is smoothly stretched to ensure an accurate resolution of the high

TABLE 1 The input space for the ROM construction [37].

Input variables	Symbol	Units	Specification
Operating variables	$\mathbf{X}^{(1)}$		
Velocity of inlet air	v^a	m/s	U(1.0, 2.5)
Mass flowrate of spray water	m^{sw}	kg/s	U(0, 1.0)
Temperature of spray water	T^{sw}	K	U(285, 310)
Volume flow rate of circulating water	m^{cw}	m³/h	U(0, 500)
Temperature of inlet circulating water	T^{cw}	K	U(313, 330)
Uncertain variables	$\mathbf{X}^{(2)}$		
Dry bulb temperature of inlet air	T^{db}	K	Uncertain distribution
Relative humidity of inlet air	φ^{rh}	%	Uncertain distribution
Design variables	$\mathbf{X}^{(3)}$		
Inner diameter of tube	d^{tube}	m	(0.01, 0.02, ..., 0.1)
Numbers of tube bundle	N^{tube}	/	(200, 201, ..., 1000)

gradient regions of the fluid fields. The quality of the mesh structure is measured by the skewness value which should be less than the required level (1-inacceptable, 0-excellent). Next, multiple CFD simulations are performed on the meshed geometry to calculate the coupled heat and mass transfer, as well as the mass and energy balances of PDAEs. In particular, the Eulerian-Lagrangian modeling approach is applied whereby the air phase is treated as continuous and the spray water particles are handled using the discrete phase model. The particle trajectories, along with mass and energy transfer to the particles, are computed with a Lagrangian formulation. The air-water two-phase flow employs the realizable k-ε model. The governing equations with boundaries are solved by the finite volume method, and the convective terms in governing equations are discretized by the QUICK scheme with second-order precision.

5 Reduced models construction

5.1 Level 1: Physics-based ROM

5.1.1 Step 1: Singular value decomposition method

PCA decomposition method based on SVD can map a vector from n-dimensional space to a k-dimensional space ($k<<n$) by transforming the snapshot data to a new orthogonal coordinate system without significant loss of information. The essence of SVD is to find a set of orthogonal bases, which is still orthogonal after transformation. Assuming $\mathbf{Z}$ is a $m \times n$ snapshot matrix, it represents all the output variables in the nodes of the mesh at a special state such as temperature and velocity fields. The matrix $\mathbf{Z}$ with rank order k maps a set of orthogonal basis $\mathbf{V} = [v_1, v_2, \ldots, v_n]$ to another set of orthogonal basis $\mathbf{ZV}$ that must be satisfy the conditions $v_a \cdot v_b = v_a^T v_b = 0$ and $\mathbf{Z}v_a \cdot \mathbf{Z}v_b = (\mathbf{Z}v_a)^T \mathbf{Z}v_b = v_a^T \mathbf{Z}^T \mathbf{Z}v_b = 0$, where $a \neq b = [1, \ 2, \ \ldots, \ n]$. Since $\mathbf{Z}^T\mathbf{Z}$ is a symmetric matrix, its Eigenvectors of different Eigenvalues are orthogonal to each other and the orthogonal basis $\mathbf{V}$ can be set the Eigenvectors of $\mathbf{Z}^T\mathbf{Z}$, $|\mathbf{V}| = k = rank(\mathbf{Z})$. Thus, there is $v_a^T \mathbf{Z}^T \mathbf{Z}v_b = v_a^T \lambda_b v_b = \lambda_b v_a \cdot v_b = 0$, where $1 \le a, \ b \le k, \ a \neq b$. The orthogonal basis $\mathbf{ZV}$ can be unitized as $u_a = \mathbf{Z}v_a/| \mathbf{Z}v_a| = \lambda_a^{-1/2}\mathbf{Z}v_a$, where $\mathbf{Z}v_a = \sigma_a u_a$, $\sigma_a = \lambda_a^{-1/2}$, $0 \le a \le k$ and σ_a is the singular value. The up-dated orthogonal basis $[u_1, u_2, \ldots, u_k]$ and $[v_1, v_2, \ldots, v_k]$ needs to be extended to another orthogonal basis $\mathbf{U} = [u_1, u_2, \ldots, u_m]$ if $k < m$, and $\mathbf{V} = [v_1, v_2, \ldots, v_n]$ if $k < n$, respectively. In addition, the dataset $[v_{k+1}, v_{k+2}, \ldots, v_n]$ can be set to the null space of $\mathbf{Z}$ ($\mathbf{Z}\mathbf{V}_a = 0, a > k, \sigma_a = 0$). Finally, matrix $\mathbf{Z}$ is decomposed into three matrixes after eliminating zero.

$$\mathbf{Z} = \mathbf{U}\boldsymbol{\Sigma}\mathbf{V}^T \tag{6}$$

where $\boldsymbol{\Sigma} = diag(\sigma_1, \ldots, \sigma_k, 0)$. Note that σ declines rapidly with the rank order of $\sigma_1 \ge \sigma_2 \ldots \ge \sigma_k$, in most cases, the sum of the top 10% of σ can account for more than 95% of the total singular values. Reducing $\mathbf{Z}$ to rank k resulting from this cutoff criterion, the reduced order dataset is obtained as given by:

$$\mathbf{Z} \approx \mathbf{Z}^{(k)} = \mathbf{U}^{(k)} \mathbf{\Sigma}^{(k)} \left(\mathbf{V}^{(k)}\right)^{T} \tag{7}$$

where the superscript (k) indicates that the first k columns are taken from the original matrix to formulate a new matrix and k is far less than m or n. Then, the matrix $\mathbf{Z}$ can be expressed as its PC matrix (δ) and score matrix (Φ).

$$\mathbf{Z} = \delta \Phi \tag{8}$$

where $\delta = \mathbf{U}^{(k)}$ is an $m \times k$ matrix; $\Phi = \mathbf{\Sigma}^{(k)} (\mathbf{V}^{(k)})^{T}$ is a $k \times n$ diagonal matrix.

5.1.2 Step 2: Kriging interpolation

The PCs are unchanged for any given input variables $\mathbf{X}$ because they represent the coordinate in the transformed system via PCA decomposition of the original dataset. For any input variable $\mathbf{X}$ restricted in the domain, it only needs to obtain the score Φ to calculate the output variables $\mathbf{Z}$ through the linear correlation defined in Eq. (8). Note that the only varying components in this equation are the score Φ obtained from PCA with the potential correlation between Φ and $\mathbf{X}$. Additional functions between them can be built with Kriging interpolation according to the complexity of the CFD model.

The Kriging predictor $\Phi(\mathbf{X})$ consists of polynomial term $pt(\mathbf{X})$ and residual term $rt(\mathbf{X})$, which can be used to substitute for the PDAEs model. Due to the stochastic assumption in the Kriging interpolation, the error in the predicted value is also a function of the input variables $\mathbf{X}$.

$$\Phi(\mathbf{X}) = pt(\mathbf{X}) + rt(\mathbf{X}) \tag{9}$$

where $pt(\mathbf{X})$ is a constant μ, $rt(\mathbf{X})$ is a stochastic Gaussian process denoting the uncertainty on the mean of $\Phi(\mathbf{X})$.

To keep the predictor unbiased, the expected value of the residual term $rt(\mathbf{X})$ is zero, $E[rt(\mathbf{X})] = 0$. The covariance between points $(\mathbf{X}_{l1}, \mathbf{X}_{l2})$ of this term can be calculated by $cov[rt(\mathbf{X}_{l1}, \mathbf{X}_{l2})] = \sigma^2 \Psi(\mathbf{X}_{l1}, \mathbf{X}_{l2})$, where σ^2 is the process variance, $\Psi(\mathbf{X}_{l1}, \mathbf{X}_{l2})$ is the spatial correlation function as follows:

$$\psi(\mathbf{X}_{l1}, \mathbf{X}_{l2}) = \exp\left[-d(\mathbf{X}_{l1}, \mathbf{X}_{l2})\right] \tag{10}$$

$$\widehat{\psi}(\mathbf{X}_{l1}, \mathbf{X}) = \exp\left[-d(\mathbf{X}_{l1}, \mathbf{X})\right] \tag{11}$$

$$d(\mathbf{X}_{l1}, \mathbf{X}_{l2}) = \prod_{h=1}^{g} \exp\left[-\theta_h (\mathbf{X}_{l,h} - \mathbf{X}_{l2,h})^{p_h}\right], \quad \forall h \in \{1, \ldots, g\} \tag{12}$$

where the Gauss correlation is employed due to the continuously differentiable of the underlying phenomenon; θ is the Kriging regression parameter; p_h is the smoothness parameter which is equal to 2 and provides a smooth infinitely differentiable correlation function. The values of parameters (μ, σ^2, θ) are fit by applying the maximum likelihood parameter estimation method, and the estimations of μ and σ^2 are $\mu = (1^T \Psi^{-1} Z)/(1^T \Psi^{-1} 1)$ and $\sigma^2 = (Z - 1\mu)^T \Psi^{-1} (Z - 1\mu)^T n^{-1}$. The physics-based ROM can be obtained by calculating the likelihood function of the original dataset (X, Φ) augmented with the new interpolating point (X^{new}, Φ^{new}).

$$\Phi^{new}(X^{new}) = \mu + r^T \psi^{-1}(\Phi - 1\mu) \qquad (13)$$

where r is the $n \times 1$ vector of correlations $\Psi(X^{new}, X_{/1})$ between the point (X^{new}) to be predicted and the sample design points. By substituting Φ^{new} into Eq. (8), the state Z^{new} can be obtained at all nodes of the mesh system corresponding to any given X^{new}.

5.2   Level 2: Data-driven ROM

5.2.1   High-dimensional model representation

HDMR is a quantitative assessment and analysis tool for improving the efficiency of deducing high dimensional input-output behavior of a system. HDMR can take into account the inherent uncertainty on input parameters and also presents the potential nonlinearity and contribution due to the interaction between input parameters. HDMR expresses the model output $y \in Y$ as a finite hierarchical cooperative function expansion in terms of its input variable $x \in X$, as given by:

$$y = f_0 + \sum_{i=1}^{NH} f_i(x_i) + \sum_{i=1}^{NH} \sum_{j=i+1}^{NH} f_{ij}(x_i, x_j) + \cdots + f_{12\cdots NH}(x_1, x_2, \cdots, x_{NH}) \qquad (14)$$

where NH is the size of input parameters, i and j index any two input parameters, and $f_0 = E[f(x)]$.

For most practical applications, the functions $f(x)$ containing more than two input parameters can often be ignored due to their fewer contributions compared to the former terms. Therefore, Eq. (14) can be simplified to a finite number of terms to calculate the predictor for any input variables. The data-driven ROM can be obtained by:

$$y = C + \sum_{i=1}^{NH} \sum_{kh=1}^{KH} A_{i,kh} \times x_i^{kh} + \sum_{i=1}^{NH} \sum_{j=i+1}^{NH} \sum_{kh=1}^{KH} \sum_{nh=1}^{KH} B_{i,j,kh,nh} \times x_i^{kh} \times x_j^{nh} \qquad (15)$$

$$i, j \in \{1, 2, 3, \cdots, NH\}, \quad kh, nh \in \{1, 2, 3, \cdots, KH\}$$

where C is a constant term, $A_{i,kh}$ and $B_{i,j,kh,nh}$ are the first and second-order coefficients, KH is the highest degree of the input variables. They can be regressed by using a specialized linear programming model, which is used to minimize the cumulative relative errors $\sum_m^M |y_m - \tilde{y}_m|$, $m \in (1, 2, \ldots, M)$ between the outputs of ROM (y_m) and CFD simulation ($\tilde{y}_m$), where M is the number of fitting datasets.

5.3 Model evaluation

The performance of the constructed models can be thoroughly evaluated by both the training and generation errors (average relative error, avgRE) between the outputs of bi-level ROMs and the Fluent model. Herein, the training error is used to evaluate the reconstruction behavior of the ROMs at the design points, while the generalization error is to assess the prediction capability at interpolated points that are not included in the sample set. The generalization performance that truly reflects the prediction ability is important for the application of developed models, which are validated by using the cross-validation (CV) method. The latter randomly divides the observations into two mutually exclusive subsets. The training subset is used to construct the model, while the test subset unseen by the model during training is used to compute a prediction error. The original sample is repeatedly cut into a test subset and a training subset to assess and to obtain the reliable average error terms for the prediction of the left-out points. The optimal solution is considered as a good approximation of the original distribution.

$$
\begin{aligned}
avgRE_t &= \frac{1}{TE} \sum_{t=1}^{TE} RE_t \\
&= \frac{1}{TE} \sum_{t=1}^{TE} \frac{\left| O_t^{RM}(X^{test}) - O_t^{CFD}(X^{test}) \right|}{O_t^{CFD}(X^{test})}, \forall O \in \{Y, Z\}, t \in \{1, ..., TE\} \quad (16)
\end{aligned}
$$

where TE is the number of training sets.

6 Illustrative example

In this chapter, a small-size circulating water system is illustrated to introduce the detailed constructions of the bi-level ROMs (Example I) and the multiscale stochastic optimization model (Example I). The CFD simulation and stochastic programming problem associated with the multiscale models are implemented via Fluent 16.0 and GAMS 24.7.1 modeling environment, respectively. Both models are solved on a workstation with Intel four processors Xeon e5 CPU @2.5 GHz and 32 GB RAM. In CFD modeling, the iterative default under-relaxation factors of 0.3 and 0.6 are used for pressure and momentum, respectively. Besides, the normalized residuals used for checking convergence are less than 10^{-3} for the momentum equations and less than 10^{-6} for the energy equations. For the two-stage stochastic optimization problem, it takes 10–90 min to obtain a feasible solution with a DICOPT [41] solver and the optimality gap of 10^{-9} at each iteration.

6.1 Example I: Model reduction

It is assumed the cooling tower system is located in Jieyang City, Guangdong province, southern China. The original statistics of uncertain environmental parameters during the years 2015–2017 are retrieved from the government database (http://www. noaa.gov). All seasons within the chosen years are taken into account. With an initial sample size of 50, the optimal number of samples generated by HSS is 300 for spring, 200 for summer, 200 for autumn, and 200 for winter. Fig. 5 shows the PDFs of dry-bulb temperature and relative humidity for

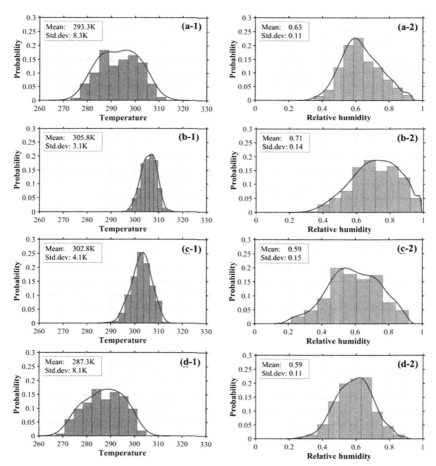

FIG. 5 PDFs of dry-bulb temperature and relative humidity. (A) spring; (B) summer; (C) autumn; (D) winter [37].

the four seasons, along with their mean value and standard derivation. As shown, it is apparent that both uncertain variables have remarkably different statistical moments among the seasons and range over a large span throughout the year.

The results of the reconstruction behavior of the developed physics-based ROM at the design points of the CWCT are presented in Table 2. Here, we systemically compare the avgRE of four monitored states concerning the temperature, pressure, velocity, and H_2O mass fraction between the outputs of physics-based ROM and Fluent model in four seasons. The statistical values of avgRE are very similar, with a tiny range between 2.44×10^{-17}–5.73×10^{-17} for temperature, 3.81×10^{-12}–5.48×10^{-11} for pressure, 2.02×10^{-8}–5.39×10^{-7} for velocity, and 7.11×10^{-16}–1.84×10^{-14} for H_2O mass fraction. From these results, it is concluded that the outputs of physics-based ROM maintain high accuracy and no observable difference at the design points for all cases.

The reconstruction behavior of the physics-based ROM is further assessed at unknown points. Four typical cases that correspond to the mean values of the input variables in each season are shown in Figs. 6–9. For each case, four monitored field profiles of state variables (temperature, pressure, velocity, and H_2O mass fraction) inside the tower are used to visually represent the reconstruction behavior of physics-based ROM. It can be seen from these figures that the contours of the field profiles predicted using physics-based ROM are comparable to those from Fluent model and the distortions of the contours in some figures (i.e., velocity profiles) are even hard to observe. Herein, it is worth noting that the developed ROMs are capable of sharply reducing the computational time and resources as compared to the Fluent model. For example, completing a Fluent model requires approximately 3 days for an investigated case, while the developed ROMs take less than 2 s.

Fig. 10 shows the relative errors of temperature and H_2O mass fraction by implementing the CV of the data-driven ROM with Fluent results. As shown,

TABLE 2 Results of avgRE for the physics-based ROM [37].

Season	Temperature	Pressure	Velocity	H_2O mass fraction
Spring	3.18×10^{-17}	5.48×10^{-11}	2.03×10^{-8}	1.84×10^{-14}
Summer	2.44×10^{-17}	2.70×10^{-11}	7.88×10^{-8}	7.11×10^{-16}
Autumn	2.89×10^{-17}	1.43×10^{-11}	5.39×10^{-7}	2.98×10^{-15}
Winter	5.73×10^{-17}	3.81×10^{-12}	2.02×10^{-8}	1.83×10^{-14}

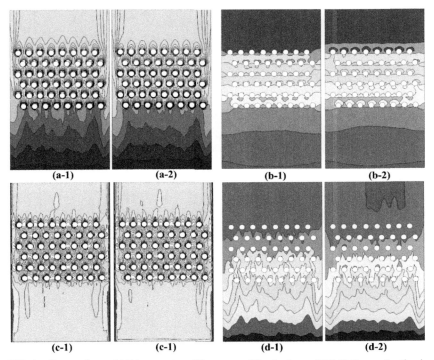

FIG. 6   Contour lines of (A) temperature, (B) pressure (C) velocity, and (D) H₂O mass fraction in the case of Spring predicted by Fluent model (*left*) and physics-based ROM (*right*) [37].

the relative errors of outlet air temperature for all 900 samples are less than 0.02, while more than 70% of the relative errors of H_2O mass fraction are less than 0.1. Also, both avgRE and interquartile range (IQR) of the relative errors are presented in Fig. 10 The former indicator ranges from 0.0038 to 0.0040 for outlet air temperature, and from 0.0436 to 0.0636 for H_2O mass fraction. The latter range from 0.0028 to 0.0037 for the outlet air temperature, and from 0.0355 to 0.0636 for H_2O mass fraction. Hence, it can be concluded that the developed data-driven ROM for the CWCT offers reasonable and sufficient confidence for further use for integration within the cooling tower systems models for optimization.

6.2   Example II: Multiscale optimization

The developed bi-level ROMs of CWCTs, together with the shortcut models of other process units are further embedded in the multiscale models of the cooling water network superstructure. In Fig. 11, the superstructure includes a set of

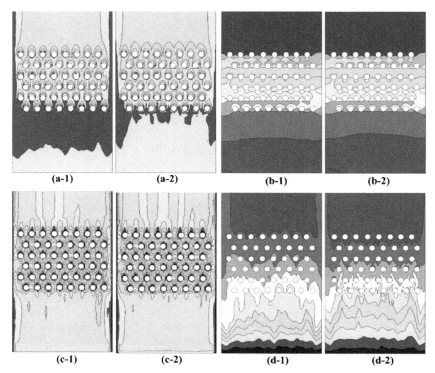

FIG. 7   Contour lines of (A) temperature, (B) pressure (C) velocity, and (D) H_2O mass fraction in the case of Summer predicted by Fluent model (*left*) and physics-based ROM (*right*) [37].

circulating water streams with known inlet temperature and flow rates, which have the potential to be split into multiple substreams for recycling or reuse. A set of CWCTs are in a parallel arrangement and each CWCT has the same cooling capacity and feeding flow rate. To cope with the variability of environmental parameters, each CWCT is equipped with a set of circulating pumps, intake fans, and spray pumps for flexibly adjustment of the flow rates of circulating water, air, and spray water, respectively. The cooled substreams of circulating water are merged into one and then mixed with make-up water and recycled circulating water. Finally, the circulating water streams at a satisfying temperature, exchange heat with process streams through the heat exchangers. To improve the model formulation, the developed data-driven ROM is integrated with the overall process model of the system. A two-stage stochastic optimization problem is used to minimize the expected total annual cost $E(TAC)$ of the system, as given by:

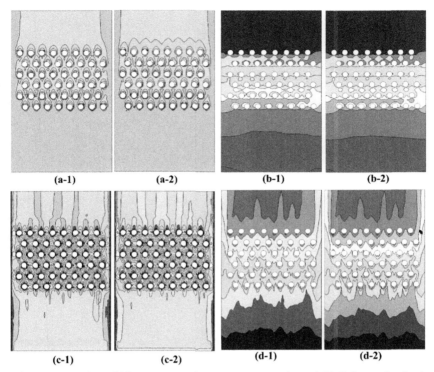

FIG. 8   Contour lines of (A) temperature, (B) pressure (C) velocity, and (D) H$_2$O mass fraction in the case of Autumn predicted by Fluent model (*left*) and physics-based ROM (*right*) [37].

$$\text{Min}\quad E(TAC) = TC^{1\text{st}}\left(\mathbf{X}^{(1)}, \mathbf{X}^{(3),1\text{st}}, \mathbf{IX}\right)$$

$$+ \sum_{s=1}^{4}\sum_{n=1}^{N} Pro_{s,n} TC_{s,n}^{2\text{nd}}\left[\mathbf{X}^{(1)}, \mathbf{IX}, \mathbf{X}_{s,n}^{(2),2\text{nd}}, \mathbf{Y}_{s,n}\left(\mathbf{X}^{(1)}, \mathbf{X}_{s,n}^{(2),2\text{nd}}\right)\right]$$

$$\begin{aligned}
s.t.\quad & Eq^{1\text{st}}\left(\mathbf{X}^{(1)}, \mathbf{X}^{(3),1\text{st}}, \mathbf{IX}\right) = 0 \\
& IEq^{1\text{st}}\left(\mathbf{X}^{(1)}, \mathbf{X}^{(3),1\text{st}}, \mathbf{IX}\right) \leq 0 \\
& \mathbf{X}^{(1),lb} \leq \mathbf{X}^{(1)} \leq \mathbf{X}^{(1),ub} \\
& \mathbf{X}^{(3),1\text{st},lb} \leq \mathbf{X}^{(3),1\text{st}} \leq \mathbf{X}^{(3),1\text{st},ub} \\
& \mathbf{IX}^{lb} \leq \mathbf{IX} \leq \mathbf{IX}^{ub} \\
& Eq_{s,n}^{2\text{nd}}\left[\mathbf{X}^{(1)}, \mathbf{IX}, \mathbf{X}_{s,n}^{(2),2\text{nd}}, \mathbf{Y}_{s,n}\left(\mathbf{X}^{(1)}, \mathbf{X}_{s,n}^{(2),2\text{nd}}\right)\right] = 0, \forall s \in \{1, ..., 4\}, \forall n \in N^{2\text{nd}} \\
& IEq_{s,n}^{2\text{nd}}\left[\mathbf{X}^{(1)}, \mathbf{IX}, \mathbf{X}_{s,n}^{(2),2\text{nd}}, \mathbf{Y}_{s,n}\left(\mathbf{X}^{(1)}, \mathbf{X}_{s,n}^{(2),2\text{nd}}\right)\right] \leq 0, \forall s \in \{1, ..., 4\}, \forall n \in N^{2\text{nd}} \\
& \mathbf{Y}^{lb} \leq \mathbf{Y}_{s,n}\left(\mathbf{X}^{(1)}, \mathbf{X}_{s,n}^{(2),2\text{nd}}\right) \leq \mathbf{Y}^{ub}, \forall s \in \{1, ..., 4\}, \forall n \in N^{2\text{nd}}
\end{aligned} \tag{17}$$

where $TC^{1\text{st}}$ and $TC^{2\text{nd}}$ are part of the objective function, the former only depends on design variables (e.g., a function of capital cost), while $TC^{2\text{nd}}$ relies on both

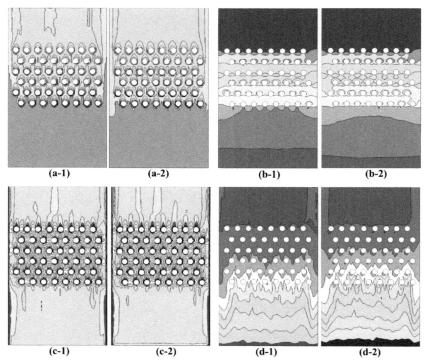

FIG. 9 Contour lines of (A) temperature, (B) pressure (C) velocity, and (D) H_2O mass fraction in the case of Winter predicted by Fluent model (*left*) and physics-based ROM (*right*) [37].

design and operating variables (e.g., a function related to water consumption). For a stochastic programming problem, the equipment capacities (e.g., CWCTs, pumps, and fans) should be the same for all samples and thus the capital cost belongs to the sample-independent variable which are determined in the first stage by design equality Eq^{1st} and inequality constraint IEq^{1st}. The operating expenses determined in the second stage are formulated by stochastic functions (operational equality Eq^{2nd} and inequality constraint IEq^{2nd} on the feasibility of the process) to capture the variability in uncertain space. $\mathbf{X}^{(1)}$ and $\mathbf{X}^{(2),2nd}$ represent the deterministic and uncertain variables to the function of ROMs; $\mathbf{IX}$ represents the remaining variables, $\mathbf{Y}(\mathbf{X}^{(1)}, \mathbf{X}^{(2),2nd})$ represents the outputs of the ROM as a function of inputs $\mathbf{X}^{(1)}$ and $\mathbf{X}^{(2),2nd}$; $\mathbf{X}^{(3),1st}$ represents the design variables; $\mathbf{X}^{(1),lb}/\mathbf{X}^{(1),ub}$, $\mathbf{X}^{(3),1st,lb}/\mathbf{X}^{(3),1st,ub}$, $\mathbf{IX}^{lb}/\mathbf{IX}^{ub}$, and $\mathbf{Y}^{lb}/\mathbf{Y}^{ub}$ are corresponding lower/upper bounds of these variables; $Pro_{s,n}$ is the probability related to the occurrence of a specific sample n in season s, N^{2nd} is the set of samples.

To solve the stochastic optimization problem, stochastic samples for each season are first generated by setting an initial size of 50, with a 95% confidence interval. The optimal sizes of the samples were determined as 300 for spring,

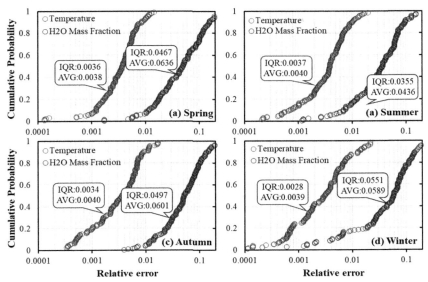

FIG. 10 Relative errors between data-driven ROM and Fluent results. (A) spring; (B) summer; (C) autumn; and (D) winter [37].

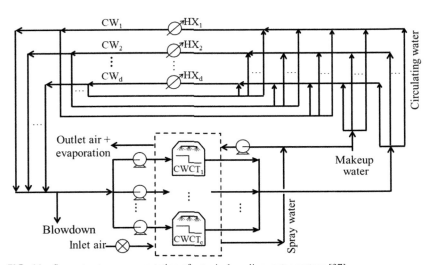

FIG. 11 Superstructure representation of a typical cooling water system [37].

100 for summer, 200 for autumn, and 200 for winter, and all the generated samples have passed the Chi-squared test. The design and operating variables and the corresponding ranges used in the optimization problem are listed in Table 1, which are consistent with those used in building the data-driven ROM. To simplify the optimization model stated in Eq. (17), the inner diameter of tubes installed in the CWCT is assumed to be a fixed value ($d^{tube} = 0.01$ m) and the cooling water system is assumed to consist of three circulating water streams, and their corresponding flow rates, inlet, and target temperatures are listed in Table 3.

In this study, a deterministic optimization model is also developed to emphasize the importance of the stochastic approach. It should be highlighted that the deterministic model can be easily obtained from the proposed stochastic approach by considering a representative sample. This single sample corresponds to the mean value of uncertain environmental parameters for the whole year (297.3 K for the dry-bulb temperature, 63.0% for the relative humidity of inlet air). Next, the equipment capacities obtained from this single sample are fixed in the model and compare with the stochastic model to further assess the impact of uncertainty on the techno-economic performances of the CWCTs and the whole cooling water system.

The optimal solutions obtained from the deterministic and stochastic approaches dispatches the inlet circulating water streams for their appropriate sinks in the cooling water network. As shown in Fig. 12, it is notable that the stochastic and deterministic solutions have the same distributed network structure, which contains four units of CWCTs, circulating pumps, spray pumps, eight intake fans, and three heat exchangers. In this figure, the inlet circulating water streams denoted by the gray lines exchange heat with process streams via heat exchangers, and their flow rates stay constant as those listed in Table 3. The distributed loads of other streams in the cooling water network (i.e., split and recycled streams), as well as the operating conditions of the CWCTs (i.e., inlet air and spray water) are constantly changing with the variation of environmental parameters in each season. The detailed model formulation of this network are provided in the supporting information of [37].

TABLE 3 Basic data of circulating water streams [37].

Stream	Flow rate (m³/h)	Inlet temperature (K)	Target temperature (K)
CW_1	216	324	[293, 315]
CW_2	548	316	[293, 311]
CW_3	468	311	[293, 309]

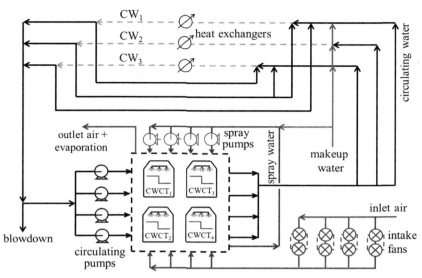

FIG. 12   The optimal cooling water network obtained from the stochastic and deterministic solutions [37].

Fig. 13A compares the breakdown of expected TAC of the cooling water system for both deterministic and stochastic solutions. The annual capital cost based on the samples of one season is equal to that based on the other one due to the constraints on the design variables for the same solution. However, note that the two optimal solutions lead to different annual capital costs, though they yield the same cooling water network. This is mainly attributed to the various design capacities and geometry sizes of the installed CWCTs and other process equipment in the same cooling water network. For instance, in the case of the same inner diameter ($d^{tube} = 0.01$ m), the optimal number of tube bundles assembled in each CWCT has grown from 407 for the deterministic solution to 519 for the stochastic solution. The corresponding design capacities of auxiliary equipment such as circulating pumps, spray pumps, and intake fan increase by 27.4%, 32.6%, and 29.4%, respectively. Thus, the annual capital cost increases from $\$3.76 \times 10^5$ for the deterministic solution to $\$4.04 \times 10^5$ for the stochastic solution, which accounts for 59.4% and 67.9% of the expected TAC, respectively.

Though the stochastic solution results in a greater annual capital cost, the expected TAC based on all 800 samples (detonated by AVG in Fig. 13A) obtained from this solution is $\$5.94 \times 10^5$, which is still lower than that of the deterministic one ($\$6.33 \times 10^5$). This reduction mainly stems from the significant reduction of operating cost, which is $\$2.57 \times 10^5$ for the deterministic solution and $\$1.91 \times 10^5$ for the stochastic solution. To be more specific, Fig. 13B shows that the mean value of operating expenses for utility usage cuts

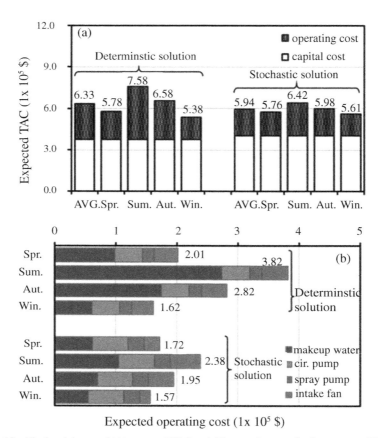

FIG. 13 The breakdowns of (A) expected TAC and (B) operating cost for four seasons [37].

down from 2.01×10^5 to 1.72×10^5 for spring, 3.82×10^5 to 2.38×10^5 for summer, 2.82×10^5 to 1.95×10^5 for autumn, and 1.62×10^5 to 1.57×10^5 for winter. Among them, it is obvious that the hot seasons would lead to higher operating expenses mainly due to the significantly increased consumption of makeup water. Especially, in summer the operating expense of makeup water had a sharp increase from 1.05×10^5 for the stochastic solution to 2.74×10^5 for the deterministic solution, which accounts for 44.1% and 71.7% of the operating cost, respectively. From the results, it can be concluded that for the stochastic solution, the reduction in the capital cost achieved by decreasing the equipment capacities is less significant and not enough to make up the loss due to the increased operating costs of CWCTs. The stochastic solution allows for a certain design margin of CWCTs to reduce the impacts of the variability of environmental conditions and to avoid the abnormal running conditions in extreme scenarios. This highlights the importance of considering the

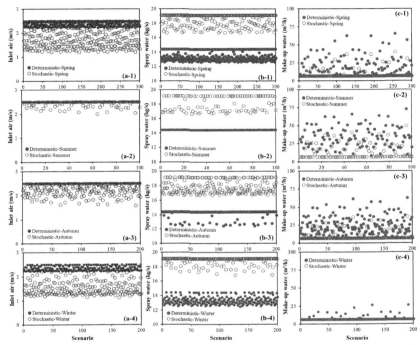

FIG. 14 The optimal values of key operating variables of the system obtained from the deterministic and stochastic solutions [37].

uncertainty of environmental parameters in the robust design of the cooling water system.

Fig. 14 shows the varying inputs of inlet air, spray water, and makeup water that corresponds to CWCTs for all 800 samples. From Fig. 14A, it is seen that the velocity of inlet air for the stochastic solution is mostly lower than that of the deterministic one, especially in cold seasons like spring and winter. For example, the velocity of inlet air for the deterministic solution has mostly reached the upper bound, 2.5 m/s, indicating the CWCT operates in a full-load state. The main reason is that the increased cooling capacity of CWCTs obtained from the stochastic solution has a greater quantity of tube bundles and larger heat exchanger areas, which reduce the required amount of inlet air as compared to the deterministic solution. However, a larger cooling capacity also means that more spray water is required to form the liquid film on the surface of the coil tubes. From Fig. 14B, it is seen that for all generated samples, the average usage amount of spray water has increased by about one-third, that is, from 13.6 (deterministic solution) to 18.6 kg/s (stochastic solution). Besides, note that the makeup water consumption is equal to the sum of the blowdown and evaporation losses according to mass balance, and the blowdown contributes

60%–80% of the total. Thus, though the increased usage of spray water leads to greater evaporation losses, the stochastic solution has remarkable advantages in reducing the amounts of circulating water load and blowdown by properly increasing the cooling capacity of CWCT. On the contrary, the CWCT within the constraints is almost operated at full load for the deterministic solution, particularly in summer, no matter how the operating variables (inlet air, spray water, and inlet stream) are manipulated. As shown in Fig. 14C, to meet the cooling target of circulating water, the amount of makeup water greatly reduces from 6.67–72.9 m^3/h (deterministic solution) to 6.67–49.5 m^3/h (stochastic solution).

It is challenging to assess the accuracy of the solution optimality and the underlying model in the presence of model reduction and assumptions. In this study, we randomly selected Sample 28 in the Spring case (dry-bulb temperature 285.5 K and relative humidity 57.7% of inlet air, see Fig. 14A) to validate the data-driven ROM of the CWCT at the optimal solution. Are listed in Table 4, compared with the FLUENT results, the relative errors of temperature and H_2O mass fraction of the outlet air are only 0.0019 and 0.0287, respectively. The magnitudes of these relative errors fall within the range of 0.0001–0.01 and 0.001–0.11. The error evaluation based on Sample 28 indicates that the ROM-based optimization model achieves an optimal result that retains the optimality of the original high-fidelity model for the same input variables.

As aforementioned, the proposed multiscale optimization approach can not only present the optimal values of objective and the corresponding operating and design variables for the whole system, but also reconstructs the high-fidelity fluid dynamics and reveals the complex thermal and flow behaviors via physics-based ROM. For figuring out of the influence of optimal solutions on heat-mass transfer processes inside CWCTs, we still selected sample 28 in spring for better physical inspection of field profiles. For this selected sample, Fig. 15 indicates that as the inlet air flows upwards and contacts with falling spray water, it is gradually saturated with water vapor and the longitudinal temperature gradient appears more evident between the coil tubes. Interestingly, the two optimal

TABLE 4 Error evaluation for Sample 28 in the Spring case [37].

Method	Temperature of outlet air (K)	Relative error	H_2O mass fraction of outlet air	Relative error
Data-driven ROM	311.8	0.0019	0.0338	0.0287
FLUENT model	312.4		0.0348	

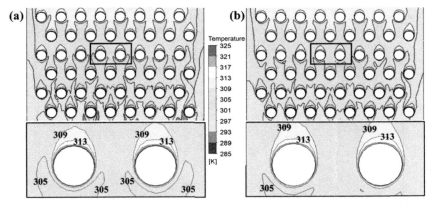

FIG. 15 Contours of fluid temperature inside CWCTs obtained from (A) deterministic model and (B) stochastic model [37].

solutions yield almost the same temperature profiles. The amplification of the selected temperature field further indicates that they have the same maximum and minimum contour lines of temperature, which are 313 and 305 K at the top and bottom of the tubes, respectively. This consistency in temperature profile is mainly because the overall heat transfer process is mainly controlled by the maximum heat transfer potential of saturated moist air under the same operating temperature of inlet air, spray water, and circulating water.

As the inlet air flows upward and encounters the coil tubes at the stagnation points, most of its kinetic energy is converted into pressure energy according to the Bernoulli equation. Thus, the pressure profile is closely related to the velocity profile inside a cooling tower. As shown in Figs. 16 and 17, the contours of

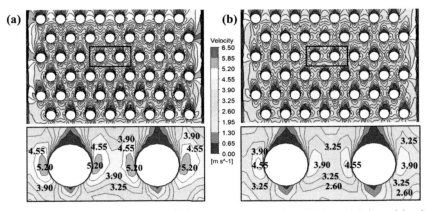

FIG. 16 Contours of inlet air velocity inside CWCTs obtained from (A) deterministic model and (B) stochastic model [37].

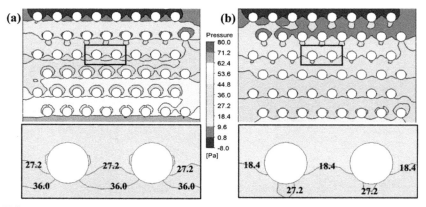

FIG. 17 Contours of fluid pressure inside CWCTs obtained from (A) deterministic model and (B) stochastic model [37].

inlet air velocity are significantly reduced, while the contours of fluid pressure increase accordingly in the surrounding region of the stagnation points. As the inlet air continues to flow upward, the pressure gradient along the flow direction of inlet air becomes negative due to the gradual shrinkage of the flow path, making the airflow at the boundary layer to be in an accelerated speed state. The maximum value of the contours of inlet air velocity can be found at the left and right ends of each tube. Thereafter, under the combined effects of the shear stress and pressure gradient, the contours of velocity at the boundary layer decrease rapidly and reach their lowest value at the upper-point of the tubes.

In Fig. 17, it is seen from the amplification of the selected field that the pressure contour lines have about one-third growth from 27.2–36.0 Pa for the stochastic solution to 18.4–27.2 Pa for the deterministic solution. Similarly, it is seen from Fig. 16 that the velocity contour lines of the inlet air at the same location increased from 3.90–4.55 m/s for the stochastic solution to 3.25–5.20 m/s for the deterministic solution. It should be noted that in practice the increased air velocity and fluid pressure are adverse to the long-term operation of cooling towers. The selected Sample 28 corresponds to normal weather conditions over the whole year. As the CWCTs operates at more extreme weather conditions, such as in summer, the cooling capacity of CWCTs would be significantly weakened mainly due to the increased temperature and relative humidity of the inlet air. For the deterministic solution, it is insufficient to cool down the inlet circulating water by only manipulating the variables of CWCT within the specified variable space, unless the consumption of makeup water is increased substantially. Instead, the stochastic solution is capable of reducing the operational and economic risks of the whole cooling water system through fully accommodating the uncertainty of environmental parameters and flexibly increasing the capacity of the equipment.

7 Conclusion

In this chapter, we developed a tailored framework based on model reduction for multiscale stochastic optimization of the cooling water system with counter-current CWCTs under the uncertainty on environmental parameters. The model reduction methodology used for generating bi-level ROMs for CWCTs included four statistical steps: optimal DoE, multisample CFD simulation, reduced model construction, and model evaluation. In particular, based on rigorous CFD simulation, a fast data-driven and physics-based ROMs was constructed by utilizing HDMR and combined PCA-Kriging interpolation, respectively, to closely approximate the high-fidelity CWCTs models. The developed bi-level ROMs were embedded in a multiscale optimization model for performing integrated design and management of the CWCTs and cooling water system. This optimization model employed sampling-based stochastic programming and the heterogeneous integration of unit-specific shortcut models and detailed models of CWCTs.

Two examples were presented to illustrate the effectiveness of the proposed methodology. In Example I, the results showed that the developed bi-level ROMs provided a good approximation of CWCTs and resulted in a significant reduction in computational resources and time. In Example II, the developed bi-level ROMs were used in a superstructure-based cooling water system for multiscale stochastic optimization. The results showed that the stochastic solution can reduce makeup consumption and save costs as compared with the deterministic solution, though it properly increased the capital costs of CWCTs and other auxiliary equipment. Besides, the stochastic solution can effectively reduce the impact of the variability of weather conditions and avoid the abnormal running of the cooling water systems in extreme scenarios. This indicated the importance of consideration of the stochastic approach and uncertainty in the robust design of cooling water systems.

References

[1] G. Gan, S.B. Riffat, Numerical simulation of closed wet cooling towers for chilled ceiling systems, Appl. Therm. Eng. 19 (1999) 1279–1296.

[2] X.C. Xie, C. He, T. Xu, B.J. Zhang, M. Pan, Q.L. Chen, Deciphering the thermal and hydraulic performances of closed wet cooling towers with plain, oval and longitudinal fin tubes, Appl. Therm. Eng. 120 (2017) 203–218.

[3] A. Hasan, K. Siren, Theoretical and computational analysis of closed wet cooling towers and its applications in cooling of buildings, Energy Build. 34 (2002) 477–486.

[4] X. Xie, H. Liu, C. He, B. Zhang, Q. Chen, M. Pan, Deciphering the heat and mass transfer behaviors of staggered tube bundles in a closed wet cooling tower using a 3-D VOF model, Appl. Therm. Eng. 161 (2019) 114202.

[5] S. Bodjona, M. Girault, E. Videcoq, Y. Bertin, Reduced order model of a two-phase loop thermosyphon by modal identification method, Int. J. Heat Mass Trans. 123 (2018) 637–654.

[6] M.M.F. Hasan, R.C. Baliban, J.A. Elia, C.A. Floudas, Modeling, simulation, and optimization of postcombustion CO_2 capture for variable feed concentration and flow rate. 1. Chemical absorption and membrane processes, Ind. Eng. Chem. Res. 51 (2012) 15642–15664.

[7] F. Boukouvala, Y.J. Gao, F. Muzzio, M.G. Ierapetritou, Reduced-order discrete element method modeling, Chem. Eng. Sci. 95 (2013) 12–26.

[8] D. Barrasso, A. Tamrakar, R. Ramachandran, A reduced order PBM-ANN model of a multi-scale PBM-DEM description of a wet granulation process, Chem. Eng. Sci. 119 (2014) 319–329.

[9] A. Rogers, M.G. Ierapetritou, Discrete element reduced-order modeling of dynamic particulate systems, AICHE J. 60 (2014) 3184–3194.

[10] M.Z. Yu, D.C. Miller, L.T. Biegler, Dynamic reduced order models for simulating bubbling fluidized bed adsorbers, Ind. Eng. Chem. Res. 54 (2015) 6959–6974.

[11] N. Quirante, J. Javaloyes, J.A. Caballero, Rigorous design of distillation columns using surrogate models based on Kriging interpolation, AICHE J. 61 (2015) 2169–2187.

[12] A. Narasingam, J.S.-I. Kwon, Development of local dynamic mode decomposition with control: application to model predictive control of hydraulic fracturing, Comput. Chem. Eng. 106 (2017) 501–511.

[13] H.S. Sidhu, A. Narasingam, P. Siddhamshetty, J.S.-I. Kwon, Model order reduction of non-linear parabolic PDE systems with moving boundaries using sparse proper orthogonal decomposition: application to hydraulic fracturing, Comput. Chem. Eng. 112 (2018) 92–100.

[14] A. Narasingam, P. Siddhamshetty, K.J. Sang-Il, Temporal clustering for order reduction of nonlinear parabolic PDE systems with time-dependent spatial domains: application to a hydraulic fracturing process, AICHE J. 63 (2017) 3818–3831.

[15] C.W. Rowley, I. Mezić, S. Bagheri, P. Schlatter, D.S. Henningson, Spectral analysis of non-linear flows, J. Fluid Mech. 641 (2009) 115–127.

[16] K. Palmer, M. Realff, Optimization and validation of steady-state flowsheet simulation meta-models, Chem. Eng. Res. Des. 80 (2002) 773–782.

[17] K. Palmer, M. Realff, Metamodeling approach to optimization of steady-state flowsheet simulations: model generation, Chem. Eng. Res. Des. 80 (2002) 760–772.

[18] F.A.N. Fernandes, Optimization of Fischer-Tropsch synthesis using neural networks, Chem. Eng. Technol. 29 (2006) 449–453.

[19] Y.D. Lang, A. Malacina, L.T. Biegler, S. Munteanu, J.I. Madsen, S.E. Zitney, Reduced order model based on principal component analysis for process simulation and optimization, Energy Fuel. 23 (2009) 1695–1706.

[20] Y.D. Lang, S.E. Zitney, L.T. Biegler, Optimization of IGCC processes with reduced order CFD models, Comput. Chem. Eng. 35 (2011) 1705–1717.

[21] M.M.F. Hasan, R.C. Baliban, J.A. Elia, C.A. Floudas, Modeling, simulation, and optimization of postcombustion CO_2 capture for variable feed concentration and flow rate. 2. Pressure swing adsorption and vacuum swing adsorption processes, Ind. Eng. Chem. Res. 51 (2012) 15665–15682.

[22] N. Quirante, J. Javaloyes-Anton, J.A. Caballero, Hybrid simulation-equation based synthesis of chemical processes, Chem. Eng. Res. Des. 132 (2018) 766–784.

[23] W.H. Ye, F.Q. You, A computationally efficient simulation-based optimization method with region-wise surrogate modeling for stochastic inventory management of supply chains with general network structures, Comput. Chem. Eng. 87 (2016) 164–179.

[24] J.A. Caballero, I.E. Grossmann, An algorithm for the use of surrogate models in modular flow-sheet optimization, AICHE J. 54 (2008) 2633–2650.

[25] F. Boukouvala, M.G. Ierapetritou, Surrogate-based optimization of expensive flowsheet modeling for continuous pharmaceutical manufacturing, J. Pharm. Innov. 8 (2013) 131–145.

[26] A. Cozad, N.V. Sahinidis, D.C. Miller, Learning surrogate models for simulation-based optimization, AICHE J. 60 (2014) 2211–2227.

[27] T.F. Yee, I.E. Grossmann, Simultaneous optimization models for heat integration—II. Heat exchanger network synthesis, Comput. Chem. Eng. 14 (1990) 1165–1184.

[28] E. Ahmetović, N. Ibrić, Z. Kravanja, I.E. Grossmann, Water and energy integration: a comprehensive literature review of non-isothermal water network synthesis, Comput. Chem. Eng. 82 (2015) 144–171.

[29] L.T. Biegler, I.E. Grossmann, A.W. Westerberg, Systematic Methods For Chemical Process Design, Prentice Hall, United States, 1997.

[30] C.A. Floudas, A.M. Niziolek, O. Onel, L.R. Matthews, Multi-scale systems engineering for energy and the environment: challenges and opportunities, AICHE J. 62 (2016) 602–623.

[31] L.T. Biegler, New nonlinear programming paradigms for the future of process optimization, AICHE J. 63 (2017) 1178–1193.

[32] M. Gao, F.Z. Sun, S.J. Zhou, Y.T. Shi, Y.B. Zhao, N.H. Wang, Performance prediction of wet cooling tower using artificial neural network under cross-wind conditions, Int. J. Therm. Sci. 48 (2009) 583–589.

[33] X.Q. Wei, N.P. Li, J.Q. Peng, J.L. Cheng, J.H. Hu, M. Wang, Performance analyses of counterflow closed wet cooling towers based on a simplified calculation method, Energies 10 (2017).

[34] S.M. Qasim, M.J. Hayder, Parametric study of closed wet cooling tower thermal performance, IOP Conf. Ser. Mat. Sci. (2017) 227.

[35] J.M. Salazar, S.E. Zitney, U.M. Dhvekar, Minimization of water consumption under uncertainty for a pulverized coal power plant, Environ. Sci. Technol. 45 (2011) 4645–4651.

[36] J.M. Salazar, U.M. Diwekar, S.E. Zitney, Stochastic simulation of pulverized coal (PC) processes, Energy Fuel (2010).

[37] Q. Zhu, B. Zhang, Q. Chen, C. He, D.C.Y. Foo, J. Ren, et al., Model reductions for multiscale stochastic optimization of cooling water system equipped with closed wet cooling towers, Chem. Eng. Sci. 224 (2020) 115773.

[38] H. Chen, C. Yang, K.J. Deng, N.N. Zhou, H.C. Wu, Multi-objective optimization of the hybrid wind/solar/fuel cell distributed generation system using Hammersley sequence sampling, Int. J. Hydrogen Energ. 42 (2017) 7836–7846.

[39] J.M. Salazar, U.M. Diwekar, S.E. Zitney, Stochastic simulation of pulverized coal (PC) processes, Energ Fuel. 24 (2010) 4961–4970.

[40] R.L. Iman, W.J. Conover, A distribution-free approach to inducing rank correlation among input variables, Commun. Stat. Simul. Comput. 11 (1982) 311–334.

[41] M.A. Duran, I.E. Grossmann, An outer-approximation algorithm for a class of mixed-integer nonlinear programs, Math. Program. 36 (1986) 307–339.

Chapter 16

Artificial intelligence algorithm-based multi-objective optimization model of flexible flow shop smart scheduling

Huanhuan Zhang, Jigeng Li, Mengna Hong, and Yi Man
State Key Laboratory of Pulp and Paper Engineering, School of Light Industry and Engineering, South China University of Technology, Guangzhou, People's Republic of China

1 Introduction

With the slowdown increasing of market demand, the manufacturing industry is facing fierce competition in recent years. To enhance the competitiveness, most of the manufacturing enterprises, especially the enterprises with flexible flow shops, choose to increase the supply of customized and personalized products. However, this will increase the complexity of the production process and rise the production costs. To address this issue, production schedule should be optimized to deal with the constraints such as the delivery term, on-hand inventory control, technological conditions, resource and energy consumption and wastage, production costs, profits, etc., and optimally determine the processing time or the sequence of the production jobs.

The conventional solution for flow shop scheduling process is to define the flow shop by only one machine at each processing stage. The purpose is to design a set of optimization schemes, reasonably arrange the processing schedule of each job, and achieve the optimization of the final processing results. Although the classic shop scheduling problem has made significant progress in theory [1, 2], there is a substitute for research on how to allocate machines. At present, internal manufacturing enterprises mainly rely on assembly line production methods. That is a production method that processes continuously and sequentially according to a certain process route and a unified production speed. Therefore, as an important branch of flexible scheduling, the problem of flexible flow shop production scheduling (flexible flow shop scheduling problem, FFSP) has attracted extensive attention from scholars at home and abroad.

Applications of Artificial Intelligence in Process Systems Engineering
https://doi.org/10.1016/B978-0-12-821092-5.00008-5

447

FFSP is a generalization of flow shops and parallel machines. It is a scheduling problem aiming at optimizing a certain index and multiple indexes. It is also called a hybrid FFSP. The problem of flexible flow shop scheduling widely exists in chemical, metallurgy, textile, machinery, semiconductor, logistics, construction, papermaking, and other fields. It was first proposed by Salvador in 1973 based on the background of the petroleum industry [3]. In terms of the FFSP model, Mousavi et al. established an FFSP mathematical model that minimized makespan and total delays [4]. Zandieh et al. studied the scheduling problem of flexible flow shop groups with sequence-related preparation time and established simultaneous minimization mathematical model of weighted delay and makespan and use a multi-group genetic algorithm to find Pareto optimal solution [5]. Defersha and Chen established a mathematical model for a batch flexible flow shop that can skip a certain stage, and the results show that parallel Genetic Algorithm (GA) has better computing efficiency [6]. Ying et al. established a mathematical model for a flow shop that cannot wait for sequence-related preparation time [7]. Bruzzone et al. established a MILP model of flexible FFSP constrained by peak power load. Its goal is to minimize the weighted sum of total delay and makespan [8]. Moon et al. proposed a discrete-time integer programming model and a hybrid genetic algorithm for unrelated parallel machine problems to minimize the weighted sum of makespan and power cost [9].

In terms of FFSP solution, the FFSP problem has been proved to be an NP-hard problem. As the scale of production scheduling problems continues to increase, the difficulty of solving them also increases [10]. Researchers have conducted a lot of research on the solution of production scheduling problems and applied various methods to solve the problem of production scheduling. The details are as follows [11]: (1) Accurate solution method. Such as linear programming, the solution to this kind of method is suitable for small-scale production scheduling problems. When the scale of the problem gradually increases, the solution time will increase exponentially. (2) Approximate solution method. The approximate solution methods can be divided into constructive methods and iterative methods. The advantage of using constructive methods (common constructive methods such as dispatch rules) to solve production scheduling problems is that the solution speed is fast. But the optimality of the solution is difficult to guarantee. Iterative methods include domain search algorithms and intelligent algorithms. The advantage of this kind of method is that the algorithm has a low dependence on the problem and strong versatility. Commonly used intelligent optimization algorithms include GA, ant colony optimization, and so on.

Although domestic and foreign scholars have made some achievements in the theoretical research of FFSP, most of the relevant research has focused on the problem of having the same processing equipment [12, 13]. In practice, it is often found that machines with different efficiencies run side by side. Besides, FFSP needs to consider not only the optimization of job sequencing, but also the allocation of machines, which increases the space of feasible solutions. When the scale of the problem reaches a certain level, the scheduling problem of the flexible flow shop is an NP problem. How to satisfy various constraints and efficiently find the best feasible solution in algorithm decoding is an issue that must be considered when studying FFSP. Therefore, how to construct fast intelligent algorithm for FFSP to seek the optimal scheduling of flexible flow shop has become the basis of agile manufacturing technology and an important component of the modern advanced manufacturing systems. Because of this, this paper studies the production scheduling problem of the flexible flow shop and establishes a classical mathematical model of production scheduling of flexible flow shop. And built an intelligent optimization algorithm (GA) for solving FFSP problem. Finally, part of the shops in the production process of ceramics and cement (such as the ball mill shop, the production method of which is a flexible flow shop scheduling) is selected as a research instance to verify the feasibility and effectiveness of the production scheduling model of the flexible flow shop and the GA's performance in solving the production scheduling problem of the flexible flow shop.

2 Literature review

There are many types of scheduling problems, which can be divided into single machine scheduling problems, parallel machine scheduling problems, job shop scheduling problems, and FFSPs. The flexible flow shop is one of the most extensive manufacturing systems. The FFSP problem widely exists in the fields of chemical industry, metallurgy, textile, machinery, semiconductor, logistics, construction, papermaking, etc. To solve the FFSP problem, the researchers conducted a lot of research on the solution of FFSP and applied various methods to solve the problem of production scheduling. At present, the methods for solving the production scheduling problem of flexible flow shops are divided into precise scheduling method and approximate scheduling method. When the problem size is small, an accurate solution algorithm can be used. The mixed linear programming and branch definition method is the most commonly used accurate solution method for solving flexible flow shop [2]. As the scale of the problem continues to increase, the solution time of the exact solution algorithm will increase exponentially, making it difficult to find a solution to the problem within an acceptable time frame.

Intelligent optimization algorithms are a type of optimization algorithms that is inspired by artificial intelligence or the social or natural phenomena of biological groups and simulates their laws. It can be roughly divided into evolutionary algorithms and swarm intelligence algorithms. Because of its global optimization performance, strong versatility, and suitability for parallel processing. At present, the intelligent optimization algorithm that has been applied to the production scheduling problem of the flexible flow shop mainly includes four categories. (1) Evolutionary algorithms: genetic algorithm [14–16], differential evolution algorithm [17]. (2) swarm intelligence algorithm: ant colony algorithm [18], particle swarm [19, 20], artificial bee colony algorithm [21]. (3) neighborhood search algorithm: simulated annealing [22], variable neighborhood search [23], tabu search algorithm [11, 24], and (4) neural network [25].

The evolutionary algorithm is to simulate the collective learning process of a group of individuals, and it is not easy to fall into local optimality in the search process. Second, because of their inherent parallelism, the algorithm is very suitable for massively parallel machines. However, evolutionary algorithms often have problems such as large search space, long search time, and easy premature convergence when solving large combinatorial optimization problems. The swarm intelligence optimization algorithm has the advantages of swarm intelligence distribution, robustness, indirect communication, and simplicity. The main disadvantage of swarm intelligence optimization algorithm is that it is prone to premature convergence and poor local optimization ability [2]. The neighborhood search algorithm can obtain the local optimal solution. It applies to various optimization problems and can be embedded in other intelligent optimization algorithms. However, the search performance of the neighborhood search algorithm depends entirely on the neighborhood structure and the initial solution. The advantages of neural networks are large storage space, strong storage capacity, strong self-learning ability, good fault tolerance, and easy classification. It has problems such as low learning efficiency, slow learning speed, and difficulty in expressing knowledge. How to use intelligent optimization algorithms to effectively solve the scheduling problem has always been a challenging problem and a research hotspot.

3 Flexible flow shop production scheduling model

3.1 Problem description

The FFSP is also called job scheduling problem with the same sequence. It is a simplified model of many practical assembly line production scheduling problems. The classic FFSP can be described as. There are n independent jobs processed on m machines according to the same process route. Each job needs to go through n procedures. These processes require different machines, and the processing of each process cannot be interrupted. The flexible flow shop model is a summary of the classic flow shop model, which is more in line with the actual production environment. The flexible flow shop model assumes that at least one stage must have multiple machines. The optimization problem includes two, one is to assign the job to the machine at each stage. The second is to sort jobs assigned to the same machine to minimize certain optimization goals [11]. The limits of its jobs are as follows: (1) A machine can process at most one job at a time. (2) A job can only be processed by one machine at a time. (3) Preemption processing is not allowed. (4) Jobs are not allowed to be split. (5) The operations of the jobs must be performed in sequence, and the processing stages must not overlap.

3.2 Production scheduling model

In the production scheduling problem of flexible flow shop, the establishment time can be divided into two types. One is related to the machine. That is, if the start time of a job depends on the machine to which the job is assigned, the creation time of the job depends on the machine. The second is when the job creation time on the machine depends on the job just completed on the machine. Then the establishment time of the job depends on the job sequence to be processed on the machine. In addition, the objective function of FFSP is usually determined according to its specific production environment and constraints. The common objective functions include minimizing the makespan, the number of delayed jobs, energy consumption, and so on. The classic model of flexible flow shop scheduling is as follows:

$$\text{Minimize } C_{max} \tag{1}$$

$$\text{Subject to}: \sum_{i=1}^{n} A_{i,l} = 1, l = 1, 2, \ldots, n \tag{2}$$

$$\sum_{k=1}^{m_j} B_{i,j,k} = 1, i = 1, 2, \ldots, n; j = 1, 2, \ldots, S \tag{3}$$

$$A_{i,l} = \begin{cases} 1, \text{if the } i\text{th job is scheduled at the position } l\text{th} \\ 0, \text{if not} \end{cases} \tag{4}$$

$$B_{i,j,k} = \begin{cases} 1, \text{if the } j\text{th operatioin of the } i\text{th job is assigned to the } k\text{th machine} \\ 0, \text{if not} \end{cases} \tag{5}$$

$$e_{i,j} = st_{i,j} + p_{i,j}, i = 1, 2, \ldots, n; j = 1, 2, \ldots, S \tag{6}$$

$$e_{i,j} \leq st_{i,j+1}, i = 1, 2, \ldots, n; j = 1, 2, \ldots, S - 1 \tag{7}$$

$$\sum_{i=1}^{n} e_{i,j} A_{i,l} \leq st_{i,j}, j = 1, 2, \ldots, S; l = 1, 2, \ldots, n \tag{8}$$

$$\sum_{i=1}^{n} st_{i,j} B_{i,j,k} \leq e_{i,j} B_{i-1,j,k}, j = 1, 2, \ldots, S; k = 1, 2, \ldots, m_j \tag{9}$$

where i represents the ith job, j represents the jth process, m_j represents the jth process has m parallel machines, k represents the kth machine, l represents the lth processing position, e represents the processing completion time of the job, st represents the job start processing time, p represents the job processing time. Eq. (1) represents the objective function that needs to be optimized. Eq. (2) indicates that all jobs must be scheduled. Eq. (3) indicates that each job can only be processed once in each process. Eqs. (4) and (5) are binary functions. Eq. (6) shows that the completion time of the jth process of the ith job is equal to the sum of the start time of the jth process of the ith job and the processing time of the jth process of the ith job. Eq. (7) indicates that the end time of the jth process of the ith job is not greater than the start time of the $(j+1)$th process of the ith job. Eq. (8) indicates that the start time of the $(i+1)$th process is not less than the end time of the ith job. Eq. (9) indicates that on the kth machine, the start time of the ith job is less than or equal to the end time of the ith job.

3.3 GA method introduction

GA was first proposed by Professor Holland in 1975 and is a method to simulate the natural evolution of living things [26]. It selects, crosses, and mutates to express the decoded problem as a "chromosome" population that evolves from generation to generation until it reaches the stopping criterion set by the algorithm. To find the optimal solution to the problem. GA has the advantages of simple principle, strong versatility, unlimited constraints, and good global solution searchability. Therefore, GA has been widely used in combinatorial optimization problems. Davis was the first scholar to apply the GA to solve production scheduling [27]. At present, GA has been applied to the FFSP problem and has achieved good results. Kurz and Askin proposed a random key encoding method, and the designed single-point cross genetic algorithm has stable performance [28]. Akrami et al. used GA to solve the batch FFSP problem with limited buffers [29]. Bellkadi et al. designed a parallel genetic algorithm and successfully solved the multi-stage FFSP problem [30]. They also designed a variety of genetic algorithms for different machine allocation rules for the production scheduling of ceramic tiles with constraints [31].

The GA process designed in this paper is shown in Fig. 1. It can be briefly described as: (1) The input data part needs to preprocess the data (including the elimination of outliers and the filling of missing values) and import the data into the model. (2) Set some parameters of GA (such as crossover probability, mutation probability, population size, number of iterations, etc.). (3) Initialize the population according to the set job amount and size and use the input data to calculate the fitness value. (4) Sort the individuals in the population according to the fitness value and save the individuals with the largest fitness value. Then, randomly select some individuals from the population to perform crossover operation and mutation operation. Repeat this process until the set stopping criterion is reached. (5) Update the population. And repeat steps (3) and (4) until the maximum number of GA iterations is satisfied.

Monte Carlo method is used to initialize the population. According to the size of the job and the size of the set population, a corresponding number of populations is automatically generated. Then encode all jobs. The calculation of fitness value can be summarized as follows: (1) Set related variables and set intermediate variables for recording process data. (2) Select the earliest available machine. Arrange the job to the machine for processing and update the earliest available time of the machine. Repeat this process until all jobs are scheduled. (3) According to the recorded intermediate data, the processing route and processing time of each job are counted and calculated. In the

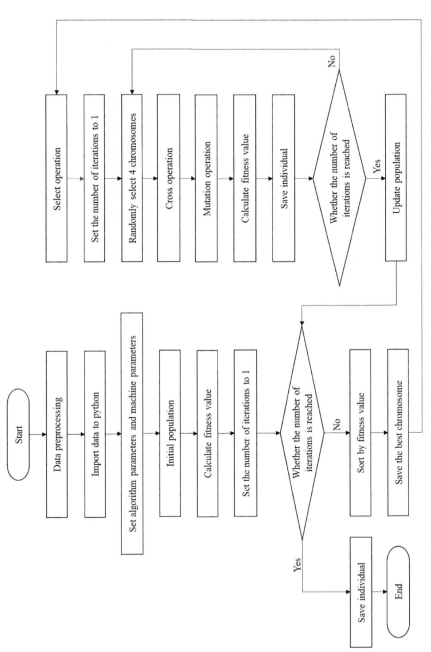

FIG. 1 The flowchart of GA.

selection operation, four chromosomes are randomly selected from the population (one chromosome represents a processing sequence for all jobs), and then the two optimal ones are selected. In the crossover operation, first select two chromosomes. Then select half of the chromosome length as the exchange gene (a gene on the chromosome represents a processing job), and finally exchange the exchange genes of the two chromosomes. In the mutation operation, a chromosome is randomly selected, and then two genes are randomly selected to exchange them.

4 Case study

A ball mill is a type of heavy mechanical equipment that relies on its own rotation to drive the steel balls inside to impact and grind material loads with high reliability. However, the ball mill has the disadvantages of low working efficiency and high energy consumption [32]. The ball mill is a key equipment for regrinding after industrial materials are crushed. The application fields of the ball mill are very wide, such as cement, new building materials, refractory materials, fertilizer and metal beneficiation, and glass ceramics. The production process of the ball mill shop can be simplified to flexible flow shop scheduling, which is an *NP* problem.

Model analysis and verification were carried out using data from a ball mill shop of a ceramic factory in Guangdong. Three sets of experimental data have been selected for a certain three-day production schedule of the ball mill shop of the ceramic factory. Each set of experiments includes a production scheduling problem with production cost minimization, which is solved using GA. Then analyze and compare with the results of manual scheduling.

4.1 Production process of ball mill shop

The production process of the ball mill shop is shown in Fig. 2, which mainly includes three stages, namely the feeding stage, feeding stage, and ball milling stage. The equipment used includes forklifts, feeders, and ball mills. In general, one feeding opportunity corresponds to multiple ball mills. But one feeder can only feed one ball mill at a time. In the feeding stage, the forklift takes the material from the warehouse and loads it into the feeder. In the feeding stage, the feeder transports the material to the ball mill through a conveyor belt. After the feeding is completed, enter the ball milling stage for ball milling. The crushing effect of the ball mill on the material is mainly the impact and grinding of the grinding body on the material. After the ball milling is finished, the ball milling effect of the material needs to be inspected, that is, quality inspection. If the quality inspection is qualified, proceed to the pulping step and cleaning step. If the quality inspection fails, continue ball milling.

The production scheduling problem of the ball mill shop can be reduced to a three-stage flexible flow workshop production scheduling model. Taking the

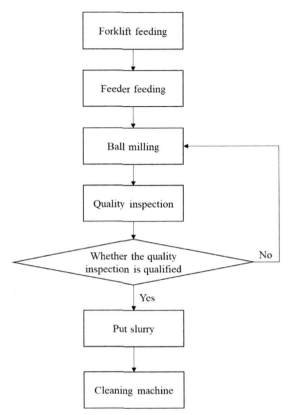

FIG. 2 Production flow chart of ball mill shop.

three-stage flexible flow shop production scheduling problem as the research object, the established production scheduling model is shown below:

$$\sum_{i=1}^{n} \sum_{j=1}^{m_1} O_{i,j} = 1 \tag{10}$$

$$\sum_{i=1}^{n} \sum_{j=1}^{m_2} O_{i,j} = 1 \tag{11}$$

$$\sum_{i=1}^{n} \sum_{j=1}^{m_3} O_{i,j} = 1 \tag{12}$$

$$I_{i,j} \geq F_{j,l-1} + TS_{l,l-1,j} \forall j \in 1,2,3\ldots,m; \forall l \in 2,3,4\ldots,N_j \tag{13}$$

$$B_{i,3} \geq E_{i,2} + f(ST), \tag{14}$$

$$f(QT, FT) = QT + FT \tag{15}$$

$$m_1 + m_2 + m_3 = m \tag{16}$$

where m is the number of production lines, and n is the number of the jobs. $O_{i,j}$ indicates whether the ith job is produced on the jth production lines. When $O_{i,j}$ is equal to 1, it means that the ith job is produced on the jth production line. When $O_{i,j}$ is equal to 0, it means that the ith job is not produced on the jth production line. Eq. (10) indicates that each job must be processed in the first stage and can only be processed once. Eq. (11) indicates that each job must be processed in the second stage and can only be processed once. Eq. (12) indicates that each job must be processed in the third stage and can only be processed once. Inequality (13) indicates that each production line can only process one job at a time. Where $I_{i,j}$ represents the start processing time of the ith job on the jth production line. On the same production line, the start time of the next job must be greater than or equal to the end time of the previous job plus the set-up time. Inequality (14) shows the relationship between the end time of each job in the second stage and the start processing time of the third stage. There is a time interval between $E_{i,2}$ and $B_{i,3}$, and this time interval is $f(QT, FT)$. Eq. (15) shows the size of the time interval. Eq. (16) represents the number of all machines.

4.2   Optimal object

Due to the pressure of the environment and energy prices, the manufacturing industry is facing huge pressure for energy saving and emission reduction. Electricity is one of the most important energy sources in the manufacturing industry. According to the difference in electricity demand, the electricity demand can be divided into three time periods in 1 day, which are the electricity peak period, the electricity valley period, and the electricity normal period. Due to the imbalance of electricity consumption, it is easy to produce waste in the valley of electricity consumption, and the supply and demand will be tight in the peak section. The implementation of time-of-use electricity prices has given enterprises the space to optimize production costs. In the ball mill shop, the energy consumption in different processing stages is different. Under the time-of-use electricity price mechanism, production scheduling that minimizes production costs requires that high-energy-consumption ball mill stages be arranged as much as possible within the electricity valley. And the feeding stage and feeding stage are arranged in the peak period of electricity consumption. Thereby reducing the total energy consumption cost. Among them, the processing energy consumption of the ball mill is the main energy consumption. Under the mechanism of time-of-use electricity price, production can be reasonably arranged through production scheduling to reduce the processing energy cost of the ball mill shop. In the ball mill shop, when switching between different types of jobs, the ball mill needs to be cleaned. In this process, the ball mill will run at lower power and will consume part of the standby energy consumption. Besides, cleaning the ball mill requires additional cleaning time. This may cause the earliest available time of the ball mill to become longer, which may affect the completion time of the job. In the process of production scheduling, reducing unnecessary production cleaning can reduce production costs for enterprises and may also shorten the completion time. Similarly, under the mechanism of time-of-use electricity prices, it is also an optimization problem at which period the production cleaning is scheduled.

Many factors affect the processing energy consumption of the ball mill. The type of job is one of the important factors that affect processing energy consumption. When producing different types of jobs, the energy consumption in the feeding stage, feeding stage, and ball milling stage is different. In addition, the energy consumption generated in different stages is also different, in which the ball milling stage is the main energy consumption stage. In the production schedule, the ball mill stage with large energy consumption is arranged as much as possible in the valley period or the normal period. In the peak period, the feeding stage and the ball washing stage with low energy consumption are carried out as much as possible. In the ball mill shop, in the process of production conversion (because the formula types of the two jobs are different, the ball mill needs to be cleaned) not only consumes part of the power consumption, but also prolongs the earliest available time of the ball mill, which may affect the completion of the job time. When jobs of the same type are scheduled to be produced in sequence, there is no set-up power consumption. Therefore, this paper regards set-up power consumption as one of the optimization goals.

4.3 Production cost model

To implement production scheduling based on the time-of-use electricity price minimization of the production cost of the ball shop in the ceramic industry, it is necessary to establish the power consumption model of the ball shop. This section mainly considers two types of power consumption, which are the processing power consumption of the ball mill and the set-up power consumption. Processing power consumption is related to equipment and product types, while processing power consumption cost is related to processing power consumption and electricity prices. Under the rules of time-of-use electricity prices, the price of electricity is different for each period of the day. This section establishes production cost based on time-of-use electricity prices, of which the specific calculation formula for processing power consumption costs is as follows:

$$EP = \sum_{j=1}^{m} \sum_{i=1}^{n} (EP1 + EP2 + EP3) \times O_{i,j} \tag{17}$$

$$EP1 = \sum_{j=1}^{m} \sum_{i=1}^{n} U_{i,j} \times P_1 \times PT_{i,j,1} \tag{18}$$

$$PT_{i,j,k} = \sum_{k=1}^{K} Q([ST_{i,j}, ST_{i,j} + T_{i,j}], k), \tag{19}$$

$$EP2 = \sum_{j=1}^{m} \sum_{i=1}^{n} U_{i,j} \times P_2 \times PT_{i,j,2} \tag{20}$$

$$EP3 = \sum_{j=1}^{m} \sum_{i=1}^{n} U_{i,j} \times P_3 \times PT_{i,j,3} \tag{21}$$

$$PT_{i,j,k} = \sum_{k=1}^{k} \sum_{j=1}^{m} \sum_{i=1}^{n} T\left(st_{i,j}, et_{i,j}\right) \tag{22}$$

where EP is the processing power cost consumed by the machine, $EP1$ is the processing power cost of all jobs in the peak period, $EP2$ is the processing power cost of all jobs in the valley period, and $EP3$ is the processing power cost of all jobs in the normal period. $U_{i,j}$ represents the power consumption per unit time of the ith job on the jth machine. $P1$, $P2$, and $P3$ are the electricity prices during peak, valley, and normal periods, respectively. $PT_{i,j,k}$ represents the processing time of the ith job on the jth machine. $ST_{i,j}$ indicates the starting time of the ith job on the jth machine. $T_{i,j}$ represents the processing time of the ith job on the jth machine. Eq. (22) indicates that the processing time of the job is counted according to the peak period, valley period and normal period, where $st_{i,j}$ indicates the start time of the ith job on the jth machine, $et_{i,j}$ represents the end time of the ith job on the jth machine.

Set-up energy consumption is the energy consumed in the process of cleaning the ball mill when the types of two adjacent processing jobs are inconsistent. In the context of time-of-use electricity prices, the cleaning stage of equipment occurs at different times, and the energy price per unit time is different. When the cleaning stage of the machine occurs at the peak period, and the energy consumption cost of cleaning will be higher than the valley period or the normal period. The formula for set-up the power consumption cost model is shown below:

$$ES = \sum_{j=1}^{m} \sum_{i=1}^{n} (ES1 + ES2 + ES3) \times O_{i,j} \tag{23}$$

$$ES1 = \sum_{j=1}^{m} \sum_{i=1}^{n} U_{i,j} \times P_1 \times ST_{i,j,1} \tag{24}$$

$$ES2 = \sum_{j=1}^{m} \sum_{i=1}^{n} U_{i,j} \times P_2 \times ST_{i,j,2} \tag{25}$$

$$ES3 = \sum_{j=1}^{m} \sum_{i=1}^{n} U_{i,j} \times P_3 \times ST_{i,j,3} \tag{26}$$

where ES is the total set-up cost, $ES1$ is the production cost of cleaning the ball mill that occurred during the peak period, $ES2$ is the set-up production cost that occurred during the valley period, and $ES3$ is the set-up production cost that occurred during the normal period. $ST_{i,j,1}$ is the set-up time during the peak period. $ST_{i,j,2}$ is the set-up time in the valley period, and $ST_{i,j,3}$ is the set-up time in the normal period.

$$\min(EPC) \tag{27}$$

$$EPC = \sum_{j=1}^{m} \sum_{i=1}^{n} (EP + SP) \times O_{i,j} \tag{28}$$

where Eq. (27) represents the optimization goal, which is the cost of energy consumption. Formula (28) represents the total production energy consumption, including processing energy consumption and set-up energy consumption.

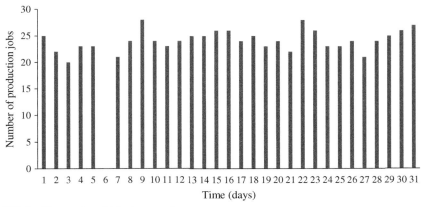

FIG. 3 The number of jobs in a month.

4.4 Experimental data

The data used in this chapter is extracted from the database of a ball mill system of a ceramic factory in Guangdong province. Data preprocessing is mainly to eliminate outliers, fill in missing values according to the set values, and match the data of daily production jobs according to the number of ball mills.

Fig. 3 shows the number of production tasks in a daily production plan of a ball mill in a ceramic factory in Guangdong Province (No. 6 has no production job data due to system abnormalities). As can be seen from Fig. 3, the number of production jobs in a day is at least 20 and at most 28. To test the performance of the GA algorithm on production scheduling problems with different numbers of jobs, this paper selects a certain 3-day scheduling data for verification. Three sets of experiments were designed, and each experiment included an example of production scheduling with production cost minimization. The three production costs minimize the name of the production scheduling instance and the number of jobs, as shown in Table 1. Table 2 is the number of equipment in the three stages of the ball mill shop, of which there is two forklift equipment of the same specification in the first stage (feeding stage). The second stage

TABLE 1 Names and number of jobs of three production scheduling instances with minimum production cost.

Instances	The number of jobs
Job1	21
Job2	24
Job3	28

TABLE 2 Number of equipment in ball mill shop.

Forklift	Feeding machine	Ball mill
2	3	18

(feeding stage) has three feeders with the same specifications. Feeder No. 1 can only feed the ball mills No. 1–6. No. 2 feeder can only feed 7–12 ball mills. Feeder 3 can only feed the ball mills 13–18. The third stage (ball milling stage) has 18 ball mills with the same capacity. Tables 3–5 correspond to the processing time of Job1, Job2, and Job3 in three stages, respectively.

Ball washing energy consumption is related to the ball washing time and the energy consumption per unit time. The ball washing time is set to 60 min (the specifications of all ball mills in this ball mill are the same). The equipment power of the ball mill during cleaning is shown in Table 6. In this chapter, the quality inspection is set to 30 min, and the slurry discharge time is set to 30 min. Also, the time division of the time-of-use electricity price adopted in this paper and the electricity price is shown in Table 7.

4.5 Parameter setting

In this paper, GA is used to solve the production cost minimization scheduling problem. In GA algorithm, the important parameters include population size, maximum iteration times, crossover probability, and mutation probability. To find the optimal parameters, this chapter designs some comparative experiments. First, the maximum number of iterations, crossover probability, and mutation probability is set as fixed values to optimize the population size. Figs. 4 and 5 show the corresponding objective function value and the running time of GA algorithm when the population number is 50, 60, 70, 80, 90, 100, 110, 120, 130, 140, 150, 160, 170, 180, 190, and 200. As can be seen from Fig. 4, when the population size is 150, the production cost is the minimum. As can be seen from Fig. 5, the running time of GA algorithm increases with the increase of population. Considering the performance and efficiency of GA, 150 is the optimal population size. Second, the population size is set to 150, and the crossover probability and mutation probability are set as fixed values to optimize the iteration times. Fig. 6 shows the change in the value of the objective function as the number of iterations increases from 10 to 80. It can be concluded that when the number of iterations is set to 50, the production cost is the lowest. Fig. 7 shows how the run time of the algorithm varies with the number of iterations. Then, the population size is set to 150, the number of iterations is set to 50, and the mutation probability is set as a fixed value to optimize the crossover probability. As can be seen from Fig. 8, when the crossover probability is 0.85, the

TABLE 3 Processing schedule of all jobs in case Job1 in three stages.

Job number	First stage processing time (min)	Stage 2 processing time (min)	Stage 3 processing time (min)
1	106	14.4	443
2	104	14.4	656
3	83	14.4	180
4	92	14.4	633
5	86	14.4	640
6	137	14.4	642
7	76	14.4	640
8	61	14.4	180
9	131	14.4	651
10	83	14.4	645
11	188	14.4	1026
12	144	14.4	480
13	192	14.4	235
14	155	14.4	659
15	97	14.4	663
16	125	14.4	447
17	80	14.4	137
18	102	14.4	660
19	100	14.4	671
20	135	14.4	1023
21	145	14.4	444

minimum value of the objective function is obtained. Finally, the population size of 150, iteration times of 50, and crossover probability of 0.85 are set to optimize the mutation probability. As shown in Fig. 9, when the mutation probability is 0.125, the objective function value is the minimum.

To sum up, the important parameters of GA in this experiment are set as follows: the population size is set to 150, the maximum iteration number is set to 50, the crossover probability is set to 0.85, and the mutation probability is set to 0.125.

TABLE 4 Processing schedule of all jobs in case Job2 in three stages.

Job number	First stage processing time (min)	Stage 2 processing time (min)	Stage 3 processing time (min)
1	79	14.4	658
2	74	14.4	175
3	131	14.4	792
4	82	14.4	637
5	185	14.4	389
6	119	14.4	154
7	103	14.4	660
8	100	14.4	645
9	192	14.4	750
10	175	14.4	438
11	149	14.4	432
12	151	14.4	449
13	171	14.4	1021
14	116	14.4	443
15	115	14.4	655
16	70	14.4	640
17	121	14.4	658
18	119	14.4	766
19	82	14.4	655
20	102	14.4	600
21	113	14.4	138
22	99	14.4	443
23	143	14.4	1011
24	131	14.4	435

TABLE 5 Processing schedule of all jobs in case Job3 in three stages.

Job number	First stage processing time (min)	Stage 2 processing time (min)	Stage 3 processing time (min)
1	97	14.4	655
2	101	14.4	673
3	98	14.4	554
4	86	14.4	444
5	96	14.4	443
6	136	14.4	783
7	127	14.4	446
8	92	14.4	647
9	95	14.4	645
10	121	14.4	653
11	154	14.4	438
12	146	14.4	462
13	226	14.4	1002
14	106	14.4	442
15	201	14.4	772
16	133	14.4	780
17	103	14.4	473
18	102	14.4	655
19	104	14.4	432
20	86	14.4	633
21	103	14.4	469
22	80	14.4	644
23	130	14.4	1012
24	117	14.4	482
25	94	14.4	482
26	135	14.4	662
27	98	14.4	639
28	87	14.4	644

TABLE 6 Ball mill equipment power table.

Ball mill power during ball milling (kW/h)	Power of ball mill when washing ball (kW/h)
180	162

TABLE 7 Time-sharing of time-of-use electricity prices and electricity prices.

Stage	Period	Electricity price
Valley period	23:00–05:00	0.43351
Normal period	05:00–17:00	0.6193
Peak period	17:00–23:00	0.80509

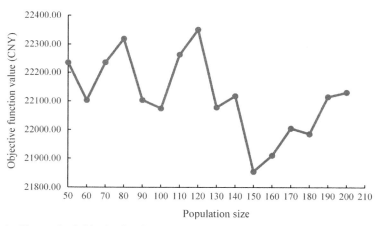

FIG. 4 The graph of objective function changing with population size.

4.6   Result analysis

This section mainly solves the production scheduling problem of minimizing the production cost under the time-of-use electricity price, and the optimization goal is the production power cost.

To this end, this study designed three sets of production scheduling examples. Using GA to solve each group of electricity costs to minimize production scheduling instances. It is compared with the results of manual scheduling to verify the performance of GA in solving the production scheduling problem.

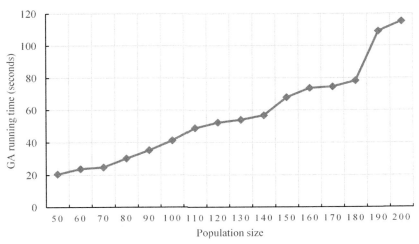

FIG. 5 GA algorithm running time changes with population size.

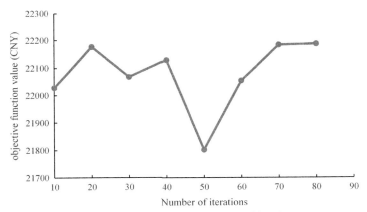

FIG. 6 The graph of the objective function with the number of iterations.

Considering the randomness of the evolutionary algorithm, when GA is used to solve each group of production scheduling job cases, it runs independently four times. Figs. 10–12 show iterative graphs of the production cost using GA, where the horizontal axis represents the number of iterations and the vertical axis represents the production cost. From the above three figures, it can be seen that the GA has good optimization ability and fast convergence speed.

Table 8 shows the results of the three sets of production scheduling instances running 4 times, respectively, and Table 9 shows the results of manual

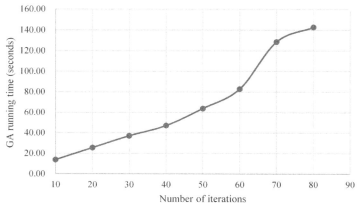

FIG. 7 GA running time changes with the number of iterations.

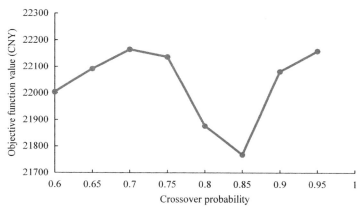

FIG. 8 The graph of the change of objective function value with cross probability.

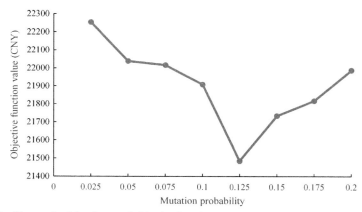

FIG. 9 The graph of the change of objective function value with mutation probability.

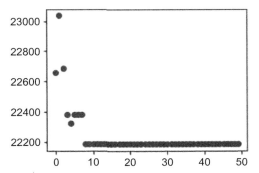

FIG. 10 Job1 iterative diagram.

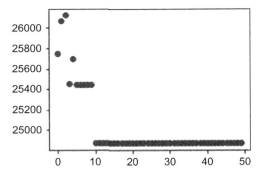

FIG. 11 Job2 iterative diagram.

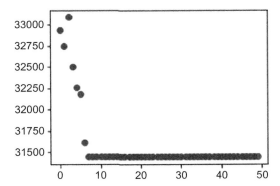

FIG. 12 Job3 iterative diagram.

scheduling. As can be seen from Tables 8 and 9, in each group of cases, the total processing time of the scheduling plan obtained by GA is the same as that obtained by manual scheduling. This illustrates the feasibility of using GA to solve the production scheduling model with the production cost minimized. It can also be seen from Tables 8 and 9 that the production cost value of each

TABLE 8 The scheduling results solved by GA.

Instance	Processing time at peak (min)	Processing time of valley period (min)	Processing time of normal period (min)	Total processing time (min)	Production cost (CNY)
Job1–1	2826.1	3187.7	5741.2	11,755	21,082
Job1–2	2659.2	3648.9	5446.9	11,755	21,099.4
Job1–3	2750.8	3240.6	5763.6	11,755	21,058.3
Job1–4	2607	3331.5	5816.5	11,755	20,832.7
Job2–1	2703.8	4236.2	6704	13,644	24,495.1
Job2–2	2662.8	3755	7226.2	13,644	24,740.4
Job2–3	2708.2	3508.2	7427.6	13,644	24,903.3
Job2–4	2630.8	4097.2	6916	13,644	24,531.9
Job3–1	3010.8	4266.2	9789	17,066	31,007.2
Job3–2	3072.4	3810.10	10,183.5	17,066	31,295.8
Job3–3	3120.6	4215.3	9730.1	17,066	31,096.8
Job3–4	3422.6	4422.7	9220.7	17,066	31,149.5

TABLE 9 Results of manual scheduling.

Instance	Processing time at peak (min)	Processing time of valley period (min)	Processing time of normal period (min)	Total processing time (min)	Production cost (CNY)
Job1	2896.5	3341	5517.5	11,755	21,916
Job2	3456.6	3818.4	6339	13,644	25,471.5
Job3	4106.1	4848.9	8111	17,066	31,616.8

group obtained by GA is less than the result obtained by manual scheduling, which shows that the GA has a good solution performance and further illustrates the effectiveness of the GA solution.

Fig. 13 shows the results of the lowest production cost value, the highest production cost value, the average production cost value in the instances Job1–1, Job1–2, Job1–3, Job1–4 obtained by GA and the results of the manually obtained instance Job1, where blue represents the result obtained by GA and orange represents the result of manual scheduling. It can be obtained from Fig. 13 that in the instance Job1 (the number of jobs is 21), the production cost obtained by GA saves about 1083 CNY, 817 CNY the lowest, and 898 CNY on average. Similarly, as can be seen from Fig. 14, in the instance Job2 (the number of jobs is 24), the production cost obtained by GA is about 976.4 CNY higher than manual scheduling, the lowest is 568.2 CNY, and the average can be 803.8

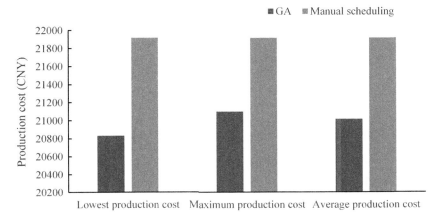

FIG. 13 Job1 solution obtained by GA and manual scheduling.

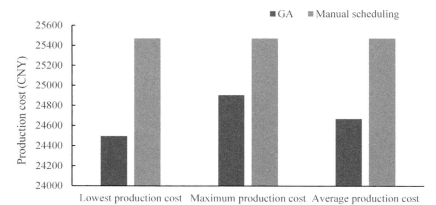

FIG. 14 Job2 solution obtained by GA and manual scheduling.

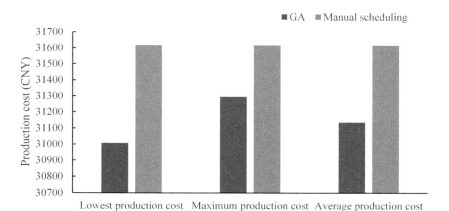

FIG. 15 Job3 solution obtained by GA and manual scheduling.

CNY. It can be seen from Fig. 15 that in the instance Job3 (the number of jobs is 28), the production cost obtained by GA is about 609.6 CNY higher than manual scheduling, the lowest is 321.1 CNY, and the average saving is 479.5 CNY. As can be seen from Figs. 13–15, when the number of jobs in a day is small (such as 21 jobs in Job1), it can save more money. This is because there are only 24 h in a day, and the larger the number of production jobs, the longer the total time required. The less time–space that can be optimized within 24 h, the less space is saved.

5 Conclusions

This chapter studies the production scheduling problem of the flexible flow shop and establishes a classical mathematical model of the production scheduling problem of the flexible flow shop. The intelligent optimization algorithm is used to solve the established production scheduling model. The results show that GA has a good ability to seek optimization and a fast convergence speed when solving the model. The ball mill shop of a ceramic factory in Guangdong Province was used as an instance to verify the feasibility and effectiveness of the model. The results show that the model established in this paper can save up to 4.9% of the production cost compared with manual scheduling. Finally, the analysis of the instance shows that the production scheduling model of flexible flow shop based on GA can effectively improve production efficiency and reduce production costs.

References

[1] Z. WHM, K. HJJ, The integration of process planning and shop floor scheduling in small batch part manufacturing, Cirp Ann. Manufact. Technol. 44 (1) (1995) 429–432.

[2] J. Mohan, K.A. Lanka, N. Rao, A review of dynamic job shop scheduling techniques, Procedia Manufact. 30 (2019) 34–39.

[3] M.S. Salvador, A solution of a special class of flow shop scheduling problems, in: Process of the symposium on the Theory of Scheduling and Its Application. Berlin, 1973, pp. 83–91.

[4] S.M. Mousavi, M. Zandich, M. Yazdani, A simulated annealing/local search to minimize the makespan and total tardiness on a hybrid flow shop, Int. J. Adv. Manuf. Technol. 64 (2013) 364–388.

[5] M. Zandieh, N. Karimi, An adaptive multi-populartion genetic algorithm to solve the multi-objective group scheduling problem in hybrid flexible flow shop with sequence-dependent setup times, J. Intell. Manuf. 22 (2011) 979–989.

[6] F.M. Defersha, M.Y. Chen, Mathematical model and parallel genetic algorithm for hybrid flexible flow shop lot streaming problem, Int. J. Adv. Manuf. Technol. 62 (2012) 249–265.

[7] K.C. Ying, S.W. Lin, Scheduling multistage hybrid flow shop with multiprocessor tasks by all effective heuristic, Int. J. Prod. Res. 47 (13) (2009) 3525–3538.

[8] A.A.G. Bruzzone, D. Anghinolfi, M. Paolucci, et al., Energy-aware scheduling for improving manufacturing process sustainability: a mathematical model for flexible flow shops, Cirp Ann. Manufact. Technol. 61 (1) (2012) 459–462.

[9] J.-Y. Moon, K. Shin, J. Park, Optimization of production scheduling with time dependent and machine-dependent electricity cost for industrial energy efficiency, Int. J. Adv. Manuf. Technol. 2013 (68) (2013) 523–535.

[10] P.G. Georgios, M.P. Borja, A.C. Daniel, et al., Optimal production scheduling of food process industries, Comput. Chem. Eng. 134 (2020).

[11] J. Jitti, R. Manop, C. Paveena, et al., A comparison of scheduling algorithms for flexible flow shop problems with unrelated parallel machines, setup times, and dual criteria, Comput. Oper. Res. 36 (2) (2009) 358–378.

[12] D. Alisantoso, L.P. Khoo, P.Y. Jiang, An immune algorithm approach to the scheduling of a flexible PCB flow shop, Int. J. Adv. Manuf. Technol. 22 (2003) 819–827.

[13] H.T. Lin, C.J. Liao, A case study in a two-stage hybrid flow shop with setup time and dedicated machines, Int. J. Prod. Econ. 86 (2) (2003) 133–143.

[14] K. Kianfar, G. Fatemi, J. Oroojlooy, Study of stochastic sequence-dependent flexible flow shop via developing a dispatching rule and a hybrid GA, Eng. Appl. Artif. Intel. 25 (3) (2012) 494–506.

[15] W. Sukkerd, T. Wuttipornpun, Hybrid genetic algorithm and tabu search for finite capacity material requirement planning system in flexible flow shop with assembly operations, Comput. Ind. Eng. 97 (2016) 157–169.

[16] J. Luo, S. Fujimura, E.B. Didier, et al., GPU based parallel genetic algorithm for solving an energy efficient dynamic flexible flow shop scheduling problem, J. Parallel Distribut. Comput. 133 (2019) 244–257.

[17] M. Mehdi, T.M. Reza, B. Armand, et al., Flexible job shop scheduling problem with reconfigurable machine tools: an improved differential evolution algorithm, Appl. Soft Comput. 94 (2020).

[18] S.C. Zhang, X. Li, B.W. Zhang, et al., Multi-objective optimization in flexible assembly job shop scheduling using a distributed ant colony system, Eur. J. Oper. Res. 283 (2) (2020) 441–460.

[19] D.B. Tang, M. Dai, A. Miguel, et al., Energy-efficient dynamic scheduling for a flexible flow shop using an improved particle swarm optimization, Comput. Ind. 81 (2016) 82–95.

[20] M. Nouiri, A. Bekrar, et al., An effective and distributed particle swarm optimization algorithm for flexible job-shop scheduling problem, J. Intell. Manuf. (2018) 1–13.

[21] X.X. Li, Z. Peng, B.G. Du, et al., Hybrid artificial bee colony algorithm with a rescheduling strategy for solving flexible job shop scheduling problems, Comput. Ind. Eng. 113 (2017) 10–26.

[22] F. Jolai, H. Asefi, M. Rabiee, et al., Bi-objective simulated annealing approaches for no-wait two-stage flexible flow shop scheduling problem, Scientia Iranica 20 (3) (2013) 861–872.

[23] G. Gong, C. Raymond, Q.W. Deng, et al., Energy-efficient flexible flow shop scheduling with worker flexibility, Expert Syst. Appl. 141 (2020).

[24] Z.Q. Zeng, M.N. Hong, Y. Man, et al., Multi-object optimization of flexible flow shop scheduling with batch process—consideration total electricity consumption and material wastage, J. Clean. Prod. 183 (2018) 925–939.

[25] K. Wang, S.H. Choi, A holonic approach to flexible flow shop scheduling under stochastic processing times, Comput. Oper. Res. 43 (2014) 157–168.

[26] X.H. Shi, Y.C. Liang, H.P. Lee, et al., An improved GA and a novel PSO-GA-based hybrid algorithm, Inf. Process. Lett. 93 (5) (2004) 255–261.

[27] K. Deb, K. Mitra, D. Rinku, et al., Towards a better understanding of the epoxy-polymerization process using multi-objective evolutionary computation, Chem. Eng. Sci. 59 (20) (2004) 4261–4277.

[28] M.E. Kurz, R.G. Askin, Scheduling flexible flow lines with sequence-dependent setup times, Eur. J. Oper. Res. 159 (1) (2004) 66–82.

[29] B. Akrami, B. Karlmi, S.M.M. Hosseini, Two meta-heuristic methods for the common cycle economic lot sizing and scheduling in flexible flow shops with limited intermediate buffers: the finite horizon case, Appl. Math Comput. 183 (1) (2006) 634–645.

[30] K. Bekadi, M. Gourgand, M. Benyettou, Parallel genetic algorithms with migration for the hybrid flow shop scheduling problem, J. Appl. Math. Decis. Sci. 7 (1) (2006) 1–17.

[31] K. Belkadi, M. Gourangd, M. Benyettou, et al., Sequence and parallel genetic algorithm for the hybrid flow shop scheduling problem, J. Appl. Sci. 6 (4) (2006) 775–788.

[32] J. Tang, J.F. Qiao, Z. Liu, et al., Mechanism characteristic analysis and soft measuring method review for ball mill load based on mechanical vibration and acoustic signals in the grinding process, Miner. Eng. 128 (2018) 294–311.

Chapter 17

Machine learning-based intermittent equipment scheduling model for flexible production process

Zifei Wang and Yi Man

State Key Laboratory of Pulp and Paper Engineering, School of Light Industry and Engineering, South China University of Technology, Guangzhou, People's Republic of China

1 Introduction

Production scheduling is one of the main factors affecting productivity and efficiency [1]. Effective scheduling plans greatly improve the performance of manufacturing systems and reduce the cost of production. Therefore, many industrial engineers have recently focused on making effective and reasonable scheduling schemes for different demands.

Scheduling is an important decision-making process in most manufacturing and service industries [2]. Dynamic scheduling methods include traditional and intelligent methods. Traditional methods include optimization (mathematical programming, branch and bound, and elimination methods), simulation, heuristic, and intelligent methods, including expert systems, artificial neural networks, and intelligent search algorithms [3]. Zhang et al. [4] proposed a schedule adjustment method combining fuzzy optimization theory and dynamic programming. Johnson [5] and Panwalker [6] studied the scheduling rules separately. ISIS [7] is one of the earliest AI-based scheduling systems. When the production environment changes or is in conflict, the scheduling mechanism of this system selectively relaxes certain constraints and reschedules the affected artifacts or orders. OPIS [8] is a workshop scheduling system based on ISIS that uses dynamic scheduling strategies for scheduling according to production conditions. SONIA [9] is a scheduling system that includes prediction and feedback scheduling. OPAL [10] is a scheduling system that uses a heuristic method. Numao et al. [11] designed an expert system for steel smelters using artificial intelligence and human–computer interaction. Xi et al. [12] proposed a neural

Applications of Artificial Intelligence in Process Systems Engineering
https://doi.org/10.1016/B978-0-12-821092-5.00018-8

network FMS dynamic scheduling method based on a rule-based backpropagation learning algorithm. Lee et al. [13] solved the scheduling problem by using a combination of a genetic algorithm and machine learning. Jian et al. [14] proposed an FMS scheduling and rescheduling algorithm that can greatly shorten the rescheduling time. Luo et al. [15] proposed a priority-based hybrid parallel genetic algorithm with a fully rescheduling strategy that predicts reactivity. Zhou and Yang [16] proposed four automatic design methods based on multi-objective genetic programming that perform better than manual scheduling. In summary, research on dynamic scheduling methods has become increasingly mature and has laid the foundation for the establishment of an equipment optimization scheduling model.

Nayak et al. [17] proposed a smart algorithm framework with manufacturing plants as the research object. The framework aimed to minimize the total power cost so that the total order delay can be controlled. Wu and Che [18] solved the replacement flow workshop scheduling problem to minimize the maximum completion time and total energy consumption. The results showed that the processing speed of the machine can be dynamically adjusted according to different jobs. Dorfeshan et al. [19] proposed a new flow shop job scheduling method based on weighted distance approximation and verified the effectiveness of the method. Shao et al. [20] studied distributed pipeline operation scheduling to minimize the maximum completion time and proposed an improved hybrid discrete fruit fly optimization algorithm. Experimental results showed that this method is more effective than existing methods. Feng et al. [21] proposed a comprehensive method of intelligent green scheduling for edge computing in flexible workshops and verified that the method is effective. Kim et al. [22] established a dynamic scheduling model for precast concrete machines and verified that the model is effective through simulation experiments.

In summary, scholars worldwide have conducted a great amount of research on the dynamic scheduling method and its application, which have provided many research ideas and methods. Therefore, this study establishes an optimized scheduling model for intermittent production equipment that utilizes these mature dynamic scheduling methods.

2 Problem description and solution

In this section, we first provide a description of the intermittent production equipment scheduling problem. Then, an optimal scheduling model based on the NSGA-II algorithm is proposed. Finally, an optimal scheduling model is established.

2.1 Dispatching status of intermittent production equipment

After inspection, research, and study, it was found that workers have the following disadvantages in the scheduling of intermittent production equipment: (1) Workers cannot make the best use of the time when electricity prices are at their lowest or avoid the peak electricity price times and (2) workers are unstable factors. If the factory needs to improve the effect of optimized scheduling, it needs to invest in more manpower. Therefore, production costs will increase greatly, which will put pressure on the enterprise. (3) The manufacturing shop is unstable. In the face of different working conditions, workers cannot change the scheduling of intermittent production equipment in time; therefore, workers miss the optimal scheduling times of machines.

2.2   NSGA-II algorithm overview

The essence of mathematical modeling is to transform a multi-objective optimization problem into a single-objective optimization problem through a weighting method [23–28]. A single-objective optimization theory is then used to solve the problem; however, the single-objective optimization theory often has shortcomings in practical applications. For example, subjectively setting the weight of each goal cannot lead to the optimization of each goal. Multi-objective optimization is a process of finding decision variables that satisfy constraints and assigning values to all objective functions [29]. In theory, such a problem will produce a set of optimal solutions, called Pareto optimal sets, rather than a single solution [30]. Different algorithms, such as the strength Pareto evolutionary algorithm, nondominated sorting genetic algorithm (NSGA), ant colony, and particle swarm optimization (PSO) are often used to solve complex multi-objective problems [31–34]. The meta-heuristic algorithm is prone to premature convergence, especially when dealing with complex multi-peak search problems, and the local optimization ability is poor. NSGA-II is one of the most popular multi-objective algorithms in meta-heuristic algorithms. It not only has the advantages of a fast running speed and good convergence to the solution set but also reduces the complexity of the NSGA algorithm [35]. Because NSGA-II is more competitive in solving multi-objective optimization problems [36], the proposed scheduling model uses NSGA-II to find solutions.

Zhang et al. [37] studied the configuration and operation of a hybrid photovoltaic power generation system using the NSGA-II algorithm. This study utilized a solar-wind-storage hybrid power system model to determine the optimal hybrid power system composition. Hu et al. [38] solved a multi-objective optimization problem of gas and electricity combined network expansion planning using the NSGA-II algorithm. Jain et al. [39] established a multi-objective optimized vapor compression-absorption cascade refrigeration system using the NSGA-II algorithm. The research results showed that multi-objective optimization is better than a single-objective optimization alone. Martínez et al. [40] proposed a multi-objective optimization model using the NSGA-II algorithm to optimize the economic performance of micro-generation systems.

Deb et al. [41] proposed the NSGA-II algorithm for the first time. The algorithm can maximize the independence of each optimization goal, and it has a fast convergence speed and strong global optimization ability. The calculation processes of both the traditional standard genetic and NSGA-II algorithms have selection, crossover, and mutation operations. The NSGA-II algorithm performs a nondominated sorting and stratification of individuals before selection and judges the differences in each nondominated solution in the same group through the crowding strategy. The main steps of the NSGA-II algorithm are as follows: (1) The algorithm uses a fast nondominated sorting method to find a set of equally optimal solutions that are closest to the Pareto optimal frontier. (2) The algorithm maintains the diversity of the Pareto optimal solutions by calculating the crowded distance. (3) The introduction of elitism in the algorithm allows the parent and child groups to compete and enhances the convergence of the optimization theory [42]. The NSGA-II algorithm process is shown in Fig. 1.

The key steps of the NSGA-II algorithm are as follows:

(1) Decode and generate initial populations. The group size is set, and the number of individuals included in the group is N. The range of random numbers is defined in advance, and the initial population P_0 is randomly produced by the program.

(2) Fast nondominated sorting. Zitzler et al. [43] proposed a fast nondominated sorting method. This is a complex algorithm $O(gN^2)$, where g represents the number of goals, and N represents the number of individuals in the population. The nondominant ranking is based on the dominant number of solutions in the entire solution space, and the dominant number is finally determined to be the first layer. Then, an individual is selected from the first layer, its corresponding dominating set is traversed, and one is subtracted from each individual's dominating number. If the advantage number is 0, the second layer is obtained. In this way, the algorithm can obtain all the layers. When the algorithm compares different results, it prefers a low-ranked solution.

(3) Calculate the crowding distance. A congestion distance of ∞ is assigned to the boundary individuals. For the remaining individuals, the crowding degree can be expressed as [44, 45]:

$$\delta_d(l) = \sum_{s=1}^{g} \frac{|E_s(l+1) - E_s(l-1)|}{E_s^{max} - E_s^{min}}, l \in \{2, 3, \ldots, n-1\}, \tag{1}$$

where d represents the number level of the lth individual. n represents the number of individuals with rank d. g represents the number of objective functions. $\delta_d(l)$ represents the crowding distance of the lth individual. E_s represents the value of the sth objective function.

(4) Selection, crossover, and mutation. Each individual in the population has two properties that are determined by calculating the crowding degree and

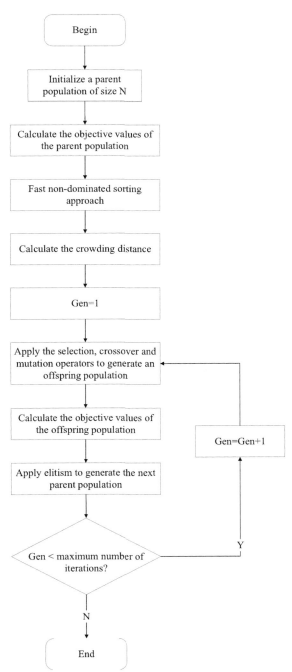

FIG. 1   NSGA-II procedure.

sorting: the crowding distance and nondominated order. When the nondominated orders of two individuals are equal and the former has a higher crowding degree than the latter, the former is considered to be better than the latter.

    a. A crossover is an operation in genetic algorithms that is also known as a reorganization. It chooses a point between two randomly selected individuals and divides it into two parts. Then, the left (or right) parts of the two individuals are swapped, and new individuals are created.

    b. A mutation is an operation in genetic algorithms. New individuals are generated by randomly selecting certain parts of individuals in the current population and turning them over.

(5) Elitism operator. Elitism is a strategy to compare the parent population with the offspring population. [46, 47] The steps of the elitism operator are as follows:

    a. The parent and offspring populations are combined to create a temporary population of $2N$.

    b. The temporary populations are sorted by a fast nondominant sorting method, and the crowding distance is calculated.

    c. The best N individuals are selected as the next parent population. In the selection, if two individuals have different levels, the individual with the smaller level is chosen; if two individuals have the same rank, the individual with the larger crowded distance is selected.

2.3 Objective function and constraints

The optimal scheduling model of intermittent production equipment is designed to reduce the cost of electricity by scheduling processing times. The objective function of this model is the power cost and processing time of the equipment.

    The power cost is used to describe the electricity consumption of intermittent production equipment. The machine processing time is used to ensure that the processing time can meet the order requirements. The objective function is shown in Eqs. (2) and (3):

$$\min function1 = \sum P \times t_K \times f_k, \tag{2}$$

where P represents the power of the intermittent production equipment, kW. k represents the different periods. f_k represents the time-of-use price, yuan/kWh. t_k represents the working time of the equipment, min.

$$\max function2 = T_{m,k}, \tag{3}$$

where T represents the cumulative processing time of the equipment (min).

    The constraints are the processing conditions that need to be met in the production process.

3   Case study

In 2016, the energy costs of papermaking enterprises accounted for approximately 6% of the total plant costs, and electricity accounted for 51% of the energy costs [48, 49]. According to data from the National Bureau of Statistics, in 2018, there were 2657 paper enterprises above the designated size with a main business revenue of 815.2 billion yuan, and tissue paper output accounted for 6.61% of the total revenue. In 2018, 9.7 million tons of tissue were produced, which was an increase of 1.04% over the previous year. In that same year, 9.01 million tons of tissue were consumed, which was an increase of 1.24% over the previous year. In summary, the tissue overcapacity has led to fierce market competition. To improve the competitiveness of enterprises, it is necessary to start with the enterprises themselves and reduce the cost of production. According to survey statistics, the power consumption cost of tissue pulping lines accounts for 25%–29% of the entire production line, and the power consumption cost in pulping lines for other paper production processes accounts for 15%–20%. This is because the production line for tissue paper is shorter than that for other paper products. A Yangke dryer is used in the tissue paper process, while other paper production processes use multi-cylinder dryers. Therefore, the pulping process of household paper was selected as the research object.

The papermaking process production equipment includes intermittent and nonintermittent equipment. Intermittent production equipment includes primarily pulpers, grinders, and pumps for moving pulp. Nonintermittent production equipment includes equipment such as paper machines. The electricity consumption in many industrially developed areas is charged based on peak and valley times, and the price of industrial electricity in different regions varies according to local conditions. Therefore, the period and peak and valley time prices in different regions will differ. It is assumed that most paper mills have no restrictions on the start and stop times of intermittent power equipment, which leads to high electricity costs for enterprises. It is possible to reduce electricity costs by optimizing the scheduling of the above equipment and reasonably arranging its processing time.

In summary, the intermittent production equipment in the tissue pulping workshop is taken as the research object and the analysis case of this paper.

3.1   The pulping process

The pulping line of a papermaking workshop is shown in Fig. 2. The pulp board is transported to the pulper by the conveyor, and after the water and chemicals are added, it is pumped into the pre-grinding tank. The slurry in the pre-grinding

FIG. 2   Pulping line of papermaking workshop.

tank requires slag-removal and is then pumped into the grinder, after which the processed slurry is pumped into the after-grinding tank for further processing. The slurry in the after-grinding tank is then pumped into the mixing tank. Throughout the pulping process, the liquid levels of the three slurry tanks must be within the normal range; otherwise, the slurry overflows from the pulp pond to cause waste, or an insufficient slurry production causes the paper machine to stop, which causes economic losses.

Based on the current electricity prices for peak and valley periods, tissue companies actively respond to the government's call for equipment optimization scheduling. The main scheduling object is the intermittent pulping production equipment. Currently, the scheduling of pulpers and grinders by workers is based on the liquid level of the pulp tank and the periodic electricity price.

3.2 Technical route

The technical route of the research includes the liquid level calculation and scheduling plan models. The liquid level calculation model (Fig. 3A) refers to the slurry flow velocity and tank liquid level established based on the paper-making and pulping line data. The model includes data collection and preprocessing such as outlier removal and missing value filling. The processed data is used as the initial input of the slurry flow rate tank liquid level models. The scheduling plan model (Fig. 3B) is established by using the NSGA-II algorithm for the pulping equipment based on the pulp speed and slurry tank liquid level models. The objective function of this model is the electricity cost and running time of the equipment. Workers control the pulper and grinder according to the optimal dispatch plan derived from the equipment dispatch plan model.

3.3 Data collection

Data collection is used to obtain the relevant data from the papermaking and pulping lines. Data preprocessing is used to clean the collected real-time production data and consists of deleting outliers and filling in missing values. Deleting the outliers prevents abnormal data from affecting the model, and by filling in the missing data from the equipment and slurry, data diversity is guaranteed.

Table 1 presents the variable information from the collected data. These data were all collected under real working conditions and include related data such as pulp concentration and information from pulping process equipment and paper machines.

3.4 Establishment of a liquid level calculation model

The research object of this study is the tissue paper pulping process equipment. In this study, the purpose of cutting energy consumption at peak times and

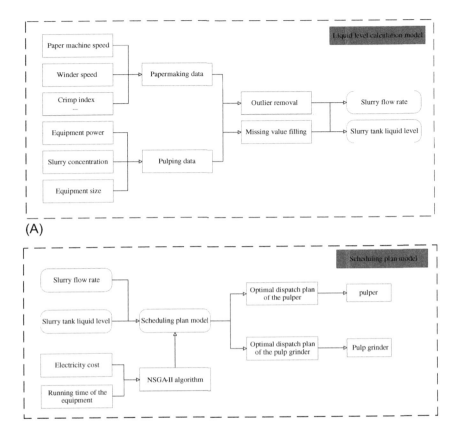

(A)

(B)

FIG. 3   (A) Liquid level calculation model and (B) scheduling plan model.

TABLE 1 Variable information from data collection.

Number	Style	Unit
1	Pulper power	kW
2	Grinder power	kW
3	Concentration of pre-grinding tank	%
4	Concentration of after-grinding tank	%
5	Concentration of mixing tank	%
6	Flow of grinder outlet	m³/h
7	Consumption of pulp board	t

Continued

TABLE 1 Variable information from data collection—cont'd

Number	Style	Unit
8	Paper machine speed	m/min
9	Winder speed	m/min
10	Crimp index	%
11	Grammage	g/m^2
12	Breadth	mm
13	Volume of pre-grinding tank	m^3
14	Volume of after-grinding tank	m^3
15	Volume of mixing tank	m^3

increasing it when energy costs are lower is achieved by establishing an optimal scheduling model for the equipment. The optimal scheduling model needs to be established based on a quantitative model of the pulping process. The quantitative model uses mathematical methods to describe the material flow process of the pulping process. The level of the slurry tank is the key factor that determines the start and stop times of the pulping process equipment. Therefore, quantifying the liquid level of the slurry tank is the basis for the optimal scheduling of equipment. In addition, the liquid level of the slurry tank depends on the slurry inflow and outflow.

3.4.1 Slurry flow rate

There are three types of pulp ponds in the pulping line: pre-grinding, after-grinding, and mixing ponds. Because the pump between the after-grinding and mixing tanks is a variable power device, this study considers the grinding and sizing tanks as a whole.

The velocity of the slurry flowing into the pre-grinding pool is given by Eqs. (4) and (5):

$$v_{ss,j} = \frac{C_{pu} \times p}{c_q \times T} \tag{4}$$

$$p = CS \times JQL \times F \times D \times 1.05 \times 0.97, \tag{5}$$

where $v_{ss,j}$ represents the production speed of the slurry before grinding in different time periods, t/h. p represents the consumption of the pulpboard, t. C_{pu} represents the fiber content in the pulp board. T represents the time of slurry extraction from the pulper, and c_q represents the concentration in the pre-grinding tank, %. CS represents the speed of the paper machine, m/min. JQL

represents the crimp index, %. F represents the breadth of the paper machine, mm. D represents grammage, g/m^2. 1.05 is the ratio of raw materials such as pulp board to paper. 0.97 is the proportion of damaged paper.

The outflow speed of slurry in the pre-grinding pool, speed of slurry flowing into the after-grinding pool, and outflow speed of slurry in the after-grinding pool are shown in Eqs. (6) and (7):

$$v_{sx,j} = v_{os,j} = L_c \tag{6}$$

$$v_{ax,j} = \frac{V \times D \times F \times C_{pa}}{c_s}, \tag{7}$$

where $v_{sx,j}$ represents the consumption rate of the pre-grinding slurry in different periods, t/h. $v_{os,j}$ represents the speed of the slurry flowing into the after-grinding tank in different periods, t/h. L_c represents the flow at the grinder outlet, m^3/h. $v_{ax,j}$ represents the consumption rate of the slurry in the tank after grinding, t/h. V represents the winder speed, m/min. c_s represents the concentration in the mixing tank, %. C_{pa} represents the fiber content in the finished paper.

3.4.2 Slurry tank liquid level

The slurry tank consists of three parts: the bottom (cylinder), middle (pillar), and top (cylinder). The level of the slurry tank refers to the position of the slurry inside the tank. Therefore, this study represents the dynamic liquid level in the slurry tank according to the height occupied by the slurry per unit time. The quantitative model of the slurry tank liquid level is shown in Eqs. (8) and (9):

$$v = v_{in} \times t_{in} - v_{out} \times t_{out} \tag{8}$$

$$h(v) = \begin{cases} \dfrac{v}{\pi \times r_1{}^2}, & v \leq v_1 \\[2ex] \dfrac{v + \dfrac{1}{3\pi} \times r_1{}^3 \times \tan\alpha - \pi \times r_1{}^2 \times h_1}{\dfrac{1}{3\pi} \times r_2{}^2} + h_1 - r_1 \times \tan\alpha, & v_2 \geq v \geq v_1 \quad (9) \\[2ex] \dfrac{v - v_1 - v_2}{\pi \times r_2{}^2} + h_1 + h_2, & v \geq v_2 \end{cases}$$

where h represents the height of the slurry in the tank, %. v represents the volume of slurry in the tank, m^3. v_{in} represents the speed of the slurry flowing into the tank, t/h. t_{in} represents the time required for the slurry to flow into the tank, min. v_{out} represents the speed of slurry outflow from the tank, t/h. t_{out} represents the time required for the slurry to flow out of the tank, min. v_1 represents the volume of the bottom of the tank, m^3. v_2 represents the volume of the middle of the tank, m^3. r_1 represents the radius of the bottom of the tank, m. r_2 represents the radius of the top of the tank, m. α represents the inclination angle of the middle of the pulp tank, (°). h_1 represents the liquid level at the bottom of the slurry tank, %. h_2 represents the liquid level at the middle of the slurry tank, %.

The papermaking workshop produces different types of paper according to the order requirements and prepares different types of paper that require pulp. Suppose that during the current shift, it is necessary to produce two different types of paper, A and B. When the processing of type A slurry is completed, a period of time needs to be set aside. The calculation method is shown in Eqs. (10) and (11):

$$t = \frac{v(h_{now}) - v(h_{min})}{v_{ax,j}} \tag{10}$$

$$v(h) = \begin{cases} h \times \pi \times r_1^2 \ , & h \leq h_1 \\ \dfrac{1}{3\pi} \times \left[r_2^2 \times (r_1 \times tan\alpha + h - h_1) - r_1^3 \times tan\alpha \right] + h_1 \times \pi \times r_1^2, & h_2 \geq h \geq h_1 \\ (h - h_1 - h_2) \times \pi \times r_2^2 + v_1 + v_2 \ , & h \geq h_2 \end{cases} \tag{11}$$

where t represents the reserved time, min. $v(h_{now})$ represents the volume of slurry in the slurry tank, m³. $v(h_{min})$ represents the corresponding volume of slurry when the slurry tank liquid level is the lowest, m³. $v(h)$ represents the piecewise function of the slurry tank liquid level and slurry volume.

3.5 Schedule planning model of pulping equipment based on the NSGA-II algorithm

The scheduling model is built on a quantitative model using the NSGA-II algorithm. The result of the model is the optimal dispatch plan obtained by using the electricity cost and running time of the pulping process equipment as the objective function. Workers control the pulper and grinder according to the optimal scheduling plan obtained by the model optimization.

The scheduling model includes two working conditions: normal and abnormal paper machine speeds. The normal paper machine speed also includes two working conditions: the new pulp is processed, or it is not processed and unqualified paper is processed. Therefore, the scheduling plan model includes four types of working conditions: (1) *Condition 1*: paper machine speed is abnormal; (2) *Condition 2*: paper machine speed is normal, and no new pulp is processed; (3) *Condition 3*: paper machine speed is normal, and new pulp is processed; (4) *Condition 4*: speed of the paper machine is normal, and unqualified paper is processed. Fig. 4 shows the technical route of the scheduling plan model.

(1) Condition 1: Paper machine speed is abnormal

An abnormal speed indicates that the paper machine is in a shutdown state; therefore, no slurry preparation is required. After the model is

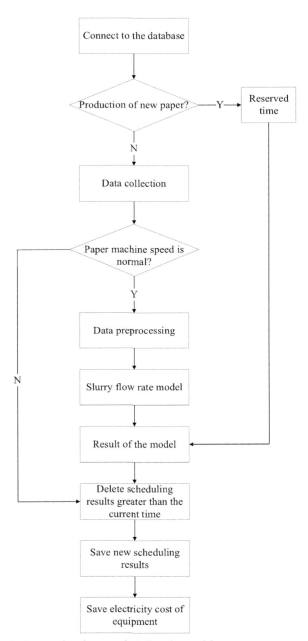

FIG. 4 Technical route of equipment scheduling plan model.

triggered, the scheduling plan in the database that is greater than the current time should be removed, and the scheduling plan is not displayed in the workshop. When the speed of the paper machine increases, the worker needs to input the slurry processing signal. When the model receives this signal, it generates a new scheduling plan that is then displayed.

(2) Condition 2: Paper machine speed is normal, and no new pulp is processed
    After the model is triggered, the model connects to a database to collect production data for pulping and papermaking. Because there are a few outliers and null values in the production data, the collected data must first be cleansed as discussed above. After the data are pre-processed, the quantitative model calculates the liquid level of the pulp pond at the current moment and every minute within the next 8 h and finally iterates the optimal scheduling plan of the pulping process equipment using the multi-objective solution algorithm. After the model is triggered, the scheduling plan in the database, which is greater than the current time, is removed. The main reason for this is to avoid a duplication of scheduling results. Then, according to the power of the equipment, the electricity cost generated by the pulping process equipment under the dispatch plan is calculated.

(3) Condition 3: Paper machine speed is normal, and new pulp is processed
    When the speed of the paper machine is normal and new pulp is processed, the process of establishing the scheduling plan model is as follows. After the model is triggered, the scheduling plan in the database, which is greater than the current time, is removed. Next, the model calculates the reserved time for the grinder and collects the production data from the pulping and papermaking. Because there are a few outliers and null values in the production data, the collected data must be cleansed of extreme and missing values. After the data are pre-processed, the quantitative model calculates the liquid level of the pulp pond at the current moment and every minute during the next 8 h and finally determines the optimal scheduling plan for the pulping process equipment using the multi-objective solution algorithm. The reserved time calculated by the model is the time that the grinder needs to wait before being scheduled. The scheduling of the pulper is not affected. Then, according to the power of the equipment, the electricity cost generated by the pulping process equipment under the dispatch plan is calculated.

(4) Condition 4: Paper machine speed is normal, and unqualified paper is processed
    The time and duration of unqualified paper processing are confirmed by many indicators, such as onsite conditions and production needs. When the speed of the paper machine is normal and unqualified paper is processed, the process of establishing the scheduling plan model is as follows. After the model is triggered, the scheduling plan in the database, which is greater than the current time, is removed. Because there are a few outliers and nulls

in the production data, the collected data must be cleansed of extreme and missing values. After the data are pre-processed, the quantitative model calculates the liquid level of the pulp pond at the current moment and every minute during the next 8 h and finally determines the optimal scheduling plan for the pulping process equipment using the multi-objective solution algorithm. The unqualified processing paper time is the time that the pulper needs to work first. The grinder scheduling is not affected. Then, according to the power of the equipment, the electricity cost generated by the pulping process equipment under the dispatch plan is calculated.

3.6 Industrial verification

The scheduling plan model was deployed in a tissue company with an annual output of 80,000 tons. The model used real-time data for industrial verification. The electrovalence based on the time period is 1.0906 yuan/kWh (10:00–12:00 and 18:00–22:00), 0.9086 yuan/kWh (08:00–10:00, 12:00–18:00, and 22:00–24:00), and 0.4246 yuan/kWh (00:00–08:00) [50]. The model reasonably allocates the processing time of the intermittent production equipment and calculates its electricity cost. The configuration of the pulping line is as follows: a pulper, pre-grinding tank, grinder, after-grinding tank, and sizing tank. The model collects the relevant data of the pulping line, and the frequency of data collection was 1 min.

3.6.1 Adjustment of scheduling plan model parameters

The parameters of the NSGA-II algorithm include population size, number of iterations, crossover rate, and mutation rate. Determining the parameters of the model and selecting the optimal parameters is the key to the scheduling model. The process and results of the parameter adjustments are as follows:

First, we collect the relevant data during a shift (8 h) to establish a scheduling model. Next, values for the iteration number, crossover rate, and mutation rate are fixed, and the population size is changed according to the electricity cost under the optimal results of different population size models. The value range of the population size is [30,350], which is incremented by one. When the population size is 100, the electricity cost is the lowest, and as the population size increases, the electricity cost does not change; therefore, the optimal value of the population size is 100, as shown in Fig. 5. Similarly, the number of iterations is determined. The value range of the iteration number is [30,170], which increases gradually. When the number of iterations is 70, the electricity cost is the lowest, and as the number of iterations increases, the electricity cost does not change; therefore, the optimal number of iterations is 70, as shown in Fig. 6. The crossover rate is also determined in a similar manner. The value range of the crossover rate is [0.05, 0.95], and 0.01 is added successively. When the

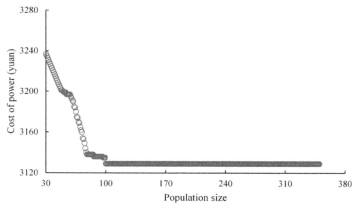

FIG. 5 Optimization of population size.

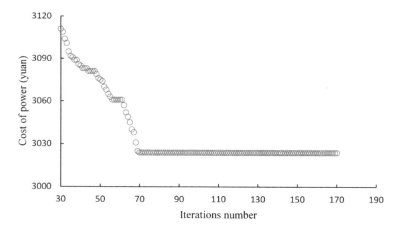

FIG. 6 Optimization of iteration number.

crossover rate is 0.35, the electricity cost is the lowest, and as the crossover rate increases, the electricity cost does not change; therefore, the best value for the crossover rate is 0.35, as shown in Fig. 7. Then, the mutation rate is determined. The value range of the mutation rate is [0.05, 0.1], and 0.01 is added successively. When the mutation rate is 0.75, the electricity cost is the lowest, and as the value of the mutation rate increases, the electricity cost does not change. Therefore, the optimal mutation rate is 0.75, as shown in Fig. 8. Finally, the parameter settings of the model are listed in Table 2.

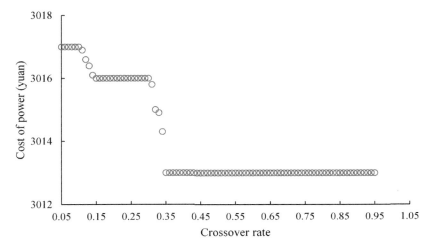

FIG. 7 Optimization of crossover rate.

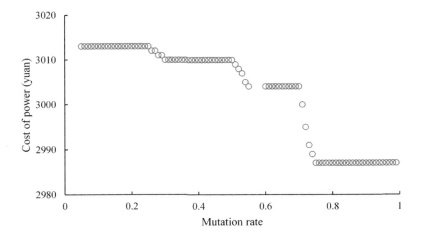

FIG. 8 Optimization of mutation rate.

3.6.2 Results of industrial verification

In this study, an offline verification was performed. The results are shown in Fig. 9A and B. In these figures, PART represents the actual running time of the pulper, PRRT represents the recommended running time of the pulper by model, GART represents the actual running time of the pulp grinder and GRRT represents the recommended running time of the pulp grinder by model.

TABLE 2 Parameter settings for the pulping scheduling model.

Number	Style	Setting
1	Population size	100
2	Iterations number	70
3	Crossover rate	0.35
4	Mutation rate	0.75

Fig. 9A is based on the data of a production line in the enterprise. Peak period, flat period, and valley period represent the electricity price of that period in the region. The peak price is the highest, the flat price is second, and the valley price is the lowest. It can be seen from Fig. 9A that the electricity price of the shift (08:00–16:00) experienced changes from the flat period to the valley period and then to the flat period. The recommended running result of pulper is better than the actual running result. The recommended running time of the pulper is longer in the valley so the running cost of the pulper is lower. Although the running time of the recommended running result and the actual running result is same during the valley period. But the recommended running result realizes fewer machine starts and stops.

It can be seen from 16 to 9(b) that the electricity price of this shift (16:00–24:00) changes from the flat period, peak period, flat period, and valley period. The results show that the running time of pulper is longer in the peak period and shorter in the valley period. The recommended operation result of pulper optimizes the problems existing in actual operation. The recommended running results increase the running time in the valley period and decrease the running time in the peak period. In the actual operation, the running time of the pulp grinder in peak period is too long. This problem can be optimized by the recommended running results.

The results of offline verification prove that the model is effective, so the model is used to guide the actual production. Table 3 compares the unoptimized results with those optimized using the model.

It can be seen from Table 3 that the total power cost of the day after model optimization was reduced by 234 yuan. The weighted electricity price has also fallen from 0.64 to 0.60 yuan/kWh. Since the enterprise is a high energy consumption enterprise, every 0.01 yuan/kWh reduction in the weighted electricity price will have a great impact on the production cost of the enterprise. The model proposed in this study can reduce the weighted electricity price by about 0.04 yuan/kWh. This is a clear step forward, and it will have huge economic benefits.

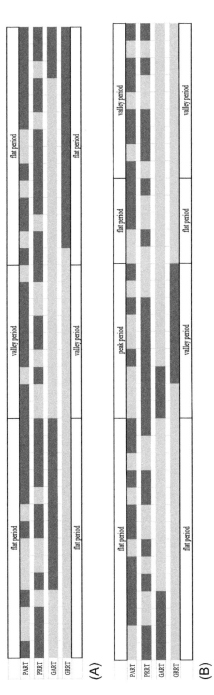

FIG. 9 (A) Equipment running time (08:00–16:00). (B) Equipment running time (16:00–24:00).

TABLE 3 Analysis of electricity consumption.

Period	Electricity consumption (kWh)		Total power cost (yuan)		Weighted electricity price (yuan/kWh)	
	Unoptimized	Optimized	Unoptimized	Optimized	Unoptimized	Optimized
Valley period	2309.33	2617.33	2755.85	2521.03	0.64	0.60
Flat period	1734.67	1426				
Peak period	214.67	136				

4 Conclusion

The scheduling model is designed based on scientific and objective principles, and the basis of the calculation is objective data from the production process. The data guarantee the objectivity of the scheduling results and prevent the influence of subjective human factors. The scheduling of this model is realized by computer software, and the final scheduling result is accurate and scientific.

In this study, the decision objectives of the scheduling plan model were determined by analyzing the current situation and existing problems of the pulping process equipment operation, and a multi-objective decision model based on the NSGA-II algorithm was established. To achieve the decision objective, the model integrates two quantifiable objectives determined through the analysis of influencing factors, so as to be suitable for actual factory operation and have practical applicability. The equipment scheduling model of the pulping process established in this study has a great practical value and economic benefit for actual plant operations.

In summary, the dispatching model based on the NSGA-II algorithm has an obvious effect of peak-shaving and valley-filling the power cost intervals, and it can reasonably allocate the working time of equipment to reduce electricity costs. In addition, the model maximizes the time workers spend on the production line and reduces unnecessary operations, such as workers who constantly monitor the liquid level of the pulp pond. The industrial verification shows that the electricity cost generated by the scheduling plan can save 234 yuan per day on average compared with the actual scheduling results. The dispatch generated by this model can reduce the weighted electricity price by 0.04 yuan/kWh.

References

[1] X. Chen, Y. An, Z. Zhang, Y. Li, An approximate nondominated sorting genetic algorithm to integrate optimization of production scheduling and accurate maintenance based on reliability intervals, J. Manuf. Syst. 54 (January) (2020) 227–241.

[2] M. Pinedo, X. Chao, Operations Scheduling With Applications in Manufacturing and Services, Irwin/McGraw-Hill, Boston, 1999, ISBN: 0-07-289779-1.

[3] Q. Xiaolong, T. Lixin, L. Wenxin, Dynamic scheduling: a survey of research methods, Control Decis. 016 (002) (2001) 141–145 (in Chinese).

[4] Z. Jie, L. Yuan, Z. Kaifu, H. Yang, Fuzzy dynamic programming method for progress adjustment of project management, Comput. Integr. Manuf. Syst. 08 (2006) 108–112 (in Chinese).

[5] S.M. Johnson, Optimal two and three stage production schedules with setup times included, Naval Res. Log Quart. 1 (1) (1954) 61–68.

[6] P.W. Iskander, A survey of scheduling rules, Oper. Res. 25 (1) (1977) 45–61.

[7] M.S. Fox, ISIS; a knowledge-based system for factory scheduling, Expert Syst. 1 (1) (1984) 25–49.

[8] S.F. Smith, J.E. Hynyen, Integrated decentralization of production management for factory scheduling, Symp. Integ. Intel. Manuf. Boston (1987).

[9] A. Collinot, C.L. Pape, G. Pinoteau, SONIA: a knowledge-based scheduling system, Artif. Intell. Eng. 3 (2) (1988) 86–94.

[10] E. Bensana, G. Bel, D. Dubois, OPAL: a multi-knowledge-based system for industrial job-shop scheduling? Int. J. Prod. Res. 26 (5) (1988) 795–819.
[11] M. Numao, S.I. Morishita, Cooperative scheduling and its application to steelmaking processes, IEEE Trans. Ind. Electron. 38 (2) (1991) 150–155.
[12] X. Lifeng, Z. Binghai, C. Jianguo, F. Kun, FMS dynamic scheduling based on rules artificial neural network, Mach. Tool Hydraul. 4 (2002) 32–34 (in Chinese).
[13] C.Y. Lee, S. Piramuthu, Y.K. Tsai, Job shop scheduling with a genetic algorithm and machine learning, Int. J. Prod. Res. 35 (4) (1997) 1171–1191.
[14] A.K. Jain, H.A. Elmaraghy, Production scheduling/rescheduling in flexible manufacturing, Int. J. Prod. Res. 35 (1) (1997) 281–309.
[15] J.P.D. Comput, J. Luo, S. Fujimura, D. El, B. Plazolles, GPU based parallel genetic algorithm for solving an energy efficient dynamic flexible flow shop scheduling problem, J. Par. Distrib. Comput. 133 (2019) 244–257.
[16] Y. Zhou, J.-j. Yang, Automatic design of scheduling policies for dynamic flexible job shop scheduling by multi-objective programming based hyper-heuristic, Procedia CIRP 79 (2019) 439–444.
[17] W. Sutherland, W. West, M. Dsm, P. Stief, J. Dantan, A. Etienne, A. Siadat, ScienceDirect dynamic load scheduling for energy generation, Procedia CIRP 80 (2018) 197–202.
[18] X. Wu, A. Che, Energy-efficient no-wait permutation flow shop scheduling by adaptive multi-objective variable neighborhood search R, Omega 102117 (2019).
[19] Y. Dorfeshan, R. Tavakkoli-moghaddam, S.M. Mousavi, B. Vahedi-nouri, A new weighted distance-based approximation methodology for flow shop scheduling group decisions under the interval-valued fuzzy processing time, Appl. Soft Comput. J. 91 (2020) 106248.
[20] Z. Shao, D. Pi, W. Shao, Hybrid enhanced discrete fruit fly optimization algorithm for scheduling blocking flow-shop in distributed environment, Expert Syst. Appl. 145 (2020) 113147.
[21] Y. Feng, Z. Hong, Z. Li, H. Zheng, J. Tan, Integrated intelligent green scheduling of sustainable flexible workshop with edge computing considering uncertain machine state, J. Clean. Prod. 246 (2020) 119070.
[22] T. Kim, Y. Kim, H. Cho, Dynamic production scheduling model under due date uncertainty in precast concrete construction, J. Clean. Prod. 257 (2020) 120527.
[23] F.S. Wen, Z.X. Han, Fault section estimation in power systems using genetic algorithm and simulated annealing, Proc. CSEE 14 (3) (1994) 29–35.
[24] F.S. Wen, Z.X. Han, Fault section estimation in power systems using a genetic algorithm, Electr. Pow. Syst. Res. 34 (3) (1995) 165–172.
[25] W.X. Guo, F.S. Wen, G. Ledwich, Z.W. Liao, X.Z. He, J.H. Liang, An analytic model for fault diagnosis in power systems considering malfunctions of protective relays and circuit breakers, IEEE Trans. Power Deliv. 25 (3) (2010) 1393–1401.
[26] D.B. Liu, X.P. Gu, H.P. Li, A complete analytic model for fault diagnosis of power systems, Proc. CSEE 31 (34) (2011) 85–93.
[27] D.M. Zhao, X. Zhang, J. Wei, W.C. Liang, D.Y. Zhang, Power grid fault diagnosis aiming at reproducing the fault process, Proc. CSEE 34 (13) (2014) 2116–2123.
[28] Y. Zhang, C.Y. Chung, F.S. Wen, J.Y. Zhong, An analytic model for fault diagnosis in power systems utilizing redundancy and temporal information of alarm messages, IEEE Trans. Power Syst. 31 (6) (2016) 4877–4886.
[29] C. Coello, A.D. Christiansen, Multiobjective optimization of trusses using genetic algorithms, Comput. Struct. 75 (2000) 647–660.
[30] S. Lotfan, R.A. Ghiasi, M. Fallah, M.H. Sadeghi, ANN-based modeling and reducing dual-fuel engine's challenging emissions by multi-objective evolutionary algorithm NSGA-II, Appl. Energy 175 (2016) 91–99.

[31] T. Hiroyasu, S. Nakayama, M. Miki, Comparison study of SPEA2+, SPEA2, and NSGA-II in diesel engine emissions and fuel economy problem, in: The 2005 IEEE Congress on Evolutionary Computation. IEEE, 2005, pp. 236–242.

[32] J. Mohammadhassani, A. Dadvand, S. Khalilarya, M. Solimanpur, Prediction and reduction of diesel engine emissions using a combined ANN–ACO method, Appl. Soft Comput. 34 (2015) 139–150.

[33] M.M. Etghani, M.H. Shojaeefard, A. Khalkhali, M. Akbari, A hybrid method of modified NSGA-II and TOPSIS to optimize performance and emissions of a diesel engine using biodiesel, Appl. Therm. Eng. 59 (2013) 309–315.

[34] Q. Zhang, R.M. Ogren, S.-C. Kong, A comparative study of biodiesel engine performance optimization using enhanced hybrid PSO–GA and basic GA, Appl. Energy 165 (2016) 676–684.

[35] K. Deb, A. Pratap, S. Agarwal, T. Meyarivan, A fast and elitist multi-objective genetic algorithm: NSGA-II, vol. 6, 2000. KanGAL Report No. 2000-01.

[36] X. Sun, L. Zhao, P. Zhang, L. Bao, Y. Chen, Enhanced NSGA-II with evolving directions prediction for interval multi-objective optimization, Swarm Evol. Comput. 49 (April 2018) (2019) 124–133.

[37] D. Zhang, J. Liu, S. Jiao, H. Tian, C. Lou, Z. Zhou, J. Zuo, Research on the con figuration and operation effect of the hybrid solar-wind-battery power generation system based on NSGA-II, Energy 189 (2019) 116–121.

[38] Y. Hu, Z. Bie, T. Ding, Y. Lin, An NSGA-II based multi-objective optimization for combined gas and electricity network expansion planning, Appl. Energy 167 (2016) 280–293.

[39] V. Jain, G. Sachdeva, S. Singh, B. Patel, Thermo-economic and environmental analyses based multi-objective optimization of vapor compression—absorption cascaded refrigeration system using NSGA-II technique, Energ. Conver. Manage. 113 (2016) 230–242.

[40] S. Martínez, E. Pérez, P. Eguía, A. Erkoreka, Model calibration and exergoeconomic optimization with NSGA-II applied to a residential cogeneration, Appl. Therm. Eng. 114916 (2020).

[41] S. Wanxing, Y. Xueshun, L. Keyan, Optimal allocation between distributed generations and microgrid based on NSGA-II algorithm, Proc. CSEE 18 (2015) 4655–4662 (in Chinese).

[42] S. Wang, Application of NSGA-II algorithm for fault diagnosis in power system, Electr. Pow. Syst. Res. 175 (May) (2019) 105893.

[43] Z. Eckart, L. Marco, T. Lothar, SPEA2: improving the strength Pareto evolutionary algorithm for multi-objective optimization, in: K.C. Giannakoglou, et al. (Eds.), Proceedings of EUROGEN 2001-Evolutionary Methods for Design. Optimization and Control with Applications to Industrial Problems, 2001, pp. 95–100.

[44] K. Deb, A. Pratap, S. Agarwal, T. Meyarivan, A fast and elitist multiobjective geneticalgorithm: NSGA-II, IEEE Trans. Evol. Comput. 6 (2) (2002) 182–197.

[45] K. Deb, S. Agrawal, A. Pratap, T. Meyarivan, A fast elitist non-dominated sorting genetic algorithm for multi-objective optimization: NSGA-II, in: Parallel Problem Solving From Nature VI, Springer, Paris, France, 2000, pp. 849–858.

[46] K. Deb, Multi-Objective Optimization Using Evolutionary Algorithms, Wiley, Singapore, 2001, pp. 1–24.

[47] S. Ramesh, S. Kannan, S. Baskar, Application of modified NSGA-II algorithm to multi-objective reactive power planning, Soft Computer 12 (2) (2012) 741–753.

[48] China Paper Association, China Technical Association of Paper Industry, White paper on sustainable development of Chinese paper industry, 2019.

[49] National Bureau of Statistics, Statistical Yearbook of China—Energy Consumption by Industry, 2016. http://www.stats.gov.cn/tjsj/ndsj/2018/indexch.htm. (in Chinese).

[50] Y. Yang, M. Wang, Y. Liu, et al., Peak-off-peak load shifting: are public willing to accept the peak and off-peak time of use electricity price? J. Clean. Prod. 199 (2018) 1066–1071.

Chapter 18

Artificial intelligence algorithms for proactive dynamic vehicle routing problem

Xianlong Ge[a,b] and Yuanzhi Jin[a,c]

[a]*School of Economics and Management, Chongqing Jiaotong University, Chongqing, People's Republic of China*, [b]*Key Laboratory of Intelligent Logistics Network, Chongqing Jiaotong University, Chongqing, China*, [c]*Department of Computer Technology and Information Engineering, Sanmenxia Polytechnic, Sanmenxia, People's Republic of China*

1 Introduction

The earliest work of vehicle routing problems can be dated from 1959 introduced by Dantzig and Ramser [1] and expanding rapidly over the last two decades. Many deterministic constraints have been added to make the issues closer to real life. In recent decades, the Deterministic Vehicle Routing Problem (DVRP) has been conducted under the assumption that all the information necessary to formulate the problems is known and readily available in advance (i.e., one is in a deterministic setting). In practical applications, this assumption is usually not verified given the presence of uncertainty affecting the parameters of the problem. Uncertainty may come from different sources, both from expected variations and unexpected events. Such variations can affect various aspects of the problem under study, which may cause service time delays or even service termination. Once such situations occur, the operation enterprises could not give timely and appropriate responses. This may result in the decrease of customer satisfaction and even customer churn, which is not conducive to the development and progress of the enterprise. The same is true for the scheduling of vehicle routing problems. If the distribution center cannot respond to and service their customer in a reasonable time, customer satisfaction will be reduced. However, temporary deliveries for small batches will increase vehicle scheduling costs.

In this chapter, the proactive method can solve this problem to a certain extent. Proactive Dynamic Vehicle Routing Problem (PDVRP) is an extend of the classical VRP, which is based on the analysis of the customers'

Applications of Artificial Intelligence in Process Systems Engineering
https://doi.org/10.1016/B978-0-12-821092-5.00011-5

multi-dimensional historical information and uses certain forecasting methods to predict the possibility of customer demand in a certain period, which is an "ex ante" optimization method before delivery. A cluster first–route second heuristic is utilized to solve the PDVRP. First, all customers (e.g., stores) are divided into static customers (with deterministic demand) and dynamic customers (with uncertain demand); second, calculating the prospect value of each dynamic customer in order to determine whether they need to be delivered according to a threshold. Third, a clustering algorithm is performed to generate several clustering sub-regions for static customers and dynamic customers that need to be delivered in advance, and the center of each sub-region is proactive distribution center. Finally, a hybrid heuristic algorithm is applied to solve the problem within each cluster.

In order to enable an accurate mapping of vehicle tours on real roads and to calculate the resulting travel time between customers accordingly, a real-world urban road network is used. On this road network, we use the information returned by the Baidu map API related to the best driving path from one customer to its neighbor. The returned information includes travel time, travel distance, optional paths, and other formatted data, and may be varying during different time intervals. In this chapter, this function is applied to obtain the distance matrix and time matrix of all customers and the final detailed distribution scheme. Unfortunately, the number of queries is limited for free users per day. At the same time, because of the different network conditions, it often takes long time to query the required dada of large distance matrix and time matrix.

2 Main approaches for PDVRP

Proactive vehicle routing problem is relevant to the Dynamical demand Vehicle Routing Problem (DVRP). Since the deterministic VRP may not fit real-life scenarios when variables or parameters are uncertain, these problems can be solved with DVRP, which can be divided into the following categories according to its research objects: (1) dynamical demand VRP that serves a set of customers with uncertain demand (Tillman [2], Stewart and Golden [3], Bertsimas [4]). (2) VRP with dynamic customers, whom needs services or not is unknown (Gendreau et al. [5], Azi et al. [6], Abdallah et al. [7]). (3) VRP with random travel time (Tas et al.[8], Li and Tian [9], Lecluyse et al. [10]).

DVRPs are typically deal with requirements that randomly generated or real-time update. Montemanni et al. [11] designed an open road network for updating online orders so that vehicles can address new orders after leaving the distribution center. Haghani and Jung [12] proposed a time-dependent dynamic vehicle routing problem and solved it by a genetic algorithm. Branchini et al. [13] designed a structural heuristic algorithm based on the dynamic changes of customer location and demand during vehicle service time. Wohlgemuth et al. [14] synthetically analyzed the continuous travel time and short-term demand in dynamical VRP and concluded that the dynamic scheduling strategy can improve vehicle utilization ratio. Bopardikar et al. [15] used a directed acyclic graph structure to describe the vehicle routing problem with dynamic demand and designs a route optimization strategy for vehicle operators.

There are little studies on proactive research. The earlier one applying proactive theory to vehicle routing problem is presented by Ferrucci et al. [16]. They proposed a new proactive real-time control method to solve the dynamic vehicle routing problem centering on emergency distribution. This method uses historical request information to predict future requirements. Combined with tabu search algorithm, the vehicle can be guided to the possible area before the request arrives. Several solutions of intelligent transportation system are proposed to identify traffic congestion and change the road of vehicles afterwards [17]. This work introduces a fully distributed, proactive intelligent traffic information system GARUDA, which can avoid traffic congestion, reroute new routes, and actively inform the driver to follow the new route. Congestion reduces the efficiency of infrastructure and increases travel time, air pollution and fuel consumption. Martuscelli et al. [18] proposed a protocol set that adopted a packet routing strategy based on vehicle density and location to improve the performance of packet routing. Furthermore, the advance prediction time of Vehicular Ad Hoc networks (VANETs) for real-time road conditions is increased. Due to the large number of red signalized intersections on the road, frequent stops can lead to high fuel consumption. To solve this problem, Soon et al. [19] proposed a proactive pheromone-based green vehicle routing model, which makes full use of the information provided by the traffic information system to guide vehicles to avoid congestion and reduce the number of stops, thus reducing fuel consumption. At the same time, an improved dynamic k-shortest path algorithm is introduced in the path allocation process to reduce the time complexity of the algorithm.

In real case, the customer's demands will show certain regularity in a time cycle (such as 12 months) according to their consumption habits. Therefore, the customer's historical requests and shopping habits may be used to roughly predict their future needs. Here, we introduce a new hybrid Simulated Annealing— Genetic Algorithm (SA-GA) instead of using the single one to solve the PDVRP.

The rest of this chapter is organized as follows. In Section 3, we describe the problem to be solved in detail. In Section 4, we propose a formulation of Time-Dependent Vehicle Routing Problem with Soft Time Windows (TDVRPSTW) based on proactive prediction. In addition, the detailed implementation of the SA-GA for solving the model is described in Section 5. The computational results are reported in Section 6. Finally, Section 7 contains a summary and future works.

3 Problem description

Proactive vehicle routing problems can be described as: There are several customers with constant or dynamic demands (referred as static customers and

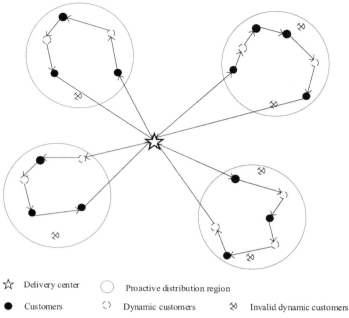

☆ Delivery center   ◯ Proactive distribution region

● Customers   ◌ Dynamic customers   ⊗ Invalid dynamic customers

FIG. 1 Proactive distribution network.

dynamic customers respectively) within a defined service range. We first use prospect theory to calculate the prospect value of each dynamic customer as the dynamic customer's acceptance degree in future, and then combine the dynamic customers possessing high prospect value and all static customers into a new group of targeted customers.

The service area is parted by the proactive clustering method according to the geographical location of the targeted customers. They are several non-overlapping sub-regions and proactive distribution centers, as is shown in Fig. 1. Finally, a proactive distribution model is adopted to serve all targeted customers.

3.1 The vehicle speed of time-dependence

The road capacity is divided into different time intervals according to the characteristics of urban traffic flow, and the average speed of vehicles in each time period is predictable (regardless of the influence of natural factors, such as sudden traffic accidents, extreme weather, and so on). Fig. 2 shows the average speed of vehicles in different time intervals within 24 h of a day. We assume that the travel time between customer i to customer j is denoted as t_{ij}. The time

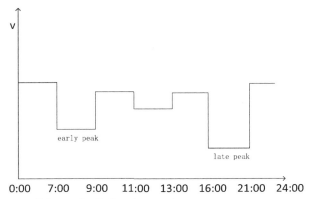

FIG. 2 Average vehicle speed with time-dependent.

$t_{ij} = t_{ij}^m$ if the vehicle k arrives at the customer j from the customer i in a time interval m; Otherwise, if the travel time of vehicle k span n time intervals, the arrival time $t_{ij} = t_{ij}^m + t_{ij}^{m+1} + \ldots + t_{ij}^{m+n}$.

3.2 Penalty cost of soft time windows

A penalty function is adopted separately according to the degree of violation of the customer's time windows. As is illustrated in Fig. 3, $[e, l]$ is the time window accepted by the customer, and the penalty coefficient is 0 if the service vehicle arriving at that time interval. $[e', e]$ and $[l, l']$ is a time interval that the customer can accept the extension of the time windows, and the penalty function is relatively small on slope. During the rest time, violating the time windows is unacceptable for customers every time. The penalty cost is so large that the vehicle does not service as much as possible during the time interval.

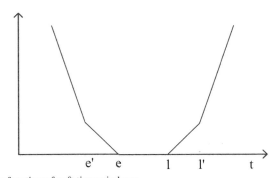

FIG. 3 Penalty function of soft time windows.

The penalty function divides the time that the vehicle arrives at the customer into five categories according to the time range and the length of the lateness:

$$P(a_i) = \begin{cases} p_1\left(e_i' - a_i\right) + p_2\left(e_i - e_i'\right) & a_i < e_i' \\ p_2(e_i - a_i) & e_i' < a_i < e_i \\ 0 & e_i < a_i < l_i \,, \\ p_3\left(l_i' - a_i\right) & l_i < a_i < l_i' \\ p_3\left(l_i' - l_i\right) + p_4\left(a_i - l_i'\right) & l' < a_i \end{cases} \tag{1}$$

where $p_1 > p_2$, $p_4 > p_3$, and $p_i (i \in 1,2,3,4)$ is four penalty coefficients of cost as the vehicle violates a time window, that is, p_1 denote the penalty coefficient that the vehicle k reaches the customer node before e_i'; p_2 denote that the vehicle reaches the customer node at $[e_i', e_i]$, and so on.

3.3 The conception of proactive distribution

In order to meet the dynamic customer demands, the conception of proactive distribution is introduced, which is divided into three stages: (1) proactive dynamic customer evaluation based on prospect theory [20], (2) proactive clustering algorithm to determine cluster center, and (3) proactive distribution with dynamic demands. The first stage is decomposing the dynamic customers' historical demand data and determine whether provide proactive services to customers or not based on the customer's prospect value. The second stage clusters the targeted customers according to their geographic location and determines their proactive distribution centers. The third stage is to plan routes for each proactive distribution center based on the result of the previous two phases.

3.3.1 Calculating prospect value

We use the prospect theory to make a proactive evaluation of the dynamic customer's requests, and then judge whether serve a dynamic customer or not based on the predicted value p_{ik}, denote the customer's demand, and a_{ik} denote the mean of distance matrix for demand of the historical demand data.

Assuming that the customer's historical demand data is m dimension, and all customers n can be expressed as matrix $A = (a_{11}, a_{12}, ..., a_{1m}; a_{21}, a_{22}, ..., a_{2m}; a_{n1}, a_{n2}, ..., a_{nm},)$. In order to describe the dynamic customer's needs more accurately, the deterministic attribute and the triangular fuzzy number [21] are used to represent the certain and the uncertain attributes in the demand matrix.

When the customer's attributes can be represented by a certain value, the logical distance between the predicted value of the customer's demand and the average historical data is:

$$d_{ik} = |p_{ik} - a_{ik}| \tag{2}$$

When a customer's attributes need to be represented by a triangular fuzzy number, the logical distance [22] is:

$$d_{ik} = \sqrt{\frac{\left|\left(p_{ik}^1 - m_{ik}^1\right)^2\right| + \left|\left(p_{ik}^2 - m_{ik}^2\right)^2\right| + \left|\left(p_{ik}^3 - m_{ik}^3\right)^2\right|}{3}} \tag{3}$$

where $(p_{ik}^1, p_{ik}^2, p_{ik}^3)$ is the customer demand predict value and $(m_{ik}^1, m_{ik}^2, m_{ik}^3)$ is the mean of customer's historical data. The formula for calculating the triangular fuzzy number is:

$$p_{ik} = \left(\max\left(\frac{l-1}{n}, 0\right), \frac{l}{n}, \min\left(\frac{l+1}{n}, 1\right)\right), l \in [1, n] \tag{4}$$

where l is the value range of the customer's attributes. A threshold is set to select dynamic customers that meet preference of decision makers, and then we combine the selected dynamic customers with static customers into a new group to be served.

For dynamic customers, the distribution center determines whether to deliver it or not according to their prospect value. Detailed calculation method is shown in Section 6.2.

3.3.2   Proactive clustering algorithm

Since dynamic customers are derived from methods of proactive evaluation, it is no guarantee that they will appear during the delivery process. Clustering only with the actual demand as the constraint will lead to the invalid expansion of the cluster from the boundary of the region. Therefore, the service radius expansion factor μ and the load expansion factor υ are added in the clustering algorithm. Cluster area demand (static customer demand and possible dynamic customer demand) can exceed the vehicle load, but the vehicle is not allowed to overload when leaving the distribution center. The specific steps of the proactive partition clustering method designed in this chapter are as follows:

Step 1. Calculate the adjacency matrix between distribution center and its customers according to known coordinate data.
Step 2. Choose some non-overlapping square, center is node i and radius R, counting the client nodes falling within the range, and then select the area covered by the square containing the most client nodes as the proactive service sub-region.
Step 3. Generate a proactive service sub-region:
Step 3.1. Determine the proactive scheduling center of the sub-region using gravity method.
Step 3.2. Determine whether the total customer demand $QD(iter)$ in an initial service sub-region is greater than the vehicle capacity Q, if true, execute Step 3.3, otherwise, execute Step 3.4.
Step 3.3. Calculate the distance between customers that fall within the sub-region and their proactive distribution center, get the customer sequence MD

sorted in descending order, and remove the first element of sequence *MD*, obtain $QD(iter+1)$, execute Step 3.5.

Step 3.4. Expand the search radius to μR, if $\mu R \leq R_{\max}$ is true, execute Step 3.2. Otherwise, go to step 4.

Step 3.5. If $QD(iter) < Q < QD(iter+1)$ is true, execute Step 4, otherwise execute Step 3.2.

Step 4. Remove the client included in the service sub-region generated in Step 3.

Step 5. Go to Step 2 if each customer has a cluster. Otherwise, stop the algorithm.

4 TDVRPSTW formulation

This study addresses a dynamic VRP with homogenous vehicles that have identical capacity Q and other parameters. The problem is defined as an undirected graph $G = (A, N)$, where $N = \{0, 1, \cdots n\}$ is the set of nodes. Node 0 represents the depot that serves customers, whereas nodes $i \in N\backslash\{0\}$ correspond to the customers. Each customer has a demand q_i that need to be served at once by a vehicle. Each vehicle route starts and ends in the depot. We also define the following quantities (Table 1).

Therefore, our mathematical model can be presented as follows:

$$\min z = \sum_{k\in K}\sum_{i\in N} fy_{ik} + \sum_{k\in K}\sum_{i\in N}\sum_{j\in N} cd_{ij}x_{ijk} + \sum_{i\in N} P(a_i) \tag{5}$$

subject to:

$$\sum_{k\in K}\sum_{i\in N} x_{ijk} \leq 1, \forall j\in N \tag{6}$$

$$\sum_{k\in K}\sum_{j\in N} x_{ijk} \leq 1, \forall i\in N \tag{7}$$

$$\sum_{k\in K} y_{ik} \leq K, \forall i\in N \tag{8}$$

$$\sum_{i,j\in N} x_{ijk} \leq |N_s| - 1, \ 2 \leq |N_s| \leq n-1, \forall k\in K \tag{9}$$

$$\sum_{i\in N} q_i y_{ik} \leq Q, \ \forall k\in K \tag{10}$$

$$\sum_{i\in N} x_{0ik} = \sum_{j\in N} x_{j0k} \leq 1, \forall k\in K \tag{11}$$

$$\sum_{i\in N} x_{ipk} - \sum_{j\in N} x_{pjk} = 0, \ \forall p\in N, \ \forall k\in K \tag{12}$$

$$a_j = u_i + t_{ij}, \ \forall i,j\in N \tag{13}$$

TABLE 1 Notations used in this chapter.

Parameter	Meaning
$G=(A, N)$	Distribution network
0	Depot
$C=\{1,2,\cdots,n\}$	Customer's set
$N=\{0,1,\cdots n\}$	Node's set
$A=\{(i,j)\mid \forall i,j\in N, i\neq j\}$	Arc's set
$\mid N_s\mid$	The number of nodes in $\forall S, S\subseteq N$
Q	The capacity of vehicle
$K=\{1,\ldots,k\}$	Vehicle 's set
q_i	The demand of customer i
$[e_i, l_i]$	The time window of customer i
t_{ij}	The travel time from i to j
s_i	The service time of customer i
a_i	The arrival time of customer i
u_i	The depart time of customer i
d_{ij}	The distance from customer i to customer j
c	The cost of per unit distance
f	The fixed cost of vehicle
P	The penalty cost for validation time windows
d_{ik}	The prospect value for predicting customer demand

$$u_i = a_i + s_i, \quad \forall i\in N \tag{14}$$

$$x_{ijk} = 0, \ 1, \quad \forall i,j\in N, i\neq j, \forall k\in K \tag{15}$$

$$y_{ik} = 0, \ 1, \quad \forall i\in N, \ \forall k\in K \tag{16}$$

Eq. (5) indicates that the objective function of the model minimizes the total cost, where the total cost includes the fixed cost of the vehicle, the distance cost, and the penalty cost of violating the time windows. Eqs. (6) and (7) indicate that each customer is visited and left no more than once because some dynamic customers may not have requests to generate. Eq. (8) ensures that the number of enabled vehicles does not exceed the total number of fleet vehicles. Eq. (9) eliminates subtours in the solution, and Eq. (10) represents that the vehicle is not overloaded. Eq. (11) means that each vehicle must depart from the distribution center and eventually return to it. Eq. (12) ensures the flow balance of the incoming and outgoing customers. Eqs. (13) and (14) represent the time constraint that the vehicle arrives at the next customer and leaves the previous customer. Binary variables x_{ijk} in Eq. (15) equal to 1 if and only if vehicle $k \in K$ directly travels from customer i to customer $j(i, j \in C, i \neq j)$, and y_{ik} in Eq. (16) also describe a set of binary variables, which equal to 1 if and only if vehicle $k \in K$ visits customer i ($i \in C$).

5 The algorithm

Although a large body of heuristic algorithms has been generated to solving vehicle routing problems, TDVRPSTW considering four factors (real distance, speed, time windows, and cost) has higher requirements on the efficiency and solution quality. Therefore, a hybrid simulated annealing algorithm with genetic operators is designed. The standard Simulated Annealing (SA) has the ability of escaping local optimum, but its global search ability is inadequate. Combined with the characteristics of genetic algorithm that has a good global searching ability, it will make the SA-GA more efficient.

5.1 Coding structure

In this chapter, the natural number coding is used to represent the solution of the route. The customer number is used as an initial solution which is generated by the forward insertion algorithm that can be divided into three stages: (1) insert the customer into one route, (2) update the vehicle capacity, and (3), insert the customer into the next route if the vehicle is overloaded, otherwise, insert to current route until all the customers are inserted into the route. The distribution center between the routes is indicated by 0, as shown in Fig. 4.

5.2 Genetic operator design

5.2.1 Crossover operator

An improved position-based crossover approach is employed in the crossover process. In order to improve the algorithm's search ability, some better genic segments are retained in the original solution when selecting an intersection point and judging its quality according to the vehicle's load rate in the route. The details are as follows:

FIG. 4 Chromosome coding.

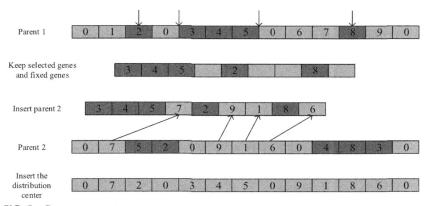

FIG. 5 Crossover operator.

Step 1 Random select two chromosomes as parent 1 and parent 2. Find the path of the vehicle with the largest load in parent 1 and use it as the selected gene in advance. Then select two other random genes on the chromosome except the distribution center as the fixed gene.

Step 2 Exclude other genes in parent 1 and retain only the fixed gene.

Step 3 Exclude the same gene in the parent 2 from the parent 1 and insert the remaining genes in the parent 2 into the parent 1 in turn except the distribution center.

Step 4 Reinsert the distribution center 0 according to the constraints of the vehicle load. The crossover operation is completed, as is shown in Fig. 5.

5.2.2 Mutation operator

In order to accelerate the convergence of the solution, new chromosomes are obtained by exchanging genes in chromosomes. The steps are as follows:

Step 1 Select the parent chromosome for the mutation operation, and randomly generates two numbers.

Step 2 Swap two genes at the two positions.

Step 3 Perform the decoding operation and insert 0 to update when it is overloaded. As is shown in Fig. 6.

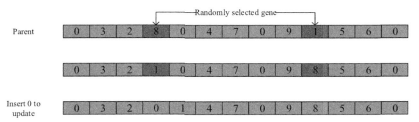

FIG. 6 Mutation operator.

5.2.3 Pseudo code of the SA-GA

Pseudo-code for the SA-GA

1: **Input**: problem
2: Preprocessing
3: Set the parameters of the algorithm
4: Generate initial solution X by random
5: **While** $T_{curr} \geq T_{min}$
6:    Initialize counter: iteration, stop
7:    **While** iteration $< MAX_{iter}$ **and** stop $< MAX_{stop}$
8:          **For** $i = 0$ to M
9:                Calculate objective value
10:                Randomly select a move operator and generate a new solution X^*
11:                Update the objective value
12:          **End For**
13:          Calculate fitness of X^*
14:          Perform select, crossover, mutate operator
15:          Find the best individuals and migrate the elites to each subpopulation
16:          **If** objective value does not change **Then**
17:                stop $=$ stop $+1$
18:          **End If**
19:          iteration $=$ iteration $+1$
20:       **End While**
21:          $T_{curr} = T_{curr} * C_i$
22: **End While**
23: **Output** X^*

---

6 Numerical experiment

This section can be divided into two parts. The first part evaluates the algorithm's performance, and the second step is to verify the application of proactive distribution scheme in a supermarket's distribution system.

6.1 Evaluation of the algorithm's performance

The well-known Solomon's instances are selected as benchmark to test the performance of the SA-GA. The travel speed of the path is based on the information obtained by the Baidu map API and converts to the average speed per unit time, 30 min (Table 2). The remaining parameters are as follows: the working time of the distribution center is 8:00–18:00 every day, the fixed cost of the vehicle is 200, the cost per unit of distance is 5, and the vehicle max capacity is 200. Unit penalty cost for violation of time windows is $p_1 = 10$,

TABLE 2 Average velocity in each time interval (km/h).

t	8:00–8:30	8:30–9:00	9:00–9:30	9:30–10:00	10:00–10:30	10:30–11:00	11:00–11:30	11:30–12:00	12:00–12:30	12:30–13:00
v	12.7	7.8	11.8	18.8	21.1	22.5	22.8	38.1	44.5	46.0
t	13:00–13:30	13:30–14:00	14:00–14:30	14:30–15:00	15:00–15:30	15:30–16:00	16:00–16:30	16:30–17:00	17:00–17:30	17:30–18:00
v	43.1	41.9	41.9	39.5	35.9	29.7	31.7	41.9	40.7	30.3

$p_2 = 5$, $p_3 = 8$, $p_4 = 16$ respectively. Each instance is tested 10 times, and the detailed results are shown in Table 3.

It can be seen from Table 3 that the results of the SA-GA proposed in this chapter are better than other the two algorithms. The best optimal value (Task R102) is 11.1% more efficient than GA, and 13.1% than SA. At the same time, the best mean value (Task RC103) is 9.5% more cost saving than GA, and 14.4% than SA. Moreover, the solution quality for these tasks is relatively stable.

The experiment done in this section is only to test that the SA-GA has the ability to solve medium-scale instances. Because the algorithm adopts the method of clustering first–route second heuristics, it can also deal with large-scale instances. In addition, the Euclidean distance is employed for testing Solomon instances in this section. At the same time, the travel time and distance between customers are symmetric and customer requirements are stochastic data without unit, which cannot be directly applied to solve practical problems. The BD09 coordinate system is employed in the next section and a real case of the supermarket is to be studied.

6.2 A real case study

We take a supermarket distribution as a case to validate the proactive distribution method. The supermarket includes one distribution center (DC) and 46 customer points (i.e., stores, numbered C01-C46). Some properties, such as the customer coordinates, time windows, and the penalty cost of violating the time windows, are known in advance. The geographical location distribution of all customers is shown in Fig. 7. Since we use the BD09 coordinate system, it may cause subtle deviation to display the location information of all customers with Google Map as the base.

6.2.1 Customer classification

All customers can be divided into two categories. One is static customers with fixed needs each day, the other is dynamic customers with uncertain needs. That is, the static customers here represent stores with fixed demands every day, while the dynamic customers indicate stores with uncertain demands. Such uncertainty of demand is often caused by online or offline promotion activities of stores. We can use the historical demand data of these stores to predict their future demand and carry out corresponding distribution in advance according to the predicted results. Such operation makes the users who would go to the nearest store to pick up their online order waiting shorter and increases their satisfaction. In this case, the information of customer classification is shown in Table 4.

TABLE 3 The results of Solomon's instances.

Task	GA		SA			SA-GA				
	Opt.	Mean	Opt.	Mean	Opt.	opt_diff_GA	opt_diff_SA	mean_diff_GA	mean_diff_SA
C101	2061	2178	2116	2212	1911	2015	7.9%	10.8%	8.1%	9.8%
C102	2148	2224	2181	2280	2024	2194	6.1%	7.7%	1.4%	3.9%
C103	1983	2157	2038	2185	1904	1987	4.1%	7.0%	8.6%	10.0%
R101	2565	2892	2593	2917	2350	2651	9.2%	10.4%	9.1%	10.1%
R102	2489	2648	2532	2800	2240	2618	**11.1%**	**13.1%**	1.1%	7.0%
R103	2202	2494	2251	2530	2062	2350	6.8%	9.2%	6.1%	7.6%
RC101	2648	2697	2743	2796	2462	2524	7.6%	11.4%	6.9%	10.8%
RC102	2383	2564	2417	2648	2285	2480	4.3%	5.8%	3.4%	6.8%
RC103	2263	2376	2220	2481	2063	2169	9.7%	7.6%	**9.5%**	**14.4%**

The bold values are the biggest improvement of the results obtained by our SA-GA compared with the standard GA or SA.

FIG. 7   The geographical location of DC and customers.

TABLE 4 Customer category.

Type	Customer no.	Count
Static	C01,C02,C03,C04,C05,C08,C09,C11,C12,C13,C15,C16, C17,C18,C19,C20,C21,C23,C24,C26,C30,C31,C32,C33, C35,C36,C38,C40,C41, C43	30
Dynamic	C06,C07,C10,C14,C22,C25,C27,C28,C29,C34,C37,C39, C42,C44,C45, C46	16

Through the analysis of the large amount of historical data of the supermarket, we get the average demand and standard deviation of each store in a delivery cycle (1 day). Detailed data are shown in Table 5.

Note that the above statistics do not include beverages, vegetables, meat, etc., as these goods are regularly replenished by specific suppliers. The following section predicts which dynamic customers need to be served using prospect theory mentioned above.

TABLE 5 The analyzed result of customer demand, the unit of mean and std. is kg.

No.	Mean	Std.	No.	Mean	Std.
C01	206.0	46.4	C24	156.9	41.9
C02	102.1	38.3	C25	124.7	45.2
C03	204.0	54.0	C26	265.7	47.3
C04	152.7	52.1	C27	154.9	53.0
C05	169.0	32.9	C28	137.5	53.6
C06	178.0	55.6	C29	170.0	32.4
C07	161.7	47.3	C30	291.5	53.7
C08	163.0	64.2	C31	291.0	55.8
C09	250.0	39.3	C32	248.6	58.5
C10	141.0	67.8	C33	188.4	54.0
C11	109.0	46.0	C34	142.8	65.0
C12	248.0	63.2	C35	281.6	47.7
C13	274.0	31.6	C36	255.1	39.3
C14	134.3	58.8	C37	150.2	48.5
C15	121.0	67.3	C38	138.1	52.8
C16	286.0	37.0	C39	124.0	64.0
C17	160.0	31.2	C40	250.5	42.3
C18	138.4	45.9	C41	120.1	56.8
C19	152.5	43.4	C42	200.0	63.9
C20	192.0	52.5	C43	180.0	39.4
C21	228.0	58.7	C44	113.0	56.4
C22	133.0	55.4	C45	149.0	38.7
C23	238.0	55.2	C46	161.3	32.1

6.2.2 Evaluation of dynamic customers

In order to make proactive evaluation of customer's future demand, the four parameters: average sales speed of goods a_1, user's dependence on the store a_2, waiting time between user 's order conforming and pick up its goods in the store a_3, and customer's historical demand a_4 are used to calculate its prospect value, each of them have a weight coefficient $\lambda_i, i \in \{1, 2, 3, 4\}$ respectively.

The attributes of the first three fuzzy parameters are classified into five grades: excellent, good, satisfactory, fair, and poor. The specific values are calculated according to Eq. (17), which is the special case of Eq. (4).

$$a_i = \left(\max \left(\frac{l-1}{5}, 0 \right), \frac{l}{5}, \min \left(\frac{l+1}{5}, 0 \right) \right), l \in [1, 5] \quad (17)$$

With the customer evaluation indicator, the prospect value of each customer can be calculated by Eq. 18:

$$pv = \sum_{i=1}^{4} \lambda_i \alpha_i \quad (18)$$

According to the above four indicators to calculate the prospect value of dynamic customers to judge whether to choose it as target customers. The calculated results are normalized in Table 6.

The four weight coefficients $\lambda_1 = \lambda_2 = 0.15, \lambda_3 = 0.1$ and $\lambda_4 = 0.6$, which indicates that the customers historical demand is the most important factor. It can be seen from Table 5 that all the 11 customers (C10, C14, C25, C27, C28, C29, C39, C42, C44, C45, C46) have a positive prospect value, respectively. Hence, these customers are used in the next section for service targets.

6.2.3 Clustering for the targeted customers

The static customers and the dynamic customers with high prospect value are put together; the proactive clustering algorithm is executed according to geographical location to generate several clustering sub-regions, that is, proactive distribution areas. The result is present in Fig. 8.

The clustering algorithm divides all target customers into four sub-regions; each of them is a proactive distribution region. Theoretically, selecting the center point of each sub-region (DC1–DC4) as the proactive distribution center is the best choice. However, considering the high cost of urban warehousing, we select stores closest to each corresponding sub-regional center as proactive distribution centers. The details for each sub-region are shown in Table 7.

There are two scenarios that the prediction fails: (1) if a dynamic customer with a positive prospect value, we will consider it as a target customer to serve, but in fact it has no demand at all. (2) If a dynamic customer does not need service according to its prospect value, but actually it does, in this case, the store can be regarded as a static customer in the next distribution cycle. These two conditions resulted in an average increase of about 3.5 km per customer.

6.2.4 Proactive tour plan

The SA-GA proposed in this chapter is respectively performed within each subregion to generate a preliminary tour plan. The parameters used in this real case are different from those used in the Solomon instance above. The biggest

TABLE 6 Predict the prospect values for dynamic customers.

Customer no.	a_1		a_2		a_3		a_4		Prospect value
	Expected value	Past mean	Expected value	Past mean	Expected value	Past mean	Expected value	Past mean	
C06	4	4	4	5	4	4	187.0	178.0	−0.4546
C07	4	4	4	4	4	3	162.0	161.7	−0.3735
C10	4	4	2	1	4	5	151.0	141.0	**0.3187**
C14	4	3	4	4	4	5	123.0	134.3	**0.0789**
C22	5	4	5	5	4	4	136.0	133.0	−0.2026
C25	5	5	4	4	3	2	135.0	124.7	**0.1587**
C27	4	4	4	3	3	3	181.9	154.9	**0.7556**
C28	4	4	3	3	2	2	156.5	137.5	**0.2896**
C29	4	3	4	2	2	3	173.0	170.0	**0.1618**
C34	1	2	3	3	2	3	126.5	142.8	−0.5227
C37	4	4	3	4	4	3	151.0	150.2	−0.5902
C39	3	4	2	2	4	3	149.0	124.0	**0.6098**
C42	4	3	4	4	4	5	194.0	180.0	**0.0407**
C44	4	4	4	3	3	4	136.0	113.0	**0.7672**
C45	3	2	2	2	4	3	153.0	149.0	**0.1449**
C46	4	3	4	4	5	4	174.5	161.3	**0.3485**

The bold values represent the corresponding customers with positive prospect value, which means that the customers have a high probability of replenishment demand.

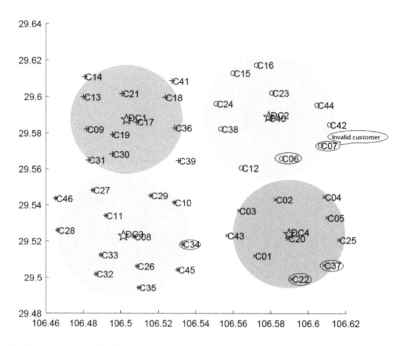

FIG. 8 Proactive clustering for targeting customers.

TABLE 7 The information of four sub-regions.

	DC1	DC2	DC3	DC4
Clustering radius	0.03	0.035	0.035	0.03
Coordinates of clustering center	29.587397	29.589027	29.5231	29.524271
	106.50308	106.57899	106.50145	106.58977
Proactive distribution center	C17	C40	C08	C20
Number of static customers	10	7	6	7
Number of dynamic customers	2	4	7	3
Number of invalid customers	0	2	1	2
Number of targeted customers	12	9	12	8

change is the method of calculating the distance between two customers. The Euclidean plane distance formula is adopted in the Solomon instance for evaluating the performance of SA-GA, while the Haversine formula is adopted here, and the main parameters are shown in Table 8. Then, the results of SA-GA are shown in Table 9.

As can be seen from Table 9, each sub-region can be served by one truck. In effect, the route inside each region is a TSP tour that meets the time window

TABLE 8 Parameter settings of the SA-GA.

Type	Parameters	Description	Values
Vehicle	c	Cost per kilometer	2 (¥)
	f	Fixed cost of each vehicle	300 (¥)
	Q	Maximum load	2.5 (t)
GA	P_{ic}	Crossover probability	0.9
	P_{im}	Mutation probability	0.05
	$GGAP$	Generation gap	0.9
	M	Population size	300
SA	T_i	Initial temperature	100
	C_i	Annealing coefficient	0.9
	T_{min}	Stop temperature	0.01
	MAX_{stop}	The number that limits the objective without improvement	20

TABLE 9 Tour plan of each sub-region.

Sub-regions	Tour	Cost (¥)
DC1	C17->C18->C41->C21->C14->C13->C09->C19->C31->C30->C39->C36->C17	355
DC2	C40->C42->C44->C23->C16->C15->C24->C28->C38->C12->C40	342
DC3	C08->C26->C35->C45->C10->C29->C27->C46->C28->C32->C33->C11->C08	368
DC4	C20->C01->C43->C03->C02->C04->C05->C25->C20	350

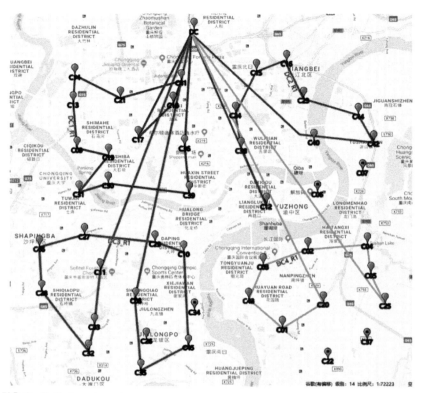

FIG. 9 Preliminary tour plan.

and load constraints of its inner customers. During each delivery cycle, vehicles start from the distribution center, reach the center of each sub-region, and then serve customers one by one along with each TSP tour before returning to the distribution center (see Fig. 9). If the total demand for each sub-region is very high, it is possible to save costs by sending the goods to the center of each sub-region via a larger van and then switching to a smaller vehicle to serve their internal customer.

As can be seen from Fig. 9, the tour plan obtained by inserting the distribution center into each TSP tour in the central pointcut of each sub-region is not optimal. Obviously, we need to choose the nearest and the second nearest customer node from each TSP tour to the distribution center, disconnect the TSP tour, insert the distribution center into it, and finally get the VRP tour (see Fig. 10).

Therefore, the latter can save 50 min in time, 34.5 km in distance and 69¥ in cost.

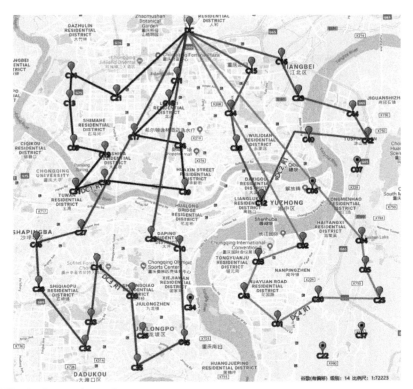

FIG. 10 Improved tour plan.

6.2.5 Route fitting

Although the routes shown in Fig. 10 may satisfy the needs of the supermarket decision maker, it is not practical for the drivers, because they only have the point order for customers. In order to solve this problem, we make additionally route fitting for the distribution scheme (see Fig. 11).

Combined with the JavaScript technology, we put each route into a single HTML file to precisely guide the corresponding driver for their delivery activity (see Fig. 12).

7 Conclusions and future research directions

According to changing travel speed in time-varying road network, a mathematical model TDVRPSTW with minimizing total cost is established. In addition, a proactive distribution scheme is designed to meet dynamic customer's needs in advance, which reduces the waiting time of customers. At the same time, the new hybrid heuristic algorithm is proposed and tested on Solomon benchmark,

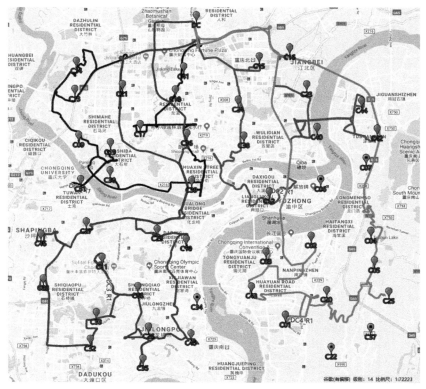

FIG. 11 Fitted routes.

which evaluates the performance of the system. And then a real case of a supermarket in Chongqing is taken to verify the proactive distribution scheme. Finally, a route fitting method is utilized to improve the display results for decision makers.

In this chapter, the time-dependent speed is a series of average speed within 30 min queried from Baidu map API, and the speed between each time interval is not smooth. In the future, time and distance data can be obtained from Baidu at any time during the operation of the algorithm, so as to make full use of real-time road information. However, if the customer nodes that you service is more than 80, you may apply for a paid AK from Baidu, and the multithreading technology is also needed to speed up the queries. Moreover, the influence of customer type and geographical distribution on clustering is not investigated when considering the dynamic customers. These two aspects are valuable for future research.

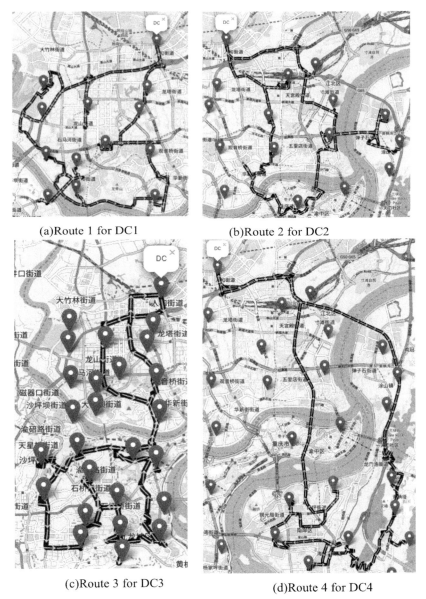

(a)Route 1 for DC1   (b)Route 2 for DC2

(c)Route 3 for DC3   (d)Route 4 for DC4

FIG. 12   Precise routes. (A) Route 1 for DC1. (B) Route 2 for DC2. (C) Route 3 for DC3. (D) Route 4 for DC4.

References

[1] G.B. Dantzig, J.H. Ramser, The truck dispatching problem, Management Sci. 6 (1) (1959) 80–91.

[2] F.A. Tillman, The multiple terminal delivery problem with probabilistic demands, Trans. Sci. 3 (3) (1969) 192–204.

[3] W.R. Stewart, B.L. Golden, Stochastic vehicle routing: a comprehensive approach, Eur. J. Operat. Res. 14 (4) (1983) 371–385.

[4] D.J. Bertsimas, A vehicle routing problem with stochastic demand, Oper. Res. 40 (3) (1992) 574–585.

[5] M. Gendreau, G. Laporte, R. Séguin, A tabu search heuristic for the vehicle routing problem with stochastic demands and customers, Oper. Res. 44 (3) (1996) 469–477.

[6] N. Azi, M. Gendreau, J.-Y. Potvin, A dynamic vehicle routing problem with multiple delivery routes, Ann. Oper. Res. 199 (1) (2012) 103–112.

[7] A.M.F.M. AbdAllah, D.L. Essam, R.A. Sarker, On solving periodic re-optimization dynamic vehicle routing problems, Appl. Soft Comput. 55 (2017) 1–12.

[8] D. Taş, et al., Vehicle routing problem with stochastic travel times including soft time windows and service costs, Comput. Oper. Res. 40 (1) (2013) 214–224.

[9] X. Li, P. Tian, S.C.H. Leung, Vehicle routing problems with time windows and stochastic travel and service times: models and algorithm, Int. J. Prod. Econ. 125 (1) (2010) 137–145.

[10] C. Lecluyse, T. Woensel, H. Peremans, Vehicle routing with stochastic time-dependent travel times, 4OR 7 (4) (2009) 363–377.

[11] R. Montemanni, et al., Ant colony system for a dynamic vehicle routing problem, J. Combinat. Optim. 10 (4) (2005) 327–343.

[12] A. Haghani, S. Jung, A dynamic vehicle routing problem with time-dependent travel times, Comput. Oper. Res. 32 (11) (2005) 2959–2986.

[13] R. Moretti Branchini, V.A. Armentano, A. Løkketangen, Adaptive granular local search heuristic for a dynamic vehicle routing problem, Comput. Oper. Res. 36 (11) (2009) 2955–2968.

[14] S. Wohlgemuth, R. Oloruntoba, U. Clausen, Dynamic vehicle routing with anticipation in disaster relief, Socio-Econ. Plan. Sci. 46 (4) (2012) 261–271.

[15] S.D. Bopardikar, S.L. Smith, F. Bullo, On dynamic vehicle routing with time constraints, IEEE Trans. Robot. 30 (6) (2014) 1524–1532.

[16] F. Ferrucci, S. Bock, M. Gendreau, A pro-active real-time control approach for dynamic vehicle routing problems dealing with the delivery of urgent goods, Eur. J. Oper. Res. 225 (1) (2013) 130–141.

[17] A.M.D. Souza, et al., GARUDA: a new geographical accident aware solution to reduce urban congestion, in: 2015 IEEE International Conference on Computer and Information Technology; Ubiquitous Computing and Communications; Dependable, Autonomic and Secure Computing; Pervasive Intelligence and Computing, 2015.

[18] G. Martuscelli, et al., V2V protocols for traffic congestion discovery along routes of interest in VANETs: a quantitative study, 16 (17) (2016) 2907–2923.

[19] K.L. Soon, et al., Proactive eco-friendly pheromone-based green vehicle routing for multi-agent systems, Exp. Syst. Appl. 121 (2019) 324–337.

[20] D. Kahneman, A. Tversky, Prospect theory: an analysis of decision under risk, in: Handbook of the Fundamentals of Financial Decision Making: Part I, World Scientific, 2013, pp. 99–127.

[21] S. Datta, N. Sahu, S. Mahapatra, Robot selection based on gray—MULTIMOORA approach, Gray Syst. Theory Appl. 3 (2) (2013) 201–232.

[22] W.-F. Dai, Q.-Y. Zhong, C.-Z. Qi, Multi-stage multi-attribute decision-making method based on the prospect theory and triangular fuzzy MULTIMOORA, Soft Comput. 24 (13) (2020) 9429–9440.

9780128210925